Remote Sensing

Remote Sensing
Models and Methods for Image Processing

Third Edition

Robert A. Schowengerdt

Professor Emeritus
Department of Electrical and Computer Engineering,
College of Optical Sciences, and Office of Arid Lands Studies
University of Arizona
Tucson, Arizona

Amsterdam • Boston • Heidelberg • London • New York
Oxford • Paris • San Diego • San Francisco • Singapore
Sydney • Tokyo

ELSEVIER

Academic Press is an imprint of Elsevier
30 Corporate Drive, Suite 400, Burlington, MA 01803, USA
525 B Street, Suite 1900, San Diego, California 92101-4495, USA
84 Theobald's Road, London WC1X 8RR, UK

This book is printed on acid-free paper. ∞

Library of Congress Cataloging-in-Publication Data
Application submitted.

British Library Cataloguing-in-Publication Data
A catalogue record for this book is available from the British Library.

ISBN 13: 978-0-12-369407-2
ISBN 10: 0-12-369407-8

For information on all Academic Press publications
visit our Web site at www.books.elsevier.com

Printed and bound by CPI Group (UK) Ltd, Croydon, CR0 4YY
Transferred to Digital Print 2011

To my teachers and students

Contents

Figures .... *xvii*

Tables .... *xxxiii*

Preface to the Third Edition .... *xxxvii*

Preface to the Second Edition .... *xxxix*

CHAPTER 1 The Nature of Remote Sensing *1*

   1.1 Introduction 1

   1.2 Remote Sensing 2

       1.2.1 Information Extraction from Remote-Sensing Images 7

       1.2.2 Spectral Factors in Remote Sensing 8

   1.3 Spectral Signatures 13

   1.4 Remote-Sensing Systems 16

       1.4.1 Spatial and Radiometric Characteristics 16

       1.4.2 Spectral Characteristics 30

       1.4.3 Temporal Characteristics 32

       1.4.4 Multi-Sensor Formation Flying 35

   1.5 Image Display Systems 36

   1.6 Data Systems 39

   1.7 Summary 42

   1.8 Exercises 44

CHAPTER 2 Optical Radiation Models *45*

   2.1 Introduction 45

2.2 Visible to Shortwave Infrared Region 46
   2.2.1 Solar Radiation 46
   2.2.2 Radiation Components 47
      Surface-reflected, unscattered component 48
      Surface-reflected, atmosphere-scattered component......................53
      Path-scattered component 54
      Total at-sensor, solar radiance 55
   2.2.3 Image Examples in the Solar Region 58
      Terrain shading 58
      Shadowing 58
      Atmospheric correction 61
2.3 Midwave to Thermal Infrared Region 61
   2.3.1 Thermal Radiation 61
   2.3.2 Radiation Components 63
      Surface-emitted component 64
      Surface-reflected, atmosphere-emitted component........................66
      Path-emitted component...67
      Total at-sensor, emitted radiance 68
   2.3.3 Total Solar and Thermal Upwelling Radiance 68
   2.3.4 Image Examples in the Thermal Region 69
2.4 Summary 72
2.5 Exercises...73

CHAPTER 3 Sensor Models 75

3.1 Introduction 75
3.2 Overall Sensor Model..76
3.3 Resolution..76
   3.3.1 The Instrument Response 77
   3.3.2 Spatial Resolution 77
   3.3.3 Spectral Resolution 82
3.4 Spatial Response..85
   3.4.1 Optical PSF_{opt} 86
   3.4.2 Detector PSF_{det}...88
   3.4.3 Image Motion PSF_{IM} 88
   3.4.4 Electronics PSF_{el} 90
   3.4.5 Net PSF_{net}..90
   3.4.6 Comparison of Sensor PSFs..90

3.4.7 Imaging System Simulation 91
3.4.8 Measuring the *PSF* 95
    ALI LSF measurement 98
    QuickBird LSF measurement 101
3.5 Spectral Response..104
3.6 Signal Amplification 106
3.7 Sampling and Quantization 107
3.8 Simplified Sensor Model 109
3.9 Geometric Distortion 110
    3.9.1 Sensor Location Models..110
    3.9.2 Sensor Attitude Models 110
    3.9.3 Scanner Models 113
    3.9.4 Earth Model...114
    3.9.5 Line and Whiskbroom Scan Geometry 119
    3.9.6 Pushbroom Scan Geometry...119
    3.9.7 Topographic Distortion 121
3.10 Summary 125
3.11 Exercises..125

CHAPTER 4 Data Models *127*

4.1 Introduction 127
4.2 A Word on Notation 128
4.3 Univariate Image Statistics...128
    4.3.1 Histogram 129
    Normal distribution 130
    4.3.2 Cumulative Histogram 131
    4.3.3 Statistical Parameters 131
4.4 Multivariate Image Statistics..133
    4.4.1 Reduction to Univariate Statistics 140
4.5 Noise Models..140
    4.5.1 Statistical Measures of Image Quality 146
    Contrast 146
    Modulation 146
    Signal-to-Noise Ratio (SNR) 147
    National Imagery Interpretability Scale (NIIRS) 149
    4.5.2 Noise Equivalent Signal 152
4.6 Spatial Statistics 152

4.6.1 Visualization of Spatial Covariance 153
4.6.2 Covariance and Semivariogram 153
        Separability and anisotropy 160
4.6.3 Power Spectral Density 162
4.6.4 Co-Occurrence Matrix 164
4.6.5 Fractal Geometry 166
4.7 Topographic and Sensor Effects 169
4.7.1 Topography and Spectral Scattergrams 169
4.7.2 Sensor Characteristics and Spatial Statistics 174
4.7.3 Sensor Characteristics and Spectral Scattergrams 178
4.8 Summary 181
4.9 Exercises 182

CHAPTER 5 Spectral Transforms *183*

5.1 Introduction 183
5.2 Feature Space 184
5.3 Multispectral Ratios 186
5.3.1 Vegetation Indexes 188
5.3.2 Image Examples 191
5.4 Principal Components 193
5.4.1 Standardized Principal Components (SPC) 199
5.4.2 Maximum Noise Fraction (MNF) 199
5.5 Tasseled-Cap Components 202
5.6 Contrast Enhancement 206
5.6.1 Global Transforms 208
        Linear stretch 209
        Nonlinear stretch 209
        Normalization stretch 210
        Reference stretch 210
        Thresholding 216
5.6.2 Local Transforms 217
5.6.3 Color Images 219
        Min-max stretch 221
        Normalization stretch 221
        Reference stretch 221
        Decorrelation stretch 222
        Color-space transforms 223

Spatial domain blending...224
5.7 Summary 227
5.8 Exercises...227

CHAPTER 6 Spatial Transforms *229*

6.1 Introduction 229
6.2 An Image Model for Spatial Filtering 230
6.3 Convolution Filters..230
    6.3.1 Linear Filters 232
        Convolution..232
        Low-pass and high-pass filters (LPF, HPF) 233
        High-boost filters (HBF) 234
        Band-pass filters (BPF) 235
        Directional filters 236
        The border region 237
        Characteristics of filtered images 239
        Application of the blending algorithm to spatial filtering...............239
        The box-filter algorithm 240
        Cascaded linear filters 241
    6.3.2 Statistical Filters...242
        Morphological filters 244
    6.3.3 Gradient Filters...245
6.4 Fourier Transforms..246
    6.4.1 Fourier Analysis and Synthesis 246
    6.4.2 Discrete Fourier Transforms in 2-D 249
    6.4.3 The Fourier Components..253
    6.4.4 Filtering with the Fourier Transform...255
        Transfer functions 257
    6.4.5 System Modeling Using the Fourier Transform.................................259
    6.4.6 The Power Spectrum 263
6.5 Scale-Space Transforms 263
    6.5.1 Image Resolution Pyramids 265
    6.5.2 Zero-Crossing Filters 267
        Laplacian-of-Gaussian (LoG) filters 269
        Difference-of-Gaussians (DoG) filters..274
    6.5.3 Wavelet Transforms 278
6.6 Summary 282
6.7 Exercises...282

CHAPTER 7 Correction and Calibration *285*

7.1 Introduction 285
7.2 Distortion Correction...286
    7.2.1 Polynomial Distortion Models 287
        Ground Control Points (GCPs) 291
    7.2.2 Coordinate Transformation 298
        Map projections...299
    7.2.3 Resampling..300
7.3 Sensor MTF Compensation..309
    7.3.1 Examples of MTF compensation 311
7.4 Noise Reduction 315
    7.4.1 Global Noise..315
        Sigma filter 315
        Nagao-Matsuyama filter 317
    7.4.2 Local Noise 318
        Detection by spectral correlation 318
    7.4.3 Periodic Noise 320
    7.4.4 Detector Striping 323
        Global, linear detector matching 325
        Nonlinear detector matching 325
        Statistical modification..325
        Spatial filter masking 327
        Debanding 328
7.5 Radiometric Calibration 332
    7.5.1 Multispectral Sensors and Imagery 334
        Sensor calibration 334
        Atmospheric correction 337
        Solar and topographic correction 339
        Image examples...340
    7.5.2 Hyperspectral Sensors and Imagery 341
        Sensor calibration 341
        Atmospheric correction 343
        Normalization techniques...344
        Image examples...350
7.6 Summary 352
7.7 Exercises..353

CHAPTER 8 Registration and Fusion *355*

8.1 Introduction 355
8.2 What Is Registration? 356
8.3 Automated GCP Location 357
    8.3.1 Area Correlation 357
        Relation to spatial statistics 362
    8.3.2 Other Spatial Features for Registration 362
8.4 Orthorectification 363
    8.4.1 Low-Resolution DEM 363
    8.4.2 High-Resolution DEM 364
        Hierarchical warp stereo 366
8.5 Multi-Image Fusion 371
    8.5.1 Feature Space Fusion 374
    8.5.2 Spatial Domain Fusion 375
        High frequency modulation 376
        Filter design for HFM 378
        Sharpening with a sensor model 378
    8.5.3 Scale-Space Fusion 380
    8.5.4 Image Fusion Examples 380
8.6 Summary 384
8.7 Exercises 384

CHAPTER 9 Thematic Classification *387*

9.1 Introduction 387
9.2 The Classification Process 388
    9.2.1 The Importance of Image Scale and Resolution 390
    9.2.2 The Notion of Similarity 391
    9.2.3 Hard Versus Soft Classification 393
9.3 Feature Extraction 395
9.4 Training the Classifier 395
    9.4.1 Supervised Training 396
        Separability analysis 396
    9.4.2 Unsupervised Training 399
        K-means clustering algorithm 400
        Clustering examples 400
    9.4.3 Hybrid Supervised/Unsupervised Training 402

9.5 Nonparametric Classification 405
    9.5.1 Level-Slice Classifier 405
    9.5.2 Histogram Estimation Classifier 406
    9.5.3 Nearest-Neighbors Classifier 407
    9.5.4 Artificial Neural Network (ANN) Classifier 407
          Back-propagation algorithm...409
    9.5.5 Nonparametric Classification Examples 413
9.6 Parametric Classification..417
    9.6.1 Estimation of Model Parameters 417
    9.6.2 Discriminant Functions 418
    9.6.3 The Normal Distribution Model 418
    9.6.4 The Nearest-Mean Classifier 421
    9.6.5 Parametric Classification Examples 422
9.7 Spatial-Spectral Segmentation 427
    9.7.1 Region Growing 427
9.8 Subpixel Classification...430
    9.8.1 The Linear Mixing Model 434
          Unmixing examples 437
          Relation of fractions to neural network output 440
          Endmember specification..441
    9.8.2 Fuzzy Set Classification 442
          Fuzzy C-Means (FCM) clustering 442
          Fuzzy supervised classification 443
9.9 Hyperspectral Image Analysis..445
    9.9.1 Visualization of the Image Cube 445
    9.9.2 Training for Classification..447
    9.9.3 Feature Extraction from Hyperspectral Data 447
          Image residuals..447
          Absorption-band parameters 448
          Spectral derivative ratios..448
          Spectral fingerprints 449
    9.9.4 Classification Algorithms for Hyperspectral Data 450
          Binary encoding 452
          Spectral-angle mapping..452
          Orthogonal Subspace Projection (OSP) 454
9.10 Summary 455
9.11 Exercises..456

APPENDIX A Sensor Acronyms *457*

APPENDIX B 1-D and 2-D Functions *461*

References *467*

Index *509*

Figures

CHAPTER 1 *The Nature of Remote Sensing*

FIGURE 1-1.   A plot of some remote-sensing systems in a two-dimensional parameter space. ... 4

FIGURE 1-2.   An example of how maps and imagery complement each other. 9

FIGURE 1-3.   This airborne Thematic Mapper Simulator (TMS) image of a devastating wildfire in Yellowstone National Park, Wyoming, was acquired on September 2, 1988...11

FIGURE 1-4.   A nomograph for finding wavelength given frequency, or vice versa, in the microwave spectral region. 13

FIGURE 1-5.   Spaceborne imaging radar image of Isla Isabella in the western Galapagos Islands taken by the L-band radar in HH polarization from the Spaceborne Imaging Radar C/X-Band Synthetic Aperture Radar on the 40th orbit of the space shuttle Endeavour on April 15, 1994. 14

FIGURE 1-6.   Exo-atmospheric (i.e., arriving at the top of the atmosphere) solar spectral irradiance and the daylight-adapted response of the human eye. 15

FIGURE 1-7.   Example vegetation spectral reflectance curves (Bowker et al., 1985). 17

FIGURE 1-8.   Example mineral spectral reflectance curves for clay (top) and several alteration minerals (bottom) (Bowker et al., 1985; Clark et al., 1993)................................ 18

FIGURE 1-9.   Comparison of the spatial and spectral sampling of the Landsat TM and AVIRIS in the VNIR spectral range. 19

FIGURE 1-10. Definition of basic scanner parameters and depiction of three scanning methods, with specific examples of whiskbroom and pushbroom scanners. 20

FIGURE 1-11. Simple geometric description of a single detector element in the focal plane of an optical sensor. 21

FIGURE 1-12. The relationship between GIFOV and GSI for most scanning sensors and the particular relationship for the Landsat MSS and AVHRR. 22

FIGURE 1-13. Focal plane detector layout of ETM+. 24

FIGURE 1-14. Detector layout of the MODIS uncooled VIS and NIR focal planes................... 25

FIGURE 1-15. Detector layout of the MODIS cooled focal planes... 26

FIGURE 1-16. The four sensor chip assemblies (SCAs) in the ALI. 27

FIGURE 1-17. The pushbroom 2-D array concept used in the HYDICE, Hyperion, and MERIS hyperspectral sensors. 28

FIGURE 1-18. The 2-D detector array layout used in HYDICE. 29

FIGURE 1-19. Visual comparison of ETM+ whiskbroom and ALI pushbroom imagery acquired on July 27, 2001, of center-pivot irrigated agricultural fields near Maricopa, Arizona. 31

FIGURE 1-20. Visual comparison of ETM+ whiskbroom and ALI pushbroom panchromatic imagery of an area in Alaska acquired by ETM+ on November 27, 1999, and ALI on November 25, 2000. 32

FIGURE 1-21. Spectral ranges for the 36 MODIS bands. 33

FIGURE 1-22. Four MODIS image bands collected on March 2, 2006, showing James Bay, Canada, at the top, the Great Lakes in the middle, and Florida at the bottom. 34

FIGURE 1-23. The three most common multispectral image formats: BIS, BSQ, and BIL, illustrated with an 8 sample-by-8 line-by-7 band TM image. 37

FIGURE 1-24. The conversion from DN to GL to color in a 24 bits/pixel digital video display...38

CHAPTER 2 *Optical Radiation Models*

FIGURE 2-1. Comparison of the exo-atmospheric (top-of-the-atmosphere) solar spectral irradiance as used in the atmospheric modeling code MODTRAN (Berk et al., 1989) to the blackbody curve for T = 5900 K. 47

FIGURE 2-2. Spectral distributions at the top-of-the-atmosphere for the two radiation sources in the visible through thermal infrared spectral regions. 48

FIGURE 2-3. The most significant radiation components seen by the sensor in solar reflective remote sensing are the "direct" component, the "skylight" component, and the "path radiance" component (commonly called "haze"). 49

FIGURE 2-4. Transmittance of the atmosphere as calculated by the program MODTRAN 50

FIGURE 2-5. Solar irradiance in the visible and shortwave IR regions (for a solar elevation angle of 45°), above the atmosphere and at the earth's surface. 51

FIGURE 2-6. The geometry of solar direct irradiance on the earth's surface............................. 52

FIGURE 2-7. Atmospheric path transmittance as viewed by a satellite sensor......................... 54

FIGURE 2-8. The path-scattered and ground-reflected components of the total upwelling radiance seen by a satellite sensor for a surface reflectance of Kentucky Bluegrass. 56

FIGURE 2-9. The spectral response of Kentucky Bluegrass as predicted by the MODTRAN model and a plot of a mixed grass and trees response from the AVIRIS image of Palo Alto (Plate 1-3). 57

FIGURE 2-10. Comparison of the reflectance and remotely-sensed radiance spectral signals for grass. 57

FIGURE 2-11. The influence of terrain relief on image structure is depicted with a co-registered DEM and TM band 4 image near Berkeley, California....................................... 59

FIGURE 2-12. Maps of the self-shadowed pixels and projected shadows for the solar angles and DEM of Fig. 2-11. 60

FIGURE 2-13. Landsat MSS images of the Grand Canyon, Arizona, acquired on two dates.......60

FIGURE 2-14. TM band 1 through band 4 images of the San Pablo (left) and Briones Reservoirs (right) north of Berkeley, California (part of the same TM scene used in Fig. 2-11). 62

FIGURE 2-15. The reflected and scattered components of the at-sensor radiance (Fig. 2-3) and the analogous emitted components.. 63

FIGURE 2-16. The dependence of radiant exitance from a blackbody on its temperature at three wavelengths. 65

FIGURE 2-17. Atmospheric transmittance (solar elevation = 45°) in the midwave IR and thermal IR spectral regions. 66

FIGURE 2-18. Solar irradiance in the midwave and thermal IR regions (solar elevation = 45°). 67

FIGURE 2-19. The at-sensor radiance above the atmosphere in the middle and thermal IR regions. 69

FIGURE 2-20. Lake Anna, Virginia, viewed by the Landsat MSS band 4 (June 12, 1978, at 9:30 A.M.) and the HCMM thermal band (June 11, 1978, at 3:30 A.M.) and Landsat ETM+ (September 30, 1999, at 10:30 A.M.). 70

FIGURE 2-21. TM band 4 (30 m GSI) and band 6 (120 m GSI) images of the San Francisco area. 71

FIGURE 2-22. Landsat TM band 2 and band 6 images of New Orleans, Louisiana, including Lake Pontchartrain and the Mississippi River (September 16, 1982). 71

FIGURE 2-23. The Santa Rita Mountains, south of Tucson, Arizona, viewed by Landsat TM on January 8, 1983.. 72

CHAPTER 3 Sensor Models

FIGURE 3-1. The primary components in an electro-optical remote-sensing system................ 76

FIGURE 3-2. Example of subpixel object detection.. 78

FIGURE 3-3. More examples of subpixel object detection. 79

FIGURE 3-4. Detectability analysis for a single target at two different contrasts to the surrounding background and for an idealized sensor... 80

FIGURE 3-5. The effect of spatial phasing between the pixel grid and the ground target. 82

FIGURE 3-6.   A contrast-enhanced enlargement of Fig. 3-2 and DN profile plots along three adjacent scanlines near the center of the pier, illustrating the sample-scene phase effect. 83

FIGURE 3-7.   The effective reflectance of alunite as measured by a multispectral sensor. 84

FIGURE 3-8.   The at-sensor radiance for Kentucky Bluegrass (solid curve), the weighted spectral distribution seen by each TM band (dotted curves), and the total (integrated over the weighted spectral distribution) effective radiance (solid circles). 84

FIGURE 3-9.   The diffraction-limited optical PSF and its radial profile. 87

FIGURE 3-10. The MODIS scan mirror sweeps the image continuously across the focal plane. 89

FIGURE 3-11. This normalized detector PSF is used for comparison of different sensor PSFs.. 91

FIGURE 3-12. Perspective views of the total system PSF for several remote sensing systems... 92

FIGURE 3-13.   Simulation of the effect of GSI and GIFOV on visual image quality. 93

FIGURE 3-14. The individual component PSFs for TM. 94

FIGURE 3-15. Scanned aerial photograph of an area in Tucson, Arizona, used in spatial simulation of TM imagery. 95

FIGURE 3-16. A sequential series of images produced by the components of the TM spatial response are shown at the top, with the individual PSF components shown to scale. 96

FIGURE 3-17. The mathematical relations among various representations of optical spatial response. 97

FIGURE 3-18. Example transects of agriculture field berms used to measure the ALI LSF are shown in white in these ALI band 3 images. 100

FIGURE 3-19. The effect of line target width on the measured LSF. 101

FIGURE 3-20. QuickBird image collected on November 4, 2002, of the Tucson Bargain Center parking lot. 102

FIGURE 3-21. The QuickBird LSF analysis using parking lot stripes. 103

FIGURE 3-22. Normalized spectral response functions for several sensors. 105

FIGURE 3-23. Comparison of the VSWIR spectral bands for MODIS, ASTER, and ETM+. .. 106

FIGURE 3-24. Spectral characteristics of the AVIRIS and Hyperion sensors. 107

FIGURE 3-25. Relation between the amplified signal, a_b, and the raw signal out of the detector, e_b. 108

FIGURE 3-26. Relation between the signal level at the A/D converter, a_b, and the DN output. 108

FIGURE 3-27. Conventional definitions for the three attitude axes of a sensor platform. 112

FIGURE 3-28. Airborne ASAS imagery of Maricopa Farm near Phoenix, Arizona, one taken at-nadir and the other at +30° off-nadir. 112

FIGURE 3-29. The key parameters for modeling the imaging geometry for an earth-orbiting satellite in a near-polar descending orbit on the sunlit side, such as used by Landsat and Terra. 116

FIGURE 3-30. Line and whiskbroom scanner geometry in the cross-track direction used to derive Eq. (3-39) and Eq. (3-40)...................... 120

FIGURE 3-31. The "bow-tie" effect in AVHRR data, characteristic of line and whiskbroom scanners........................ 121

FIGURE 3-32. Cross-track pushbroom scanner geometry for Eq. (3-42) and Eq. (3-44). 122

FIGURE 3-33. Geometry for a pushbroom scanner imaging a topographic feature off-nadir in the cross-track direction.......................... 123

FIGURE 3-34. Geometry of stereo imaging. 124

CHAPTER 4 Data Models

FIGURE 4-1. Data models form the link between physical radiation and sensor models and image processing algorithms. 127

FIGURE 4-2. A discrete matrix (row, column) notation compared to a continuous, Cartesian (x,y) notation. 129

FIGURE 4-3. An example image histogram compared to a Gaussian distribution with the same mean and variance. 130

FIGURE 4-4. Comparison of an example image cumulative histogram to a Gaussian cumulative histogram using the same data as in Fig. 4-3................... 131

FIGURE 4-5. Visualization of a three-band multispectral image pixel $\mathbf{DN}_p$ as a vector in three-dimensional space......................... 133

FIGURE 4-6. Three-band scatterplots of bands 2, 3, and 4 of a TM image, viewed from three different directions............................ 134

FIGURE 4-7. Reduction of 3-D scatterplots to 2-D scatterplots by projections onto the three bounding planes. 135

FIGURE 4-8. A full set of band-pair scatterplots for a seven-band TM image. 136

FIGURE 4-9. Two-dimensional scattergrams with density coded as grey levels (inverted for better visibility) and displayed as surfaces. 137

FIGURE 4-10. The correlation coefficient indicates the shape of the scatterplot (or scattergram) of a multispectral image............................ 139

FIGURE 4-11. The normal probability density function in 2-D, displayed as an inverted greyscale image with overlaying contours of equal probability. 139

FIGURE 4-12. Band-to-band scattergrams for a TM image of an agricultural area (Plate 4-1). 141

FIGURE 4-13. An example of photographic granularity.......................... 142

FIGURE 4-14. A gallery of scanner noise. 143

FIGURE 4-15. More examples of scanner noise........................ 144

FIGURE 4-16. Example of spectral crosstalk in ASTER data...................... 145

FIGURE 4-17. A 1-D periodic function of spatial coordinates, with a period of 100 spatial units, a mean of 3 units, an amplitude of 1.2 units, and a phase of 5.8 radians, corresponding to an offset of 92.3 spatial units.. 148

FIGURE 4-18. The effect of random noise on image quality is simulated in this aerial photograph of a residential neighborhood in Portland, Oregon, by adding a normally-distributed random number to each pixel.. 150

FIGURE 4-19. Comparison of the visual effect of equal amounts of random and striping noise. 151

FIGURE 4-20. A TM band 4 image of agricultural fields (top) and a series of scattergrams between pairs of horizontally-separated pixels.. 154

FIGURE 4-21. Graphs of the exponential covariance model (top) and three normalized semivariogram models (bottom)... 156

FIGURE 4-22. Scanned aerial image used to illustrate spatial statistics.................................... 157

FIGURE 4-23. Covariance functions and semivariograms for the transects of Fig. 4-22. 158

FIGURE 4-24. Normalized covariance functions. In effect, the data variance is set to one for each class. 158

FIGURE 4-25. Exponential model fitted to covariance functions from Fig. 4-24..................... 159

FIGURE 4-26. Three possible forms for 2-D exponential covariance models, in perspective and as contour maps... 161

FIGURE 4-27. Power spectra calculated from the data in Fig. 4-24.. 162

FIGURE 4-28. Landsat TM band 4 image of sand dunes and irrigated agriculture near Yuma, Arizona, partitioned into 16 128-by-128 pixel blocks and the corresponding power spectrum of each block. 163

FIGURE 4-29. Power spectra calculated by the Fourier transform of the covariance functions in Fig. 4-25, compared to the model of Eq. (4-46). 164

FIGURE 4-30. Three sample TM images used for co-occurrence matrix analysis. 166

FIGURE 4-31. The contrast and entropy spatial features from the sample images of Fig. 4-30. 167

FIGURE 4-32. Contrast and entropy CM features after equalization of the image histograms.. 167

FIGURE 4-33. A TM band 4 image (the same area as Fig. 4-28) and the 32-by-32 pixel block fractal dimension map... 169

FIGURE 4-34. The original image and fractal dimension map histograms............................... 170

FIGURE 4-35. The shaded relief image generated from a DEM, part of the same area shown in Fig. 2-11, and the synthesized soil and vegetation masks. 171

FIGURE 4-36. The simulated red and NIR images, generated by multiplying the shaded relief image by the simulated class pixels in each class and band, are added together to produce the final simulated images. 172

FIGURE 4-37. NIR versus red scattergrams for the original soil and vegetation data, before and after merger with the terrain model. 173

FIGURE 4-38. The two simulated images used to investigate the influence of the sensor on spatial statistics measured from its imagery. 174

FIGURE 4-39. Covariance and semivariogram functions for the original image and with added spatially-uncorrelated noise. 175

FIGURE 4-40. Covariance functions for the original image with a simulated GIFOV of 5-by-5 pixels. 176

FIGURE 4-41. Normalized covariance and semivariogram functions for the original image with a simulated GIFOV of 5-by-5 pixels. 177

FIGURE 4-42. Creation of a simulated two-band scene with three classes and different amounts of spatial "texture," and the associated scattergrams. 179

FIGURE 4-43. The simulated images resulting from spatial averaging of the two textured-scene cases in Fig. 4-42 and the associated scattergrams. 180

FIGURE 4-44. Simulation of a 5 x 5 pixel GIFOV on the synthetic images of Fig. 4-36 and the resulting scattergram, which shows both topographic-induced correlation and spectral mixing. 181

CHAPTER 5 *Spectral Transforms*

FIGURE 5-1. Some properties of a linear feature space transform in 2-D. 185

FIGURE 5-2. The DN scattergram for band 4 versus band 3 of a TM image of an agricultural scene near Marana, Arizona (Plate 9-1). 187

FIGURE 5-3. Plot of the modulation ratio as a function of the simple ratio............................ 188

FIGURE 5-4. Isolines for the PVI. The data are the same as that in Fig. 5-2 and are uncalibrated DNs. 190

FIGURE 5-5. The RVI and NDVI index images for the TM agriculture image of Plate 4-1... 191

FIGURE 5-6. Spectral band ratios for a Landsat TM image of Cuprite, Nevada, acquired on October 4, 1984. 192

FIGURE 5-7. Two-band scattergrams of different images from different sensors. 194

FIGURE 5-8. 2-D PCTs of two highly-correlated (ρ = 0.97) and two nearly-uncorrelated (ρ = 0.13) TM bands from a nonvegetated desert scene and a partially-vegetated agriculture scene, respectively... 196

FIGURE 5-9. Comparison of the 2-D PC axes from Fig. 5-8 (dashed arrows) to the 6-D PC1 and PC2 axes (solid arrows) projected to the band 4-band 3 data plane....................197

FIGURE 5-10. Distribution of total image variance across the original spectral bands and across the principal components... 197

FIGURE 5-11. The dependence of the percentage of total image variance captured by each eigenvalue. 198

FIGURE 5-12. PC transformation of a nonvegetated TM scene. PC components are individually contrast stretched. 200

FIGURE 5-13. PC transformation of a vegetated TM scene. PC components are individually contrast stretched. 201

FIGURE 5-14. The first three eigenvectors for the nonvegetated and vegetated scenes. 202

FIGURE 5-15. Standardized principal components; compare to the conventional PCs for the two scenes in Fig. 5-12 and Fig. 5-13. 203

FIGURE 5-16. Projection of two of the component axes from 6-D PC and TC transformations onto the TM band 4 versus band 3 data plane. 206

FIGURE 5-17. The PC and TC components for the Yuma agricultural scene. 207

FIGURE 5-18. Some types of DN-to-GL transformations for contrast enhancement. 208

FIGURE 5-19. Examples of contrast enhancement using point transformations and global statistics. 211

FIGURE 5-20. Examples of the normalization stretch..212

FIGURE 5-21. The procedure to match the CDFs of two images. 212

FIGURE 5-22. Contrast matching of TM band 3 images of San Jose, California. 214

FIGURE 5-23. DN histograms of original multitemporal images and the contrast transformations and resulting histograms for the December image after matching to the August image... 215

FIGURE 5-24. Examples of thresholding the GOES image of Fig. 5-19. 216

FIGURE 5-25. Blocking parameters for the LRM adaptive contrast enhancement.................... 218

FIGURE 5-26. Adaptive contrast enhancement with blockwise stretching and the LRM algorithm, which does not introduce block discontinuities. 220

FIGURE 5-27. The normalization algorithm for color images. 221

FIGURE 5-28. The PCT decorrelation contrast stretch algorithm. 222

FIGURE 5-29. The use of a Color-Space Transform (CST) to modify the perceptual color characteristics of an image... 223

FIGURE 5-30. Generation of the hexcone CST... 225

FIGURE 5-31. Four examples of the HSI space components for a test image. 226

CHAPTER 6 Spatial Transforms

FIGURE 6-1. Examples of the global spatial frequency image model at two scales................ 231

FIGURE 6-2. A moving window for spatial filtering. 231

FIGURE 6-3. 1-D signal processed with 13 and 17 LPFs and HPFs....................................... 234

FIGURE 6-4. Example application of 33 HB filters from Table 6-3. 236

FIGURE 6-5. Examples of directional enhancement using derivative filters. 238

FIGURE 6-6.   The border region for a 33 filter. 238

FIGURE 6-7.   Histograms of a TM image, and its LP and HP components............................. 239

FIGURE 6-8.   Application of the blending algorithm to variable spatial filtering. 240

FIGURE 6-9.   Depiction of the box-filter algorithm applied to neighboring regions along an image row. 241

FIGURE 6-10. Example processing by 33 statistical filters... 243

FIGURE 6-11. Application of the median filter to the 1-D signal used earlier......................... 243

FIGURE 6-12. Examples of 33 morphological filter processing of the image in Fig. 6-10. 244

FIGURE 6-13. Vector geometry for calculating image gradients.. 245

FIGURE 6-14. Comparison of the gradient magnitude images produced by common gradient filters. 247

FIGURE 6-15. Fourier synthesis of a 1-D square wave signal by superposition of sine wave signals. 248

FIGURE 6-16. Fourier synthesis of a 2-D square wave..251

FIGURE 6-17. Fourier synthesis of a portion of a TM image. 251

FIGURE 6-18. The implied periodicity of the discrete Fourier transform extends infinitely in both directions... 252

FIGURE 6-19. Coordinate geometry in the Fourier plane for an NN array.............................. 253

FIGURE 6-20. The Fourier components of an image. 254

FIGURE 6-21. Evidence for the importance of spatial phase information. 255

FIGURE 6-22. A filtering algorithm that uses the Fourier transform to compute a spatial domain convolution. 256

FIGURE 6-23. The 2-D MTFs for the 33 box filters of Table 6-2 and Table 6-3. 258

FIGURE 6-24. The MTF of a Gaussian filter with a spatial radius of 1.5 pixels and truncated to 33 pixels... 259

FIGURE 6-25. The continuous-discrete-continuous model for image acquisition, processing, and display is shown in this simplification and expansion of Fig. 3-1...................... 260

FIGURE 6-26. Expansion of Fig. 3-17 to include the Fourier component, the Optical Transfer Function (OTF, or TF$_{opt}$ in our notation). 261

FIGURE 6-27. The ALI in-track and cross-track model MTFs from Table 6-8........................ 262

FIGURE 6-28. The model and measured ALI LSFs. 263

FIGURE 6-29. The dependence of power spectra on image spatial structure. 264

FIGURE 6-30. Construction of a 2 2 box pyramid and an example of six levels, starting with a 256 × 256 image.. 266

FIGURE 6-31. Gaussian pyramid construction with the 5×5 pixel weighting function of Eq. (6-37). 267

FIGURE 6-32.   The REDUCE and EXPAND procedures as defined in Burt and Adelson (1983).
.... 268

FIGURE 6-33.   The construction of level 1 of the Gaussian pyramid and level 0 of the Laplacian
pyramid. 269

FIGURE 6-34.   Level 3 and level 1 images compared for the box and Gaussian pyramids. 270

FIGURE 6-35.   The links between a pixel in level 3 of the Gaussian pyramid and pixels at lower
levels. 271

FIGURE 6-36.   Levels 1 through 3 of the Gaussian pyramid, without down-sampling. 272

FIGURE 6-37.   Example of the relation between slope changes in a function and zero-crossings in
its second derivative (marked with small circles). 273

FIGURE 6-38.   First and second derivatives of the 1-D Gaussian function g(x)........................ 273

FIGURE 6-39.   The process used here to find zero-crossings. 274

FIGURE 6-40.   Profiles of the DoG filter for various size ratios between the two Gaussian functions
in Eq. (6-45)... 275

FIGURE 6-41.   Zero-crossing maps for different size ratios in the DoG filter............................ 276

FIGURE 6-42.   Profiles of the DoG filter for different overall sizes, but the same size ratio between
the two Gaussians. 276

FIGURE 6-43.   Zero-crossing maps for constant size ratio, but different overall DoG filter size.
.... 277

FIGURE 6-44.   Comparison of the Roberts thresholded gradient edge map with a zero-crossing map
for the image on the left... 278

FIGURE 6-45.   Wavelet decomposition from one pyramid level to the next...............................279

FIGURE 6-46.   Calculation of one of the four wavelet components in level 1. 280

FIGURE 6-47.   The four wavelet transform components of level 1 produced by the filter bank in
Fig. 6-45... 281

CHAPTER 7 Correction and Calibration

FIGURE 7-1.   Rectification of a Landsat TM band 4 image of Tucson, Arizona. 287

FIGURE 7-2.   Geometric processing data flow for the common two-stage process and a single-
stage process for rectification. 288

FIGURE 7-3.   Contributions of each quadratic polynomial term to the total warp. 289

FIGURE 7-4.   Polynomial geometric warps. 290

FIGURE 7-5.   GCP location for rectifying the aerial photograph (top) to the scanned map (bottom).
.... 294

FIGURE 7-6.   Direct mapping of the image GCPs (black arrowheads) and GPs (white arrowheads)
to those in the map... 295

FIGURE 7-7.   RMS deviations between the GCP and GP locations as predicted by the fitted-polynomial and their actual locations for different numbers of terms in the polynomial. 295

FIGURE 7-8.   Refinement of GCPs. 296

FIGURE 7-9.   Comparison of actual scanner panoramic distortion and a polynomial model over the FOVs of the AVHRR and Landsat sensors. 297

FIGURE 7-10. Piecewise polynomial mapping. 298

FIGURE 7-11. Rectification of the airborne ASAS image of Fig. 3-28 using six GCPs from a synthesized orthogonal grid and using global polynomial models is shown above. 299

FIGURE 7-12. The two-way relationship between the reference and distorted coordinate systems, expressed as rows and columns in the digital data. 300

FIGURE 7-13. Geometry for resampling a new pixel at (x,y). 302

FIGURE 7-14. Nearest-neighbor, linear, and PCC resampling spatial-weighting functions are compared at the top. 303

FIGURE 7-15. Image magnification using nearest-neighbor and bilinear resampling. 304

FIGURE 7-16. Surface plots of DN(x,y) from 4X-zoomed images for three resampling functions. 305

FIGURE 7-17. A portion of the rectified TM image of Fig. 7-1, as obtained with four different resampling functions. 307

FIGURE 7-18. Demonstration of the effect of resampling on the spectral scattergram. 308

FIGURE 7-19. The MTF, inverse MTF and a constrained inverse MTF are shown, using the ALI as an example system (Fig. 6-27). 310

FIGURE 7-20. An IKONOS panchromatic band image of Big Spring, Texas, collected on August 5, 2001, is shown. 312

FIGURE 7-21. A TM image of Sierra Vista, Arizona, acquired in 1987, is shown.................... 313

FIGURE 7-22. Landsat-7 ETM+ processing with MTFC is shown for a band 8 pan image of Basra, Iraq, acquired on February 23, 2000. 314

FIGURE 7-23. The behavior of the sigma filter near edges and lines. 316

FIGURE 7-24. The nine subwindows used to calculate local DN variance in the Nagao-Matsuyama algorithm. 317

FIGURE 7-25. Speckle noise filtering of a SLAR image of Deming, New Mexico, acquired in the X-band with HH polarization on July 1, 1991, from 22,000 feet altitude. 319

FIGURE 7-26. Local line noise in an MSS image. 320

FIGURE 7-27. Application of the PCT to noise isolation in a TM multispectral image. 321

FIGURE 7-28. A second example of the ability of the PCT to isolate noise, in this case in an ALI image. 322

FIGURE 7-29.   Detector striping and removal in Hyperion images of large copper mines and waste ponds south of Tucson, Arizona. 326

FIGURE 7-30.   Algorithm flow diagram for Fourier amplitude filtering. 327

FIGURE 7-31.   Fourier amplitude filtering to remove striping from a nearly uniform ocean area in a TM image acquired on December 31, 1982. 329

FIGURE 7-32.   Application of the filter derived from the ocean area (Fig. 7-31) to another part of the same TM image 330

FIGURE 7-33.   Automatic filter design for striping. 331

FIGURE 7-34.   The convolution filter algorithm proposed in Crippen (1989) to remove banding noise in system-corrected TM imagery. 332

FIGURE 7-35.   Landsat TM band 4 of San Francisco Bay from August 12, 1983, shows banding, probably caused by a bright fog bank just off the left edge of the image 333

FIGURE 7-36.   Data flow for calibration of remote sensing images to physical units 335

FIGURE 7-37.   Time history of sensor gain for the Landsat-5 TM sensor in the non-thermal bands. 336

FIGURE 7-38.   Example calibration and path radiance correction of TM bands 1, 2, and 3 using sensor calibration coefficients and DOS 340

FIGURE 7-39.   Sensor DN-to-scene radiance calibration for a Landsat TM scene of Cuprite, Nevada, acquired on October 4, 1984 342

FIGURE 7-40.   The at-sensor radiance of a soil area in the AVIRIS image of Plate 7-2 is shown at the top with the solar exo-atmospheric irradiance curve. 345

FIGURE 7-41.   The averaged apparent reflectance images for the two water vapor absorption bands and the surrounding background spectral regions are shown. 346

FIGURE 7-42.   Dividing the center image by the average of the two on either side in Fig. 7-41 produces these estimated atmospheric transmittance maps (top). 347

FIGURE 7-43.   AVIRIS at-sensor spectral radiances for the three mineral sites at Cuprite. 349

FIGURE 7-44.   Mineral spectra after normalization by the flat-field and IARR techniques 350

FIGURE 7-45.   Comparison of mineral reflectances and flat-field-normalized relative reflectances from the 1990 AVIRIS image 351

FIGURE 7-46.   Comparison of the bright target and average scene spectra for the Cuprite scene. 351

FIGURE 7-47.   Mineral spectra adjusted by continuum removal. 352

CHAPTER 8 Image Registration and Fusion

FIGURE 8-1.   Four ways to obtain multitemporal data of the same area from a single satellite sensor or two different sensors. 356

FIGURE 8-2.   Area correlation for image registration. A 55 pixel target area, T, and a 99 search area, S, are shown at the top. 359

FIGURE 8-3.   Example cross-correlation coefficient surfaces obtained from multitemporal TM image chips of San Jose, California, acquired on December 31, 1982 (search chip), and August 12, 1983 (target chip). 361

FIGURE 8-4.   The process used to create a digital orthorectified photograph (commonly called an "orthophoto") from aerial photography and a pre-existing low-resolution DEM.
....364

FIGURE 8-5.   The full frame as scanned from a National Aerial Photography Program (NAPP) aerial photograph of Harrisburg, Pennsylvania, acquired on April 8, 1993. 365

FIGURE 8-6.   The Harrisburg NE quarter quad DEM (left) and a shaded relief representation (right). 366

FIGURE 8-7.   The digital orthophoto and the corresponding portion of the quadrangle map. . 367

FIGURE 8-8.   The HWS algorithm. 369

FIGURE 8-9.   Stereo pair of aerial photos acquired at 12:20 P.M. on November 13, 1975, over Cuprite, Nevada. 370

FIGURE 8-10.   The raw disparity map shows the effect of uncorrected camera tilt relative to the topographic map DEM...371

FIGURE 8-11.   Subsets of the interpolated topographic map and HWS-derived DEM, corrected for tilt and z-scale as described in the text (compare to Fig. 8-9). 372

FIGURE 8-12.   Image fusion using feature space component replacement, in this case either the first principal component or the intensity component... 375

FIGURE 8-13.   The sequence of processing to fuse two images using HFM (Eq. (8-10)).......... 377

FIGURE 8-14.   The scattergram of the original pan band versus the value component of the visible bands 3, 2, and 1 shows high correlation (r = 0.817)... 381

FIGURE 8-15.   The scattergram of the original pan band versus the value component of the NIR band 4 and visible bands 3 and 2 shows much less correlation (r = 0.454) than for the visible bands alone (Fig.)... 382

FIGURE 8-16.   The original pan band and the PC1 of bands 4, 3, and 2 are highly correlated (ρ = 0.823). 383

CHAPTER 9 *Thematic Classification*

FIGURE 9-1.   The data flow in a classification process. 389

FIGURE 9-2.   Classification as a data compression technique.. 390

FIGURE 9-3.   Two possible sets of training data for three classes in feature space and candidate decision boundaries for class separation.. 393

FIGURE 9-4.   One way to view the difference between hard and soft classification................ 394

FIGURE 9-5.   The L1, L2, and ANG distance measures depicted for two vectors in 3-D........ 397

FIGURE 9-6.   An idealized data distribution during three iterations of the K-means clustering algorithm with the nearest-mean decision criterion... 401

FIGURE 9-7.   Typical behavior of the net mean migration from one iteration to the next in the K-means algorithm. 402

FIGURE 9-8.   Final cluster maps and band 4 versus band 3 scattergrams for different numbers of clusters. 403

FIGURE 9-9.   Residual magnitude error maps between the original image and the approximation given by the cluster mean DNs are shown above. 404

FIGURE 9-10. Level-slice decision boundaries for three classes in two dimensions................ 406

FIGURE 9-11. The traditional structure of a three-layer ANN, the components of a processing element, and the sigmoid activation function. 410

FIGURE 9-12. The number of hidden layer nodes required in a three-layer ANN to match the degrees-of-freedom in a maximum-likelihood classifier................................... 412

FIGURE 9-13. Level-slice classification results in the image space and the decision regions in the feature space. 413

FIGURE 9-14. The behavior of the output-node errors and learning rate and momentum parameters during training of the network. 414

FIGURE 9-15. Hard maps and decision boundaries at three intermediate stages and at the final stage (5000 cycles) of the back-propagation algorithm...................................... 415

FIGURE 9-16. Soft classification maps at four stages of the back-propagation training process. 416

FIGURE 9-17. Maximum-likelihood decision boundaries for two continuous Gaussian DN distributions in one dimension.. 420

FIGURE 9-18. Maximum-likelihood decision boundaries for three classes in two dimensions, with Gaussian distributions for each class. 421

FIGURE 9-19. Nearest-mean decision boundaries for three classes in two dimensions, using the L2-distance measure. 423

FIGURE 9-20. Nearest-mean and maximum-likelihood classification results in image and feature space. 424

FIGURE 9-21. Lake Anna Landsat MSS band 4 image with three training sites and initial classification maps...425

FIGURE 9-22. The effect of a priori probabilities on the goodness-of-fit between the class Gaussian models and the data.. 426

FIGURE 9-23. The natural log of the probability density functions of the maximum-likelihood classifer (top two rows) and the output surfaces of the neural network (bottom two rows) for four of the classes in Plate 9-2. 428

FIGURE 9-24. The spatial neighborhood and rules used for the region-growing algorithm...... 431

FIGURE 9-25. Segmentation results for two DN-difference thresholds, 2 and 5. 432

FIGURE 9-26. Simple example to illustrate spatial mixing. 433

FIGURE 9-27. The linear mixing model for a single GIFOV. 435

FIGURE 9-28. The spatial integration involved in mixing of spectral signatures. 436

FIGURE 9-29. Three possible choices for endmembers for the classes "dark soil," "light soil," and "crop" in a TM NIR-red spectral space. 436

FIGURE 9-30. Class fraction maps produced by the data-defined endmembers (middle triangle, Fig. 9-31) and "virtual" endmembers (outer triangle, Fig. 9-31). 440

FIGURE 9-31. Scatterplots between the ANN output node values and linear unmixing fractions of the image in Plate 9-1. 441

FIGURE 9-32. Comparison of hard and fuzzy clustering results. 444

FIGURE 9-33. Visualization of hyperspectral image data by spatial spectrograms. 446

FIGURE 9-34. The definition of three absorption-band parameters. 449

FIGURE 9-35. Spectral fingerprints for three mineral reflectance spectra (data from Fig. 1-8) and corresponding radiance data from the AVIRIS image of Cuprite in Plate 7-3... 451

FIGURE 9-36. Binary encoding of the spectral radiance for four classes in the AVIRIS Palo Alto scene of Plate 1-3. 453

FIGURE 9-37. The spectral-angle classifier decision boundaries. 454

APPENDIX B Function Definitions

FIGURE B-1. The 1-D square pulse, or rectangle, function. 462

FIGURE B-2. The 2-D square pulse, or box, function. 463

FIGURE B-3. The 2-D sinc function. 464

Tables

CHAPTER 1 *The Nature of Remote Sensing*

TABLE 1-1.   Some special issues of scientific journals that contain design, performance, calibration and application articles for specific sensors.. 5

TABLE 1-2.   Primary geophysical variables measurable with each spectral band of the EOS MODIS system (Salomonson et al., 1995). 6

TABLE 1-3.   The primary spectral regions used in earth remote sensing................................. 10

TABLE 1-4.   Microwave wavelengths and frequencies used in remote sensing........................ 12

TABLE 1-5.   Revisit intervals and equatorial crossing times for several satellite remote sensing systems... 35

TABLE 1-6.   Sensor band mapping to RGB display color for standard color composites. 39

TABLE 1-7.   Processing levels for NASA earth remote-sensing systems. 41

TABLE 1-8.   Example MODIS science data products. 42

TABLE 1-9.   Processing levels for commercial earth remote-sensing systems 43

CHAPTER 3 *Sensor Models*

TABLE 3-1.   Some examples of sensor spatial response measurement from operational imagery. 99

TABLE 3-2.   Parking lot stripe parameters relevant to QuickBird image analysis. 103

TABLE 3-3.   The angle between two adjacent pixels for a number of sensors........................ 111

TABLE 3-4.   Examples of sensor-specific internal distortions. 113

TABLE 3-5.   Useful parameters for the "Figure of the Earth" and its rotational velocity. 115

TABLE 3-6.   Publications on scanner and satellite orbit modeling... 117

TABLE 3-7.   Some of the published work on satellite production image geometric quality... 118

TABLE 3-8.   AVHRR rectification experiments using sensor and geoid models. 118

CHAPTER 4 Data Models

TABLE 4-1.   Example of the National Image Interpretability Scale (NIIRS) 151

TABLE 4-2.   Some 1-D continuous models for the discrete spatial covariance and semivariogram.. 155

TABLE 4-3.   Example applications of the semivariogram in remote sensing 156

TABLE 4-4.   Correlation lengths obtained from the exponential model fits in Fig. 4-25. 159

TABLE 4-5.   Some of the spatial texture features derivable from the CM, as originally proposed in Haralick et al. (1973). 165

TABLE 4-6.   Some ways to estimate fractal dimensions from images. 168

TABLE 4-7.   Examples of research on measurement and correction of topographic effects in analysis of remote sensing imagery 173

CHAPTER 5 Spectral Transforms

TABLE 5-1.   Example applications of the PCT to multitemporal imagery............................. 199

TABLE 5-2.   Tasseled-cap coefficients for several sensors.. 204

TABLE 5-3.   Summary of contrast enhancement algorithms... 217

TABLE 5-4.   Base image to be used to manipulate different color image properties with the interpolation/extrapolation blending algorithm. 226

CHAPTER 6 Spatial Transforms

TABLE 6-1.   Catalog of local filter types... 232

TABLE 6-3.   Example 3 x 3 HB box filters for different values of K..................................... 235

TABLE 6-2.   Examples of simple box filters, which have uniform weights in the *LPF* and the complementary weights in the *HPF*... 235

TABLE 6-4.   Example directional filters. 237

TABLE 6-5.   Example cascaded filters and their equivalent net filter..................................... 242

TABLE 6-6.   Example local gradient filters. See also Robinson (1977). 246

TABLE 6-7.   Examples of sensor PSF and MTF measurement and modeling. 261

TABLE 6-8.   The MTF modeling parameters used for ALI... 262

TABLE 6-9.   Descriptive relationships between the spatial and spatial frequency domains. .. 264

CHAPTER 7 *Correction and Calibration*

TABLE 7-1.   Specific affine transformations for Landsat MSS data (Anuta, 1973; Richards and Jia, 1999)... 291

TABLE 7-2.   Projection plane equations for several common map projections (Moik, 1980). 301

TABLE 7-3.   Some of the research published on MTF compensation of remote-sensing imagery. 309

TABLE 7-4.   Some research papers on reducing random image noise. 323

TABLE 7-5.   Some of the published work on sensor-specific periodic noise reduction.......... 324

TABLE 7-6.   Pre-flight measurements of the TM calibration gain and offset coefficients for Landsat-4 and -5 are calculated using the procedure provided in EOSAT (1993). 336

TABLE 7-7.   Examples of atmospheric correction techniques for multispectral remote-sensing images. 338

TABLE 7-8.   The discrete characterization of atmospheric conditions used in Chavez (1989). 339

TABLE 7-9.   Atmospheric modeling and correction software programs............................... 343

TABLE 7-10.   Empirical normalization techniques for hyperspectral imagery that has previously been calibrated to at-sensor radiances. 348

TABLE 7-11.   Comparison of hyperspectral image normalization techniques in terms of their ability to compensate for various physical radiometric factors. 348

CHAPTER 8 *Image Registration and Fusion*

TABLE 8-1.   Some examples of image registration work.. 358

TABLE 8-2.   Example approaches to determining elevation from stereo remote-sensing imagery. 368

TABLE 8-4.   Multisensor and multispectral image fusion experiments 373

TABLE 8-3.   The panchromatic band response range for several sensors. 373

CHAPTER 9 *Thematic Classification*

TABLE 9-1.   An example category in a 3-level Anderson land-cover/land-use classification scheme. 391

TABLE 9-2.   A level I and II land-use and land-cover classification hierarchy (Anderson et al., 1976). 392

TABLE 9-3.   Distance measures between the means of two distributions in feature space..... 398

TABLE 9-4.   Example applications of ANNs in remote-sensing image classification. 408

TABLE 9-5.   Some spatial-spectral segmentation algorithms, comparisons and applications 429

TABLE 9-6.   Example applications of subpixel analyses. 434

TABLE 9-7.   Endmember DN values for the 2-D unmixing example. 438

TABLE 9-8.   Augmented matrices for unmixing. 439

TABLE 9-9.   Hamming distance table for the binary-coded spectral classes of Fig. 9-38. 453

APPENDIX A Sensor Acronyms

TABLE A-1.   Some of the more common remote sensor system acronyms............................. 458

APPENDIX B Function Definitions

TABLE B-1.   2-D Fourier transform properties 464

TABLE B-2.   Spatial and frequency domain functions used to model the ALI sensor in Chapter 6.
                465

Preface to the Third Edition

With nearly ten years since publication of the Second Edition of *Remote Sensing - Models and Methods for Image Processing*, the Third Edition provides a needed and comprehensive update. The changes include:

- Sensor updates, including the NASA satellites Terra, Aqua, and EO-1, and the commercial satellites IKONOS, OrbView, and Quickbird, with many new image examples
- Research literature updates for all topics
- New and expanded sections on
    - sensor spatial response modeling and measurement
    - MTF correction
    - atmospheric correction
    - multispectral fusion
    - noise reduction techniques
- 32 color plates, including more than 20 new to this edition, over fifteen new exercises, over forty new figures, and twenty revised figures
- Many improvements to the text and figures for improved teaching and understanding

The style of the Second Edition was retained throughout, but a new wider format permitted reorganization and rearrangement of many figures to better present concepts.

In preparing the Third Edition, I was struck (but perhaps really shouldn't have been) by two things. One is the continuing and even accelerating increase in the number of operating sensors from an expanding number of countries. In a practical sense, it was quite difficult to fit all the desired contemporary information into Fig. 1-1, and yet the figure is woefully incomplete in representing the systems currently available for remote sensing of the earth and its environment. Some anticipated future systems, such as the National Polar-orbiting Operational Environmental Satellite System (NPOESS), were deliberately excluded as their specifications may still change before launch.

The other major new realization is of the impressive amount of information available on the Internet, ranging from detailed sensor technical documentation to data itself. One is quite spoiled these days, sitting at a computer and accessing nearly all that is needed to do a book of this sort. For example, in looking for a MODIS image example of multispectral sensing of land, snow and clouds, I used a MODIS Direct Broadcast site operated by the U.S. Geological Survey (*http:// modisdb.usgs.gov/*) to browse for suitable images and to download the recent images in Fig. 1-22. Another example is the Landsat-7 ETM+ image of Fig. 2-20; a search of the Internet for an ETM+ image of Lake Anna found this one at the Global Land Cover Facility at the University of Maryland (*http://glcf.umiacs.umd.edu/*). Moreover, nearly all of my "library" research for this edition was done online using journal and conference paper databases, and government and commercial satellite websites. It almost makes you wonder if books are obsolete!

As was the case for the Second Edition, I'm indebted to many colleagues for their assistance in preparing the Third Edition. Among them are Ken Ando and John Vampola of Raytheon and Bill Rappoport of Goodrich Electro-Optical Systems who provided illustrations and information on real sensor focal planes, George Lemeshewsky of the U.S. Geological Survey for image examples and technical advice on multispectral image fusion and restoration, Jim Storey of USGS EROS/SAIC for discussions and technical input on countless topics throughout the book, and James Shepherd of Manaaki Whenua Landcare Research, New Zealand, and Dr. Rudolf Richter of the DLR/DFD, Germany, for kindly providing digital image files for publication of their work. I also want to thank the reviewers of my book proposal who provided many useful comments on topics to include and organization matters, and colleagues who took the time to send corrections to the Second Edition, which I've included in the new volume. In the end, however, I accept full responsibility for the material and any errors in the Third Edition.

I also thank the editors and production personnel at Elsevier and Alan Rose and Tim Donar of Multiscience Press for their patience and collaboration in producing the Third Edition. Despite all the powerful computer tools and access available now to authors, creation of a real, "hold in your hand" book requires the teamwork of many capable and conscientious professionals.

At the completion of this Third Edition, it is especially important to me to acknowledge mentors throughout my life and career, including my parents, who set me on the right path, my graduate advisor, Prof. Phil Slater, who kept me on the right path and paved the way in many cases, inspiring teachers and students, and colleagues at the University of Arizona, NASA, USGS, and many other organizations who've made it all interesting and enjoyable. I'm especially mindful of two friends and co-authors, Prof. Steve Park and Jim Fahnestock, who died years too early. Lastly, I most appreciate my family's support through all three editions. To one and all, thank you!

Robert A. Schowengerdt
Tucson, Arizona
June 2006

Preface to the Second Edition

This book began as a rather conservative revision of my earlier textbook, *Techniques for Image Processing and Classification in Remote Sensing*. Like many "limited" endeavors, however, it soon grew to be a much larger project! When it became clear that simply a revision would not suffice, I gave considerable thought on a new way to present the subject of image processing in the context of remote sensing. After much mental wandering about, it became clear that there was a unifying theme through many of the image processing methods used in remote sensing, namely that they are based, directly or indirectly, on *models* of physical processes. In some cases there is a direct dependence, for example, on physical models that describe orbital geometry or the reflectance of radiation. In other cases, the dependence is indirect. For example, the common assumption of data similarity implies that neighboring pixels in the space or spectral domains are likely to have similar values. The origin of this similarity is in the physical processes leading up to the acquisition of the data, and in the acquisition itself. In nearly all cases, the motivation and rationale for remote sensing image processing algorithms can be traced to an assumption of one or more such models. Thus, I settled on this viewpoint for the book.

It was obvious from the beginning that the book should be an entirely digital production. The computer tools currently available for desktop publishing easily support this, and given the subject matter, seem almost obligatory. Therefore, extensive use is made of computer-generated graphics and image processing. Nearly all figures are entirely new and produced especially for this edition. Three-dimensional graphing programs were used to visualize multidimensional data, and an assortment of desktop image processing programs was used to produce the images. These include in particular, IPT, a development version of the MacSADIE image processing software from my laboratory, and MultiSpec, a multispectral classification program from David Landgrebe's laboratory at Purdue University.

To enhance the use of the book in the classroom, exercises are included for each chapter. They range from conceptual, "gedanken" experiments to mathematical derivations. The exercises are intended to promote an understanding of the material presented in the chapter. Extensive bibliographies of many of the subjects covered are provided in the form of tables to conserve space and provide a compact source for further information. In the references, I've emphasized archival journal papers, because they are generally easiest for the reader to acquire.

Chapter 1 provides an overview of remote-sensing science and technology as of 1996. The basic parameters for optical remote sensing are established here, and the main types of scanning sensors are described. In Chapter 2, the most important optical radiation processes in remote-sensing are described mathematically. These include solar radiation, atmospheric scattering, absorption and transmission, and surface reflectance. The wavelength region from 400 nm wavelength to the thermal infrared is analyzed. Sensor models for radiometric and spatial response are explained in Chapter 3. Satellite imaging geometry is also included because of its importance for image rectification and geocoding and for extraction of elevation information from stereo images.

In Chapter 4, data models provide a transition between the physical models of Chapters 2 and 3 and the image processing methods of later chapters. Spectral and spatial statistical models for remote sensing data are described. A series of imaging simulations illustrate and explain the influence of the sensor's characteristics on the data acquired by remote-sensing systems.

Chapter 5 begins the discussion of image processing methods and covers spectral transforms, including various vegetation indicies, principal components and contrast enhancement. Chapter 6 includes convolution and Fourier filtering, multiresolution image pyramids and scale-space techniques such as wavelets. The latter types of image analyses appear to have considerable potential for efficient and effective spatial information extraction. The concept of image spatial decomposition into two or more components is used here to provide a link among the different spatial transforms. In Chapter 7, several examples of the use of image processing for image radiometric and geometric calibration are given. The importance of image calibration for high spectral resolution imagery ("hyperspectral") data is also discussed. The topic of multiimage fusion is addressed in Chapter 8, with reference to the spatial decomposition concept of Chapter 6. The various approaches are explained and illustrated with Landsat TM multispectral and SPOT panchromatic image fusion. An image pyramid-based scheme for digital elevation model (DEM) extraction from a stereo image pair is also described in detail. Chapter 9 is devoted to thematic classification of remote-sensing images, including the traditional statistical approaches and newer neural network and fuzzy classification methods. Techniques specifically developed for hyperspectral imagery are also described.

Some topics that one usually finds in a remote sensing image processing book, such as classification map error analysis, were deliberately excluded. This was done not only for space reasons, but also because I felt they departed too far from the main theme of the relation of image processing methods to remote sensing physical models. Likewise, classification methods such as rule-based systems that rely on higher level abstractions of the data, although effective and promising in many cases, are not described. I also view Geographic Information Systems (GIS) as being outside the scope of this work.

I'm indebted to many colleagues for their advice and assistance. In some cases, their contributions were substantial: Phil Slater and Kurt Thome (Optical Sciences Center, University of Arizona) provided knowledgeable guidance to keep me on the right path in Chapters 2 and 3, as did Jennifer Dungan (NASA/Ames Research Center) with respect to Chapter 4. Others provided reviews of selected portions, including Chuck Hutchinson and Stuart Marsh (Office of Arid Lands Studies, University of Arizona) and Chris Hlavka (NASA/Ames Research Center). I also wish to acknowledge the insight provided by Eric Crist (Environmental Research Institute of Michigan) on the tasseled cap transformation. Many of my former and current students provided invaluable data and examples, including Dan Filiberti (Science Applications International Corporation), Steve Goisman (University of Arizona), Per Lysne (University of Arizona), Justin Paola (Oasis Research Center), and Ho-Yuen Pang (University of Arizona). Gerhard Mehldau (Photogrammetrie GMBH) supported my efforts with updated versions of IPT. Colleagues in the U. S. Geological Survey, including William Acevedo, Susan Benjamin, Brian Bennett, Rick Champion, Len Gaydos and George Lee, and Jeff Meyers of the NASA/Ames Research Center, kindly supplied image and digital elevation data that provided important examples throughout the book. I would also like to thank Peter B. Keenan, a longtime friend and colleague, for helping me collect ground truth by bicycle on a typically beautiful day in the San Francisco Bay area!

I am grateful for technical and management contributions by several editors and professional staff at Academic Press, including Lori Asbury, Sandra Lee and Bruce Washburn of the San Diego office, and Diane Grossman, Abby Heim and Zvi Ruder of the Chestnut Hill office.

Finally, I must thank my family, Amy, Andrea and Jennifer, for they sacrificed more than I did during the long hours devoted to this book. Hopefully, the result is to some degree worthy of their support.

CHAPTER 1

The Nature of Remote Sensing

1.1   Introduction

The first Landsat Multispectral Scanner System (MSS) launched in 1972, with its 4 spectral bands, each about 100nm wide, and 80m pixel size, began the modern era of land remote sensing from space. Remote-sensing systems now exhibit a diversity and range of performance that make the MSS specifications appear modest indeed. There are operational satellite systems that sample nearly all available parts of the electromagnetic spectrum with dozens of spectral bands, and with pixel sizes ranging from less than 1m to 1000m, complemented by a number of airborne hyperspectral systems with hundreds of spectral bands, each on the order of 10nm wide. The general characteristics of these remote-sensing electro-optical imaging instruments and of the images produced by them are described in this chapter.

1.2   Remote Sensing

Remote sensing is defined, for our purposes, as the measurement of object properties on the earth's surface using data acquired from aircraft and satellites. It is therefore an attempt to measure something *at a distance*, rather than *in situ*. Since we are not in direct contact with the object of interest, we must rely on propagated signals of some sort, for example optical, acoustical, or microwave. In this book, we will limit the discussion to remote sensing of the earth's surface using optical signals. While remote-sensing data can consist of discrete, point measurements or a profile along a flight path, we are most interested here in measurements over a two-dimensional spatial grid, i.e., *images*. Remote-sensing systems, particularly those deployed on satellites, provide a repetitive and consistent view of the earth that is invaluable to monitoring short-term and long-term changes and the impact of human activities. Some of the important applications of remote-sensing technology are:

- environmental assessment and monitoring (urban growth, hazardous waste)
- global change detection and monitoring (atmospheric ozone depletion, deforestation, global warming)
- agriculture (crop condition, yield prediction, soil erosion)
- nonrenewable resource exploration (minerals, oil, natural gas)
- renewable natural resources (wetlands, soils, forests, oceans)
- meteorology (atmosphere dynamics, weather prediction)
- mapping (topography, land use, civil engineering)
- military surveillance and reconnaissance (strategic policy, tactical assessment)
- news media (illustrations, analysis)

To meet the needs of different data users, many remote-sensing systems have been developed, offering a wide range of spatial, spectral, and temporal parameters. Some users may require frequent, repetitive coverage with relatively low spatial resolution (meteorology).[1] Others may desire the highest possible spatial resolution with repeat coverage only infrequently (mapping); while some users need both high spatial resolution and frequent coverage, plus rapid image delivery (military surveillance). Properly calibrated remote-sensing data can be used to initialize and validate large computer models, such as Global Climate Models (GCMs), that attempt to simulate and predict the earth's environment. In this case, high spatial resolution may be undesirable because of computational requirements, but accurate and consistent sensor calibration over time and space is essential. An example of the use of remote sensing data for global monitoring of vegetation is shown in Plate 1-1.

---

1. "Resolution" is a term that can lead to much confusion. We use the common meaning in this chapter, namely the spacing between pixel samples on the earth's surface (Fig. 1-11). The subject is discussed in detail in Chapter 3.

The modern era of earth remote sensing from satellites began when the Landsat Multispectral Scanner System (MSS) provided, for the first time, in 1972 a consistent set of synoptic, high resolution earth images to the world scientific community. The characteristics of this new sensor were multiple spectral bands (sensing four regions of the electromagnetic spectrum, each about 100 nm wide[2]—a coarse spectrometer, if you will), with reasonably high spatial resolution for the time (80 m), large area (185 km by 185 km), and repeating (every 18 days) coverage. Moreover, the MSS provided general purpose satellite image data directly in digital form. Much of the foundation of multispectral data processing was developed in the early 1970s by organizations such as the National Aeronautics and Space Administration (NASA), Jet Propulsion Laboratory (JPL), U.S. Geological Survey (USGS), Environmental Research Institute of Michigan (ERIM), and the Laboratory for Applications of Remote Sensing (LARS) at Purdue University. An excellent history and discussion of the motivations for the Landsat program and data processing are provided in (Landgrebe, David, 1997).

Since 1972, there have been four additional MSS systems, two Thematic Mapper (TM) systems, and the Enhanced Thematic Mapper Plus (ETM+) in the Landsat series. There have also been five higher resolution French SPOT systems, several lower resolution AVHRR and GOES systems, and NASA's sensor suites on the Earth Observing System (EOS) Terra and Aqua satellites, as well as a wide variety of other multispectral sensors on aircraft and satellites. Many countries, including Canada, India, Israel, Japan, South Korea, and Taiwan, and multinational agencies such as the European Space Agency (ESA) now operate remote sensing systems. A depiction of some of these optical remote sensing systems in a performance space defined by two key sensor parameters, the number of spectral bands and the *Ground-projected Sample Interval (GSI)*,[3] is shown in Fig. 1-1. An example TM image is shown in Plate 1-2. Many remote sensing systems have been described in detail in special issues of scientific journals (Table 1-1).

A so-called *hyperspectral* sensor class occupies the upper portion of Fig. 1-1. The Advanced Visible/InfraRed Imaging Spectrometer (AVIRIS) and the HyMap are airborne sensors that produce hundreds of images of the same area on the ground in spectral bands about 10 nm wide over the solar reflective portion of the spectrum from 400 to 2400 nm. The Hyperion was on NASA's Earth-Observing-1 (EO–1) satellite as the first civilian hyperspectral satellite system. Although it has relatively fewer spectral bands, the European Space Agency's MEdium Resolution Imaging Spectrometer (MERIS) is also an imaging spectrometer. The separation of spectral bands in these systems is achieved with a continuously dispersive optical element, such as a grating or prism. The MODerate Imaging Spectroradiometer (MODIS), a discrete filter-based system, on Terra and Aqua provides images in 36 spectral bands over the range 0.4 to 14 μm. Such sensors have provided large improvements in the quantity and quality of information that can be gathered about the earth's surface and near environment (Table 1-2). Example AVIRIS and Hyperion images are shown in Plates 1-3 and 1-4 and MODIS images in Fig. 1-22.

---

2. The spectral range of sensitivity is referred to as the *bandwidth* and can be defined in a number of ways (Chapter 3). It determines the spectral resolution of the sensor.

3. The *GSI* is synonymous with the simple meaning of spatial resolution used here.

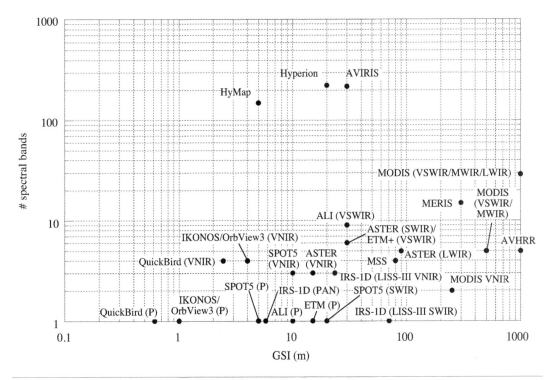

FIGURE 1-1. A plot of some remote-sensing systems in a two-dimensional parameter space. The sensor acronyms are defined in Appendix A and the notations in the graph refer to the sensor spectral regions: V = Visible, NIR = Near InfraRed, LWIR = Long Wave IR, MWIR = Mid Wave IR, SWIR = Short Wave IR, and P = Panchromatic. These terms are explained later in this chapter. All of these systems are on satellites, except AVIRIS and HyMap. There are a number of airborne simulators of satellite systems which are not shown, e.g. the MODIS Airborne Simulator (MAS), the Airborne MISR (AirMISR), and the Thematic Mapper Simulator (TMS). A list of these and other sensor acronyms is given in Appendix A. For a thorough survey of remote sensing systems, the book by Kramer is recommended (Kramer, 2002).

The increasing number and resolution of sensor systems present a continuing challenge for modern data storage and computing systems. For example, the Land Processes Distributed Active Archive Center (LP DAAC) at the USGS Center for Earth Resources Observation and Science (EROS) in Sioux Falls, South Dakota, surpassed one petabyte of data holdings in November 2003.[4] Non-commercial data have been made available for little or no charge by electronic distribution from Internet sites operated by universities and other non-profit organizations over the last few years. This form of data "sharing" is likely to continue and increase in the future.

---

4. One petabyte = 1,125,899,906,842,624 bytes.

TABLE 1-1. *Some special issues of scientific journals that contain design, performance, calibration and application articles for specific sensors.*

sensor or platform	journal issue
Aqua	IEEE Transactions on Geoscience and Remote Sensing, Vol 41, No 2, February 2003
ASTER (science results)	Remote Sensing of Environment, Vol 99, Nos 1–2, November 15, 2005
ASTER (calibration and performance)	IEEE Transactions on Geoscience and Remote Sensing, Vol 43, No 12, December 2005
EO-1	IEEE Transactions on Geoscience and Remote Sensing, Vol 41, No 6, June 2003
IKONOS	Remote Sensing of Environment, Vol 88, Nos 1–2, November 30, 2003
Landsat-4	IEEE Transactions on Geoscience and Remote Sensing, Vol GE-22, No 3, May 1984
	Photogrammetric Engineering and Remote Sensing, Vol LI, No 9, September 1985
Landsat-5, -7 (performance characterization)	IEEE Transactions on Geoscience and Remote Sensing, Vol 42, No 12, December 2004
MERIS	International Journal of Remote Sensing, Volume 20, Number 9, June 15, 1999
MODIS (land science)	Remote Sensing of Environment. Vol 83, Nos 1–2, November 2002
MTI	IEEE Transactions on Geoscience and Remote Sensing, Vol 43, No 9, September 2005
Terra	IEEE Transactions on Geoscience and Remote Sensing, Vol 36, No 4, July 1998

Although electro-optical imaging sensors and digital images dominate earth remote sensing today, earlier technologies remain viable. For instance, aerial photography, although the first remote-sensing technology, is still an important source of data because of its high spatial resolution and flexible coverage. Photographic imagery also still plays a role in remote sensing from space. Firms such as SOVINFORMSPUTNIK, SPOT Image, and GAF AG market scanned photography from the Russian KVR-1000 panchromatic film camera, with a ground resolution of 2 m, and the TK-350 camera, which provides stereo coverage with a ground resolution of 10 m. The U.S. government has declassified photography from its early national surveillance satellite systems, CORONA, ARGON, and LANYARD (McDonald, 1995a; McDonald, 1995b). This collection

TABLE 1-2. Primary geophysical variables measurable with each spectral band of the EOS MODIS system (Salomonson et al., 1995). Note the units of spectral range are nanometers (nm) for bands 1–19 and micrometers (μm) for bands 20–36.

geophysical variables		band	spectral range	GSI (m)
general	specific			
land/cloud boundaries	vegetation chlorophyll	1	620–670 nm	250
	cloud and vegetation	2	841–876	
land/cloud properties	soil, vegetation differences	3	459–479	500
	green vegetation	4	545–565	
	leaf/canopy properties	5	1230–1250	
	snow/cloud differences	6	1628–1652	
	land and cloud properties	7	2105–2155	
ocean color	chlorophyll observations	8	405–420	1000
	chlorophyll observations	9	438–448	
	chlorophyll observations	10	483–493	
	chlorophyll observations	11	526–536	
	sediments	12	546–556	
	sediments, atmosphere	13	662–672	
	cholorophyll flourescence	14	673–683	
	aerosol properties	15	743–753	
	aerosol/atmosphere properties	16	862–877	
atmosphere/ clouds	cloud/atmosphere properties	17	890–920	
	cloud/atmosphere properties	18	931–941	
	cloud/atmosphere properties	19	915–965	
thermal properties	sea surface temperatures	20	3.66–3.84 μm	
	forest fires/volcanoes	21	3.929–3.989	
	cloud/surface temperature	22	3.929–3.989	
	cloud/ surface temperature	23	4.02–4.08	
	troposphere temp/cloud fraction	24	4.433–4.498	
	troposphere temp/cloud fraction	25	4.482–4.549	
atmosphere/ clouds	cirrus clouds	26	1.36–1.39	
thermal properties	mid-troposphere humidity	27	6.535–6.895	
	upper-troposphere humidity	28	7.175–7.475	
	surface temperature	29	8.4–8.7	
	total ozone	30	9.58–9.88	
	cloud/surface temperature	31	10.78–11.28	
	cloud/surface temperature	32	11.77–12.27	
	cloud height and fraction	33	13.185–13.485	
	cloud height and fraction	34	13.485–13.785	
	cloud height and fraction	35	13.785–14.085	
	cloud height and fraction	36	14.085–14.385	

consists of more than 800,000 photographs (some scanned and digitized), mostly black and white, but some in color and stereo, over large portions of the earth at resolutions of 2 to 8 m. The imagery covers the period 1959 to 1972 and, although less systematically acquired over the whole globe than Landsat data, provides a previously unavailable, 12-year historical record that is an invaluable baseline for environmental studies. These data are available from the USGS Center for EROS.

Commercial development in the late 1990s of high-performance orbital sensors (Fritz, 1996), with resolutions of 0.5 to 1 m in panchromatic mode and 2.5 to 4 m in multispectral mode, has opened new commercial markets and public service opportunities for satellite imagery, such as real estate marketing, design of cellular telephone and wireless Personal Communications System (PCS) coverage areas (which depend on topography and building structures), urban and transportation planning, and natural and man-made disaster mapping and management. These systems also have value for military intelligence and environmental remote sensing applications. The first generation includes the IKONOS, QuickBird, and OrbView sensors, and further development is expected, particularly toward higher resolution capabilities, subject to legal regulations and security concerns of various countries. An example QuickBird image is shown in Plate 1-5.

The highest ground resolution of all is possible with airborne sensors, which have traditionally been photographic cameras. However, digital array cameras and pushbroom scanners have been developed to the point where they are beginning to compete with photography for airborne mapping projects. An example of imagery from digital airborne systems is shown in Plate 1-6.

1.2.1 Information Extraction from Remote-Sensing Images

One can view the use of remote-sensing data in two ways. The traditional approach might be called *image-centered*. Here the primary interest is in the spatial relationships among features on the ground, which follows naturally from the similarity between an aerial or satellite image and a cartographic map. In fact, the common goal of image-centered analyses is the creation of a map. Historically, aerial photographs were analyzed by *photointerpretation*. This involves a skilled and experienced human analyst who locates and identifies features of interest. For example, rivers, geologic structures, and vegetation may be mapped for environmental applications, or airports, troop convoys, and missile sites for military purposes. The analysis is done by examination of the photograph, sometimes under magnification or with a stereo viewer (when two overlapping photos are available), and transfer of the spatial coordinates and identifying attributes of ground features to a map of the area. Special instruments like the *stereoplotter* are used to extract elevation points and contours from stereo imagery. Examples of photointerpretation are provided in many textbooks on remote sensing (Colwell, 1983; Lillesand *et al.*, 2004; Sabins, 1997; Avery and Berlin, 1992; Campbell, 2002).

With most remote-sensing imagery now available in digital form, the use of computers for information extraction is standard practice. For example, images can be *enhanced* to facilitate visual interpretation or *classified* to produce a digital thematic map (Swain and Davis, 1978; Moik, 1980; Schowengerdt, 1983; Niblack, 1986; Mather, 1999; Richards and Jia, 1999; Landgrebe, 2003; Jensen, 2004). In recent years, the process of creating feature and elevation maps from remote-sensing images has been partially automated by *softcopy photogrammetry*. Although these computer tools speed and improve analysis, the end result is still a map; and in most cases, visual interpretation cannot be supplanted completely by computer techniques (Fig. 1-2).

The second view of remote sensing might be called *data-centered*. In this case, the scientist is primarily interested in the data dimension itself, rather than the spatial relationships among ground features. For example, specialized algorithms are used with hyperspectral data to measure *spectral absorption features* (Rast *et al.*, 1991; Rubin, 1993) and estimate *fractional abundances* of surface materials for each pixel (Goetz *et al.*, 1985; Vane and Goetz, 1988).[5] Atmospheric and ocean parameters can be obtained with *profile retrieval* algorithms that invert the integrated signal along the view path of the sensor. Accurate absolute or relative radiometric calibration is generally more important for data-centered analysis than for image-centered analysis. Even in data-centered analysis, however, the results and products should be presented in the context of a spatial map in order to be fully understood.

Interest in global change and in long-term monitoring of the environment and man's effect on it naturally leads to the use of remote-sensing data (Townshend *et al.*, 1991). Here the two views, image-centered and data-centered, converge. The science required for global change monitoring means that we must not only extract information from the spectral and temporal data dimensions, but also must integrate it into a spatial framework that can be understood in a global sense. It is particularly important in this context to ensure that the data are spatially and radiometrically calibrated and consistent over time and from one sensor to another. For example, imagery is *georeferenced* to a fixed spatial grid relative to the earth (geodetic coordinates) to facilitate analysis of data from different sensors acquired at different times. The data can then be "inverted" by algorithms capable of modeling the physics of remote sensing to derive *sensor-independent* geophysical variables.

1.2.2 Spectral Factors in Remote Sensing

The major optical spectral regions used for earth remote sensing are shown in Table 1-3. These particular spectral regions are of interest because they contain relatively transparent atmospheric "windows," through which (barring clouds in the non-microwave regions) the ground can be seen from above, and because there are effective radiation detectors in these regions. Between these windows, various constituents in the atmosphere absorb radiation, e.g., water vapor and carbon dioxide absorb from $2.5-3\,\mu m$ and $5-8\,\mu m$. In the microwave region given in Table 1-3, there is a minor water absorption band near 22 GHz frequency (about 1.36 cm wavelength)[6] with a transmittance of about 0.85 (Curlander and McDonough, 1991). Above 50 GHz (below 0.6 cm wavelength), there is a major oxygen absorption region to about 80 GHz (Elachi, 1988). At the frequencies of high

---

5. A *pixel* is one element of a two-dimensional digital image. It is the smallest sample unit available for processing in the original image.

6. For all electromagnetic waves, the frequency in Hertz (Hz; cycles/second) is given by $v = c/\lambda$, where c is the speed of light ($2.998 \times 10^8\,m/sec$ in vacuum) and λ is the wavelength in meters (Slater, 1980; Schott, 1996). A useful nomograph is formed by a log-log plot of λ versus v (Fig. 1-4).

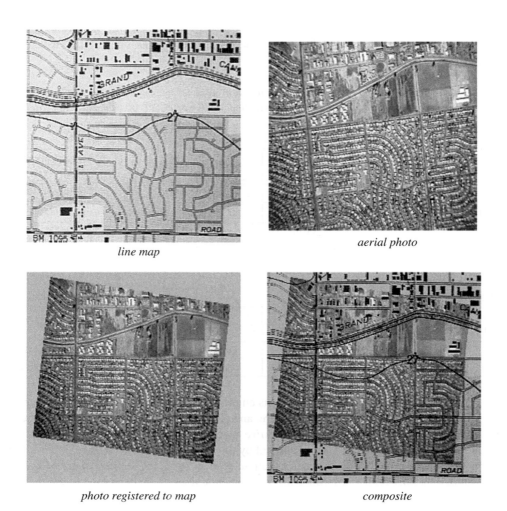

line map

aerial photo

photo registered to map

composite

FIGURE 1-2. *An example of how maps and imagery complement each other. The line map of an area in Phoenix, Arizona, produced manually from a stereo pair of aerial photographs and scanned into the digital raster form shown here, is an abstraction of the real world; it contains only the information that the cartographer intended to convey: an irrigation canal (the Grand Canal across the top), roads, elevation contours (the curved line through the center), and large public or commercial buildings. An aerial photograph can be registered to the map using image processing techniques and the two superimposed to see the differences. The aerial photo contains information about land use which is missing from the map. For example, an apartment complex that may not have existed (or was purposely ignored) when the map was made can be seen as a group of large white buildings to the left-center. Also, agricultural fields are apparent in the aerial photo to the right of the apartment complex, but are not indicated on the line map. In the lower half, the aerial photo shows many individual houses that also are not documented on the map.*

atmospheric transmittance, microwave and radar sensors are noted for their ability to penetrate clouds, fog, and rain, as well as an ability to provide nighttime reflected imaging by virtue of their own active illumination.

TABLE 1-3.   *The primary spectral regions used in earth remote sensing. The boundaries of some atmospheric windows are not distinct and one will find small variations in these values in different references.*

name	wavelength range	radiation source	surface property of interest
Visible (V)	0.4–0.7 μm	solar	reflectance
Near InfraRed (NIR)	0.7–1.1 μm	solar	reflectance
Short Wave InfraRed (SWIR)	1.1–1.35 μm 1.4–1.8 μm 2–2.5μm	solar	reflectance
MidWave InfraRed (MWIR)	3–4 μm 4.5–5 μm	solar, thermal	reflectance, temperature
Thermal or LongWave InfraRed (TIR or LWIR)	8–9.5 μm 10–14 μm	thermal	temperature
microwave, radar	1 mm–1 m	thermal (passive), artificial (active)	temperature (passive), roughness (active)

Passive remote sensing in all of these regions employs sensors that measure radiation naturally reflected or emitted from the ground, atmosphere, and clouds. The Visible, NIR, and SWIR regions (from 0.4 μm to about 3 μm) are the *solar-reflective* spectral range because the energy supplied by the sun at the earth's surface exceeds that emitted by the earth itself. The MWIR region is a transition zone from solar-reflective to thermal radiation. Above 5 μm, *self-emitted thermal radiation* from the earth generally dominates. Since this phenomenon does not depend directly on the sun as a source, TIR images can be acquired at night, as well as in the daytime. This self-emitted radiation can be sensed even in the microwave region as *microwave brightness temperature*, by passive systems such as the Special Sensor Microwave/Imager (SSM/I) (Hollinger *et al.*, 1987; Hollinger *et al.*, 1990). An example multispectral VSWIR and TIR image is shown in Fig. 1-3.

Active remote-sensing techniques employ an artificial source of radiation as a probe. The resulting signal that scatters back to the sensor characterizes either the atmosphere or the earth. For example, the radiation scattered and absorbed at a particular wavelength from a laser beam probe into the atmosphere can provide information on molecular constituents such as ozone. In the microwave spectral region, *Synthetic Aperture Radar (SAR)* is an imaging technology in which radiation is emitted in a beam from a moving sensor, and the backscattered component returned to the sensor from the ground is measured. The motion of the sensor platform creates an effectively larger antenna, thereby increasing the spatial resolution. An image of the backscatter spatial

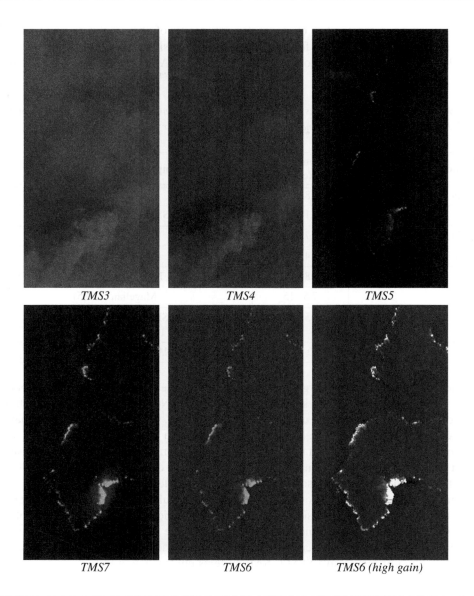

TMS3   TMS4   TMS5

TMS7   TMS6   TMS6 (high gain)

FIGURE 1-3. This airborne Thematic Mapper Simulator (TMS) image of a devastating wildfire in Yellowstone National Park, Wyoming, was acquired on September 2, 1988. The TMS bands are the same as those of TM. In the VNIR bands, TMS3 and TMS4, only the smoke from the fire is visible. The fire itself begins to be visible in TMS5 (1.55–1.75 µm) and is clearly visible in TMS7 (2.08–2.35 µm) and TMS6 (8.5–14 µm). The high gain setting in the lower right image provides higher signal level in the TIR. (Imagery courtesy of Jeffrey Myers, Aircraft Data Facility, NASA/Ames Research Center.)

distribution can be reconstructed by computer processing of the amplitude and phase of the returned signal. Wavelengths used for microwave remote sensing, active and passive, are given in Table 1-4 and graphically related to frequencies in Fig. 1-4. An example SAR image is shown in Fig. 1-5.

TABLE 1-4. Microwave wavelengths and frequencies used in remote sensing. Compiled from Sabins, 1987, Hollinger et al., 1990, Way and Smith, 1991, and Curlander and McDonough, 1991.

band	frequency (GHz)	wavelength (cm)	examples (frequency in GHz)
Ka	26.5–40	0.8–1.1	SSM/I (37.0)
K	18–26.5	1.1–1.7	SSM/I (19.35, 22.235)
Ku	12.5–18	1.7–2.4	Cassini (13.8)
X	8–12.5	2.4–3.8	X-SAR (9.6)
C	4–8	3.8–7.5	SIR-C (5.3), ERS-1 (5.25), RADARSAT (5.3)
S	2–4	7.5–15	Magellan (2.385)
L	1–2	15–30	Seasat (1.275), SIR-A (1.278), SIR-B (1.282), SIR-C (1.25), JERS-1 (1.275)
P	0.3–1	30–100	NASA/JPL DC-8 (0.44)

Figure 1-6 shows the solar energy spectrum received at the earth (above the atmosphere), with an overlaying plot of the daylight response of the human eye. Notice that what we see with our eyes actually occupies only a small part of the total solar spectrum, which in turn is only a small part of the total electromagnetic spectrum. Much remote-sensing data is therefore "non-visible," although we can, of course, display the digital imagery from any spectral region on a monitor. Visual interpretation of TIR and microwave imagery is often considered difficult, simply because we are not innately familiar with what the sensor "sees" outside the visible region.

As indicated, most optical remote-sensing systems are multispectral, acquiring images in several spectral bands, more or less simultaneously. They provide multiple "snapshots" of spectral properties which are often much more valuable than a single spectral band or broad band (a so-called "panchromatic") image. Microwave systems, on the other hand, tend to be single frequency, with the exception of the passive SSM/I. SAR systems emit radiation in two polarization planes, horizontal (H) and vertical (V), and sense the return in either the same planes (HH, VV modes) or in the orthogonal planes (HV, VH modes) (Avery and Berlin, 1992; Richards and Jia, 1999). More and more, images from different spectral regions, different sensors or different polarizations are combined for improved interpretation and analysis. Examples include composites of thermal and visible imagery (Haydn *et al.*, 1982), radar and visible imagery (Wong and Orth, 1980; Welch and Ehlers, 1988), aerial photography and hyperspectral imagery (Filiberti *et al.*, 1994), and gamma ray maps with visible imagery (Schetselaar, 2001).

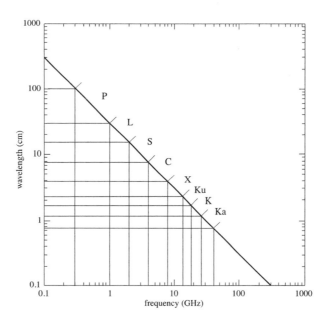

FIGURE 1-4. *A nomograph for finding wavelength given frequency, or vice versa, in the microwave spectral region. The major radar bands are indicated. A similar nomograph can be drawn for any region of the electromagnetic spectrum.*

1.3   Spectral Signatures

The spatial resolution of satellite remote-sensing systems is too low to identify many objects by their shape or spatial detail. In some cases, it is possible to identify such objects by spectral measurements. There has, therefore, been great interest in measuring the *spectral signatures* of surface materials, such as vegetation, soil, and rock, over the spectral range in Fig. 1-6. The spectral signature of a material may be defined in the solar-reflective region by its reflectance as a function of wavelength, measured at an appropriate spectral resolution. In other spectral regions, signatures of interest are temperature and emissivity (TIR) and surface roughness (radar). The motivation of multispectral remote sensing is that different types of materials can be distinguished on the basis of differences in their spectral signatures. Although this desirable situation is often reached in practice, it is also often foiled by any number of factors, including

- natural variability for a given material type
- coarse spectral quantization of many remote-sensing systems

Example pahoehoe lava flow. (Image from U.S. Geological Survey.)

FIGURE 1-5. Spaceborne imaging radar image of Isla Isabella in the western Galapagos Islands taken by the L-band radar in HH polarization from the Spaceborne Imaging Radar C/X-Band Synthetic Aperture Radar on the 40th orbit of the space shuttle Endeavour on April 15, 1994. The image is centered at about 0.5° south latitude and 91° west longitude. The radar incidence angle at the center of the image is about 20 degrees. The western Galapagos Islands, which lie about 1200km west of Ecuador in the eastern Pacific, have six active volcanoes, and since the time of Charles Darwin's visit to the area in 1835, there have been over 60 recorded eruptions. This SIR-C/X-SAR image of Alcedo and Sierra Negra volcanoes shows the rougher lava flows as bright features, while ash deposits and smooth pahoehoe lava flows appear dark. (Image and description courtesy of NASA/JPL.)

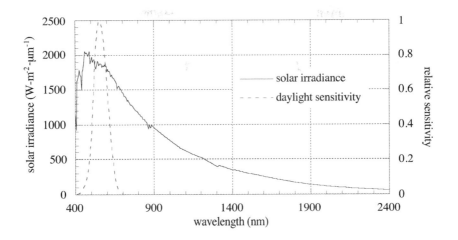

FIGURE 1-6. *Exo-atmospheric (i.e., arriving at the top of the atmosphere) solar spectral irradiance and the daylight-adapted response of the human eye.*

• modification of signatures by the atmosphere

Therefore, even though we may wish to apply different labels to different materials, *there is no guarantee that they will exhibit measurably different signatures in the natural environment.*

Figure 1-7 shows spectral reflectance curves for different types of grasses and agricultural crop types. Note that all of these vegetation "signatures" exhibit similar general characteristics, namely a low reflectance in the green–red spectrum,[7] a sharp increase in reflectance near 710 nm,[8] and strong dips in reflectance near 1400 nm and 1900 nm caused by liquid water absorption in the plant leaves. The spectral signature for vegetation is perhaps the most variable in nature since it changes completely during the seasonal life cycle of many plants, acquiring a "yellow" characteristic in senescence, with a corresponding increase in the red region reflectance caused by a loss of photosynthetic chlorophyll.

Spectral reflectance data for some geologic materials are shown in Fig. 1-8. The dry and wet clay example illustrates the overall decrease in reflectance that results from an increase in water content in the sample material. Note also the characteristic SWIR water absorption bands, similar to those seen in vegetation. The curves for the alteration minerals are high spectral resolution laboratory measurements. Each mineral shows distinguishing absorption features, in some cases "doublets." Such features will not be seen with broadband sensors, such as Landsat TM and SPOT, but can be measured with a narrowband hyperspectral sensor, such as AVIRIS, with a 10 nm spectral

---

7. The small peak in the green near 550 nm is due to low chlorophyll absorption relative to the blue and red spectral regions on either side. This peak is the reason healthy plants appear green to the human eye.

8. This so-called vegetation "red edge" is caused by the cellular structure within plant leaves.

bandwidth. Several spectral reflectance "libraries" have been published (Clark *et al.*, 1993; Hunt, 1979; Hook, 1998); these data can be used as reference spectra for matching to calibrated hyper-spectral sensor data (Chapter 9).

All spectral reflectance data are unique to the sample and the environment in which they are measured. Mineral signatures, for example, will vary from sample to sample. Vegetation is even more variable, being dependent on growth stage, plant health, and moisture content. Complicating matters further, it is impossible to duplicate field reflectance measurement conditions in the labora-tory. Even if the reference reflectance data are taken in the field, aerial and satellite imagery suffers from atmospheric, topographic, and calibration influences (Chapters 2 and 3) that alter the signa-ture as imaged and recorded by the sensor (Marsh and Lyon, 1980). Therefore, the use of labora-tory or field reflectance data should be tempered by the fact that they only approximate the "signature" in the real world, and that remote sensor data need careful calibration for comparison to laboratory or field measurements. What saves image-based analysis in many cases is that we can compare relative signatures, one material versus another within a single image acquisition, rather than absolute signatures.

1.4   Remote-Sensing Systems

The details of sensor construction and materials vary with the wavelengths of interest, and the dimensions of optical systems and detectors depend on the engineering limitations in particular spectral regions. However, all passive, scanning optical sensors (visible through thermal spectral regions) operate on the same principles of optical radiation transfer, image formation, and photon detection. Our descriptions will focus on this type of sensor. Microwave sensors, active and passive, are decidedly different in their nature and are not described.

1.4.1 Spatial and Radiometric Characteristics

Every pixel represents an average in each of three dimensions: space, wavelength, and time. The average over time is usually very small (on the order of microseconds for a whiskbroom scanner such as TM and milliseconds for a pushbroom scanner such as SPOT) and is inconsequential in most applications. The averages over space and wavelength, however, define the characteristics of the data in those critical dimensions.

If we imagine a three-dimensional continuous parameter space (x,y,λ), defined over spatial coordinates (x,y) and spectral wavelength (λ), we can visualize each pixel of a given image as rep-resenting an integration over a relatively small volume element in that continuous space (Fig. 1-9). We will see in Chapter 3 that the (x,y,λ) space is not quite as neatly divided as Fig. 1-9 indicates. Specifically, the volume of integration represented by each pixel is not a well-defined box, but overlaps in both the spatial and spectral dimensions with the integration volumes of neighboring pixels. For now, however, we will assume this convenient subdivision.

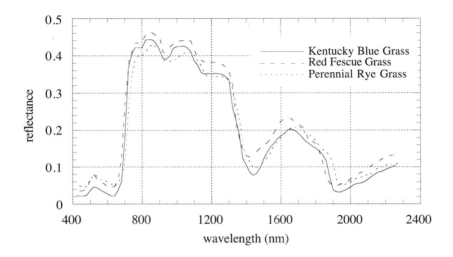

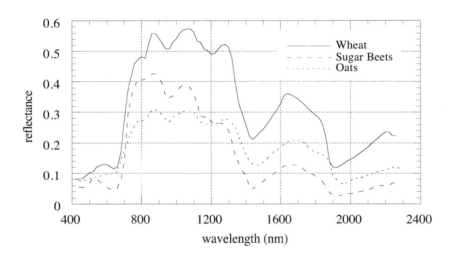

FIGURE 1-7. *Example vegetation spectral reflectance curves (Bowker et al., 1985). The curves in the upper graph show variability among three types of grasses; even under relatively well-controlled laboratory conditions, the reflectance of corn leaves has been found to vary as much as ±17% near the reflectance edge at 0.67 μm (Landgrebe, 1978).*

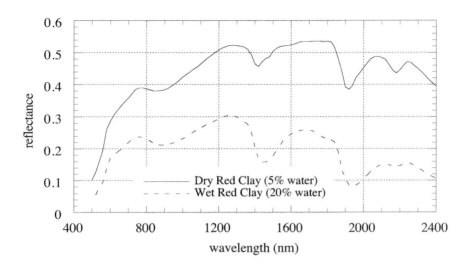

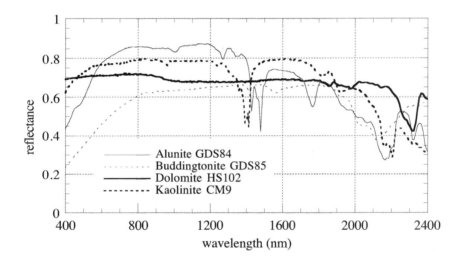

FIGURE 1-8. Example mineral spectral reflectance curves for clay (top) and several alteration minerals (bottom) (Bowker et al., 1985; Clark et al., 1993). The presence of liquid water in any material generally lowers the reflectance; for a theory of this effect, see Twomey et al., 1986.

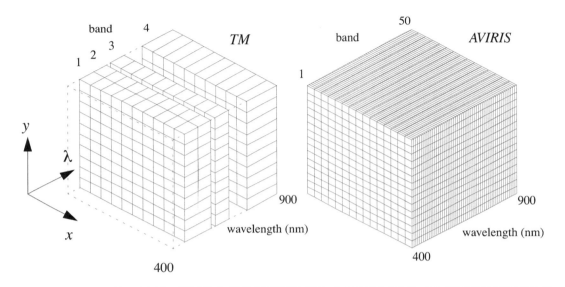

FIGURE 1-9. Comparison of the spatial and spectral sampling of the Landsat TM and AVIRIS in the VNIR spectral range. Each small rectangular box represents the spatial-spectral integration region of one image pixel. The TM samples the spectral dimension incompletely and with relatively broad spectral bands, while AVIRIS has relatively continuous spectral sampling over the VNIR range. AVIRIS also has a somewhat smaller GSI (20m) compared to TM (30m). This type of volume visualization for spatial-spectral image data is called an "image cube" (Sect. 9.9.1).

The grid of pixels that constitutes a digital image is achieved by a combination of scanning in the *cross-track* direction (orthogonal to the motion of the sensor platform) and by the platform motion along the *in-track* direction (Fig. 1-10) (Slater, 1980). A pixel is created whenever the sensor system electronically samples the continuous data stream provided by the scanning. A line scanner uses a single detector element to scan the entire scene. Whiskbroom scanners, such as the Landsat TM, use several detector elements, aligned in-track, to achieve parallel scanning during each cycle of the scan mirror. A related type of scanner is the paddlebroom, exemplified by AVHRR and MODIS, with a two-sided mirror that rotates 360°, scanning continuously cross-track. A significant difference between paddlebroom and whiskbroom scanners is that the paddlebroom always scans in the same direction, while the whiskbroom reverses direction for each scan. Pushbroom scanners, such as SPOT, have a linear array of thousands of detector elements, aligned cross-track, which scan the full width of the collected data in parallel as the platform moves. For all types of scanners, the full cross-track angular coverage is called the *Field Of View (FOV)* and the corresponding ground coverage is called the *Ground-projected Field Of View (GFOV).*[9]

---

9. Also called the *swath width*, or sometimes, the *footprint* of the sensor.

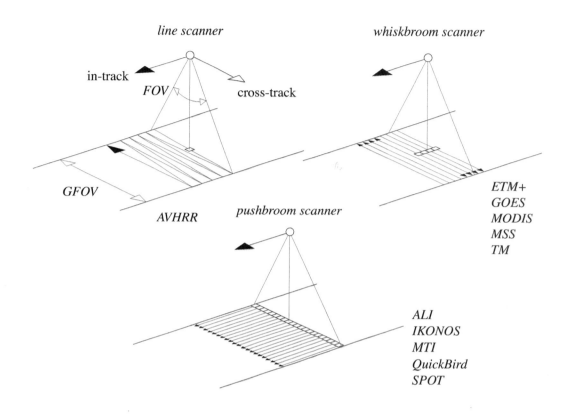

FIGURE 1-10. *Definition of basic scanner parameters and depiction of three scanning methods, with specific examples of whiskbroom and pushbroom scanners. The solid arrows represent motion relative to a stationary earth. In reality, the earth is rotating during the scanning process, approximately in the cross-track direction since most satellite remote-sensing systems are in a near-polar orbit. This results in a east-west skew in the surface coverage over the full scene.*

The spacing between pixels on the ground is the *Ground-projected Sample Interval (GSI)*. The cross-track and in-track *GSI*s are determined by the cross-track and in-track sampling rates, respectively, and the in-track platform velocity. It is common practice to design the sample rates so that the *GSI* equals the *Ground-projected Instantaneous Field of View (GIFOV)*,[10] the geometric projection of a single detector *width, w,* onto the earth's surface (Fig. 1-11 and Fig. 1-12). Thus, the *GIFOV*s of neighboring pixels will abut, both in-track and cross-track. The in-track *GSI* is determined by the necessary combination of platform velocity and sample rate (pushbroom) or scan velocity (line and whiskbroom) to match the in-track *GIFOV* at nadir. Some systems have a higher cross-track sample rate that leads to overlapping *GIFOV*s, e.g. the Landsat MSS and the AVHRR KLM models. This cross-track "over-sampling" results in some improvement in the data quality.

10. Also called the *Ground Sample Distance (GSD)*.

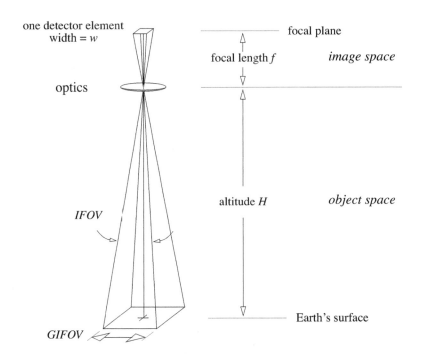

one detector element
width = w

optics

IFOV

GIFOV

focal plane

focal length f *image space*

altitude H *object space*

Earth's surface

FIGURE 1-11. *Simple geometric description of a single detector element in the focal plane of an optical sensor. The sizes of w and f are greatly exaggerated relative to H for clarity. Likewise for the optics (which, more often than not, would be a series of curved mirrors, possibly with multiple, folded optical paths). Angular parameters, such as the IFOV, are the same in image and object space in this model, but linear dimensions are related by the magnification f/H between the two spaces. Everything in this diagram is assumed stationary and in a nadir view; with scan, sensor platform, and earth motion, the GIFOV moves during the integration time of the detector, resulting in an effective GIFOV somewhat larger than shown. Also, as the scan proceeds off-nadir, the effective GIFOV increases (sometimes called "pixel growth") from oblique projection onto the earth. These effects are discussed in Chapter 3.*

The *GSI* is determined by the altitude of the sensor system H, the sensor's focal length, f, and the inter-detector spacing (or spatial sample rate as explained previously). If the sample rate is equal to one pixel per inter-detector spacing, the relation for the *GSI* at *nadir*, i.e., directly below the sensor, is simply,

$$GSI = \text{inter-detector spacing} \times \frac{H}{f} = \frac{\text{inter-detector spacing}}{m} \qquad (1\text{-}1)$$

where f/H is the *geometric magnification*, m, from the ground to the sensor focal plane.[11] As we mentioned, the inter-detector spacing is usually equal to the detector width, w.

---

11. Since $f \ll H$, m is much less than one.

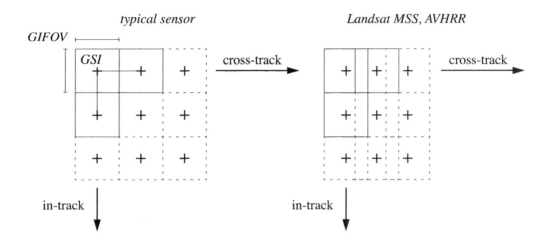

FIGURE 1-12. *The relationship between GIFOV and GSI for most scanning sensors and for the Landsat MSS and AVHRR. Each cross is a pixel. For MSS, the cross-track GSI was 57m and the GIFOV was 80m, resulting in 1.4 cross-track pixels/GIFOV. Similarly, the AVHRR KLM models have 1.36 cross-track pixels/ GIFOV. The higher cross-track sample density improves data quality, but also increases correlation between neighboring pixels and results in more data collected over the GFOV.*

The *GIFOV* depends in a similar fashion on H, f, and w. System design engineers prefer to use the *Instantaneous Field of View (IFOV)*, defined as the angle subtended by a single detector element on the axis of the optical system (Fig. 1-11),

$$IFOV = 2\operatorname{atan}\left(\frac{w}{2f}\right) \cong \frac{w}{f} . \tag{1-2}$$

The *IFOV* is independent of sensor operating altitude, H, and is the same in the image and the object space. It is a convenient parameter for airborne systems where the operating altitude may vary. For the *GIFOV*, we therefore have,

$$GIFOV = 2H\tan\left(\frac{IFOV}{2}\right) = w \times \frac{H}{f} = \frac{w}{m} . \tag{1-3}$$

The *GSI* and *GIFOV* are found by scaling the inter-detector spacing and width, respectively, by the geometric magnification, m. Users of satellite and aerial remote-sensing data generally (and justifiably) prefer to use the *GIFOV*, rather than the *IFOV*, in their analyses. Sensor engineers, on the other hand, often prefer the angular parameters *FOV* and *IFOV* because they have the same values in image and object space (Fig. 1-11).

The detector layout in the focal plane of scanners is typically not a regular row or grid arrangement. Due to sensor platform and scan mirror velocity, various sample timings for bands and pixels, the need to physically separate different spectral bands, and the limited area available on the focal plane, the detectors are often in some type of staggered pattern as depicted in Fig. 1-13 to Fig. 1-16.

The sensor collects some of the electromagnetic radiation (radiance[12]) that propagates upward from the earth and forms an image of the earth's surface on its focal plane. Each detector integrates the energy that strikes its surface (irradiance[13]) to form the measurement at each pixel. Due to several factors, the actual area integrated by each detector is somewhat larger than the *GIFOV*-squared (Chapter 3). The integrated irradiance at each pixel is converted to an electrical signal and quantized as a integer value, the *Digital Number (DN)*.[14] As with all digital data, a finite number of bits, Q, is used to code the continuous data measurements as binary numbers. The number of discrete *DN*s is given by,

$$N_{DN} = 2^Q \qquad (1\text{-}4)$$

and the *DN* can be any integer in the range,

$$DN_{range} = [0, 2^Q - 1] \ . \qquad (1\text{-}5)$$

The larger the value of Q, the more closely the quantized data approximates the original continuous signal generated by the detectors, and the higher the *radiometric resolution* of the sensor. Both SPOT and TM have 8 bits per pixel, while AVHRR has 10 bits per pixel. To achieve high radiometric precision in a number of demanding applications, the EOS MODIS is designed with 12 bits per pixel, and most hyperspectral sensors use 12 bits per pixel. Not all bits are always significant for science measurements, however, especially in bands with low signal levels or high noise levels.

In summary, a pixel is characterized, to the first order, by three quantities: *GSI*, *GIFOV*, and *Q*. These parameters are always associated with the term "pixel." Whenever there may be confusion as to what is meant between the *GSI* and *GIFOV*, "pixel" should be reserved to mean the *GSI*.

Discrete multispectral channels are typically created in an optical sensor by splitting the optical beam into multiple paths and inserting different spectral filters in each path or directly on the detectors. Some hyperspectral sensors, such as HYDICE, use a two-dimensional array of detectors in the focal plane (Fig. 1-17). A continuous spectrum is created across the array by an optical component such as a prism or diffraction grating. For HYDICE, the detector array has 320 elements cross-track and 210 in-track. The cross-track dimension serves as a line of pixels in a pushbroom mode, while the optical beam is *dispersed over wavelength* by a prism along the other direction (in-track) of the array. Therefore, as the sensor platform (an aircraft in the case of HYDICE) moves in-track, a full

---

12. Radiance is a precise scientific term used to describe the power density of radiation; it has units of $W\text{-}m^{-2}\text{-}sr^{-1}\text{-}\mu m^{-1}$, i.e., watts per unit source area, per unit solid angle, and per unit wavelength. For a thorough discussion of the role of radiometry in optical remote sensing, see Slater (1980) and Schott (1996).

13. Irradiance has units of $W\text{-}m^{-2}\text{-}\mu m^{-1}$. The relationship between radiance and irradiance is discussed in Chapter 2.

14. Also known as *Digital Count*.

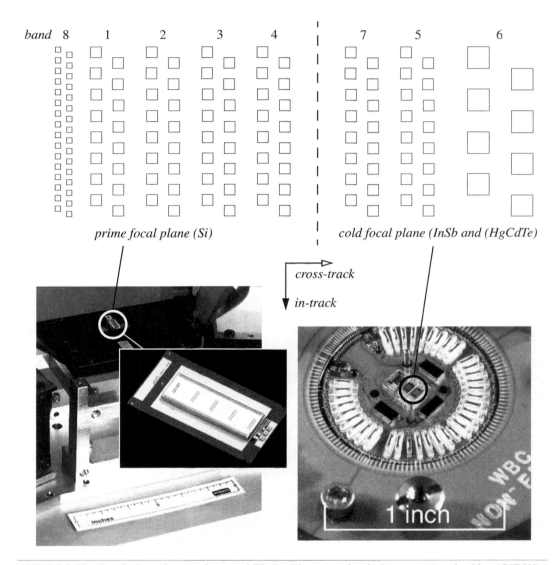

FIGURE 1-13. *Focal plane detector layout of ETM+. The prime focal plane contains the 10m (GIFOV) panchromatic and 30m VNIR bands with silicon (Si) detectors; the cold focal plane is cooled to reduce detector noise and contains the 30m SWIR bands with Indium-Antimonide (InSb) detectors and the 60m LWIR bands with Mercury-Cadmium-Telluride (HgCdTe) detectors. As the scan mirror sweeps the image across the focal planes in the cross-track direction, data are acquired in all bands simultaneously; electronic sample timing is used to correct the cross-track phase differences among pixels. The photographs of the focal plane assemblies indicate the detector array's actual size—a few millimeters. (Raytheon photographs courtesy of Ken J. Ando.)*

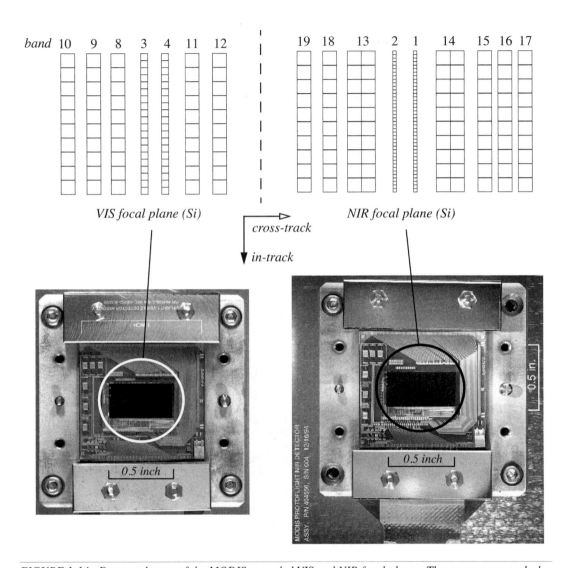

FIGURE 1-14. *Detector layout of the MODIS uncooled VIS and NIR focal planes. The arrays appear dark in both cases because of anti-reflection optical coatings. The NIR focal plane contains the 250m red and NIR bands 1 and 2 that are used mainly for land remote sensing. The double columns of detectors in bands 13 and 14 employ Time Delay Integration (TDI) to increase signal and reduce noise levels for dark scenes such as oceans. The data from one column are electronically delayed during scanning to effectively align with the data from the other column, and then the two sets of data are summed, in effect increasing the integration time at each pixel in bands 13 and 14. (Raytheon photographs courtesy of Ken J. Ando and John Vampola.)*

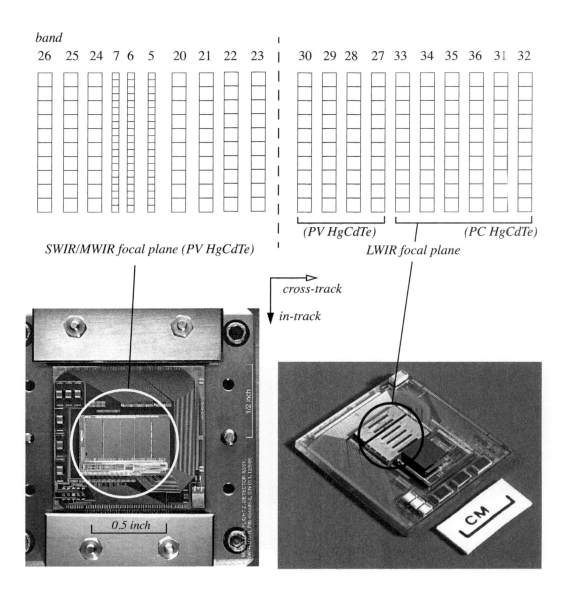

FIGURE 1-15. Detector layout of the MODIS cooled focal planes. Bands 31–36 are the longest wave IR bands and use photoconductive (PC) HgCdTe detectors, while the SWIR/MWIR bands use photovoltaic (PV) detectors. (Raytheon photographs courtesy of Ken J. Ando.)

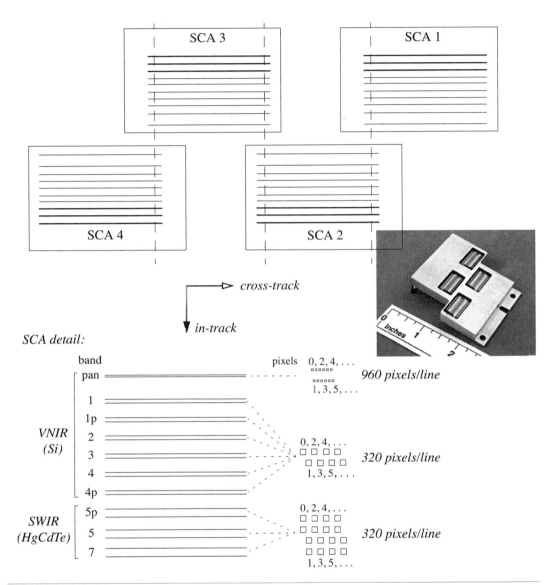

FIGURE 1-16. *The four sensor chip assemblies (SCAs) in the ALI. The even- and odd-numbered detectors in each band are in separate rows on the SCA; the phase difference is corrected by data processing. The double rows of detectors in the SWIR bands 5p (the "p" denotes a band that does not correspond to a ETM+ band), 5, and 7 allow TDI for higher signal-to-noise (SNR) (see also Fig. 1-14 for an explanation of TDI). There are a total of 3850 panchromatic detectors and 11,520 multispectral detectors on the four SCAs. The pixels resulting from overlapping detectors between SCAs are removed during geometric processing of the imagery (Storey et al., 2004).*

line of cross-track pixels is acquired *simultaneously* in all 210 spectral bands, a total of 67,200 values. These data are read out of the detector array in time for a new line to be acquired, contiguous to the previous line by virtue of the platform's ground velocity. The mapping of detector elements to spectral bands is shown in Plate 1-7 for imaging spectrometers and discrete multispectral sensors.

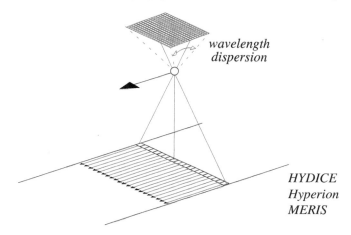

FIGURE 1-17.   The pushbroom 2-D array concept used in the HYDICE, Hyperion, and MERIS hyperspectral sensors. Each cross-track line of pixels is simultaneously dispersed in wavelength along the in-track direction of the array. Therefore, the 2-D array measures cross-track spatial variation in a large number of spectral bands simultaneously. The number of spectral bands is equal to the number of detector elements in the in-track direction. All of the spectral data for one cross-track line of pixels must be read out of the array before the sensor progresses to the next line.

The *Field of View (FOV)* of a sensor is the angular extent of data acquisition cross-track (Fig. 1-10). The corresponding cross-track ground distance is given by,[15]

$$GFOV = 2H\tan\left(\frac{FOV}{2}\right) .\qquad\qquad (1\text{-}6)$$

The *GFOV* is also called the *swath width*. While the *GSI* and *GIFOV* are specified for nadir viewing, i.e. directly below the aircraft or spacecraft, they increase toward either end of the *GFOV* as described in Chapter 3. The in-track *FOV* is not naturally defined, since data acquisition in-track is controlled by the continuous aircraft or satellite motion. Instead, it is often determined by ground processing limitations, data rate constraints, and a desire to have approximately equal coverage cross-track and in-track.

---

15. We're ignoring the curvature of the earth, which increases this distance. The error in this approximation will increase as the *FOV* or *H* increases; a more accurate description that includes earth curvature is provided in Chapter 3.

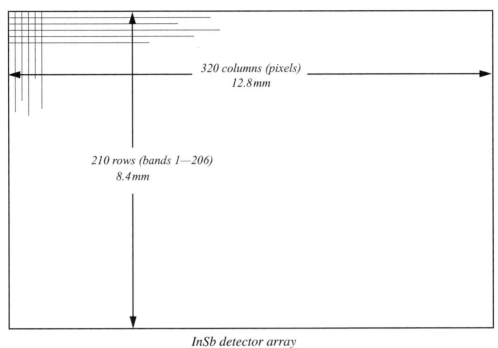

320 columns (pixels)
12.8mm

210 rows (bands 1—206)
8.4mm

InSb detector array

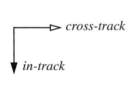

cross-track

in-track

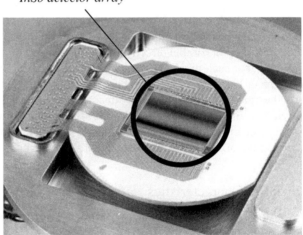

FIGURE 1-18. *The 2-D detector array layout used in HYDICE. Each detector element is 40μm on a side, and there are 67,200 detectors in the array. The array is segmented into three regions (400-1000, 1000-1900 and 1900-2500nm) with different electronic gains (Chapter 3) to compensate for different solar energy levels across the VSWIR spectrum (Fig. 1-6). (Photo courtesy of Bill Rappoport, Goodrich Electro-Optical Systems.)*

Sensors with multiple detectors per band, such as whiskbroom and pushbroom scanners, require relative radiometric calibration of each detector. The MSS had 6 detectors in each of 4 bands, for a total of 24, and the TM has 16 detectors in each of the 6 non-TIR bands (30 m *GSI*), plus 4 in band 6 (120 m *GSI*), for a total of 100 detectors. The Landsat-7 ETM+ has, in addition to the normal TM complement, 32 detectors in the panchromatic band (15 m *GSI*) and 8 in band 6 (60 m *GSI*), making a total of 136 detectors. Each detector is a discrete electronic element with its own particular responsivity characteristics, so relative radiometric calibration among detectors is particularly important. Errors in calibration lead to cross-track "striping" and "banding" noise in these systems, which can be quite visible artifacts. A pushbroom system has a very large number of cross-track detector elements (6000 in the SPOT panchromatic mode, averaged to 3000 in the multispectral mode), requiring proportionally more calibration effort. The most severe calibration requirements are for 2-D array hyperspectral sensors such as HYDICE—its focal plane has 67,200 individual detector elements within a single array (Fig. 1-18).

Line and whiskbroom scanners clearly have many dynamic motions occuring during scene acquisition (mirror rotation, earth rotation, satellite or aircraft roll/pitch/yaw) and consequently require complex post-processing to achieve accurate geometry. This can be done to a high level, however, as exemplified by the high quality of Landsat TM imagery. The geometry of a pushbroom scanner such as SPOT or ALI is comparatively simple, since scanning is achieved by the detector motion in-track. Factors such as earth rotation and satellite pointing during image acquisition affect the image geometry in any case. A comparison of ETM+ and ALI geometric performance is shown in (Fig. 1-19).

The total time required to "build" an image of a given in-track length depends on the satellite ground velocity, which is about 7 km/sec for low altitude, earth-orbiting satellites. A full TM scene therefore requires about 26 seconds to acquire, and a full SPOT scene about 9 seconds. The number of detectors scanning in parallel directly affects the amount of time available to integrate the incoming optical signal at each pixel. Pushbroom systems therefore have an advantage because all pixels in a line are recorded at the same time (Fig. 1-10). If there were no satellite motion, the radiometric quality of the image would increase with increased integration time, because the signal would increase relative to detector noise. With platform motion, however, longer integration times also imply greater smearing of the image as it is being sensed, leading to reduced spatial resolution. This effect is more than offset by the increased signal-to-noise ratio of the image (Fig. 1-20).

1.4.2 Spectral Characteristics

The spectral location of sensor bands is constrained by atmospheric absorption bands and further determined by the reflectance features to be measured. If a sensor is intended for land or ocean applications, atmospheric absorption bands are avoided in sensor band placement. On the other hand, if a sensor is intended for atmosphere applications, it may very well be desirable to put spectral bands within the absorption features. An example of a sensor designed for all three applications, MODIS collects spectral images in numerous and narrow (relative to ETM+) wavelength

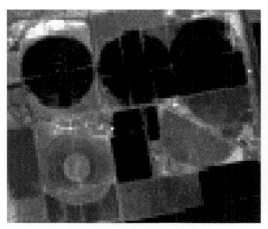
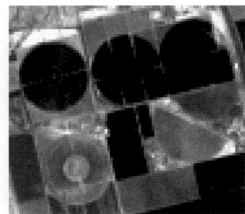

ETM+ Level 1G band 1   *ALI Level 1R band 2*

FIGURE 1-19. *Visual comparison of ETM+ whiskbroom and ALI pushbroom imagery acquired on July 27, 2001, of center-pivot irrigated agricultural fields near Maricopa, Arizona. The ALI Level 1R image, which has no geometric processing, depicts the circular shape of the center pivot irrigated fields more accurately than does the ETM+ Level 1G image, which has geometric corrections applied with nearest-neighbor resampling (Chapter 7). The pushbroom geometry of ALI is inherently better than that of a whiskbroom scanner like ETM+. Also, the superior SNR characteristics of ALI produces an image that shows somewhat better detail definition and less "noise." (ETM+ image provided by Ross Bryant and Susan Moran of USDA-ARS, Southwest Watershed Research Center, Tucson.)*

bands from the visible through the thermal infrared. The MODIS spectral band ranges are plotted in Fig. 1-21. All of the 36 bands are nominally co-registered and acquired at nearly the same time during scan mirror rotation.

The diversity of information in different bands is illustrated in Fig. 1-22 (all bands are displayed at 1 km *GIFOV*). These are Direct Broadcast (DB) data received at the USGS National Center for EROS in Sioux Falls, South Dakota. DB data are processed rapidly without the full ancillary data used for archived MODIS data. This winter image shows ice and snow in St. James Bay and Canada, thick clouds over the northeastern United States and clear skies along the eastern coast down to Florida. Band 2 is the NIR, band 6 is a band designed to distinguish ice and snow from clouds (ice and snow have low reflectance relative to clouds in this band), band 26 is designed to detect cirrus clouds, and band 27 senses mid-tropospheric water vapor. Note that the Earth's surface is not visible in bands 26 and 27 because they are in major atmospheric absorption bands (Fig. 1-21).

ETM+ pan   ALI pan

FIGURE 1-20. Visual comparison of ETM+ whiskbroom and ALI pushbroom panchromatic imagery of an area in Alaska acquired by ETM+ on November 27, 1999, and ALI on November 25, 2000 (Storey, 2005). The solar irradiance level was quite low because of the high latitude of the site and the winter acquisition dates, and both images have been strongly contrast stretched for visibility. The ALI clearly produced a better image with less "noise," largely because of the superior signal-to-noise properties of its pushbroom sensor.

1.4.3 Temporal Characteristics

One of the most valuable aspects of unmanned, near polar-orbiting satellite remote-sensing systems is their inherent repeating coverage of the same area on the earth. Regular revisits are especially important for monitoring agricultural crops; in fact, *temporal signatures* can be defined for certain crop types in different regions of the world and used to classify them from multitemporal images (Haralick *et al.*, 1980; Badhwar *et al.*, 1982). Other applications for multitemporal data are change detection of both natural and cultural features and monitoring of the atmosphere and oceans.

Many remote sensing satellites, including Landsat, AVHRR, and SPOT, are in sun-synchronous orbits which means that each always passes over the same spot on the ground at the same local time. The interval between revisits depends solely on the particulars of the satellite orbit for sensors that have a fixed view direction (nadir), such as the Landsat TM. If more than one system is in orbit

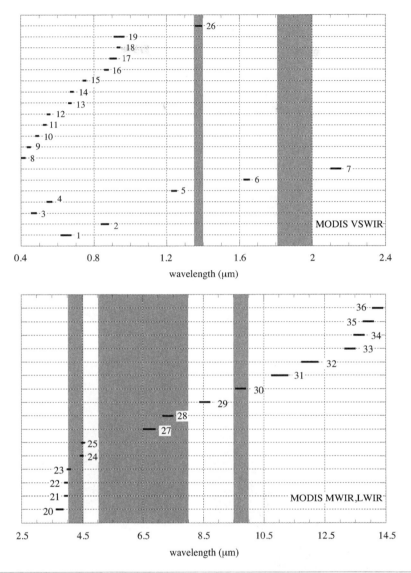

FIGURE 1-21. *Spectral ranges for the 36 MODIS bands. The shaded areas are the major atmospheric absorption bands. Note that bands 1 and 2, the 250m GSI land sensing bands, straddle the vegetation reflectance "red edge" (Fig. 1-7). Bands 26, 27, 28, and 30, all located within atmospheric absorption bands, are designed to measure atmospheric properties. All bands were located to optimize measurement of specific land, ocean, and atmospheric features (Table 1-2).*

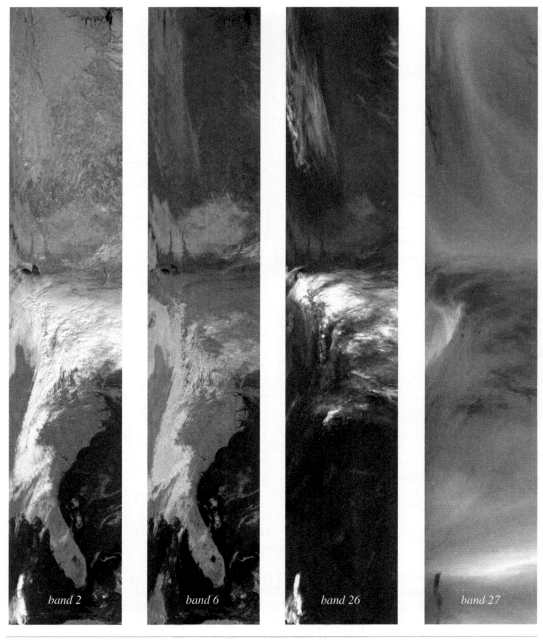

FIGURE 1-22. Four MODIS image bands collected on March 2, 2006, showing James Bay, Canada, at the top, the Great Lakes in the middle, and Florida at the bottom. All bands are displayed at 1 km GIFOV.

at the same time, it is possible to increase the revisit frequency. The SPOT systems are pointable and can be programmed to point cross-track to a maximum of $\pm 26°$ from nadir, which allows a single satellite to view the same area more frequently from different orbits. The high resolution commercial satellites, IKONOS and QuickBird, are even more "agile," i.e. they can point off-nadir within a short time. They can thus obtain two in-track views at different angles of the same area, within minutes of each other, to make a stereo pair. Without pointing, their revisit times are 1 to 3 days, depending on the latitude (more frequent at high latitudes). Manned systems, such as the Space Shuttle, which has carried several experimental SAR and other remote-sensing systems, and the International Space Station, have non-polar orbits which allow less regular revisit cycles. Some example revisit times of remote sensing satellites are given in Table 1-5. An example multitemporal image series is shown in Plate 1-8.

TABLE 1-5. *Revisit intervals and equatorial crossing times for several satellite remote sensing systems. The specifications assume only one system in operation, with the exception of the AVHRR, which normally operates in pairs, and the two MODIS systems, allowing morning and afternoon viewing of the same area in one day. The GOES is in a geostationary orbit and is always pointed at the same area of the earth. Equatorial crossing times are only approximate, as they change continually by small amounts and orbit adjustments are needed periodically.*

system	revisit interval	daylight equatorial crossing time
AVHRR	1 day (one system) 7 hours (two systems)	7:30 A.M. 2:30 P.M.
GOES	30 minutes	NA
IKONOS	minutes (same orbit), 1–3 days (pointing)	10:30 A.M.
IRS-1A, B	22 days	10:30 A.M.
Landsat	18 days (L-1, 2, 3)/16 days (L-4, 5, 7)	9:30 A.M./10:15 A.M.
MODIS	3 hours–1 day	10:30 A.M. descending (Terra) 1:30 P.M. ascending (Aqua)
QuickBird	minutes; 1–3 days	10:30 A.M.
SPOT	26 days (nadir only); 1 day or 4–5 days (pointing)	10:30 A.M.

1.4.4 Multi-Sensor Formation Flying

Since a single sensor cannot make all desired measurements of the Earth and its atmosphere, we must rely on combining data from several sensors to achieve a complete picture for scientific analysis. One of the ways to do this is to create a "train" of sensors on different satellites, traveling in the same orbit and separated by short time intervals, similar to planes flying in formation. The initial NASA demonstration of this concept was a morning formation, including Landsat-7 in the lead, EO-1 (1 minute behind Landsat-7), Terra (15 minutes behind), and the Argentine satellite SAC-C

(30 minutes behind). These satellites all descend across the equator on the daylight side of the Earth in the morning. An afternoon "A-Train" was established later with Aqua in the lead, followed by several atmospheric sensor satellites, including Cloudsat (1 minute behind Aqua) and CALIPSO (2 minutes behind), all in an ascending orbit. The close timing of data collections by the various sensors minimizes temporal changes, especially in the atmosphere, and facilitates scientific analysis. Some continual maintenance of the orbital parameters of each satellite is necessary to maintain the desired separation and the same orbit.

1.5   Image Display Systems

Remote-sensing images are stored on disk or tape in one of three formats: Band-Interleaved-by-Sample (BIS)[16], Band SeQuential (BSQ), or Band-Interleaved-by-Line (BIL). These formats are determined by different ordering of the three data dimensions (Fig. 1-23). From a data access time viewpoint, the BSQ format is preferred if one is mainly interested in working with individual spectral bands, and BIS is preferred if one is working with all spectral bands from a relatively small image area. The BIL format is a commonly used compromise.

Computer image displays convert the digital image data to a continuous, analog image for viewing. They are usually preset to display 8 bits/pixel in greyscale, or 24 bits/pixel in additive color, achieved with red, green, and blue primary screen colors. Three bands of a multispectral image are processed by three hardware *Look-Up Tables (LUTs)* to convert the integer *DN*s of the digital image to integer *Grey Levels (GLs)* in each band,

$$GL \ = \ LUT_{DN}. \tag{1-7}$$

The *DN* serves as an integer index in the *LUT*, and the *GL* is an integer index in the video memory of the display (Fig. 1-24). The range in image *DN*s is given by Eq. (1-5), while GL typically has a range,

$$GL_{range} = [0, 255] \tag{1-8}$$

in each color. The hardware *LUT* can be used to apply a "stretch" transformation to the image *DN*s to improve the displayed image's contrast or, if the *DN* range of the original image is greater than the *GL* range, the *LUT* can be used to "compress" the range for display. The output of the *LUT* will always be limited according to Eq. (1-8).

Color images are formed from composites of the triplet of *GLs* corresponding to any three bands of a multispectral image. With a 24-bit display, each band is assigned to one of three 8-bit integers corresponding to the display colors: red (R), green (G), or blue (B). There are therefore 256

---

16. The BIS format is also called *Band-Interleaved-by-Pixel (BIP)*.

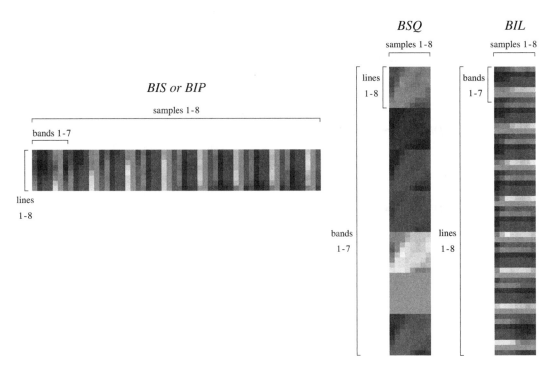

FIGURE 1-23. *The three most common multispectral image formats: BIS, BSQ, and BIL, illustrated with an 8 sample-by-8 line-by-7 band TM image. Note the very low contrast in the TIR band 6 relative to the other bands.*

*GL*s available for each band. Every displayed pixel has a color defined by a triplet of *GL*s, which we may consider a three-dimensional column vector $\boldsymbol{RGB}$,[17]

$$\boldsymbol{RGB} = \left[GL_R, GL_G, GL_B \right]^T . \qquad (1\text{-}9)$$

There are 256^3 possible $\boldsymbol{RGB}$ vectors, but fewer distinguishable display colors because there are no monitors that can display all of the colors in the color cube. The exact color displayed for a given $\boldsymbol{RGB}$ data vector depends on the phosphor characteristics and control settings of the monitor.

Computer monitors create color in an *additive* way; that is, a pixel with equal *GL*s in red, green, and blue will appear as grey on the screen (if the monitor is properly adjusted); a pixel with equal amounts of red and green, and with no blue, appears as yellow, and so forth. Certain color combinations are widely used in remote sensing (Table 1-6). However, since many commonly-used bands are not even in the visible spectrum, color assignment in the display image is arbitrary. The "best"

---

17. The triplet is conveniently written as a row vector in Eq. (1-9). The superscript *T* converts it to a column vector by a transpose operation.

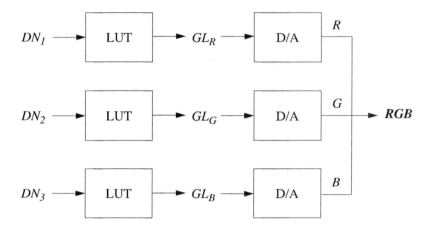

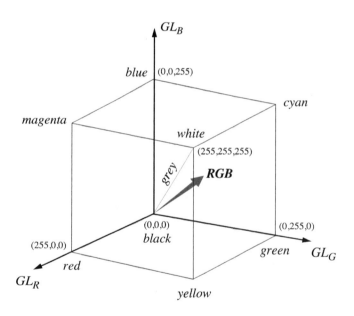

FIGURE 1-24. *The conversion from DN to GL to color in a 24 bits/pixel digital video display. Three images are converted by individual LUTs to the three display GLs that determine the amplitude of the primary display colors, red, green and blue. The last step is a digital-to-analog (D/A) conversion and combination of the three channels to achieve the color seen on the monitor. At the bottom, the color cube for a 24 bits/ pixel display is shown; the vector RGB specifies any triplet of GLs within the cube.*

colors to use are those that enhance the data of interest. The popularity of the *Color IR (CIR)* type of display derives from its emulation of color IR photography, in which vegetation appears as red because of its relatively high reflectance in the NIR and low reflectance in the visible (Fig. 1-7). Anyone with photointerpretation experience is usually accustomed to interpreting such images. The natural color composite is sometimes called a "true" color composite, but that is misleading as there is no "true" color in remote sensing—natural color is more appropriate for the colors seen by the eye. The bands used to make TM CIR and natural color composites are shown in Plate 1-9.

Single bands of a multispectral image can be displayed as a greyscale image or *pseudo-colored* by converting each *DN* or range of *DN*s to a different color using the display's *LUT*s. Pseudo-coloring makes it easier to see small differences in *DN*.

TABLE 1-6. *Sensor band mapping to RGB display color for standard color composites. A general false color composite is obtained by combining any three sensor bands.*

sensor	composite type	
	natural color	*Color IR (CIR)*
generic	red:green:blue	NIR:red:green
ALI	4:3:2	5:4:2
ASTER	NA	3:2:1
AVIRIS	30:20:9	45:30:20
Hyperion	30:21:10	43:30:21
MODIS	13:12:9	16:13:12
MSS	NA	4:2:1
SPOT	NA	3:2:1
TM, ETM+	3:2:1	4:3:2

1.6   Data Systems

The data volume and technical complexity of modern remote sensing systems dictate that some *preprocessing* of the data be performed before it is supplied to the science community. The aim of preprocessing is to create a consistent and reliable image database by

- calibrating the image radiometry
- correcting geometric distortions
- removing some types of sensor noise
- formatting to a standard prescription

The particular types of preprocessing required obviously depend on the sensor's characteristics, since the goal is to remove any undesirable image characteristics caused by the sensor. Since not all data users need, or can afford, the highest degree of correction, various levels of preprocessing are provided as options.

A taxonomy of data products has evolved for major remote-sensing systems along the lines shown in Table 1-7. A generic hierarchy that encompasses these examples is given by four major processing levels

- reformatted raw data
- sensor-corrected data
    - geometric corrections
    - radiometric corrections
- scene-corrected data
    - geometric corrections
    - radiometric corrections
- geophysical data

Each level normally requires more ancillary data and processing than the preceding level. Typically, only a small fraction of the total data collected is processed to the higher levels, and the cost to the data customer generally increases with the level of processing. NASA was responsible for Landsat TM data production until that task was assumed by the commercial firm, EOSAT, on September 27, 1985. The levels of processing produced by NASA during the pre-commercialization period are documented (Clark, 1990). In recent years, processing responsibility for Landsat-5 and -7 was returned to the government. The engineering characteristics of the Landsat TM also were described thoroughly in two special journal issues (Salomonson, 1984; Markham and Barker, 1985).

Information extraction from the preprocessed images is generally done by earth science researchers. But, since not all data are processed to the highest level at a central site, the earth scientist often must do equivalent processing before information extraction can begin, e.g., co-registration of scenes from different dates or sensors. It is, therefore, important that ancillary calibration data (primarily radiometric and geometric) be included with data supplied to users.

The approach taken in the NASA EOS program has been somewhat different from that used earlier. EOS satellite data, such as that from ASTER and MODIS, are used to produce not only lower level 1A or 1B radiance products, but also higher level products, using production algorithms developed by teams of scientists in their respective fields. Examples are the ASTER AST07 and MODIS MOD09 products for Level 2 surface reflectance and the MODIS Vegetation Index product MOD13. Many higher level EOS products use not only sensor data, but also rely on the results of other product calculations. For example, the MODIS surface reflectance product (MOD09) uses Level 1B MODIS bands 1–7 for radiance values, band 26 to detect cirrus clouds, MOD35 for cloud detection and masking, MOD05 for atmospheric water vapor correction, MOD04 for atmospheric aerosol correction, MOD07 for atmospheric ozone correction, and MOD43 for a coupling term involving the atmosphere and the surface Bi-directional Reflectance Distribution Function (BRDF).

TABLE 1-7. Processing levels for NASA earth remote-sensing systems.

system	processing level	description
NASA EOS (Asrar and Greenstone, 1995)	0	Reconstructed unprocessed instrument/payload data at full resolution; any and all communications artifacts (e.g., synchronization frames, communications headers) removed.
	1A	Reconstructed unprocessed instrument data at full resolution, time-referenced, and annotated with ancillary information, including radiometric and geometric calibration coefficients and georeferencing parameters (i.e., platform ephemeris) computed and appended, but not applied, to the Level 0 data (or if applied, applied in a manner in which Level 0 data can be fully recovered).
	1B	Level 1A data that have been processed to calibrated sensor units (not all instruments will have a Level 1B equivalent; Level 0 data is not recoverable from Level 1B data).
	2	Derived geophysical variables at the same resolution and location as the Level 1 source data.
	3	Level 1 or 2 data or other variables mapped on uniform space-time grid scales, usually with some completeness and consistency.
	4	Model output or results from analyses of lower level data, e.g., variables derived from multiple measurements.
EOSAT TM (EOSAT, 1993)	system	Geometric correction and projection to one of 20 map projections and rotation to north-up, using ephemeris data and systematic error models. Relative detector calibration non-thermal bands. UTM, SOM or polar stereographic projection with residual distortions of 200 – 300m (orbit-oriented product).
	precision	Map projection using Ground Control Points (GCPs).
	terrain	Orthographic projection using DEMs.
ETM+ (USGS, 2000)	0R	No radiometric or geometric corrections, but includes all necessary data for radiometric and geometric corrections.
	1R	Radiometrically corrected for detector gains and offsets.
	1G	Radiometrically corrected, resampled for geometric correction and registration to a geographic map projection. Includes satellite, sensor scan, and earth geometry corrections, but not topographic corrections.

In turn, the MOD09 product is used in generation of the vegetation products MOD13 and MOD15, as well as many other products. Some example MODIS science products are listed in Table 1-8, and two MODIS science products are shown in Plate 1-10.

TABLE 1-8. Example MODIS science data products. Additional published papers describing the theory and validation of these product algorithms are noted. All 44+ products are described in general terms in the EOS Data Products Handbook, Volumes 1 (NASA, 2004) and 2 (NASA, 2000) and in detail in Algorithm Theoretical Basis Documents (ATDBs) published by NASA on the Internet.

product ID	dataset	level	reference
MOD03	Geolocation	1B	Wolfe *et al.*, 2002
MOD06	Cloud Properties	2	Platnick *et al.*, 2003
MOD09	Surface Reflectance; Atmospheric Correction Algorithm Products	2	Vermote *et al.*, 1997
MOD13	Vegetation Indices	3	Huete *et al.*, 2002
MOD14	Thermal Anomalies - Fires	2, 3	Justice *et al.*, 2002
MOD15	Leaf Area Index (LAI) and Fraction of Photosynthetically Active Radiation (FPAR)-Moderate Resolution	4	Knyazikhin *et al.*, 1998
MOD16	Evapotranspiration	4	Nishida *et al.*, 2003
MOD35	Cloud Mask	2	Platnick *et al.*, 2003
MOD43	Surface Reflectance BRDF/Albedo Parameter	3	Schaaf *et al.*, 2002

The products from commercial high-resolution remote-sensing systems are generally less dependent on calibrated radiometry than products from science-oriented systems such as MODIS, but place more emphasis on calibrated sensor geometry and correction for topographic distortion, consistent with their primary application to mapping of cultural (man-made) features (Table 1-9).

1.7   Summary

We have surveyed the field of remote sensing and the types of systems used for imaging the earth. Several points can be observed:

- Remote sensing of the earth's surface is limited to spectral transmission windows in the atmosphere.
- Materials on the earth's surface may *potentially* be identified by their spectral–temporal optical reflectance signatures.
- Therefore, a key component of remote sensing is repetitive multispectral or hyperspectral imagery.

In the next three chapters, we will look in detail at the influence of the atmosphere and terrain relief on spectral "signatures" and the impact of sensor characteristics on remote-sensing

TABLE 1-9. Processing levels for commercial earth remote-sensing systems .

system	product	description
IKONOS (Space Imaging, 2004)	Geo	Geometrically corrected to a map projection.
	Standard Ortho	Geometrically corrected to a map projection and orthorectified to meet 1:100,000 National Map Accuracy Standard.
	Reference	Geometrically corrected to a map projection and orthorectified to meet 1:50,000 National Map Accuracy Standard. Collection elevation angle between 60–90°.
	Pro	Geometrically corrected to a map projection and orthorectified to meet 1:12,000 National Map Accuracy Standard. Collection elevation angle between 66–90°.
	Precision	Geometrically corrected to a map projection and orthorectified to meet 1:4,800 National Map Accuracy Standard. Collection elevation angle between 72–90°.
OrbView (Orbimage, 2005)	OrbView BASIC	Radiometrically corrected. Optionally geopositioned using ground control points.
	OrbView GEO	Radiometrically corrected. Optionally geopositioned using ground control points.
	OrbView ORTHO	Radiometrically corrected. Orthorectified to 1:50,000 or 1:24,000 National Map Accuracy Standard.
QuickBird (DigitalGlobe, 2005)	Basic	Radiometric, sensor, and geometric corrections.
	Standard	Radiometric, sensor, and geometric corrections. Mapped to a cartographic projection
	Orthorectified	Radiometric, sensor, and geometric corrections. Mapped to a cartographic projection and orthorectified to meet 1:50,000, 1:12,000, 1:5,000, or 1:4,800 National Map Accuracy Standard.
SPOT (SPOTImage, 2003)	SPOT Scene 1A	Radiometric corrections for detector normalization.
	SPOT Scene 1B	Geometric correction for earth rotation and panoramic distortion in oblique views.
	SPOT Scene 2A	Mapped to UTM WGS 84 map projection without using GCPs.
	SPOTView 2B (Precision)	Map projection using maps, Geographic Positioning System (GPS) points, or satellite ephemeris data.
	SPOTView 3 (Ortho)	Orthographic projection using 3 arc-second Digital Elevation Models (DEMs) or other models.

measurements. These interactions affect the quality and characteristics of remote-sensing data and the design and performance of image processing algorithms discussed in later chapters.

1.8   Exercises

Ex 1-1. Construct a nomograph like that of Fig. 1-4 for the full remote-sensing spectrum from 0.4 µm to 1 m and label each of the major atmospheric windows from Table 1-3.

Ex 1-2. A Landsat TM scene covers 185 km by 185 km. How many scenes are required to cover the entire earth, assuming an average 10% endlap (top and bottom) and 30% sidelap (adjacent orbital paths)? How many pixels does this correspond to? How many bytes?

Ex 1-3. We are able to visually interpret the aerial photograph of Fig. 1-2 to describe an apartment complex, roads, and agricultural fields. Think about how this interpretation might possibly be automated using computer analysis of the photograph. What physical (spatial, spectral, temporal) characteristics of these features could be used? What general algorithm characteristics would be required?

Ex 1-4. The data formats in Fig. 1-23 can be obtained by planar slices from different directions of the multispectral image cube in Fig. 1-9. Determine the slices and the order in which the data are extracted for each format.

Ex 1-5. Identify four important characteristics of a multispectral imaging system. Select one of the remote sensing applications discussed at the beginning of this chapter and explain how the sensor characteristics you identified affect the quality of the data for that application.

Ex 1-6. Suppose you are asked to design a remote-sensing system with a *GIFOV* of 1 m, an altitude of 700 km, and a focal length of 10 m. What is the required detector element size? If the sensor is a 1-D array pushbroom and has 4000 detectors, what are its *FOV* and *GFOV*, assuming a flat earth?

Ex 1-7. The ground velocity of earth-orbiting satellites in a sun synchronous orbit is about 7 km/sec. For a 185 km in-track and cross-track *GFOV* and 30 m *GSI*, what is the time between each pixel sample in TM or ETM+ multispectral bands? In the ETM+ panchromatic band? Suppose a pushbroom satellite sensor has the same *GFOV* and *GSI*—what is the time between each row of pixels?

Ex 1-8. The QuickBird satellite is at an altitude of 405 km and has a *GIFOV* of 0.6 m in the panchromatic band. Calculate the *IFOV* in radians and in degrees.

CHAPTER 2

Optical Radiation Models

2.1   Introduction

Passive remote sensing in the optical regime (visible through thermal) depends on two sources of radiation. In the visible to shortwave infrared, the radiation collected by a remote sensing system originates with the sun. Part of the radiation received by a sensor has been reflected at the earth's surface and part has been scattered by the atmosphere, without ever reaching the earth. In the thermal infrared, thermal radiation is emitted directly by materials on the earth and combines with self-emitted thermal radiation in the atmosphere as it propagates upward. In this chapter, we outline the basic models appropriate to the optical region from the visible through the thermal infrared. The science of radiometry is applied here as a means to an end, i.e., to define the major processes involved in remote sensing in this spectral region. For thorough treatments of optical radiometry in remote sensing, the books by Slater (1980) and Schott (1996) are recommended.

2.2　Visible to Shortwave Infrared Region

All materials on the earth's surface passively absorb and reflect solar radiation in the 0.4 to 3μm spectral range. Some materials also transmit solar radiation; for example, water bodies and plant canopies. At longer wavelengths, materials at normal temperatures begin to actively emit thermal radiation, which is discussed later in this chapter. A description of how solar radiation propagates and is modified prior to sensing by an optical system follows.

2.2.1 Solar Radiation

The source of energy for remote sensing in the *solar-reflective* spectral region (visible to shortwave IR) is the sun. The sun is a near-perfect *blackbody* radiator; that is, it emits radiation at nearly the maximum efficiency possible for a body at its effective temperature. The *spectral radiant exitance*, M_λ, from the sun can therefore be modeled by *Planck's blackbody equation*,

$$M_\lambda = \frac{C_1}{\lambda^5[e^{C_2/(\lambda T)}-1]}$$

(2-1)

where
  T is the blackbody's temperature in Kelvin (K),
  $C_1 = 3.74151 \times 10^8$ W-m^{-2}-μm^4, and
  $C_2 = 1.43879 \times 10^4$ μm-K.
Using these values and specifying the wavelength, λ, in μm and the temperature in Kelvin results in units for spectral radiant exitance of W-m^{-2}-μm^{-1} (Slater, 1980), i.e., power (or flux) per unit area of the sun's surface, per unit wavelength interval.

The blackbody function peaks at a wavelength given by *Wien's Law*,

$$\lambda|_{max} = 2898/T .$$

(2-2)

where the blackbody's temperature is in Kelvin and the wavelength at which the radiation is a maximum is in μm. Thus, as the temperature of a blackbody increases, the wavelength of maximum radiant exitance decreases.

We, of course, are interested in the radiation that reaches the earth. To calculate that, the spectral radiant exitance of Eq. (2-1) must first be propagated to the top of the earth's atmosphere and then converted to *spectral irradiance*, E_λ^0. This transformation is accomplished with the following equation,

$$top\text{-}of\text{-}the\ atmosphere:\ E_\lambda^0 = \frac{M_\lambda}{\pi} \times \frac{area\ solar\ disk}{(distance\text{-}to\text{-}earth)^2} .$$

(2-3)

The units of spectral irradiance are the same as those of spectral radiant exitance, namely those of *spectral flux density*, $W\text{-}m^{-2}\text{-}\mu m^{-1}$. The spectral content of the radiation does not change in the transit through space, but the magnitude of solar irradiance at the earth changes by a few percent as the sun-earth distance changes throughout the year. The exo-atmospheric flux density is plotted in Fig. 2-1 using a blackbody model and a more detailed, empirical function stored in the atmospheric modeling program MODTRAN. As can be seen from the figure, the blackbody model with a temperature of 5900K is a good approximation to measured solar radiation.

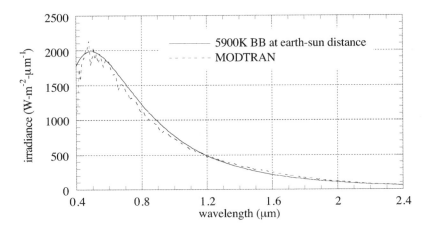

FIGURE 2-1. *Comparison of the exo-atmospheric (top-of-the-atmosphere) solar spectral irradiance as used in the atmospheric modeling code MODTRAN (Berk et al., 1989) to that produced by a blackbody object at a temperature of 5900K and at the same distance as the sun. Deviations from the blackbody model are mainly in narrow absorption lines present in the actual solar spectrum.*

As the wavelength increases to the shortwave IR, less radiation is available from the sun for signal detection by remote sensing. If we ignore atmospheric effects, the solar radiation arriving at the earth is matched by self-emitted thermal energy from the earth at about 4.5 μm (Fig. 2-2). The wavelength at which they are equal contributors to the *at-sensor radiance above the atmosphere* depends on earth surface reflectance and emissivity (see Sect. 2.3) and the atmosphere, and can range between 2.5 and 6 μm (Slater, 1980).

2.2.2 Radiation Components

The major radiation transfers of concern in the visible through SWIR spectral regions are shown in Fig. 2-3. There are generally three significant components in the upwelling at-sensor radiation:

- the unscattered, surface-reflected radiation, L_λ^{su}

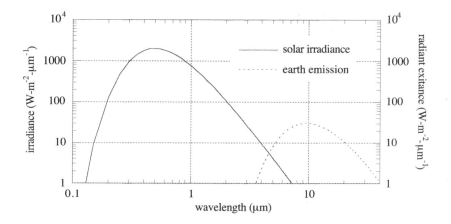

FIGURE 2-2. *Spectral distributions at the top-of-the-atmosphere for the two radiation sources in the visible through thermal infrared spectral regions. The earth is assumed to be a blackbody at T = 300K and the sun a blackbody at T = 5900K, and atmospheric effects are ignored. Blackbody curves for different temperatures cannot cross each other; it only appears to occur here because the solar radiant exitance has been scaled as described in Eq. (2-3).*

- the down-scattered, surface-reflected skylight, L_λ^{sd}
- the up-scattered path radiance, L_λ^{sp}

We can therefore write for the total upwelling radiance at a high altitude or satellite sensor,[1]

$$L_\lambda^s = L_\lambda^{su} + L_\lambda^{sd} + L_\lambda^{sp} .$$
(2-4)

In the following sections, we will look at appropriate models for each component.

Surface-reflected, unscattered component (L_λ^{su})

The atmosphere is an unavoidable influence in satellite and high altitude aerial remote sensing in the visible through shortwave IR. It scatters and absorbs radiation between the sun and the earth along the solar path, and again between the earth and the sensor along the view path. The fraction of radiation that initially arrives at the earth's surface is called the *solar path transmittance*, $\tau_s(\lambda)$, and is, by definition, between zero and one, and unitless. A typical spectral transmittance curve for the solar path, as generated from atmospheric models, is shown in Fig. 2-4.

---

1. We use the superscript *s* to denote "solar," thereby distinguishing these terms from thermal emitted radiation, described later.

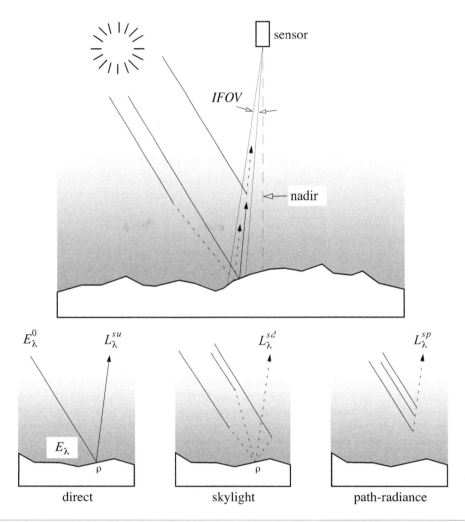

FIGURE 2-3. *The most significant radiation components seen by the sensor in solar reflective remote sensing are the "direct" component, the "skylight" component, and the "path radiance" component (commonly called "haze"). The shading in the diagram portrays the decreasing atmospheric density at higher altitudes. Other radiation can reach the sensor, such as from the "adjacency" component, consisting of a direct reflection from a nearby GIFOV, followed by either an up-scatter directly to the sensor, or a down-scatter into the GIFOV of interest, followed by a second reflection toward the sensor. The adjacency phenomenon increases local spatial correlation among pixels and reduces the contrast of dark-light boundaries, such as a shoreline. Multiple surface reflections and atmospheric scatterings generally are of less significance, because the radiation magnitude is reduced by each reflection or scattering event. Note that, because of the great distance to the sun, its rays are effectively parallel when entering the earth's atmosphere.*

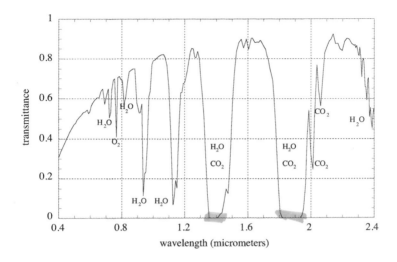

FIGURE 2-4. *Transmittance of the atmosphere as calculated by the program MODTRAN. This is the transmittance along the solar path, i.e., between the sun and the earth's surface, for a solar elevation angle of 45°. The absorption bands are primarily associated with water vapor and carbon dioxide. This figure, and many of the others in this chapter, are produced with specific atmospheric parameters in the modeling program MODTRAN (Berk et al., 1989). The atmosphere model is "mid-latitude summer" with a "visibility" of 23 km (a relatively high value, implying that Mie scattering effects are small relative to Rayleigh scattering effects). The solar elevation angle is 45° and the azimuth angle from north is 135°. Special thanks to Dr. Kurt Thome, who performed the MODTRAN runs and provided assistance in data interpretation. The reader should realize that these curves are intended to be only examples; the specifics of any particular atmospheric condition will be different, and other atmospheric modeling programs, e.g., 6S (Vermote et al., 1997), may yield somewhat, but not vastly, different curves.*

   The molecular absorption bands of water and carbon dioxide cause "deep" absorption features that, in two bands near 1.4 μm and 1.9 μm, completely block transmission of radiation. These spectral regions, therefore, are avoided for remote sensing of the earth's surface. They can be useful, however, for the detection of cirrus clouds, which are not easily distinguished from lower altitude clouds or surface features, at other wavelengths (Gao *et al.*, 1993). Since cirrus clouds are above most of the atmosphere's water vapor, they return a signal to a high-altitude sensor, while the signals from the earth's surface and lower altitude clouds are absorbed in the 1.4 μm band. This application is the motivation for the MODIS band 26 (Table 1-1). The water absorption bands near 0.9 μm and 1.1 μm are much narrower, but can still block the surface signal significantly in narrowband remote sensors and can reduce the signal in broadband sensors when the water vapor content of the atmosphere is high.

The transmittance generally decreases towards the blue spectral region. This is largely the result of light being scattered out of the solar path by air molecules, whose diameter is less than λ/π (Slater, 1980). The magnitude of scattered light is approximated by pure *Rayleigh scattering* with a λ^{-4} dependence on wavelength. The short wavelength blue light is scattered out of the direct transmitting path more than the longer wavelength red light, which explains the red sky at sunrise and sunset, when the solar path through the atmosphere is greatest. In an atmosphere containing aerosols and particulates (smoke, smog, dust, haze, fog), *Mie scattering* is also present. For particle sizes larger than about $2\lambda/\pi$, the magnitude of Mie scattering does not depend on wavelength. Thick clouds are white from above because of Mie scattering by water droplets. For particle sizes in the range from about λ/π to about $2\lambda/\pi$, there is a wavelength dependence that is less than that of Rayleigh scattering. Real atmospheres with a full range of molecular, aerosol, and particulate sizes exhibit a combination of Rayleigh and Mie scattering.

The spectral effect of atmospheric transmission on the solar radiation is shown in Fig. 2-5. The atmosphere significantly alters the spectral irradiance before it arrives at the earth. Mathematically, the irradiance E_λ on a plane perpendicular to the solar path and at the earth's surface is given by,

$$\textit{earth's surface:}\quad E_\lambda = \tau_s(\lambda)E_\lambda^0 \tag{2-5}$$

where τ_s is the solar path atmospheric transmittance. Note that, by the definition of transmittance, E_λ must be less than or equal to E_λ^0.

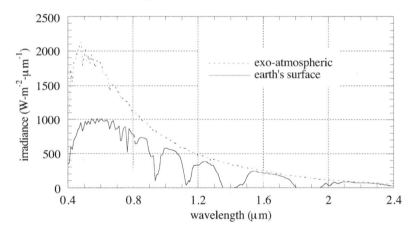

FIGURE 2-5. *Solar irradiance in the visible and shortwave IR regions (for a solar elevation angle of 45°), above the atmosphere and at the earth's surface.* The ratio of these two curves is the path transmittance depicted in Fig. 2-4.

With the exceptions of cast shadows or clouds, E_λ^0 can be assumed to be constant across the *GFOV* of a sensor such as ETM+. The irradiance at the surface depends on the incident angle, being a maximum if the surface is perpendicular to the incident angle, and less as the angle decreases. The decrease varies as the cosine of the angle, which can be calculated by a dot product of two vectors (Fig. 2-6) (Horn, 1981). The incident irradiance E_λ in Eq. (2-5) must then be modified to account for terrain shape as follows,

$$\textit{earth's surface:} \qquad \begin{aligned} E_\lambda(x, y) &= \tau_s(\lambda)E_\lambda^0\boldsymbol{n}(x, y) \bullet \boldsymbol{s} \\ &= \tau_s(\lambda)E_\lambda^0\cos[\theta(x, y)] \end{aligned} \qquad (2\text{-}6)$$

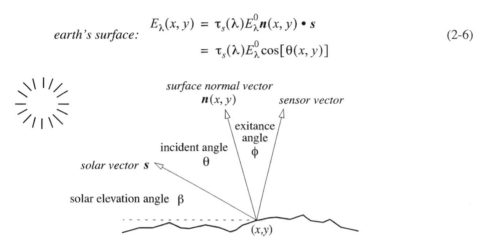

FIGURE 2-6. *The geometry of solar direct irradiance on the earth's surface. The unit length vector* **s** *points to the sun and the unit length vector* $\boldsymbol{n}(x, y)$ *is normal to the surface. The solar elevation angle is β and the solar zenith angle is 90° − β. The solar incidence angle to the surface is θ and the exitance angle from the surface normal toward the sensor is φ. The cosine of the angle θ is given by the vector dot product* $\boldsymbol{n}(x, y) \bullet \boldsymbol{s}$. *For simplicity in this diagram, the surface normal is assumed to lie in the vertical plane through the solar vector, but the dot product calculation is valid for any two vectors. Note this terrain-related effect does not involve the view angle of the sensor.*

The next energy transfer occurs upon reflectance at the earth's surface. The irradiance downward onto a *Lambertian* surface is converted to the *surface radiance* L_λ leaving the surface with the aid of a geometric factor π and a *diffuse spectral reflectance* ρ,

$$\textit{earth's surface:} \qquad \begin{aligned} L_\lambda(x, y) &= \rho(x, y, \lambda)\frac{E_\lambda(x, y)}{\pi} \\ &= \rho(x, y, \lambda)\frac{\tau_s(\lambda)E_\lambda^0}{\pi}\cos[\theta(x, y)] \; . \end{aligned} \qquad (2\text{-}7)$$

Like transmittance, reflectance is, by definition, unitless and between zero and one. Example spectral reflectance curves for some natural earth surface materials were shown in Chapter 1. Reflectance varies with wavelength and spatial location but does not depend on the view (sensor) direction for a truly Lambertian surface.[2] Deviation from this simple model by real materials is expressed in a *Bi-directional Reflectance Distribution Function (BRDF)*, which can be measured for a particular material by finding the ratio of outgoing radiance to incoming irradiance, as a function of incident and view angles. The quantity $\rho(x, y, \lambda)/\pi$ in Eq. (2-7) is then replaced by the BRDF (Schott, 1996).

The outgoing radiance from the earth's surface traverses the atmosphere to the sensor. The transmittances of the atmosphere along a nadir view path and along a view path at 40° from nadir are shown in Fig. 2-7. These curves differ from the one in Fig. 2-4 only by virtue of different path lengths through the atmosphere. Sensors such as Landsat TM with a small *FOV* show little within-scene atmospheric variation (other than localized clouds, smoke, and haze) over areas with moderate terrain relief, since the path through the atmosphere is nearly constant for all pixels. In areas of high terrain relief, even Landsat images can show pixel-to-pixel atmospheric variation due to pixel-to-pixel altitude differences (Proy *et al.*, 1989). Wide *FOV* sensors, such as the AVHRR, can show considerable scan-angle effects due to changes in the atmospheric path in the cross-scan direction. Note that the differences among the curves in Fig. 2-4 and Fig. 2-7 are relatively greater at shorter wavelengths. Visible and NIR imagery will therefore be particularly sensitive to changes in view angle across the *FOV* (Plate 2-1). Atmospheric view path and vegetation canopy radiance variation with view angle are the rationale for the MISR sensor that records images from nine cameras in four spectral bands at different in-track view angles (Plate 2-2).

We now must modify Eq. (2-7) according to the *view path* transmittance, $\tau_v(\lambda)$, to obtain the *at-sensor radiance,*

$$L_\lambda^{su} = \tau_v(\lambda)L_\lambda$$

at-sensor: $\qquad\qquad\qquad\qquad\qquad\qquad\qquad\qquad\qquad\qquad\qquad\qquad$ (2-8)

$$= \rho(x, y, \lambda)\frac{\tau_v(\lambda)\tau_s(\lambda)E_\lambda^0}{\pi}\cos[\theta(x, y)] \, .$$

This component carries the signal of interest, namely the spatial-spectral reflectance distribution, $\rho(x, y, \lambda)$.

Surface-reflected, atmosphere-scattered component (L_λ^{sd})

The sensor also sees radiance arising from radiation that is scattered *downward* by the atmosphere ("skylight") and then reflected at the earth upward into the *IFOV* of the pixel of interest. This term, $L_\lambda^{sd}(x, y)$, is responsible for the commonly-observed fact that shadows are not totally dark. The

---

2. A Lambertian surface exhibits equal radiance in all directions. Visually, we say it appears equally bright at any view angle. Such a surface is also termed *perfectly diffuse*, with no mirror-like specular reflection. Many natural surfaces are approximately Lambertian within a limited range of view angles, typically 20°–40°; as the view angle increases beyond that, most materials become non-Lambertian and show unequal reflectance in different directions. This property can be measured by sensors such as the Muli-angle Imaging SpectroRadiometer (MISR) to better characterize the surface radiative characteristics (Diner *et al.*, 1989).

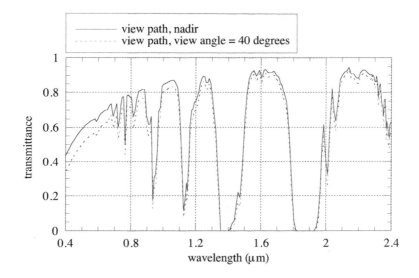

FIGURE 2-7. Atmospheric path transmittance as viewed by a satellite sensor. View angle is defined as the angle from nadir. Note the similarity in shape, but differences in magnitude, compared to Fig. 2-4. The transmittance for a view angle of 40° is less because the path through the atmosphere is longer, leading to more scattering and absorption.

reflected-skylight term is proportional to the diffuse surface reflectance, ρ, and the irradiance at the surface due to skylight, E_λ^d. This quantity is used because it is directly measurable with ground-based instruments. We accommodate the possibility that the sky may not be entirely visible from the pixel of interest due to intervening topography with a factor, $F(x, y)$,[3]

$$at\text{-}sensor: \quad L_\lambda^{sd} \; = \; F(x, y)\rho(x, y, \lambda)\frac{\tau_v(\lambda)E_\lambda^d}{\pi}. \tag{2-9}$$

Path-scattered component (L_λ^{sp})

The path radiance term is a combination of molecular *Rayleigh scattering*, which varies with wavelength as λ^{-4} , and aerosol and particulate *Mie scattering,* which depends less strongly, or not at all, on wavelength. The combined effect of Rayleigh and Mie scattering in a clear atmosphere results in a net wavelength dependence of between λ^{-2} and $\lambda^{-0.7}$ (Curcio, 1961).

---

3. *F* is the *fraction* of the sky hemisphere that is visible from the pixel of interest. For completely flat terrain, *F* equals one. A detailed description and calculated examples are given in Schott (1996).

Path radiance can vary within a scene, for example, between a rural and an urban area, or in the presence of smoke plumes from fires. It also will vary with view angle, which is particularly important for a wide *FOV* sensor (such as AVHRR) or a sensor pointed off-nadir (such as SPOT). For scenes of homogeneous landscapes and for nadir-viewing sensors with relatively small *FOV*s (such as TM or ETM+), the path radiance is reasonably assumed to be a constant over the entire scene, and we simply write this term as L_λ^{sp}.

Total at-sensor, solar radiance (L_λ^s)

The total at-sensor, solar radiation is the sum of the three components described previously,

at-sensor:

$$L_\lambda^s(x, y) = L_\lambda^{su}(x, y) + L_\lambda^{sd}(x, y) + L_\lambda^{sp}$$

$$= \rho(x, y, \lambda)\frac{\tau_v(\lambda)\tau_s(\lambda)E_\lambda^0}{\pi}\cos[\theta(x, y)] + F(x, y)\rho(x, y, \lambda)\frac{\tau_v(\lambda)E_\lambda^d}{\pi} + L_\lambda^{sp} \qquad (2\text{-}10)$$

$$= \rho(x, y, \lambda)\frac{\tau_v(\lambda)}{\pi}\{\tau_s(\lambda)E_\lambda^0\cos[\theta(x, y)] + F(x, y)E_\lambda^d\} + L_\lambda^{sp}.$$

The essence of Eq. (2-10) is that:

- the total spectral radiance received by the sensor is linearly proportional to the surface diffuse reflectance, modified by

- a multiplicative, spatially- and spectrally-variant factor that depends on the terrain shape, and

- an additive, spatially-invariant, spectrally-variant term due to view path scattering.

As an example, Fig. 2-8 shows how the at-sensor radiance components vary with wavelength for a reflecting surface described by Kentucky Bluegrass (Fig. 1-7). The sum of these two components is the total radiance seen by a satellite or high altitude aircraft sensor and is compared to the spectral profile of an AVIRIS pixel containing grass and trees in Fig. 2-9. Even though there has been no attempt to emulate the atmospheric conditions, surface reflectance magnitude, or solar angles of this particular AVIRIS scene, the data are remarkably similar in shape to the results of the MODTRAN simulation. The MODTRAN atmospheric models are evidently realistic, and if *actual* atmosphere values are used in predicting the radiances of a particular image with a program such as MODTRAN, good agreement is possible. The problem, of course, is that such calibration data are almost never available for a given scene. Measurements of the atmosphere and ground reflectance in conjunction with aircraft or satellite overpasses require substantial planning and effort, and are subject to the vagaries of weather and equipment failure. Therefore, there has been a great amount of interest in *image-based* atmospheric calibration techniques (Teillet and Fedosejevs, 1995).

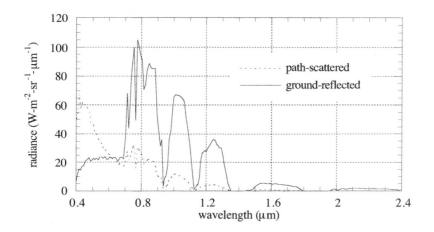

FIGURE 2-8. The path-scattered and ground-reflected components of the total upwelling radiance seen by a satellite sensor for a surface reflectance of Kentucky Bluegrass. These components, as defined in MODTRAN, are related to the terms of Eq. (2-10) as follows. The path-scattered component is L_λ^{sp}, plus radiation that is reflected by the surface in a direction other than toward the sensor (remember, we assume the surface is perfectly diffuse and reflects equally in all directions), and is then scattered into the IFOV (we have not included this term in our discussion). The strong increase in the path-scattered component below 0.7 μm is due to molecular scattering and is primarily the L_λ^{sp} term, since the surface reflectance here is relatively low. Above 0.7 μm, the influence of the reflected and then scattered component is apparent. The ground-reflected component is the sum of L_λ^{su} and L_λ^{sd}. In the ground-reflected component, little information about the grass signature is seen until above 0.7 μm, where the reflectance becomes relatively high. The ground-reflected component only exceeds the path-scattered component above 0.7 μm, but both contain information about the signal (grass reflectance). Note the atmospheric water vapor absorption bands near 0.9, 1.1, 1.4, and 1.9 μm (compare to Fig. 2-4).

The at-sensor spectral radiance and the spectral reflectance of Kentucky bluegrass are dramatically different (Fig. 2-10)! How then is it possible to even hope to use remote sensing to recognize different terrestrial materials? The major reason why this drastic alteration of spectral "signatures" is not, in fact, a disaster, is that many image processing algorithms rely only on *relative spectral differences* among pixels. It is clear, however, that *remote sensing data must be corrected for atmospheric, topographic, and solar effects if they are to be compared to a library of spectral reflectance curves*. Furthermore, *relative* atmospheric correction is needed if data signatures from one image date are to be compared to those from another date. These aspects of data calibration are discussed in detail in Chapter 7.

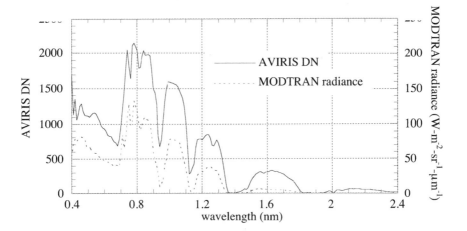

FIGURE 2-9. *The spectral response of Kentucky Bluegrass as predicted by the MODTRAN model and a plot of a mixed grass and trees response from the AVIRIS image of Palo Alto (Plate 1-3). The shapes of the curves are quite similar, even though the MODTRAN model parameters have no direct relation to this particular AVIRIS data.*

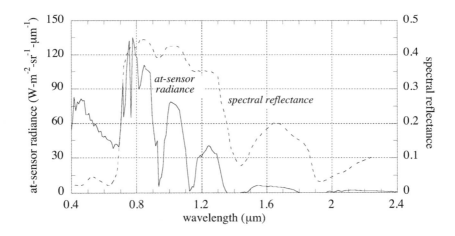

FIGURE 2-10. *Comparison of the reflectance and remotely-sensed radiance spectral signals for grass. The characteristic features in the reflectance curve are masked and modified by all the factors involved in optical remote sensing. The most significant feature however, the "red edge" near 0.71 μm, is preserved.*

2.2.3 Image Examples in the Solar Region

Some image examples are used in this section to illustrate the theory and models of optical radiation transfer in the solar reflective portion of the spectrum.

Terrain shading

The cosine factor arising from terrain relief (Eq. (2-10)) can be important to the spatial content of images as demonstrated in Fig. 2-11. The *Digital Elevation Model (DEM)*[4] has a *GSI* of 30m, matching the TM *GSI*. A *shaded relief* image is created from this DEM by calculating the $\cos[\theta(x,y)]$ term at each pixel and assuming that the surface reflectance is diffuse (i.e., the surface is Lambertian) and constant everywhere. In this case, we have set the solar elevation and azimuth angles to correspond to those for the particular TM image in Fig. 2-11. The shading caused by the terrain relief is clearly predictable with the DEM data.

The differences that remain between the contrast-stretched TM image and the shaded relief image of Fig. 2-11 are partially due to the reflectance spatial variations that are not modeled in the relief image. The modeling of vegetation reflectance from plant canopies is quite complex and is a matter of considerable research interest. The reader is referred to an extensive review (Goel, 1988) as a starting point for this subject and to Liang (2004) for more details.

The influence of terrain geometry on remote sensing measurements has been discussed in detail in Sjoberg and Horn (1983) and Proy *et al.* (1989), including the possibility of reflectance by neighboring ground elements in mountainous terrain into the *IFOV* at a pixel of interest. Looking at Eq. (2-10), we see that the image spatial content results from the product of two spatially-varying terms, the reflectance and the terrain-dependent cosine factor (assuming the down-scattered term is small). Either may be considered to *spatially modulate* the other. This relationship is useful in certain algorithms for image fusion, as described in Chapter 8.

Shadowing

Analysis of the DEM in Fig. 2-11 provides information on shadowing in the TM image (Dubayah and Dozier, 1986; Giles *et al.*, 1994). For example, the *self-shadowed* points can be found directly from the shaded relief image (see Exercise 2-1); these are terrain segments that face away from the solar irradiance direction. With a "line-of-sight" algorithm, one can also find the pixels that lie within *projected shadows* (Fig. 2-12). Because of the mild terrain relief and high solar elevation for the TM image, there are relatively few projected shadows. Both types of shadow regions may be used to estimate atmospheric path radiance if a reasonable estimate is made of the reflectance within the shadowed area.

---

4. A DEM is a regular spatial grid of elevation values. It can be created in a variety of ways, for example, indirectly via analysis of stereo imagery (see Chapter 8) or satellite Interferometric SAR (InSAR) data, or directly from airborne laser LIght Detection And Ranging (LIDAR) measurements.

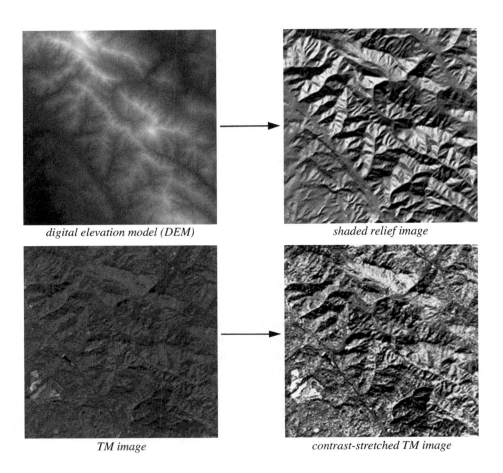

digital elevation model (DEM)   shaded relief image

TM image   contrast-stretched TM image

FIGURE 2-11. *The influence of terrain relief on image structure is depicted with a co-registered DEM and TM band 4 image near Berkeley, California. Elevation is coded as brightness in the DEM image. The solar elevation angle of 35° and azimuth angle of 151° for the October 25, 1984, TM image are used to create a shaded relief image from the DEM. The contrast of the TM image is significantly lower than that of the shaded relief image because of atmospheric contrast reduction and the fact that the displayed image contrast is controlled by the high reflectance man-made feature in the lower left. If the contrast of the TM image is adjusted to match that of the shaded relief image, the similarity of the two is evident. The residual differences are due to surface cover reflectance variation in the TM image and reflected skylight within shadows, neither of which are modeled in the shaded relief image. (The DEM is courtesy of William Acevedo and Len Gaydos of the U.S. Geological Survey.)*

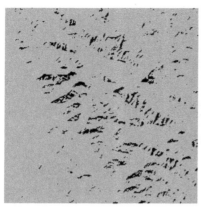

self-shadowed pixels   *projected shadows*

FIGURE 2-12. Maps of the self-shadowed pixels and projected shadows for the solar angles and DEM of Fig. 2-11. (The projected shadows map was produced by Justin Paola, Oasis Research Center.)

An image's appearance can vary dramatically with different solar angles, as illustrated in Fig. 2-13. The difference in acquisition dates of four months creates a dramatically different image because the changes in solar elevation and, to a lesser extent, azimuth cause changes in the $\cos[\theta(x,y)]$ term of Eq. (2-10) and in shadowing.

June 11, 1981   *October 20, 1980*

FIGURE 2-13. Landsat MSS images of the Grand Canyon, Arizona, acquired on two dates. The lower sun elevation of 38° for the October image dramatically increases the shadowing in the Canyon, compared to the June image with a sun elevation of 65°.

Atmospheric correction

A portion of a Landsat TM scene near Oakland, California, is shown for bands 1 through 4 in Fig. 2-14. These data, although not calibrated for sensor gain and offset (Chapters 3 and 7), nevertheless indicate the general properties of atmospheric influence as a function of wavelength. A common technique for atmospheric correction of multispectral imagery uses a "dark object" as a calibration target (Chavez, 1988). The dark object may be a region of cast shadow or, as in this case, a deep body of water. It is *assumed* that the dark object has uniformly zero radiance for all bands, and that any non-zero measured radiance must be due to atmospheric scattering into the object's pixels.[5] For the images in Fig. 2-14, the Briones Reservoir is a good choice for a dark object (it is significantly darker than the San Pablo Reservoir at shorter wavelengths). Averaging the darkest pixels in the Reservoir yields *DN*s of 53, 20, 11, and 14 for bands 1 through 4. It is tempting at this point to remove the atmospheric path scattering bias by simply subtracting these *DN* values from every pixel in the corresponding band, given that the at-sensor radiance-to-*DN* calibration is linear. However, that would not correct the data for the sensor gain and would therefore not be a complete calibration. Calibration of the raw image *DN*s to at-sensor radiance and correction to reflectance are discussed in Chapter 7.

2.3 Midwave to Thermal Infrared Region

At longer wavelengths, beyond the SWIR and into the MWIR spectral region, the importance of solar radiation declines and that of emitted thermal radiation increases for objects that are Lambertian reflectors (Fig. 2-2). At the longer wavelengths of the TIR, direct solar radiation is not a factor compared to self-emitted thermal radiation, other than solar-induced heating of the surface. The only exception to these statements is if the objects of interest are specular reflectors; in that case the solar reflected component may exceed the emitted component, even in the TIR (Slater, 1996). We will ignore this relatively rare circumstance in the remainder of this section.

2.3.1 Thermal Radiation

Every object at a temperature above absolute zero (0K) emits thermal radiation due to kinetic energy of molecules within the object. The radiation obeys Planck's equation (Eq. (2-1)) *if* the object is a perfect emitter and absorber, i.e., a blackbody. Real objects are not perfect emitters or absorbers, and Eq. (2-1) is modified by an emission efficiency factor, the *emissivity*, which is generally a function of wavelength. Real materials are sometimes referred to as *greybodies*, but their spectral radiant exitance may not follow the blackbody curve because of the wavelength dependence of emissivity.

---

5. If any data are available on the actual reflectance of the "dark object," then it should be used to obtain a better estimate of the at-sensor path radiance for this correction (Teillet and Fedosejevs, 1995).

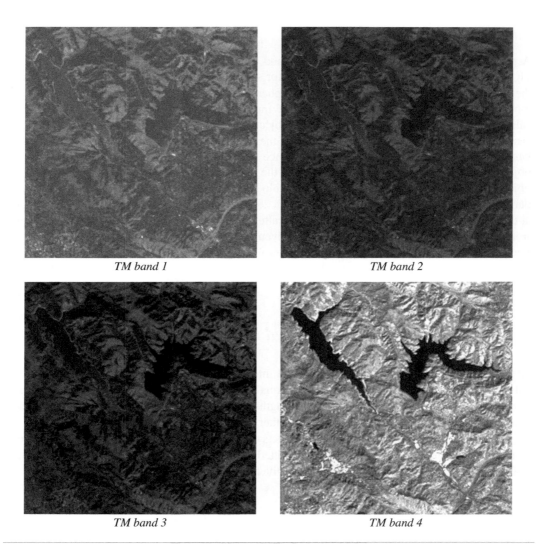

FIGURE 2-14. *TM band 1 through band 4 images of the San Pablo (left) and Briones Reservoirs (right) north of Berkeley, California (part of the same TM scene used in Fig. 2-11). The individual bands are uncalibrated and shown with their recorded relative brightness and contrast. Atmospheric scattering reduces the contrast in band 1, while bands 2 and 3 are dark, due to low vegetation reflectance and lower sensor gain than band 1. Band 4 shows high contrast between the water-filled reservoirs and surrounding vegetated and bare soil terrain. Note the Briones Reservoir is relatively darker in the shorter wavelength spectral bands than the San Pablo Reservoir. This indicates the latter may have suspended sediments and particulates in the water, which is particularly likely since it is at a lower altitude and subjected to more runoff from the surrounding terrain. In band 4, both reservoirs have little radiance because of the near zero reflectance of water in the NIR.*

2.3.2 Radiation Components

The three emitted thermal components considered here arise from:

- the surface-emitted radiation from the earth, L_λ^{eu} ;

- the down-emitted, surface-reflected radiation from the atmosphere, L_λ^{ed} ; and

- the path-emitted radiance, L_λ^{ep} .

These are shown in (Fig. 2-15), with the previously discussed solar components for comparison.

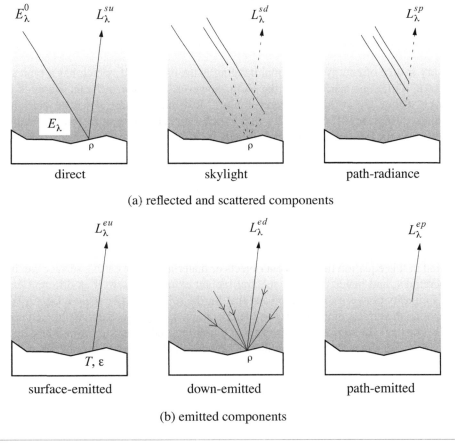

(a) reflected and scattered components

(b) emitted components

FIGURE 2-15. *The reflected and scattered components of the at-sensor radiance (Fig. 2-3) and the analogous emitted components. In the 2.5 to 6 μm spectral region, both must generally be considered; in the thermal IR region (8 to 15 μm) only the emitted components are important.*

The total at-sensor radiance contribution from emission is

$$L_\lambda^e = L_\lambda^{eu} + L_\lambda^{ed} + L_\lambda^{ep}. \tag{2-11}$$

In the MWIR, we write for the total radiance,[6]

$$at\text{-}sensor\ (MWIR)\text{:}\quad L_\lambda^{MWIR} = L_\lambda^s + L_\lambda^e \tag{2-12}$$

where L_λ^s is given by Eq. (2-10). In the 8–15 μm region, however, the contribution of solar energy is negligible compared to that of the self-emitted thermal component, so we can just write,

$$at\text{-}sensor\ (TIR)\text{:}\quad L_\lambda^{TIR} = L_\lambda^e. \tag{2-13}$$

Surface-emitted component (L_λ^{eu})

The primary source of energy for thermal imaging is the earth itself, which has a typical temperature of 300 K. Different materials on the earth, however, can emit different amounts of thermal energy even if they are at the same temperature. Most materials are not ideal blackbodies with 100 % radiative efficiency. The efficiency with which real materials emit thermal radiation at different wavelengths is determined by their *emissivity, ε*. Emissivity plays a proportionality role in the thermal region much like that of reflectance in the visible; it is defined as the ratio of the spectral radiant exitance of a *greybody* to that emitted by a blackbody (M_λ in Eq. (2-1)) at the same temperature, and is therefore unitless and between zero and one. The emitted radiance of the earth is therefore,

$$earth\text{'}s\ surface\text{:}\quad L_\lambda(x, y) = \varepsilon(x, y, \lambda)\frac{M_\lambda[T(x, y)]}{\pi}. \tag{2-14}$$

It is implied in this equation that different objects or materials on the earth's surface can have different temperatures, and hence different spectral radiant exitances, as well as different emissivities. Note the similarity between this relation and that for the solar reflective region, Eq. (2-7).

To separate the effects of emissivity and temperature, scientists usually assume one or the other is spatially constant. In thermal studies, such as aerial thermal scanning for building heat loss, the emissivity of various roof materials might be assumed equal in order to estimate temperatures. On the other hand, in some geologic applications, temperature might be assumed constant in order to estimate emissivity. The fact that both affect the radiance, however, means that one should be able to justify ignoring the variation of either parameter. The situation is particularly complicated because emissivity can vary with wavelength, and even temperature. An example of the type of analysis needed to separate T and ε effects in the MWIR is given in Mushkin *et al.* (2005).

---

6. We will use an extra superscript, *e*, on some radiance quantities in this section to distinguish emission-related terms from solar reflectance terms.

The relation between emitted radiance and the source temperature is not obvious from Eq. (2-1) and Eq. (2-14). To get a better feeling for that, we plot in Fig. 2-16 the spectral radiance as a function of temperature for three fixed wavelengths in the TIR, assuming constant emissivity. The temperature range, 250 K to 320 K, includes normal daytime and nighttime temperatures on the earth. We see that *spectral radiance is approximately linear with temperature over this range*, and for any smaller range, as might actually be encountered in a thermal image, a linear approximation is even better. Thus, for our purposes, we can approximate Eq. (2-14) by,

$$\text{earth's surface:} \quad L_\lambda(x, y) \approx \varepsilon(x, y, \lambda)\frac{[a_\lambda T(x, y) + b_\lambda]}{\pi} \tag{2-15}$$

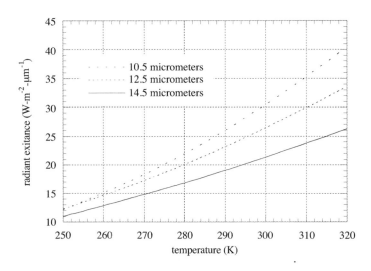

FIGURE 2-16. *The dependence of radiant exitance from a blackbody on its temperature at three wavelengths. Emissivity is held constant at one, whereas it actually can vary with temperature and wavelength for a greybody. The temperature range depicted is that for normal temperatures at the earth's surface.*

where a_λ and b_λ are relatively weak functions of wavelength λ. In practice, these coefficients would be given by their average over the spectral passband of the sensor (Chapter 3). Our intent with Eq. (2-15) is to provide a simple, easily-visualized relationship between radiance and temperature; in effect, a complex system has been approximated by a linear relationship that applies over a limited temperature range. If one's goal is to actually calculate temperatures from remote-sensing imagery, then the accurate form of Eq. (2-14) should be used.

The radiation emitted from the earth is transmitted by the atmosphere along the view path to the sensor,

$$at\text{-}sensor: \quad \begin{aligned} L_\lambda^{eu}(x, y) &= \tau_v(\lambda)L_\lambda(x, y) \\ &= \varepsilon(x, y, \lambda)\frac{\tau_v(\lambda)[a_\lambda T(x, y) + b_\lambda]}{\pi} \end{aligned} \tag{2-16}$$

The atmosphere's transmittance from $2.5\,\mu\mathrm{m}$ through the thermal IR is shown in (Fig. 2-17). There are four distinct spectral windows available for remote sensing in this spectral region, as determined by molecular absorption bands. As mentioned earlier, the solar energy contribution in this spectral region is relatively small; the rapid fall-off in solar irradiance above $2.5\ \mu\mathrm{m}$ is shown in Fig. 2-18.

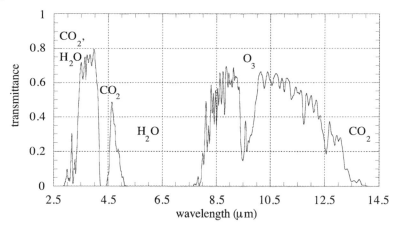

FIGURE 2-17. *Atmospheric transmittance (solar elevation = 45°) in the midwave IR and thermal IR spectral regions. This curve, like those in the visible and shortwave IR, is obtained from the atmospheric modeling program MODTRAN. The curve appears more detailed than Fig. 2-4 because the abscissa covers six times the range in wavelength with about the same spectral resolution.*

Surface-reflected, atmosphere-emitted component (L_λ^{ed})

The atmosphere also emits thermal radiation downward, which is then reflected at the earth's surface and transmitted upward through the atmosphere to the sensor. This term is analogous to the skylight component arising from scattering in the visible spectral region.

$$at\text{-}sensor: \quad L_\lambda^{ed} = F(x, y, \lambda)\rho(x, y, \lambda)\frac{\tau_v(\lambda)M_\lambda^a}{\pi} \tag{2-17}$$

where M_λ^a denotes the spectral radiant exitance of the atmosphere. The factor F is the fraction of the sky hemisphere that is seen from the surface at (x,y) and is the same function that was used for reflected skylight (compare to Eq. (2-9)).

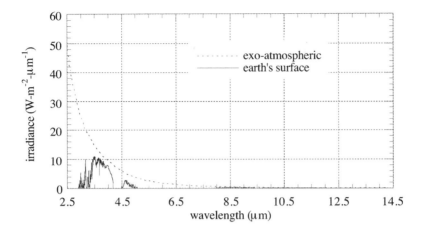

FIGURE 2-18. *Solar irradiance in the midwave and thermal IR regions (solar elevation = 45°). The ratio of these two curves is the path transmittance depicted in Fig. 2-17.*

A useful relation between an object's emissivity and its reflectance (assuming that the object is thick and does not transmit radiation) is *Kirchhoff's Law*,

$$\rho(x, y, \lambda) = 1 - \varepsilon(x, y, \lambda), \qquad\qquad (2\text{-}18)$$

which is valid at any wavelength. It implies that any object which is a good emitter of thermal energy ($\varepsilon \cong 1$) is also a poor reflector, and vice versa. This equation may be substituted in Eq. (2-17), if desired.

Path-emitted component (L_λ^{ep})

The atmosphere also emits radiation upward (according to Planck's blackbody law) as a function of the temperature at different altitudes. The total energy that arrives at the sensor is integrated over the view path from the contributions at all altitudes. We will call this the *path-emitted* component, L_λ^{ep}. The resulting spectral distribution does not particularly resemble that of a blackbody at a single temperature; it is a mixture of blackbodies over a range of temperatures. Moreover, radiation from lower altitudes is absorbed and re-emitted at higher altitudes, making the situation quite complicated.

It is reasonable to assume that this component does not vary significantly over a scene; the only exceptions are for large angles from nadir (above about ±20°), where the path-emitted term tends to increase (Schott, 1996), and in areas where the surface temperature has significant spatial variation (for example, areas of burning vegetation) that may influence the near-surface atmospheric temperature.

Total at-sensor, emitted radiance (L_λ^e)

We write the total at-sensor, emitted radiance as the sum of the three components described previously,

at-sensor:

$$L_\lambda^e(x, y) = L_\lambda^{eu} + L_\lambda^{ed} + L_\lambda^{ep}$$

$$= \varepsilon(x, y, \lambda)\frac{\tau_v(\lambda)}{\pi}[a_\lambda T(x, y) + b_\lambda] + F(x, y, \lambda)\rho(x, y, \lambda)\frac{\tau_v(\lambda)M_\lambda^a}{\pi} + L_\lambda^{ep}. \tag{2-19}$$

Similar to the solar reflectance region (Eq. (2-10)), we note that:

  · the total spectral thermal radiance received by the sensor is approximately linearly proportional to the surface temperature, modified by

  · a multiplicative, spatially- and spectrally-variant emissivity factor, and

  · an additive, spatially-invariant, spectrally-dependent term due to view path emission.

2.3.3 Total Solar and Thermal Upwelling Radiance

The sum of the non-thermal and thermal radiation contributions is,

$$\textit{at-sensor:} \quad L_\lambda(x, y) = L_\lambda^s(x, y) + L_\lambda^e(x, y) \tag{2-20}$$

where L_λ^s is given by Eq. (2-10) and L_λ^e is given by Eq. (2-19). As discussed earlier, the second term is negligible in the visible and shortwave IR, while the first term is negligible in the thermal IR. In the midwave IR, both can be important, depending on the surface reflectance, emissivity, and temperature. A general review of MWIR remote sensing is given in (Boyd and Petitcolin, 2004).

The at-sensor radiance that would be seen by a satellite thermal sensor is depicted in Fig. 2-19. In this MODTRAN simulation, the reflectance and emittance are assumed to be spectrally uniform. The direct component due to solar irradiance at the earth is small, but contributes the majority of energy between 2.5 and 5 μm. Above 5 μm, the upwelling radiance is due to the three components discussed previously. The path-emitted component shifts to longer wavelengths as the atmosphere becomes colder with increasing altitude, and the result seen by a downward-looking sensor is a mixture from all levels. For these reasons, the total at-sensor radiance of Fig. 2-19 does not resemble a blackbody curve for a single temperature source (Fig. 2-1).

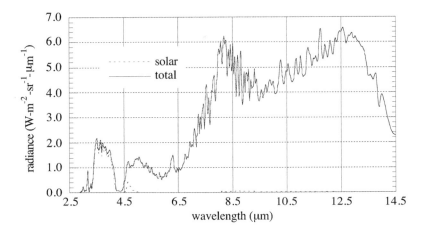

FIGURE 2-19. *The at-sensor radiance above the atmosphere in the middle and thermal IR regions. Note how the two sources of radiation, solar and thermal emission, exchange relative importance from the MWIR to the TIR. The satellite view angle is zero degrees from nadir and the surface emissivity is assumed to be one. The spectral reflectance is also assumed to be uniform in wavelength.*

2.3.4 Image Examples in the Thermal Region

Nighttime thermal imagery is particularly useful because solar heating is not present. In Fig. 2-20, a nighttime HCMM image is compared to a co-registered Landsat MSS image. Since the emissivity can safely be assumed constant across the lake, the HCMM-measured radiance is a good representation of the lake's temperature. The large *GIFOV* of the HCMM sensor, however, mixes the radiance coming from the lake and surrounding terrain, which is mostly vegetated. A detailed analysis of mixing proportions using the MSS image to separate water and land at each HCMM pixel was required to estimate lake temperatures from the relatively low resolution HCMM data (Schowengerdt, 1982). The newer ETM+ thermal band spatially resolves the temperature variations within the lake.

Although the atmosphere emits thermal radiation, only the layer near the earth is near 300 K. As altitude above the surface increases, the temperature falls off rapidly. Clouds appear cooler than land in daytime TIR imagery since they are at some altitude above the earth (Fig. 2-21).

Urban areas expectedly show a "heat island" effect in thermal imagery. In Fig. 2-22, New Orleans appears to be warmer than the surrounding undeveloped areas and water. One must know the emissivities of the different surface materials represented, however, to correctly find the temperature differences.

Thermal images do not show topographic shading for the same reasons as visible imagery. Since thermal radiation is self-emitted by the surface, there is no directional dependence due to external irradiance. However, one sees so-called "thermal shadows" caused by preferential heating of the surface depending on its orientation to the sun. The example in Fig. 2-23 clearly shows the

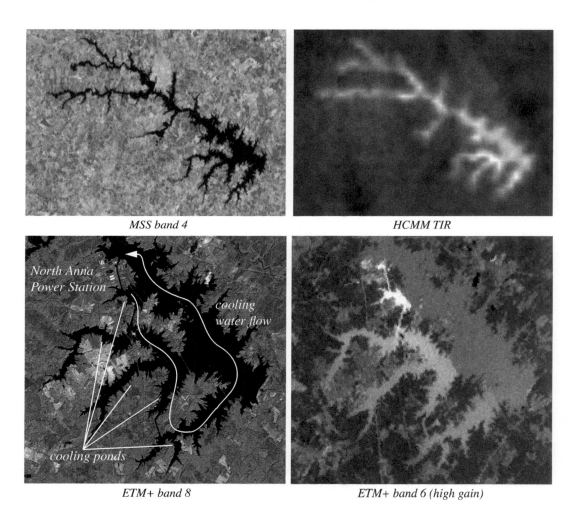

MSS band 4

HCMM TIR

ETM+ band 8

ETM+ band 6 (high gain)

FIGURE 2-20. *Lake Anna, Virginia, viewed by the Landsat MSS band 4 (June 12, 1978, at 9:30A.M.) and the HCMM thermal band (June 11, 1978, at 3:30A.M.) and Landsat ETM+ (September 30, 1999, at 10:30A.M.). Water from the center of the lake is used to cool the North Anna reactor and then pumped through a series of canals and cooling ponds, and eventually back into the main lake; the intent is to preserve the temperature and ecology of the main lake. Even though the HCMM GIFOV was only about 600m, it is clear that the water in these cooling ponds is warmer than that in the main body of the lake. The ETM+ pan band 8 more clearly shows the cooling system canals and dams, and the TIR band 6 (60m GIFOV) shows how the initially elevated temperature of the cooling water gradually decreases as the water moves through the ponds to the lake. Note, ETM+ records TIR imagery with two electronic gains, low and high. The high-gain mode is useful for low radiance scenes (Chapter 3). (MSS and HCMM imagery courtesy of Dr. Alden P. Colvocoresses, U.S. Geological Survey; ETM+ image courtesy Global Land Cover Facility, University of Maryland (http://glcf.umiacs.umd.edu/) and USGS)*

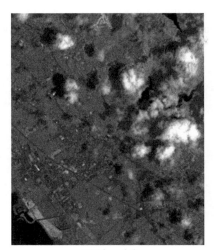
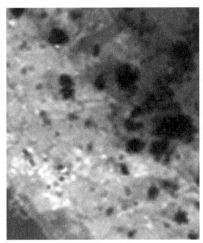

TM band 4   TM band 6

FIGURE 2-21. *TM band 4 (30m GSI) and band 6 (120m GSI) images of the San Francisco area. The bright clouds visible in band 4 appear darker than the ground in band 6, implying they are cooler than the ground in this daytime image. The cloud shadows, visible to the upper left of each cloud in band 4, appear slightly darker than surrounding areas in band 6, implying that the ground in shadow is slightly cooler than that in sunlight. Note that weather satellite TIR images shown on television often have an inverted greyscale in order to make cold clouds appear bright!*

TM band 2   TM band 6

FIGURE 2-22. *Landsat TM band 2 and band 6 images of New Orleans, Louisiana, including Lake Pontchartrain and the Mississippi River (September 16, 1982). The urban area, particularly the denser core, appears warmer than surrounding vegetation, although their respective emissivities may be a factor. The darker, rectangular feature above the city center and adjoining the lake is a city park.*

effect of solar heating. It lends a different appearance to the topography compared to the visible image; note the brighter appearance of the southeast facing slopes in the lower half of the mountain range.

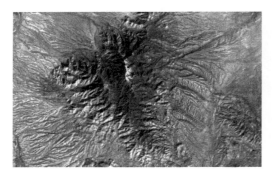
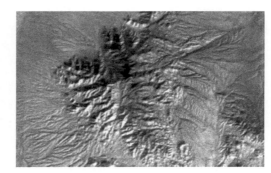

TM band 2   TM band 6

FIGURE 2-23. The Santa Rita Mountains, south of Tucson, Arizona, viewed by Landsat TM on January 8, 1983. The elevation ranges from about 1000 meters to 2880 meters at Mt. Wrightson, and the mountains are heavily vegetated at the higher elevations. Note the thermal "shadows" on slopes facing away from the direction of solar irradiance (from the lower right); these valleys are cooler than the solar-facing slopes in this mid-morning, winter image.

2.4   Summary

The basic models for optical radiation transfer at the earth's surface and through the atmosphere have been described. The most important components for remote sensing were covered and illustrated with image examples. The key elements of this chapter are:

-   The energy collected by a remote sensing system is proportional to the surface reflectance (solar reflective region) and to the surface emissivity and temperature (thermal region).

-   A spatially-invariant, but spectrally-dependent, constant bias term arising from atmospheric scattering (solar reflective region) and atmospheric emission (thermal region) is present in the sensed signal.

-   A "coupling" exists between the surface and the atmosphere; they interact as a function of the surface reflectance, emission, and topography.

In the next chapter, we will look at the influence of the sensor on remote-sensing measurements. Its primary impact is a local spatial and spectral integration of the signal, by which the "resolution" of the image data is determined.

2.5   Exercises

Ex 2-1. Suppose you have a DEM and wish to calculate a shaded relief image. The formula in Eq. (2-6) can be used, but it does not explicitly account for surface points that are self-shadowed, i.e., points that face away from the sun. What mathematical constraint needs to be imposed on Eq. (2-6) for such points so that they are correctly depicted? How would you determine points that are shadowed by other points, i.e., points that lie in a cast shadow? (This requires a rather involved mathematical solution — in your response, just indicate what needs to be done.) For simplicity, assume a 1–D problem in the vertical plane defined by the solar vector.

Ex 2-2. Suppose we want to measure the amount of liquid water contained in the leaves of an agricultural crop by remote sensing in the absorption bands at 920 nm and 1120 nm. What effect will the atmosphere have on our ability to obtain accurate measurements?

Ex 2-3. In the discussion leading to Eq. (2-6), we noted that, in the absence of cast shadows or clouds, the solar irradiance is effectively constant across the 185 km *GFOV* of Landsat ETM+. Is this also true for a sensor like AVHRR or MODIS with a *GFOV* of 2800 km?

Ex 2-4. What physical factors may contribute to the differences between the two curves in Fig. 2-9?

Ex 2-5. What physical assumption (other than altitude) is behind the statement ". . . the bright clouds appear darker . . . implying they are cooler than the ground in this daytime image" in the caption of Fig. 2-21? Why is that assumption less important to the statement that the land surface in cloud shadows is slightly cooler than the surrounding land surface?

CHAPTER 3

Sensor Models

3.1   Introduction

The sensor converts the upwelling radiance (reflected and/or emitted) into an image of the radiance spatial distribution. Several important transformations of the radiometric, spatial, and geometric properties of the radiance occur at this stage. Generally, the sensor degrades the signal of interest, i.e., the portion of the total radiance that contains information about the earth's surface. It is important to understand the nature of this degradation to properly design image processing algorithms and interpret their results.

3.2   Overall Sensor Model

An electro-optical sensor may be modeled by the processes shown in Fig. 3-1. The scanning operation (Chapter 1) converts the spatial at-sensor radiance to a continuous, time-varying optical signal on the detectors. The detectors, in turn, convert that optical signal into a continuous time-varying electronic signal, which is amplified and further processed by the sensor electronics. At the Analog/Digital (A/D) converter, the processed signal is sampled in time and quantized into discrete *DN* values representing the spatial image pixels.

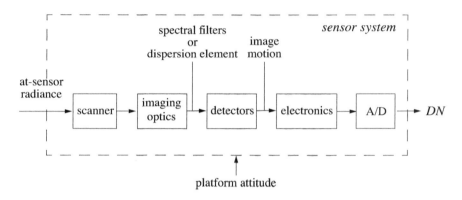

FIGURE 3-1. *The primary components in an electro-optical remote-sensing system. A whiskbroom type of system is illustrated here. Although the platform attitude is external to the sensor per se, it has important effects on the final image characteristics and quality.*

3.3   Resolution

No property of images is more widely quoted, and simultaneously misused, than *resolution*. It is a term that conveys a strong intuitive meaning, but is difficult to define quantitatively. Remote sensing systems have "resolution" in the spectral, spatial, and temporal measurement domains. Furthermore, there is a numerical resolution associated with the data itself by virtue of radiometric quantization. We will discuss the notion of resolution in this section, and in the process, introduce important concepts such as convolution, mixed pixels, and sampling.

3.3.1 The Instrument Response

No instrument, remote-sensing systems included, can measure a physical signal with infinite precision. If the signal varies in time, the instrument must average over a non-zero integration time; if the signal varies in wavelength, the instrument must average over a non-zero spectral bandwidth; or if the signal varies in space, the instrument must average over a non-zero spatial distance. In general, we can write the output of most instruments as

$$o(z_0) = \int_W i(\alpha) r(z_0 - \alpha) d\alpha \qquad\qquad (3\text{-}1)$$

where

$i(\alpha)$ = input signal,
$r(z_0 - \alpha)$ = instrument response (unit area), inverted and shifted by z_0,
$o(z_0)$ = output signal at $z = z_0$, and
W = range over which the instrument response is significant.

The physical interpretation of Eq. (3-1) is that the instrument weights the input signal in the vicinity (W) of z_0 and integrates the result. If we allow z_0 to have any continuous value, say z , then this relation is known as a *convolution*. A convenient, commonly-used shorthand notation for Eq. (3-1) is,

$$o(z) = i(z)*r(z) \qquad\qquad (3\text{-}2)$$

which reads "the output signal equals the input signal convolved with the response function." This mathematical description can be applied to a wide range of instruments.[1] In the following sections, we use it to describe the spatial and spectral response characteristics of a remote-sensing imaging system.

3.3.2 Spatial Resolution

While the spatial "resolution" of a sensor (or its image) is often quoted as the *GSI* or *GIFOV*, it is well known that it is possible to *detect* considerably smaller objects if the contrast with the surrounding background is sufficiently high. Even though such objects may be detectable, they are not necessarily *recognizable*, except by the general context of the image. Thus, in TM imagery one frequently "sees" roads or bridges over water that are 10m (one third of a pixel) or less wide (Fig. 3-2). Similarly, high contrast, linear features much smaller than a pixel are visible in MODIS and IKONOS imagery (Fig. 3-3).

---

1. Specifically, systems that are linear (the output for multiple inputs is the sum of their individual outputs) and invariant (the form of $r(z_0 - \alpha)$ does not depend on z_0).

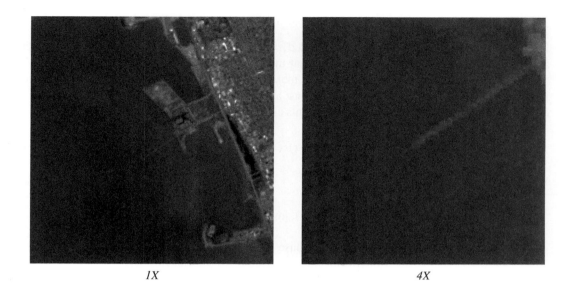

1X   *4X*

FIGURE 3-2. Example of subpixel object detection. This is part of a TM band 3 image of San Francisco showing the Berkeley Pier. The pier is 7m wide and made of concrete. An older extension of the same width, but made of wood with a lower reflectance than concrete, is barely visible. (Acknowledgments to Joseph Paola for providing details on the size and construction of the pier.)

Suppose we have a sensor that produces a linear *DN* output versus scene reflectance.[2] A ground area that is larger than *GIFOV*-squared[3] and has zero reflectance produces a zero *DN*, and an area with a reflectance of one produces a maximum *DN*, say 255. Now, suppose the area within one *GIFOV* contains two types of materials, with reflectances of zero and one. Surrounding pixels contain only the dark material and represent the background against which the pixel of interest is compared. At that pixel, the signal produced by the sensor will be the integrated effect of the mixture of the two components, target and background, within the *GIFOV*. If their relative proportion within the *GIFOV* is 50%, then the *DN* will be 128 (rounded to the nearest integer *DN*). If the brighter material occupies only 10% of the *GIFOV* area, the *DN* will be 26. Since the sensor presumably can reliably distinguish two *DNs* separated by only one unit, we could theoretically (assuming no image noise) detect a bright object against a dark background even if it occupied only 0.4% of the *GIFOV* area (Fig. 3-4).

---

2. We simplify the discussion by assuming remote-sensing measurements are a direct function of reflectance, without intervening influences.

3. The product of the in-track and cross-track *GIFOVs*. For brevity, this two-dimensional area will simply be called a *GIFOV* in the rest of this section

TM band 4

MODIS band 2

IKONOS pan band

FIGURE 3-3. More examples of subpixel object detection. At the top are TM and MODIS images of the All American Canal in California. The Canal carries water for irrigation and is about 60m wide (two TM pixels). Because of the high contrast to the Imperial Sand Dunes, it is visible in the 250m MODIS band 2 image. Even the narrower Interstate 8 highway at the top is visible. At the bottom is an IKONOS image of a parking lot and buildings in Tucson, AZ. The parking lot markings are about 10cm wide, or about 1/10 of a IKONOS 1m pan band pixel. (IKONOS image provided by Space Imaging LLC and NASA Scientific Data Purchase Program.)

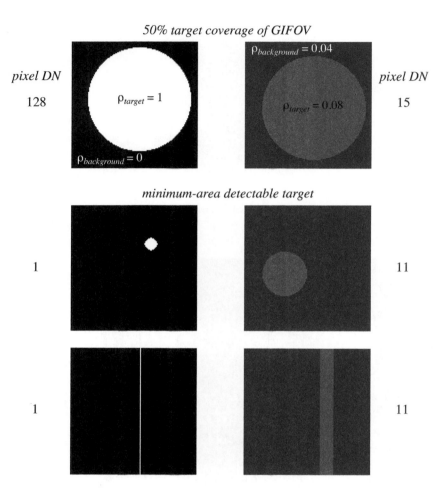

FIGURE 3-4. *Detectability analysis for a single target at two different contrasts to the surrounding background and for an idealized sensor. The squares represent the area sampled by a single GIFOV. The minimum-area detectable target results in a one DN difference from the background. Note that the target does not have to be centered in the GIFOV (if the sensor response is uniform across the GIFOV); the spatially-integrated signal will be the same for any internal location. For the same reason, the shape of the target cannot be discerned from a single pixel. If the target is linear, such as a road or bridge, then the shape may be inferred by the context of several pixels; cases in point are Fig. 3-2 and Fig. 3-3.*

Now, we will make the object reflectances more realistic. The darker background might have a reflectance of 4% and the lighter target a reflectance of 8%, yielding a *contrast ratio* of 2:1. If the target covers 50% of the *GIFOV*, the *DN* of that pixel will be 15, while the *DN* of pure background pixels will be 10. The target is still detectable since it differs from the background by five *DN*s. If the size of the target is less than 10% of a pixel, however, it falls below the threshold (one *DN*) for detection. We see, therefore, that the radiometric quantization, coupled with the target and background reflectances, and the sensor *GIFOV*, all conspire to determine the "resolution" of the image. If there is any noise in the image, the threshold for detection will be higher, i.e., a greater target-to-background contrast will be required for reliable detection.

An often neglected influence on image resolution is the *sample-scene phase*, i.e., the relative location of the pixels and the target (Park and Schowengerdt, 1982; Schowengerdt *et al.*, 1984). This relative spatial phase is unpredictable (and almost always unknown) for any given image acquisition, and varies from acquisition-to-acquisition with a uniform probability distribution between ±1/2 pixel. Two of an infinite number of possible sample-scene phases are shown in Fig. 3-5 for the low-contrast case described previously. On the left, the target appears equally in each of four *GIFOV*s, and on the right, unequally among the four. The resulting *DN*s of the four pixels containing part of the target are shown below the diagram. In the case of equal subdivision of the target, the four pixels would have a *DN* of 11 and the target would be detectable in each. If the image pixel grid happened to partition the target unequally as shown to the right in the figure, the percent coverage by the target in each *GIFOV* would be 30, 5, 2, and 13 (clockwise from the upper left). The four pixels would thus have *DN*s of 13, 11, 10, and 12, respectively. The target would remain detectable in three of the pixels (again, if there were no noise).

Sample-scene phase is important for more complex targets, as well. For example, suppose the target consists of a series of equally-spaced bright bars against a dark background. If the sensor *GIFOV* is equal to the width of one bar, the two extreme sample-scene phases result in a spatial signal that either has maximum contrast or no contrast! If the sensor *GIFOV* is exactly twice the width of one bar, the spatial signal will have zero contrast for *any* sample-scene phase.

An enlargement of the TM image in Fig. 3-2 is shown in Fig. 3-6. It can be seen that the *DN* profile along each scan line across the pier is different. This is the result of sample-scene phase variation, because the pier is not oriented at 90° to the scan. Thus, in one particular scan the pier may appear to be one pixel wide, while in another it appears to be two pixels wide. To estimate the true width of this subpixel object, an interleaved composite of many lines should be made, with careful attention to phasing them correctly to fractions of a pixel. The composite would represent the convolution, Eq. (3-2), of the sensor spatial response function and the pier radiance profile, sampled at an interval much finer than the pixel *GSI*. Just such an analysis is used later in this chapter to evaluate the ALI and QuickBird sensor spatial response functions.

With the inclusion of factors that come into play in real images, namely sensor noise, non-uniform targets and backgrounds, and variable solar angle and topography, it becomes clear that the situation with respect to image resolution is not simple! A common statement such as "the resolution of this image is 30 meters," probably refers to the sensor *GIFOV* or *GSI*, but it is not a precise statement, as we have seen. In this book, we will assume the term resolution refers simply to the *GSI*.

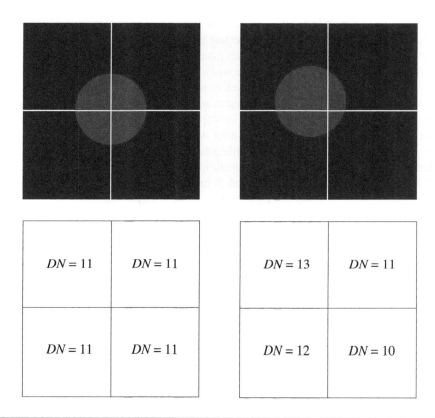

FIGURE 3-5. *The effect of spatial phasing between the pixel grid and the ground target. Four adjacent GIFOVs are shown and the target area is 50% of the GIFOV area. On the left, the target fills 12.5% of each GIFOV. On the right, the target occupies 30%, 5%, 2%, and 13% of the four GIFOVs. The location of the grid is unpredictable prior to imaging any given scene; it could be anywhere within ±1/2 pixel interval with equal probability. The imaging of long linear features can sometimes allow precise measurement of the phase to a small fraction of a pixel. The background DN is 10.*

3.3.3 Spectral Resolution

The total energy measured in each spectral band of a sensor is a spectrally-weighted sum of the image irradiance over the spectral passband, Eq. (3-26). This weighting by wavelength is the primary determinant of the sensor's capability to *resolve* details in the spectral signal. To see why this is so, we will take the reflectance data for the mineral alunite (Chapter 1) in the vicinity of the OH absorption doublet between 1350 and 1550nm wavelength, and simulate its measurement by a multispectral sensor. Each half of the doublet is only about 10 to 20nm wide, and they are separated by about 50nm. Now, imagine these data are measured with a hyperspectral sensor having many

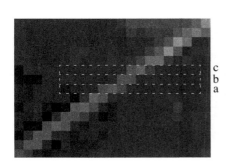
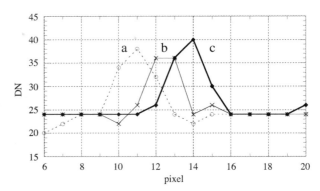

FIGURE 3-6. *A contrast-enhanced enlargement of Fig. 3-2 and DN profile plots along three adjacent scanlines near the center of the pier, illustrating the sample-scene phase effect. The pier's profile is different in each line (the linear interpolation between individual pixels is only to aid visualization of the graph).*

spectral bands, each 10 nm wide (at 50% of the peak responsivity) and spaced at 10 nm intervals. Each band sees an effective reflectance which is the weighted reflectance over the band; the weighting function is the spectral response of the sensor in each band (Fig. 3-7).[4] We will use a bell-shaped function which approximates actual spectral band responsivities. Since the spectral passband is comparable in width to the details in the signal, they are preserved. The exact reflectance minima are not found, however, because of the location of the bands along the wavelength scale.[5]

If we do a similar exercise with 50 nm-wide spectral bands, the result shows complete loss of information about the doublet; it is literally "averaged away" by the broad spectral bands (Fig. 3-7). Even if the spectral band locations were shifted, the doublet would not be "resolved." If the spectral bands do not overlap as they do in Fig. 3-7, (for example, bands 3, 4, 5, and 7 of TM), the sensor's ability to resolve even coarser features is seriously hampered. The trade-off, of course, is the increased data burden of finely-sampled spectra.

To illustrate how the actual TM spectral response modifies the at-sensor radiance, we will use the Kentucky Bluegrass spectral reflectance shown in Chapter 1 and the atmospheric propagation model of Chapter 2. The net at-sensor radiance is shown in Fig. 3-8. This is multiplied by the spectral response of each of the four VNIR TM bands (Fig. 3-22), to yield the weighted spectral distribution seen by each band. The integral of this function over wavelength then provides the total *effective* radiance in each band.

---

4. To simplify the illustration, we are not including solar irradiance or atmospheric propagation as described in Chapter 2. These factors would change the input function, but not alter the concept.

5. This is another example of sample-scene phase, but in the spectral dimension rather than the spatial dimension.

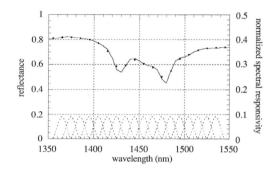
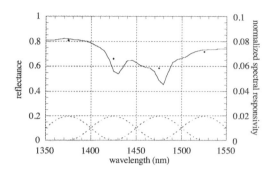

10nm spectral bandwidth   *50nm spectral bandwidth*

FIGURE 3-7. The effective reflectance of alunite as measured by a multispectral sensor. The solid line is the original reflectance sampled at 1nm and the individual band spectral responsivities are shown as dashed lines. Each solid dot is the output of the corresponding band. The graph with 10nm-wide spectral bands represents the response of a hyperspectral sensor such as AVIRIS or HYDICE. The graph with 50nm-wide spectral bands represents the response of a sensor such as TM (although TM does not actually have any spectral bands in this part of the spectrum).

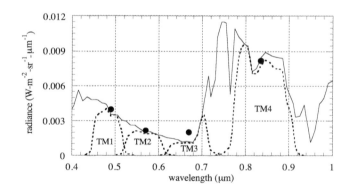

FIGURE 3-8. The at-sensor radiance for Kentucky Bluegrass (solid curve), the weighted spectral distribution seen by each TM band (dotted curves), and the total (integrated over the weighted spectral distribution) effective radiance (solid circles). Note how the broad bandwidth of bands 3 and 4, in particular, averages spectral detail present in the original radiance.

Now, in fact, the spectral reflectances of many natural materials are fairly smooth, without fine absorption lines (see the reflectance curves for soil and vegetation in Chapter 1). So the high spectral resolution of hyperspectral sensors at first seems to be of little value in those cases. With better understanding of the absorption properties and biochemical interactions among natural material components, for example, cellulose, lignin and protein in vegetation, the additional resolution afforded by imaging spectroscopy can be fully utilized (Verdebout *et al.*, 1994; Wessman, 1994). As seen in the prior examples, the *placement* of spectral bands is as important as the spectral *bandwidth* to a sensor's ability to resolve spectral features. Hyperspectral sensors offer generally contiguous bands over a wide spectral range, and are, therefore, superior to multispectral sensors with a few, albeit judiciously-placed, spectral bands.

3.4   Spatial Response

The sensor modifies the spatial properties of the scene in two ways: (1) blurring due to the sensor's optics, detectors, and electronics; and (2) distortion of the geometry. In this section, we discuss spatial blurring, which generally occurs on a smaller spatial scale (a few pixels) than distortion.

The image of the scene viewed by the sensor is not a completely faithful reproduction. Small details are blurred relative to larger features; this blurring is characterized by the net (total) sensor *Point Spread Function (PSF$_{net}$)*, which we can accurately view as the *spatial responsivity* of the sensor (just as we described a spectral responsivity in the previous section). *PSF$_{net}$* is the weighting function for a spatial convolution (refer to Eq. (3-1)), resulting in the electronic signal, e_b,

$$e_b(x, y) = \int_{\alpha_{min}}^{\alpha_{max}} \int_{\beta_{min}}^{\beta_{max}} s_b(\alpha, \beta) PSF_{net}(x - \alpha, y - \beta) d\alpha d\beta . \qquad (3-3)$$

The sensor response function weights the measured physical signal, which is then integrated over the range of the response function to produce the output value. The limits of the integral define the *spatial extent* of the *PSF* about the coordinate (x,y). Note that the left side of Eq. (3-3) still depends on the continuous spatial coordinates (x,y); we do not convert this to discrete pixel coordinates until the signal is sampled.[6]

PSF$_{net}$ consists of several components. First, the optics induce blurring by the *optical PSF*. The image formed by the optics on the detectors may in some cases move during the integration time for each pixel; this introduces an *image motion PSF*. Then the detector adds additional blurring due to the *detector PSF*. The detected signal is further degraded by the *electronics PSF*. The process is outlined in the flow diagram of Fig. 3-1. The following type of analysis has been described for the Landsat MSS (Park *et al.*, 1984), TM (Markham, 1985) and AVHRR (Reichenbach *et al.*, 1995).

---

6. Throughout this chapter and the remainder of the book, the coordinates (x,y) will represent the cross-track and in-track directions in image space, respectively. This "path-oriented" coordinate system will suffice for any sensor that points to nadir relative to the platform track. It would need to be reconsidered for describing off-nadir imaging by a pointable sensor.

The appropriate sensor parameters for the models are generally found in contractor reports, e.g., TM data are found in SBRC (1984), although such data are normally pre-flight laboratory measurements which may not be appropriate after launch and possible sensor changes.

An important assumption in this analysis is that the net 2-D sensor *PSF* is given by a product of two 1-D *PSF*s in the cross-track and in-track directions,

$$PSF_{net}(x, y) = PSF_c(x)PSF_i(y).$$ (3-4)

The net *PSF* is termed *separable* if this equation holds. Separability is generally valid for the types of scanners of interest and allows a simpler analysis with 1-D functions (see Appendix B for an elaboration on separable functions).

In the following sections, we describe the various components that contribute to a sensor's total *PSF*. Unless noted otherwise, the coordinates (x,y) and all parameters are assumed to be in image space (rather than object space; see Fig. 1-11), since most data for sensor modeling are available from the sensor engineering documents and tend to be specified in image space. The conversion factor between image and ground distances and velocities is simply the magnification of the sensor,

$$\text{ground distance} = \text{image distance} / \text{magnification}$$
$$= \frac{H}{f} \times \text{image distance}$$ (3-5)

and,

$$\text{ground velocity} = \text{image velocity} / \text{magnification}$$
$$= \frac{H}{f} \times \text{image velocity}.$$ (3-6)

3.4.1 Optical PSF_{opt}

The optical *PSF* is defined as the spatial energy distribution in the image of a point source, such as a star, or in the laboratory an illuminated pinhole. An optical system is never "perfect" in the sense of forming a point image of a point source. The energy from the source is spread over a small area in the focal plane. The extent of spreading depends on many factors, including optical diffraction, aberrations, and mechanical assembly quality.

An optical system with no degradation other than diffraction (which is unavoidable) is termed "diffraction-limited." The resulting *PSF* is called the *Airy Pattern*, with a bright central *Airy Disk* surrounded by concentric rings of decreasing brightness (Fig. 3-9). The mathematical form is given by,

$$PSF(r') = \left[2\frac{J_1(r')}{r'} \right]^2$$ (3-7)

where J_1 is a Bessel function of the first kind and the normalized radius r' is given by,

$$r' = \frac{\pi D}{\lambda f} r = \frac{\pi r}{\lambda N} \text{ , where} \quad \begin{array}{l} D = \text{aperture diameter} \\ f = \text{focal length} \\ N = \text{f-number} \\ \lambda = \text{wavelength of light ,} \end{array} \tag{3-8}$$

and the unnormalized radius to the first dark ring, i.e., the radius of the Airy Disk, is,

$$r = 1.22\left[\frac{\lambda f}{D}\right] = 1.22\lambda N, \tag{3-9}$$

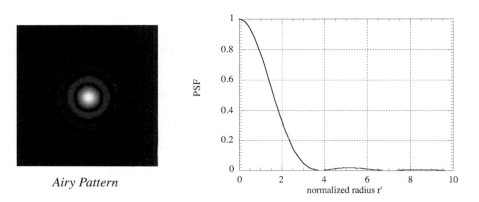

Airy Pattern

FIGURE 3-9. *The diffraction-limited optical PSF and its radial profile.*

where N is the optics *f-number*, given by the ratio of the optical focal length *f* divided by the aperture stop diameter *D*.[7]

A real optical system may not be diffraction-limited because of numerous factors, as mentioned above. Also, it may not be possible to describe the *PSF* analytically, so it must be measured after the system is built. A common generic model for a measured optical *PSF* is the 2-D Gaussian function,

$$PSF_{opt}(x, y) = \frac{1}{2\pi ab}e^{-x^2/a^2}e^{-y^2/b^2}. \tag{3-10}$$

The parameters *a* and *b* determine the width of the optics *PSF* in the cross- and in-track directions, respectively. For well-designed optics, *a* equals *b*. Notice that the Gaussian is separable, consistent with Eq. (3-4).

---

7. Actually, the diameter of the *entrance pupil* of the system, which is an image of the aperture stop, should be used. Space does not permit us to elaborate further on this topic from geometrical optics; interested readers are referred to Slater (1980) or any of a number of contemporary optics texts.

3.4.2 Detector PSF_{det}

This is the spatial blurring caused by the non-zero spatial area of each detector in the sensor. Although there can be small non-uniformities in detector response across its dimension, it is a reasonable assumption to let,

$$PSF_{det}(x, y) = rect(x/w)rect(y/w),\qquad(3\text{-}11)$$

which is a separable, uniform square pulse function (see Appendix B).

3.4.3 Image Motion PSF_{IM}

If the image moves across the detectors during the time it takes to integrate the signal for a pixel, blurring results.[8] The image motion is modeled by a square pulse *PSF* (Appendix B) in one direction only,

$$whiskbroom\ scanner:\ PSF_{IM}(x, y) = rect(x/s)\qquad(3\text{-}12)$$

$$pushbroom\ scanner:\ PSF_{IM}(x, y) = rect(y/s)\qquad(3\text{-}13)$$

where *s* is the spatial "smear" of the image in the focal plane and is given by,

$$whiskbroom\ scanner:\ \ s\ =\ \text{scan velocity} \times \text{integration time}\qquad(3\text{-}14)$$

$$pushbroom\ scanner:\ \ s\ =\ \text{platform velocity} \times \text{integration time}.\qquad(3\text{-}15)$$

For some whiskbroom scanners, such as AVHRR and TM, the integration time is negligible compared to the sample interval (the pixel spacing, or the *GSI*). Typically, the integration time of these systems causes a spatial blurring on the order of 1/10 pixel. MODIS, however, also a whiskbroom scanner, has an integration time that corresponds to image motion of nearly a whole detector width in the cross-track direction. In a pushbroom scanner, such as SPOT or ALI, the integration time causes in-track blurring. For SPOT and ALI, the in-track spatial movement during the integration time is one detector width, and is therefore a significant contribution to the total sensor *PSF* in the in-track direction.

The MODIS design is used to illustrate the interaction of the detector *PSF* and the image motion *PSF* in Fig. 3-10. The sampled area of each pixel overlaps that of its neighboring pixels by 50% in all bands, i.e., the parameters scale proportionately with each *GSI*—250m, 500m, and 1km. This design allows improved SNR with less resolution loss than would result from a larger detector.

---

8. In a conventional camera photograph, this smearing of the image is all too visible if the camera is moved during the exposure, or the object being photographed moves.

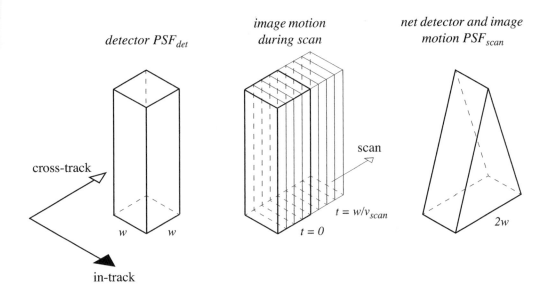

detector PSF$_{det}$

image motion
during scan

net detector and image
motion PSF$_{scan}$

cross-track

scan

$t = w/v_{scan}$

w w

$t = 0$

$2w$

in-track

pixel samples in scan direction

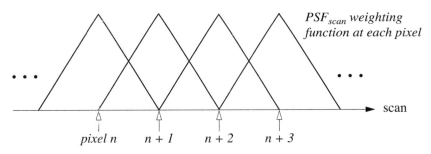

PSF$_{scan}$ weighting
function at each pixel

scan

pixel n   n + 1   n + 2   n + 3

FIGURE 3-10. *The MODIS scan mirror sweeps the image continuously across the focal plane. This is equivalent to the detectors scanning a stationary image (the perspective taken in this figure); a few intermediate locations of the detector are shown in the middle diagram to represent the motion of the image. Pixels are sampled at 83.333 μs (250 m bands), 166.667 μs (500 m bands), and 333.333 μs (1 km bands) intervals (Nishihama et al., 1997). During a sample interval, the image moves by one detector width (in all bands) and is integrated, except for 10 μs used to read out the detectors (ignored here). PSF$_{scan}$ (a convolution of PSF$_{det}$ and PSF$_{IM}$) introduces a non-uniform weighting of the local image by a triangle, whose base is twice the detector width w in the scan direction; neighboring pixels are therefore correlated because their sampled image areas overlap.*

3.4.4 Electronics PSF_{el}

The signal from the detectors is sometimes filtered electronically to reduce noise. The electronic components operate in the time domain of the signal as it is scanned and read from the detectors. The time dependence can be converted to an equivalent spatial dependence by,

$$\text{\textit{whiskbroom scanner:} } x = \text{scan velocity} \times \text{sample time interval} \tag{3-16}$$

or

$$\text{\textit{pushbroom scanner:} } y = \text{platform velocity} \times \text{sample time interval} . \tag{3-17}$$

In the AVHRR, MSS, TM, and ETM+ whiskbroom scanners, the electronic filter is a low-pass, Butterworth-type filter effective in the cross-track direction (Reichenbach *et al.*, 1995; Park *et al.*, 1984; Markham, 1985; Storey, 2001). This filter smooths the data in the cross-track direction.[9] Such electronic filters are not common in pushbroom CCD sensors. However, there are certain intrinsic sensor array properties, like spatial diffusion of photoelectrons in a 2-D array or charge transfer efficiency in a linear CCD array, that can be modeled by an electronic *PSF*. Such a component is included in the ALI model described in Chapter 6 and Appendix B.

3.4.5 Net PSF_{net}

By a simple theorem (Chapter 6), the net *PSF* is the convolution of the component *PSF*s,

$$PSF_{net}(x, y) = PSF_{opt} * PSF_{det} * PSF_{IM} * PSF_{el} . \tag{3-18}$$

The width of the net *PSF* is the sum of the widths of each of the component *PSF*s. We could write this equation as,

$$PSF_{net}(x, y) = PSF_{opt} * PSF_{scan} * PSF_{el} \tag{3-19}$$

where

$$PSF_{scan}(x, y) = PSF_{det} * PSF_{IM} \tag{3-20}$$

as explained in Fig. 3-10 for the paddlebroom MODIS scanner.

3.4.6 Comparison of Sensor PSFs

To emphasize the relative differences in *PSF* among different sensors with a wide range of *GSI*'s, it is helpful to normalize them to the same *GSI*, as shown in Fig. 3-11. The scale, then, is the same for AVHRR with a 1 km *GSI* and TM with a 30 m *GSI*. The total sensor *PSF*, modeled as described above, is shown for several satellite sensors in Fig. 3-12. Note that for the AVHRR and MSS, the amount of sensor blur is about twice as great in the cross-track direction than in the in-track

---

9. See Chapter 6 for a discussion of spatial filtering and low-pass filters.

direction. All systems, however, show a cross-track response that is *considerably broader than that of the detector*. This fact has important implications in all aspects of information extraction from remote sensing imagery, from calculations of small target signatures to spatial-spectral signature mixing at each pixel. Because of this blurring, the *effective GIFOV* of remote sensing systems is larger than the oft-quoted geometric *GIFOV*. For example, the effective TM *GIFOV* was found to be 40 to 45 m, rather than 30 m (Anuta *et al.*, 1984; Schowengerdt *et al.*, 1985).

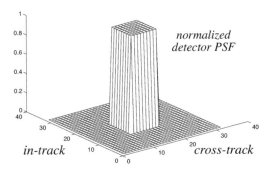

FIGURE 3-11. *This normalized detector PSF is used for comparison of different sensor PSFs. The detector is plotted with eight samples in-track and cross-track, independent of the particular sensor GSI and GIFOV.*

3.4.7 Imaging System Simulation

Simulation of imaging systems is useful in visualizing the effects of various image sensor parameters. For example, Fig. 3-13 shows at the top an aerial image with a small *GSI* that is subsampled by taking every 2nd or 4th pixel along rows and columns. It becomes increasingly difficult to recognize the man-made structures such as buildings and roads as the *GSI* increases. Magnification of the image does not help because the scene detail is not present in the image to begin with (analogous to so-called "empty magnification" in microscopy).

The *GIFOV* also affects the level of detail within the image, as seen at the bottom of Fig. 3-13. Here the original aerial image is first averaged over either 2 × 2 or 4 × 4 pixels before subsampling to simulate a sensor with the *GIFOV* equal to the *GSI*. As the *GIFOV* and *GSI* increase, more scene detail is lost in the digital representation, although, curiously, the overall visual effect does not seem as great as that incurred by the *GSI* alone. This is because the spatial averaging resulting from the increased *GIFOV* reduces "aliasing" caused by the subsampling. The sensor *GSI* is typically specified to be equal to the *GIFOV*.

To illustrate the cascading of sensor components in the image forming process, we will do a simulation of TM imaging. Three TM *PSF* components will be used in the simulation (Fig. 3-14). The simulation starts with the scanned aerial photograph of Fig. 3-15. The *GSI* and *GIFOV* of this

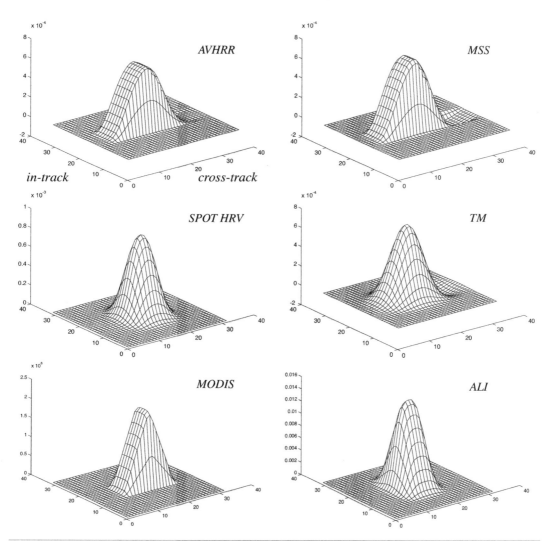

FIGURE 3-12. *Perspective views of the total system PSF for several remote sensing systems. The PSFs are normalized to the same scale as described in Fig. 3-11 and all are considerably broader than the detector PSF. Three whiskbroom scanners (AVHRR, MSS and TM) have asymmetric cross-track spatial responses, caused by the low-pass electronic filter. The same filter also causes a small negative response to one side of the PSF (which leads to the vertical offsets in the plots). The large cross-track size of the MODIS PSF is caused by the image motion blur at each pixel, as described in the text (note the similarity to Fig. 3-10). Likewise, the ALI pushbroom system has a PSF that is broader in-track than cross-track because of image motion. The AVHRR, MODIS, and MSS in-track responses are nearly that of the detector PSF alone, because their optical PSFs are relatively small compared to the detector. The SPOT and TM have smaller detectors and the optical system PSF causes blurring comparable to that of the detector in both directions.*

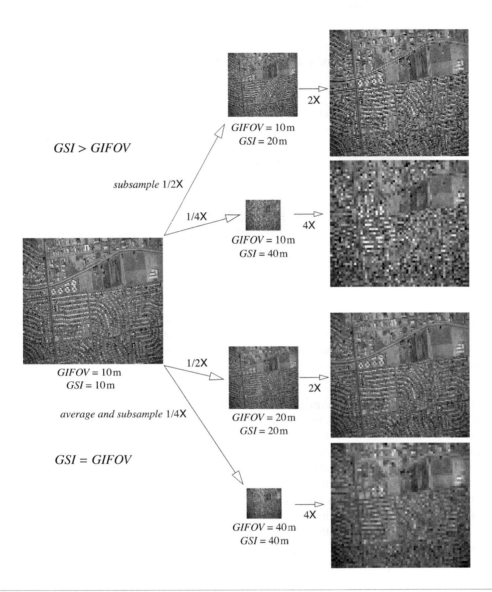

FIGURE 3-13. *Simulation of the effect of GSI and GIFOV on visual image quality. The upper two examples are for undersampled ("aliased") conditions with GSI > GIFOV, and the lower two examples are for equal GSI and GIFOV. The latter case results in better image quality even though the GIFOV is larger.*

image are each about 2m. First, the image is rotated to align with the in-track and cross-track directions of the Landsat orbit. This represents the radiance spatial distribution seen by the TM that is to be convolved with the TM *PSF* components.

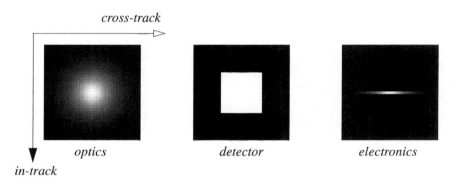

FIGURE 3-14. *The individual component PSFs for TM. All are displayed at the same scale; the detector width corresponds to a 30m GIFOV. The background around each PSF is set to zero. These 31 × 31 sample arrays (2m/sample) were used for digital convolution (Chapter 6) of the image in Fig. 3-15 to simulate TM imaging.*

   The resulting simulated intermediate images in the overall imaging process are shown in Fig. 3-16. The first image depicts the effect of the optical *PSF*; it represents the irradiance spatial distribution on the detectors in the TM focal plane. Next, the blurring induced by the scanning detector *IFOV* is shown. If we were also simulating detector noise (which is quite small for TM), it would be added at this stage. Finally, the electronic filter is applied in the cross-track direction. The original 2m *GSI* has been maintained to this point for simulation accuracy, but must be reduced to the 30m TM *GSI* in the middle image of Fig. 3-16.

   Finally, an enlargement of the simulated TM image is compared to an actual TM image acquired about four months later than the date of the aerial photograph used in the simulation. The simulation has clearly produced an image with a similar spatial resolution to that of the TM. Although the TM image's contrast was adjusted to account for unspecified radiometric differences, no attempt was made to simulate radiometric properties, including different sun angles, shadows, atmospheric conditions, and spectral response, which probably account for the residual differences.

   A spatial degradation process similar to that above has been described in Justice *et al*. (1989) for the simulation of lower resolution data from Landsat MSS imagery. In that work, the original transfer function of the MSS was taken as the "source" for further degradation to a "target" transfer function representing a sensor with lower spatial resolution, such as the AVHRR.

*rotate
and trim*

FIGURE 3-15. *Scanned aerial photograph of an area in Tucson, Arizona, used in spatial simulation of TM imagery. The scanned photograph has a GSI of about 2m and is panchromatic, with a spectral response similar to the average of TM bands 2 and 3. The dark feature at the bottom is a focal plane fiducial mark used for distortion calibration in aerial cameras. The right image is rotated to align its rows and columns with the TM scan and orbit directions, respectively, and then trimmed to the final size used in the simulation.*

3.4.8 Measuring the *PSF*

Traditionally, the optical *PSF* is measured in a laboratory using controlled conditions. A special target, such as a "point source," "line source," or "edge source" is imaged by an optical system, and the image is scanned to produce the *PSF* in 2-D or the *Line Spread Function (LSF)* or *Edge Spread Function (ESF)* in 1-D. The *LSF* and *ESF* are often easier to measure because of their greater light level, but must be measured in multiple directions if there is any asymmetry in the *PSF*. Other representations of the spatial response can be calculated via the Fourier transform of the *PSF*, but a detailed description of the Fourier transform is inappropriate at this point. We'll focus here on the spatial domain representations of spatial response; the Fourier domain equivalents are discussed in Chapter 6. The relations among the *PSF*, *LSF*, and *ESF* are shown in Fig. 3-17.

In optical terminology, the *PSF* is the image of a point source, the *LSF* is the image of a line source, and the *ESF* is the image of an edge source. The imperfect imaging behavior of any optical system results in image blur in each of these cases. The *LSF* is the derivative of the *ESF*, both being 1-D functions. We can write the *LSF* in two orthogonal directions in terms of the *PSF* as follows,

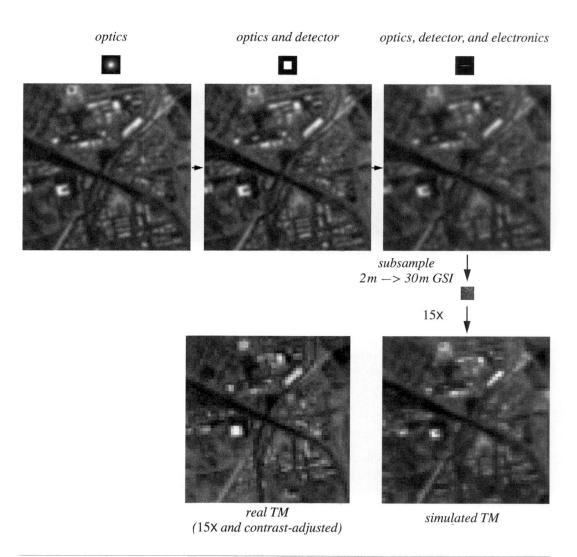

FIGURE 3-16. *A sequential series of images produced by the components of the TM spatial response are shown at the top, with the individual PSF components shown to scale. Each component introduces additional blurring. To complete the simulation, the image must be subsampled to a 30m GSI. Enlargement of this final, simulated TM image and an actual TM image are shown at the bottom. The TM image is the average of bands 2 and 3, which approximates the spectral response of the panchromatic photo used in the simulation. Contrast adjustment of the TM image compensates for radiometric differences between it and the simulated TM image due to atmospheric effects and sensor gain and bias, but does not alter the spatial detail properties for comparison to the simulated TM.*

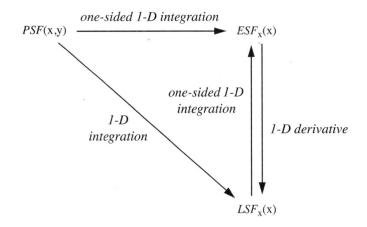

FIGURE 3-17. *The mathematical relations among various representations of optical spatial response. Reduction of the 2-D PSF to the 1-D ESF or LSF is irreversible, i.e., it is not possible to recover the PSF from the ESF or LSF. Thus, the latter are often measured in at least two directions to establish any asymmetry.*

$$LSF_c(x) = \int_{-\infty}^{\infty} PSF(x, y)dy\,, \quad LSF_i(y) = \int_{-\infty}^{\infty} PSF(x, y)dx\,. \qquad (3\text{-}21)$$

Furthermore, we can write the *ESF* in terms of the *LSF* as follows,

$$ESF_c(x) = \int_{-\infty}^{x} LSF_c(\alpha)d\alpha\,, \quad ESF_i(y) = \int_{-\infty}^{y} LSF_i(\alpha)d\alpha \qquad (3\text{-}22)$$

or equivalently,

$$LSF_c(x) = \frac{d}{dx}ESF_c(x), \quad LSF_i(y) = \frac{d}{dy}ESF_i(y)\,. \qquad (3\text{-}23)$$

In practice, the line or edge response is usually measured rather than the point response. If the edge response is measured, the *LSF* can be readily calculated as the derivative of the *ESF*, although that calculation is sensitive to any measurement noise in the *ESF*.

Measuring the total system *PSF* for digital sensors is more complicated, as the effect of pixel sampling has to be included (Park *et al*., 1984). There are also components such as electronic filters and image motion during detector integration. It is often difficult to simulate all of these effects in a pre-launch laboratory measurement. Furthermore, there is the possibility for change in system *PSF* after launch arising from thermal focus change or sensor outgassing[10] in the space environment. So, measuring and monitoring the sensor *PSF* on-orbit (after launch) is important.

A number of techniques have been used to measure the *PSF*, *ESF*, and *LSF* from operational imagery (Table 3-1). Generally, they involve either man-made targets, such as mirrors or geometric patterns, or targets-of-opportunity, such as bridges. Although not intended to measure the *PSF* but rather to mark specific ground locations in a satellite image, there were early successful experiments with specular mirrors as subpixel targets, in which a flat mirror of only 56cm diameter was successfully detected in the 80m *GIFOV* of the first Landsat MSS (Evans, 1974).

A methodology to include the effect of pixel sampling and reconstruction by resampling in a spatial response specifically for digital sensors was developed by Park *et al.* (1984). However, most measurement analyses attempt to avoid pixel sampling effects (see Fig. 3-5) and to produce an estimate of the continuous, unsampled *PSF*, *LSF*, or *ESF*. A "phased-array" of effective point sources was used to measure the TM *PSF* (Rauchmiller and Schowengerdt, 1988), and a phased array of small mirrors has been used to measure the QuickBird spatial response (Helder *et al.*, 2004). The "phased-array" concept is an array of point sources, spaced at non-integer multiples of the *GSI*. With sufficient spacing, each point source produces an independent, *sampled PSF* image. They can be averaged, maintaining their original relative phasing, to achieve the sample-scene phase-averaged response proposed by Park *et al.* (1984). Alternatively, by interleaving the samples from each *PSF* image, one can achieve subpixel sampling of the *PSF*. The same effect can be achieved with a line or edge target that is at a small angle to the pixel sample grid. Such a target is known as a "phased-line" or "phased-edge" target, referring to the changing sample-scene phase along the target (Park and Schowengerdt, 1982).

Two examples of sensor *LSF* measurement from operational imagery are presented next. One uses a "phased-line" target and the other uses a "phased-array" of line targets to produce subpixel samples and avoid the sampling effect on the measured *LSF*.

ALI LSF measurement

An analysis for the ALI *LSF* is the first example (Schowengerdt, 2002). An ALI image of the Maricopa agricultural area near Phoenix, Arizona, was acquired on July 27, 2001, and two areas used for *LSF* analysis are shown in Fig. 3-18. The fields in this area are level and periodically irrigated. Between the fields are berms, sometimes with a dirt road, to contain the irrigation water. The fields are laid in a north-south and east-west pattern, and make angles of 13.08° (as measured from the ALI image) with the ALI in-track and cross-track directions.

An IKONOS image (1m panchromatic and 4m multispectral) acquired on July 26, 2001, showed that the berms are typically 7 to 10m wide. Where the two fields on either side of a berm have a full crop canopy, the berm forms a linear, high-contrast (in the visible bands), subpixel

---

10. Outgassing can result in deposition of material onto optical surfaces, in effect "clouding" them. The *PSF* and image *SNR* are then degraded substantially. It can be reduced or controlled by "baking" of the sensor with thermal cycling. Just such a situation arose early in the ALI mission and was corrected by thermal cycling (Mendenhall *et al.*, 2002). Outgassing has also been blamed for radiometric calibration error in the Landsat-5 TM and corrected by periodic warming (Helder and Micijevic, 2004).

TABLE 3-1. *Some examples of sensor spatial response measurement from operational imagery. The ADAR System 1000 is an airborne multispectral sensor.*

sensor	target type	reference
ADAR System 1000	edge	Blonski *et al.*, 2002
ALI	agriculture field berms	Schowengerdt, 2002
ETM+	bridge	Storey, 2001
HYDICE	bridge	Schowengerdt *et al.*, 1996
Hyperion	ice shelf edge, bridge	Nelson and Barry, 2001
IKONOS	edge	Blonski *et al.*, 2002
	parking lot stripes	Xu and Schowengerdt, 2003
	edge	Ryan *et al.*, 2003
	line	Helder *et al.*, 2004
MODIS	higher-resolution imagery	Rojas *et al.*, 2002
MSS	higher-resolution imagery	Schowengerdt and Slater, 1972
OrbView-3	edge	Kohm, 2004
QuickBird	phased-array of mirrors, edge	Helder *et al.*, 2004
simulated	non-specific	Delvit *et al.*, 2004
SPOT4 (simulated)	point source array	Robinet *et al.*, 1991
SPOT5	edge, spotlight	Leger *et al.*, 2003
TM	bridge, higher-resolution imagery	Schowengerdt *et al.*, 1985
	phased-array of subpixel targets	Rauchmiller and Schowengerdt, 1988

feature that can be used to measure the multispectral band *LSF* in the in- and cross-track directions. Because of the angle between the berms and the two cardinal scanner directions, it is possible to select pixels that represent subpixel samples across the berms. For example, a column-wise set of pixels across a north-south berm is equivalent to a subpixel set of samples along the in-track direction (assuming the fields on either side of the berm and the berm itself are spatially uniform in radiance).

Several transects were extracted, registered, and averaged in both cases (columns and rows) to reduce noise. A linear trend was also removed to account for a small difference in radiance in the two adjoining fields on either side of the berm. The data were then normalized by the area, producing an estimate of the *LSF* as shown in Fig. 3-18. There is some distortion of the derived *LSF*s because of the small angle of the target to the scan directions; ideally the target should be exactly aligned with the rows and columns of pixels, but then subpixel sampling would not be achieved.

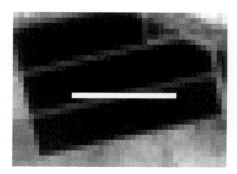

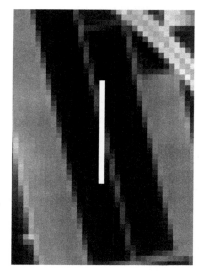

transect used to extract subpixel
samples for the in-track LSF

transect used to extract subpixel
samples for the cross-track LSF

cross-track and in-track band 3 LSFs

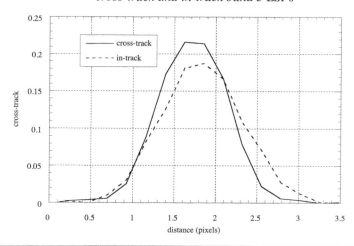

FIGURE 3-18. *Example transects of agriculture field berms used to measure the ALI LSF are shown in white in these ALI band 3 images (Schowengerdt, 2002). The pixels along columns are, in effect, cross-track subpixel samples across the berm because of its small angle to the column direction. Similarly, pixels along rows represent in-track subpixel samples across the berm. This is only true if the target fields are uniform on both sides of the berm and the berm itself is uniform. Note, the in-track LSF is wider than the cross-track LSF as expected for the ALI pushbroom sytem because of in-track sample integration, similar to the effect of cross-track sample integration in the MODIS whiskbroom system (Fig. 3-10). (© 2002 IEEE)*

A concern in such analysis is the width of the target (the berm in this case) relative to the *LSF* being measured. If the target is too narrow, it will not produce a good result because of low signal level; if it is too wide, the estimated *LSF* will be broadened by the target itself. A simple test is shown in Fig. 3-19 with targets of different size and a detector *LSF*, i.e., a rectangle function the width of one detector. A target width of 1/5 to 1/3 the *LSF* width does not dramatically broaden the measured *LSF*, a condition that is consistent with guidelines for optical *LSF* measurement using line sources.

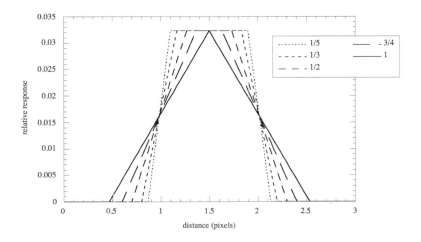

FIGURE 3-19.   *The effect of line target width on the measured LSF. The fractions are the ratio of target size to LSF width. Note that a target equal in width to the LSF results in a triangular measured LSF, similar to the image motion and detector LSF of Fig. 3-10. This is because of the underlying convolution of two rectangle functions in both cases. It is not easy to correct for the target width in spatial domain measurements, but a correction is relatively easy in the Fourier domain for simple geometric target shapes.*

QuickBird LSF measurement

There are many trade-offs between using special man-made targets and targets-of-opportunity that are already present on the ground. The former need to be installed and maintained during image collection periods, while the latter do not. However, targets-of-opportunity may not be optimal in their size and shape, and their condition may not be known at the time of imaging. The high spatial resolution of commercial remote sensing systems allows for many more potential targets-of-opportunity for imaging performance measurement. In particular, abundant man-made features in urban areas are good candidates. In this example, the white painted stripes in an asphalt vehicle parking lot are used to measure the QuickBird *LSF* from the image shown in Fig. 3-20 (Xu and Schowengerdt, 2003).

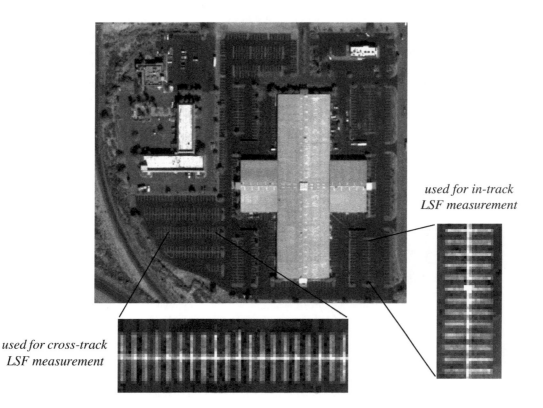

used for in-track LSF measurement

used for cross-track LSF measurement

FIGURE 3-20. QuickBird image collected on November 4, 2002, of the Tucson Bargain Center parking lot. The facility is open only from Friday through Sunday after noon, so the parking lot is empty in this late-morning image on a Monday. The target areas were chosen for their relatively clean asphalt backgrounds and large extent. Note how the image of each stripe is different due to different sample-scene phase. (Imagery provided by NASA Scientific Data Purchase program. Contains material © DigitalGlobe. All rights reserved.)

The target parameters in this case are given in Table 3-2. The individual parking stripes are only about 1/7 *GIFOV* in width, but sufficiently high contrast to the background to yield a good signal level. Their spacing is a non-integer number of pixels, meaning that each stripe is imaged with a different sample-scene phase. Finally, their length is several pixels, allowing averaging of several rows or columns of data to reduce noise in the measurement.

An algorithm was developed to interleave samples from each stripe at their correct relative spatial location, producing a composite profile with subpixel sampling. After averaging several lines or columns of data for the cross-track or in-track *LSF*, respectively, the estimated *LSF* is obtained (Fig. 3-21). These results are comparable to those from edge analysis of building roof edges (Xu and Schowengerdt, 2003).

TABLE 3-2. Parking lot stripe parameters relevant to QuickBird image analysis.

parameter	size		pixels (resampled)	
	inches	meters	in-track	cross-track
width	4	0.102	0.148	0.145
spacing	120	3.05	4.42	4.40
length	240	6.10	8.84	8.80

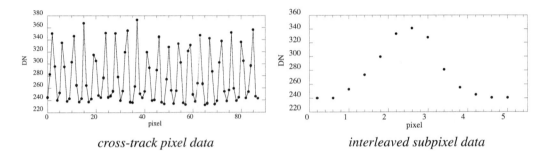

cross-track pixel data   interleaved subpixel data

cross-track and in-track pan band LSFs

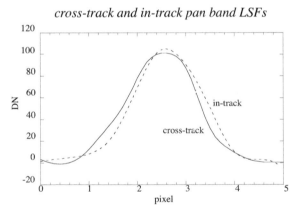

FIGURE 3-21. *The QuickBird LSF analysis using parking lot stripes. The upper left graph shows the image pixel values along a series of stripes. Each black dot is a pixel value; the intermediate values are linearly interpolated in the plot. The varying sample-scene phase is evident; each stripe is sampled at a slightly different phase along the scan. If these different sample-scene phases are carefully accounted for, it is possible to interleave the data to obtain a composite data profile of the average stripe with a subpixel sample interval. The radiance of the stripes and background are assumed to not change along the target scan. Several scan lines are averaged and a smooth interpolating curve is placed through the data points for the final LSF estimates (Xu and Schowengerdt, 2003). Note the LSF goes slightly negative on the left because of the numerical techniques used; there are no physical reasons to expect negative values for ALI.*

3.5   Spectral Response

The radiance arriving at the sensor, as described in Chapter 2, is transferred by the sensor optics to the detector focal plane where an image is formed. The spectral irradiance on a detector located on the optical axis is related to the at-sensor radiance L_λ by the *camera equation* (Slater, 1980)

$$at\text{-}image\ plane: \quad E_\lambda^i(x, y) = \frac{\pi \tau_o(\lambda)}{4N^2} L_\lambda(x, y) \quad (\text{W-m}^{-2}\text{-}\mu\text{m}^{-1}) \tag{3-24}$$

where N is the optics *f-number* defined earlier. For notational simplicity, we are assuming that the geometric magnification between the ground and the image plane (Chapter 1),

$$m = f/H, \tag{3-25}$$

is one, and we therefore use the same (x,y) coordinate system for both the scene and image. The optical system transmittance $\tau_o(\lambda)$, excluding any filters, is reasonably high (often 90% or more) for most all-reflective optical systems and is nearly spectrally flat, so little, if any, signature modification is caused by the optics per se.

At this point, multispectral filters or wavelength dispersion elements, such as prisms, are introduced to separate the energy into different wavelength bands. For example, in the MSS, TM, ETM+, and MODIS the optical path is split into multiple paths, each with different spectral band filters. If we denote the product of the spectral filter transmittance and detector spectral sensitivity as the *spectral responsivity $R_b(\lambda)$*, the signal s_b measured by the sensor in band b is,

$$s_b(x, y) = \int_{\lambda_{min}}^{\lambda_{max}} R_b(\lambda) E_\lambda^i(x, y) d\lambda \quad (\text{W-m}^{-2}) \tag{3-26}$$

where λ_{min} and λ_{max} define the range of sensitivity for the band. Although this equation is not a convolution, the spectral integration of Eq. (3-26) is much like the spatial integration of Eq. (3-3). In both cases, the measured signal is weighted by the sensor response function and integrated over the range of that response.

Equation (3-26) defines the spectral integration of the received radiance and subsequent conversion to detector current, the units of R being amps-W^{-1} or volts-W^{-1} (Dereniak and Boreman, 1996). The spectral responsivities for some multispectral sensors are shown in Fig. 3-22. The spectral characteristics of the Landsat MSS have been documented in detail; in particular, it has been shown that small differences in the responsivities among individual detectors (recall the MSS has 6 in each band and the TM has 16) can explain some striping noise in the imagery (Slater, 1979; Markham and Barker, 1983). The ETM+ and TM bands are nearly the same by design; the panchromatic band 8, however, is only on ETM+. Note the SPOT5 pan band covers the green and red bands, while the ETM+ pan band covers the green, red, and NIR bands.

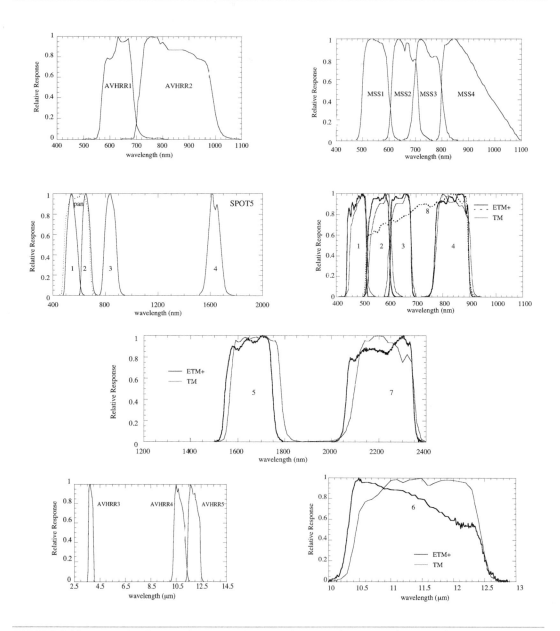

FIGURE 3-22. *Normalized spectral response functions for several sensors. These graphs are meant to be representative; individual detectors within a band can vary by several percent. Note the passbands are not rectangular and neighboring bands often overlap. An absolute vertical scale is determined by the responsivity of the detectors measured in electron units, e.g., amps or volts per unit irradiance.*

Spectral band resolution has generally improved over time, in part because of better detectors with lower noise characteristics and also because improvements in data storage and transmission allow more bands on a sensor. The VSWIR bands of MODIS, ASTER, and ETM+ are compared in Fig. 3-23. Hyperspectral sensors have a narrow spectral sensitivity in each band that can usually be considered Gaussian in shape, but with a bandwidth that varies across the full wavelength range of the sensor. The individual bands are also approximately linearly spaced over wavelength, as shown in Fig. 3-24. In cross-calibration analysis and sensor simulations, the broader spectral response of multispectral sensors can be synthesized by convolution of a hyperspectral sensor response with the response of the multispectral sensor.

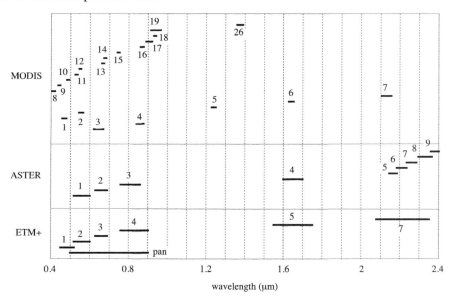

FIGURE 3-23. *Comparison of the VSWIR spectral bands for MODIS, ASTER, and ETM+. Note the relatively small spectral bandwidth of the MODIS bands and the fine spectral sampling of the ASTER SWIR bands. The latter is to provide mineral discrimination by measuring SWIR absorption features, as mentioned in Chapter 1.*

3.6   *Signal Amplification*

The electronic signal, e_b, in band b is amplified electronically and filtered by the electronics *PSF*. We included the filtering in the previous section for convenience. The amplification stage is designed to provide sufficient signal level to the A/D for quantization, without incurring saturation. This is done at the sensor design stage by estimating the maximum scene radiance range and the corresponding detector output range. The electronics gain and offset values are then set to yield a

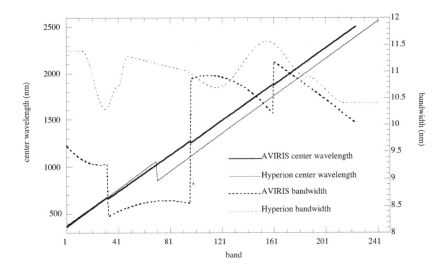

FIGURE 3-24.   *Spectral characteristics of the AVIRIS and Hyperion sensors. AVIRIS spectral characteristics have changed over the years of operation because of engineering changes in the spectrometers, and the appropriate sensor spectral characteristics included with the dataset of interest should always be used. The data plotted here are from a 1997 image collect. AVIRIS consists of four distinct spectrometers; the three discontinuities in the center wavelength and bandwidth curves mark the spectral ranges of the four spectrometers. The Hyperion has two spectrometers, one covering 360 to 1060nm and the other, 850 to 2580nm. The overlapping region from 850 to 1060nm (note the step in spectral characteristics above) allows cross-calibration of the two instruments (Pearlman et al., 2004).*

full *DN* range out of the A/D converter (Fig. 3-25). Some saturation under relatively infrequent conditions (high solar elevation over sand, for example) may be accepted in order to achieve higher gain (and therefore higher radiometric resolution) for the majority of scenes. The radiometric resolution of the sensor is, in part, controlled by the gain setting. The amplified signal, a_b, is given by,

$$a_b = gain_b \times e_b(x, y) + offset_b.$$ (3-27)

3.7 Sampling and Quantization

Finally, the amplified and filtered signal is sampled and quantized into *DN*s, usually with a linear quantizer (Fig. 3-26). This can be expressed as an *int[ ]* operator that converts the output of the signal amplifier to the nearest integer value. The final *DN* at pixel *p* in band *b* is therefore,

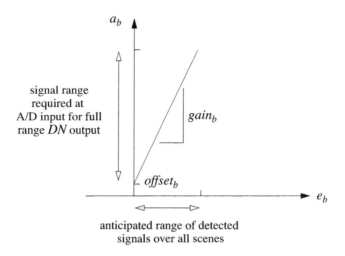

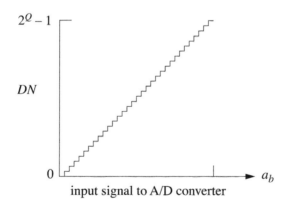

$$DN_{pb} = int[a_b]$$
$$= int[gain_b \times e_b(x, y) + offset_b]$$

(3-28)

The spatial-temporal sampling that occurs at the A/D is implicit in this equation in the conversion from continuous to discrete spatial coordinates.[11] Although the *int*[ ] operation in the A/D converter makes this a non-linear relationship between *DN* and e_b, the non-linearity can be ignored for mid-range or higher signal levels because the quantization error is a small percentage of the total signal. At lower *DN*s, however, the quantization error is more important, and should be considered in a detailed analysis. The number of discrete *DN*s, determined by the number of bits/pixel *Q*, defines the *radiometric resolution* of the system. One can express the radiometric resolution as 2^{-Q} times the dynamic range in radiance.

3.8 Simplified Sensor Model

Although Eq. (3-28) appears simple, it contains three integrations, one over the system spectral response in λ (Eq. (3-26) and two over the system spatial response in *x* and *y* (Eq. (3-3)). A simplification of Eq. (3-28) is often made in practice as follows. The spectral response of the sensor, $R_b(\lambda)$, is assumed to be an average constant over an *effective spectral band* (Palmer, 1984). Similarly, the spatial response of the sensor, *PSF(x,y)*, is assumed to be an average constant over an *effective GIFOV*. Thus, both functions can be removed from the integrals, and we can write,

$$DN_{pb} = int\left[K_b \iiint L_\lambda(x, y)d\lambda dxdy + \text{offset}_b\right],$$

(3-29)

where the sensor response functions are combined with the various other constants into the single constant K_b. The integrals in Eq. (3-29) are over the effective spectral band and the effective *GIFOV*. We therefore have, notwithstanding quantization, a *linear* relationship between *DN* and at-sensor radiance. If we further simplify the notation by calling L_{pb} the *band- and space-integrated at-sensor radiance* at a particular pixel *p* in band *b*, we can write,

$$DN_{pb} = K_b L_{pb} + offset_b.$$

(3-30)

This simplification relates the image *DN*s directly to the at-sensor radiances, integrated over the effective spectral passband and *GIFOV*. The inversion of Eq. (3-30) to obtain band radiance values from image *DN*s is known as *sensor calibration* or "calibration-to-radiance." If we go further, using the models and equations of Chapter 2, and convert radiance to reflectance, a *scene calibration,* or

---

11. Spatial-temporal sampling can be made mathematically explicit, but we have chosen not to do so in order to keep the discussion as short and direct as possible. For a mathematical treatment of sampling see, for example, Park and Schowengerdt (1982) and Park *et al.*(1984).

"calibration-to-reflectance," is achieved. Scene calibration is more difficult, of course, because it requires knowledge or assumptions about atmospheric conditions and surface terrain; a more detailed discussion appears in Chapter 7.

3.9   Geometric Distortion

The previous sections of this chapter dealt with sensor characteristics that affect the radiometric quality of the imagery, which is important in answering the question "what are we looking at?" Another relevant question is "where are we looking?" The answer to that question is determined by the geometric characteristics of the imagery, which in turn are set by the orbit, platform attitude, scanner properties, and earth rotation and shape.

As an ideal reference, consider conventional still frame imaging with a stationary scene. If the scene is a regular grid pattern and flat, and the camera optics have no distortion, the image will also be a regular grid, correct except for uniform scaling by virtue of the camera magnification. Now, imagine the camera is a pushbroom scanner, moving across the scene in a straight, constant altitude and velocity path. The resulting image will be geometrically identical to the still frame case. Any deviations from these conditions will result in *distortion* of the grid in the image. This non-ideal, but realistic, situation is the subject of this section.

3.9.1 Sensor Location Models

The orbits of most earth remote-sensing satellites are nearly circular because a constant image scale is desired. For precise modeling, an elliptical orbit is assumed.[12] The orbital velocity of satellites can be considered constant in time (e.g., 1.0153×10^{-3} radians/second for the Landsat-1 and -2 (Forrest, 1981)). Airborne sensors are another story, however, and variation in platform altitude and ground speed cannot be assumed negligible. In many cases, neither is recorded sufficiently often to permit post-flight correction; if sufficient data exist, a polynomial time model may be useful to interpolate between sample points.

3.9.2 Sensor Attitude Models

A small change in the platform attitude can result in a large change in the viewed location on the ground because of the long "moment arm" of high-altitude aircraft or satellite sensors. To see this, calculate the angle corresponding to the *GSI* between two neighboring pixels (Table 3-3). Any change in satellite attitude by this amount will result in a change of one pixel in the viewed location. The civilian high-resolution sensors have the highest attitude control and reporting requirements.

---

12. Even smaller pertubations are possible by changes in gravitational field and terrain elevation under the satellite orbit; these are difficult to model and can safely be ignored here.

Attitude is expressed by three angles of platform rotation: *roll*, *pitch*, and *yaw*. These are shown for one coordinate system convention in Fig. 3-27. Various schemes are used to automatically monitor and control satellite attitude within specified limits, including horizon sensors and torque flywheels. The actual values of roll, pitch, and yaw are sampled and recorded with the image data. Unfortunately, these data are not always available to the end user, and even if available, they are usually in complicated, non-standard formats. While any information on spacecraft attitude is useful, it is important that the precision and frequency of the reported values be sufficiently high to meet the geometric specifications for the imagery.

TABLE 3-3. *The angle between two adjacent pixels for a number of sensors. AVHRR and Landsat are not pointable; all the other sensors are pointable.*

system	altitude (km)	in-track GSI (m)	angle (mrad)
AVHRR	850	800	0.941
Landsat-4,-5 TM (multispectral)	705	30	0.0425
SPOT-1 to -4 (multispectral)	822	20	0.0243
Landsat-7 ETM+ (panchromatic)	705	15	0.0213
SPOT-5 (multispectral)	822	10	0.0122
SPOT-5 (panchromatic)	822	5	0.00608
OrbView-3 (panchromatic)	470	1	0.00213
IKONOS (panchromatic)	680	1	0.00147
QuickBird (panchromatic)	450	0.6	0.00133

Although the spacecraft's attitude does not behave in a predictable manner within the controlled limits of excursion, i.e., it is not systematic, it can usually be assumed to be a slowly changing function of time. Some success has been achieved by modeling the attitude variable, α (representing roll, pitch, or yaw), with a power series polynomial, over time periods of several image frames for TM (Friedmann *et al.*, 1983) and for individual SPOT frames (Chen and Lee, 1993),

$$\alpha = \alpha_0 + \alpha_1 t + \alpha_2 t^2 \ldots . \tag{3-31}$$

The attitude of aircraft sensors is subject to large changes from wind and turbulence. If a gyro-stabilized platform is not used for the sensor, the resulting imagery can contain severe distortions (Fig. 3-28).

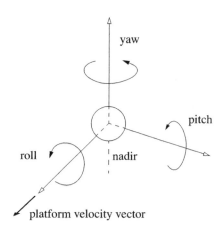

FIGURE 3-27. *Conventional definitions for the three attitude axes of a sensor platform. Although the choice of axes direction is arbitrary (Puccinelli, 1976; Moik, 1980), a right-handed coordinate system should be used (the cross product of any two vectors points in the direction of the third).*

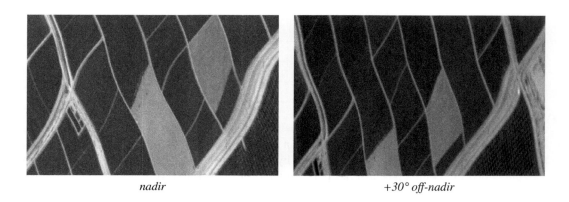

nadir   *+30° off-nadir*

FIGURE 3-28. *Airborne ASAS imagery of Maricopa Farm near Phoenix, Arizona, one taken at-nadir and the other at +30° off-nadir. Note how the distortion pattern changes within and between the two images, indicating a continuous change in the aircraft platform attitude over time. Also note the radiometric differences at the two view angles due to atmospheric differences and possibly non-Lambertian reflectance characteristics of the fields. (Imagery courtesy of Dr. Alfredo Huete, University of Arizona.)*

3.9.3 Scanner Models

Scanner-induced distortions are one of the easier factors to model, because they can usually be described by a fixed function of time. For example, the MSS has a non-linear scan mirror velocity that was well documented (Anuta, 1973; Steiner and Kirby, 1976). The effect of this was to cause a sinusoidal-like displacement of pixels across each scan, with a maximum error of about ±400 m near the midpoint of the scan on either side of nadir. As long as such distortions are consistent throughout an image and from orbit-to-orbit, they can be easily calibrated and corrected.

Whiskbroom scanners have more inherent distortions than pushbroom scanners because they have more moving parts. Pixel positioning cross-track is determined by the scan mirror motion (the MSS acquired data in one direction only, while TM and ETM+ acquire data in opposite directions on alternate scans), which couples with satellite motion to determine pixel positioning in-track. A pushbroom scanner, either of the linear or array type, on the other hand, has rigid cross-track geometry which is essentially decoupled from the in-track geometry. Some of the more important sources of scanner distortion are summarized in Table 3-4.

TABLE 3-4. *Examples of sensor-specific internal distortions. The reader is cautioned that some measurements of distortion were made from ground-processed imagery, and that a careful reading of the indicated reference is required before assuming the errors apply to all of the imagery from a given sensor. For example, the inter-focal plane misregistration in TM is given for early data; later data were found to be registered to within 0.5 pixel because of improved ground processing (Wrigley et al., 1985) .*

sensor	source	effect on imagery	maximum error	reference(s)
MSS	non-unity aspect ratio sampling	cross-track versus in-track scale differential	1.41:1	USGS/NOAA, 1984
	nonlinear scan mirror velocity	nonlinear cross-track distortion	±6 pixels	Anuta, 1973; Steiner and Kirby, 1976
	detector offset	band-to-band misregistration	2 pixels between bands	Tilton *et al.*, 1985
TM	focal plane offset	misregistration between visible (bands 1–4) and IR (bands 5–7)	–1.25 pixels	Bernstein *et al.*, 1984; Desachy *et al.*, 1985; Walker *et al.*, 1984
SPOT	detector element misalignment	in-track and cross-track pixel-to-pixel positional error	±0.2 pixels	Westin, 1992

Scanner-induced distortions can be measured pre-flight and an appropriate functional model determined for application to flight imagery, assuming there is no change in system performance during flight. If the source of the distortion can be physically modeled (for example, a motion analysis of the MSS scan mirror), it can be expected to result in the best correction possible.

The earliest Landsat MSS data were provided in an uncorrected format,[13] so there were numerous efforts to model the distortions arising from the sensor and platform motion relative to the earth (Anuta, 1973; Steiner and Kirby, 1976). Because these distortions are consistent over time, they can be corrected in a deterministic manner using orbital models and scanner calibration data.

3.9.4 Earth Model

Although the earth's geometric properties are independent of the sensor's, they interact intimately via the orbital motion of the satellite. There are two factors to consider here. One is that the earth is not an exact sphere; it is somewhat oblate, with the equatorial diameter larger than the polar diameter. In many satellite imaging models, the intersection of the sensor view vector with the earth's surface is calculated (Puccinelli, 1976; Forrest, 1981; Sawada *et al.*, 1981). Therefore, the exact shape of the earth is important. The earth ellipsoid is described by the equation,

$$\frac{p_x^2 + p_y^2}{r_{eq}^2} + \frac{p_z^2}{r_p^2} = 1 \qquad (3\text{-}32)$$

where (p_x, p_y, p_z) are the *geocentric* coordinates of any point P on the surface (Fig. 3-29), r_{eq} is the equatorial radius, and r_p is the polar radius. The *geodetic latitude* φ and *longitude* λ, as given on maps, are related to the components of $\boldsymbol{p}$ by Sawada *et al.* (1981),

$$\varphi = \text{asin}(p_z/r) \qquad (3\text{-}33)$$

and

$$\lambda = \text{atan}(p_y/p_x), \qquad (3\text{-}34)$$

where r is the local radius of the earth at the point P. The *eccentricity* of the earth ellipsoid is a useful quantity in map projections (Chapter 7),

$$\varepsilon = \frac{r_{eq}^2 - r_p^2}{r_{eq}^2}. \qquad (3\text{-}35)$$

The eccentricity of a sphere is zero. Some basic properties of the earth are listed in Table 3-5 for reference. Parameters such as the radii are updated periodically as more precise spacecraft measurements become available. Furthermore, different countries use different values (Sawada *et al.*, 1981) to produce their maps.

The second factor is that the earth rotates at a constant angular velocity, ω_e. While the satellite is moving along its orbit and scanning orthogonal to it, the earth is moving underneath from west to east. The velocity at the surface is,

---

13. The infamous *X-format*.

TABLE 3-5. Useful parameters for the "Figure of the Earth" and its rotational velocity. The dimensional values are from the Geodetic Reference System 1980 (GRS80) (Maling, 1992).

parameter	value
equatorial radius	6,378.137 km
polar radius	6,356.752 km
equatorial circumference	40,075.02 km
polar circumference	39,940.65 km
eccentricity	0.00669
angular velocity	7.2722052×10^{-5} rad/sec

$$v_0 = \omega_e r_e \cos\varphi \qquad (3\text{-}36)$$

where r_e is the earth's radius and φ is the geodetic latitude. Since satellites such as Landsat and SPOT have an orbit inclination, i, of about 9.1° to the poles (in order to set up the desired revisit period and sun-synchronism), the earth's rotation is not quite parallel to the cross-track scans. The projected earth rotational velocity in the scan direction is thus reduced,

$$v_e = v_0 \cos(i) = 0.98769 v_0 . \qquad (3\text{-}37)$$

The most important geometric parameters for earth-orbit modeling are shown in Fig. 3-29. The three vectors, s, g, and p, form the "fundamental observation triangle" (Salamonowicz, 1986) and obey the "look vector equation" (Seto, 1991),

$$p = s + g . \qquad (3\text{-}38)$$

A particularly simple algorithm to find P is given by Puccinelli (1976). The actual modeling can become quite involved mathematically, but all approaches use the above framework as a starting point.

The orbit shown in Fig. 3-29 is a *descending node* orbit, which is most common among earth remote sensing satellites. For the NASA Terra satellite, it results in a morning crossing of the equator from north to south at about 10:30 A.M. on the sunlit side of the earth. The NASA Aqua satellite is in an *ascending node* orbit, resulting in a afternoon equator crossing from south to north at about 1:30 P.M.. This combination of orbits provides same-day complementary measurements by Terra and Aqua sensors, as well as afternoon atmospheric measurements by Aqua sensors when solar heating effects are largest. The orbital paths of Terra and Aqua form a crisscross pattern as shown in Plate 3-1.

There is a long and rich history of modeling satellite orbits and image distortion (Table 3-6). There is also a parallel body of literature on *measuring* the residual distortion in production images that have been geometrically corrected (Table 3-7). Note the lack of references for AVHRR production imagery geometric quality; this is because there is no central site for production and

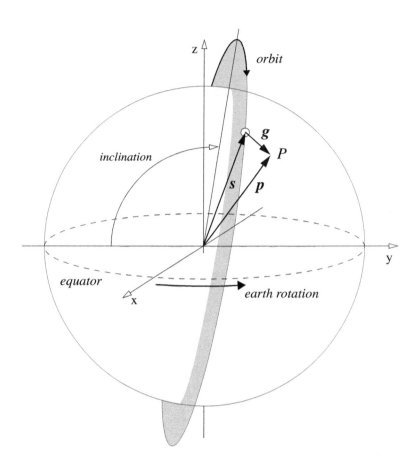

*FIGURE 3-29. The key parameters for modeling the imaging geometry for an earth-orbiting satellite in a near-polar descending orbit on the sunlit side, such as used by Landsat and Terra. The inclination is the angle between the equatorial plane and the orbital plane and is about 98° for sun-synchronous remote sensing satellites. A fixed (non-rotating) geocentric coordinate system is defined by (x,y,z). The three vectors in that system define the satellite location (**s**), the view direction to a particular point P on the earth's surface (**g**), and the location of P (**p**). Although we are ignoring topography here, it can be included in the vectors **g** and **p**. The analysis geometry is the same for an ascending orbit, such as used by Aqua, except the inclination angle has opposite sign to that shown, and the direction of the satellite along the path is reversed on the daylight side of the earth (see Plate 3-1).*

dissemination of AVHRR imagery. It can be received with a relatively low-cost station, and, therefore, users of the data have tended to create their own geometric correction software, which is evidenced by the large number of AVHRR modeling references.

TABLE 3-6. *Publications on scanner and satellite orbit modeling. The MSS geometric scanner model is essentially unchanged for all five Landsats, and the MSS orbital model for Landsat-4 and -5 is the same as that for TM. A comprehensive review of sensor geometric modeling for mapping is presented in Toutin (2004).*

sensor	reference
AVHRR	Emery and Ikeda, 1984; Brush, 1985; Ho and Asem, 1986; Brush, 1988; Emery *et al.*, 1989; Bachmann and Bendix, 1992; Moreno *et al.*, 1992; Moreno and Meliá, 1993; Krasnopolsky and Breaker, 1994
IRS-1C PAN	Radhadevi *et al.*, 1998
MISR	Jovanovic *et al.*, 2002
MODIS	Nishihama *et al.*, 1997; Wolfe *et al.*, 2002
MSS (Landsat-1, -2, -3)	Anuta, 1973; Puccinelli, 1976; Forrest, 1981; Sawada *et al.*, 1981; Friedmann *et al.*, 1983; Salamonowicz, 1986
SPOT	Kratky, 1989; Westin, 1990
TM (Landsat-4, -5)	Seto, 1991

There has been considerable interest in models for "navigation" within AVHRR imagery because of its large *FOV* and relatively low spatial resolution. A relatively simple sensor and geoid (earth shape) model incorporating ephemeris data has yielded positional accuracies of 1 to 1.5 pixels (Emery *et al.*, 1989), and with a more complex model involving TBUS ephemeris messages receivable at AVHRR ground stations, subpixel accuracies are claimed (Moreno and Meliá, 1993). A summary of AVHRR rectification work is given in Table 3-8. In all cases, the additional use of *Ground Control Points (GCPs),* whose locations are precisely known, has been recommended to obtain the highest possible accuracy. In the case of TM, rectification accuracies of less than 0.5 pixel have been reported using a bi-directional scan model (Seto, 1991). For both TM and AVHRR, these accuracies apply only in areas of low topographic relief, i.e., they assume a spherical or ellipsoidal earth surface.

The Global Positioning System (GPS) has markedly improved the measurement of GCPs in accessible areas, reducing dependency on sometimes uncertain topographic maps for orthorectification applications (Ganas *et al.*, 2002). In addition, there is increasing use of higher resolution image "chip" databases of GCP features (Parada, 2000; Wolfe *et al.*, 2002) to automate and improve the subpixel geolocation of lower resolution imagery. Spatial correlation is applied between the imagery of interest and the corresponding images in the chip database to refine geolocation (see Chapter 8).

TABLE 3-7. Some of the published work on satellite production image geometric quality. Unless otherwise noted, data are for system-corrected products, after least-squares polynomial adjustment using Ground Control Points (GCPs) obtained from maps or GPS data .

sensor	typical error	reference
ASTER	±15 m (L1B band 2 at nadir) ±30 m (L1B band 2 at 10° off-nadir)	Iwasaki and Fujisada, 2005
ETM+ (Landsat-7)	±54 m (L1G uncontrolled)	Lee *et al.*, 2004
IKONOS	±5–7 m (pan band, Standard Original product)	Helder *et al.*, 2003
IRS-1D PAN	±3 m (differentially-corrected GPS GCPs)	Turker and Gacemer, 2004
MODIS	< 50 m (all bands,TM GCP database)	Wolfe *et al.*, 2002
MSS (Landsat-1–3)	±160 m – ±320 m (bulk film product) ±40 m – ±80 m (after polynomial adjustment)	Wong, 1975
MSS (Landsat-4,-5)	±120 m	Welch and Usery, 1984
SPOT	±3 m (P band, earth-orbit model)	Westin, 1990
TM (Landsat-4,-5)	±45 m (L-4), ± 12 m (L-5) ±33 m (L-4) ±30 m (L-4) ±30 m (L-5) ± 21 m ±18 m (L-4), ±9 m (L-5) ±45 m (earth-orbit model)	Borgeson *et al.*, 1985 Walker *et al.*, 1984 Welch and Usery, 1984 Bryant *et al.*, 1985 Fusco *et al.*, 1985 Welch *et al.*, 1985 Wrigley *et al.*, 1985

TABLE 3-8. AVHRR rectification experiments using sensor and geoid models.

model and data	positional accuracy (pixels)	reference
circular orbit and earth	2–3	Legeckis and Pritchard, 1976
elliptical orbit and earth + ephemeris	1–2	Emery and Ikeda, 1984
elliptical orbit and earth + TBUS	<1	Moreno and Meliá, 1993

3.9.5 Line and Whiskbroom Scan Geometry

The cross-track pixel sampling is in fixed time increments for line or whiskbroom scanners, which, given a constant scan velocity, results in fixed angular increments, $\Delta\theta$, where θ is the scan angle from nadir. For line and whiskbroom scanners, the cross-track *GSI* therefore varies across the scan, increasing with increasing scan angle according to Richards and Jia (1999),

$$\textit{flat earth:}\quad GSI_f(\theta)/GSI(0) = [1/\cos(\theta)]^2 ,\qquad (3\text{-}39)$$

again assuming a flat earth. This approximation is accurate for a rather large scan angle, even at the altitude of the AVHRR (Fig. 3-30). However, at larger angles, the curvature of the earth must be accounted for, and the equation for the cross-track *GSI* becomes (Richards and Jia, 1999),

$$\textit{spherical earth:}\quad GSI_e(\theta)/GSI(0) = \frac{[H + r_e(1 - \cos\phi)]}{H\cos(\theta)\cos(\theta + \phi)} \qquad (3\text{-}40)$$

where ϕ is the geocentric angle corresponding to the surface point at the scan angle θ and is given in Brush (1985) as,

$$\phi = \text{asin}\{[(r_e + H)/r_e]\sin(\theta)\} - \theta . \qquad (3\text{-}41)$$

The cross-track *GIFOV* varies similarly, yielding a ground-projected "bow-tie" scan pattern (Fig. 3-31). Since the surface distance from nadir is proportional to ϕ and the pixel index from the center of a scan is proportional to θ, Eq. (3-41) also describes the ground distance from nadir at each scanline pixel.

3.9.6 Pushbroom Scan Geometry

The cross-track *GSI* for pushbroom scanners does not vary in the same way as for whiskbroom scanners, assuming the imaging system has constant magnification across the linear detector array (Fig. 3-32). In a pushbroom system (linear or area arrays; see Chapter 1), each cross-track line of the image is formed optically as in a conventional frame camera. The detector elements are equally spaced at a distance w (equal to the detector element width) across the array, and therefore the cross-track *IFOV* changes across the array, i.e., as a function of the cross-track view angle.[14] If the earth were flat, however, and the optical system exhibits no distortion, the equal spacing of w in the focal plane corresponds to a uniform cross-track *GSI* on the ground,

$$\textit{flat earth:}\quad GSI_f = w \times \frac{H}{f} ,\qquad (3\text{-}42)$$

but a changing cross-track *IFOV*,

---

14. Notice the use of "view" angle for pushbroom scanners in contrast to "scan" angle for line and whiskbroom scanners.

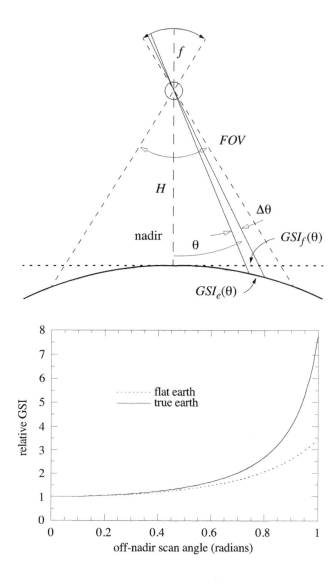

FIGURE 3-30.   Line and whiskbroom scanner geometry in the cross-track direction used to derive
    Eq. (3-39) and Eq. (3-40). The data along the scan are sampled at a fixed time interval to create pixels.
    Assuming the scan rotational velocity is constant, the fixed time interval corresponds to a fixed angular
    interval, Δθ. The cross-track GSI therefore increases with increasing θ, as shown below for an altitude of
    850 km. The flat earth approximation is good within 4% out to a scan angle of about 0.4 radians, or 23°.

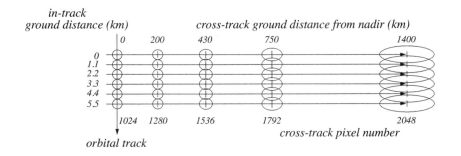

FIGURE 3-31. *The "bow-tie" effect in AVHRR data, characteristic of line and whiskbroom scanners. It is similar to distortion arising in wide-field panoramic cameras. The horizontal and vertical scales above are linear in ground distance. The ellipses represent the area of integration of the system spatial response function and the crosses represent pixel samples. The GIFOV and GSI are greatly exaggerated in size relative to the scan length, and only five cross-track samples are shown to emphasize the effect in this drawing. (Adapted from Moreno et al., 1992.)*

$$IFOV(\theta)/IFOV(0) = [\cos(\theta)]^2 . \tag{3-43}$$

With Eq. (3-43) and Eq. (3-40), it can be shown that for the true spherical earth, the cross-track *GSI* varies with view angle according to,

$$spherical\ earth: \quad GSI_e(\theta)/GSI(0) = \frac{[H + r_e(1 - \cos\phi)]\cos(\theta)}{H\cos(\theta + \phi)} \tag{3-44}$$

where ϕ is given by Eq. (3-41). This function is plotted in Fig. 3-32 with the flat earth approximation for the 832 km altitude of the SPOT satellite. Now, since the off-nadir *FOV* of the SPOT HRV sensor is only $\pm\text{atan}(30/832)$, or ±0.036 radians, the flat earth approximation is valid within 0.03%. The HRV has the capability, however, to be pointed off-nadir by up to $\pm26°$; the geometry in that configuration is considerably more complicated than the nadir-pointing condition assumed here.

3.9.7 Topographic Distortion

While the previous discussion accounts for the most significant geometric factors influencing satellite imaging, topography must be included for precise calculations. The height of any point on the ground, from a reference plane, or *datum*, affects its image location. A simple example is shown in Fig. 3-33. Intuitively, one can see that the displacement of an image point from its orthographic location depends on the off-nadir view angle and the height of the object from the reference plane; the greater the off-nadir angle and object height, the greater the image displacement.

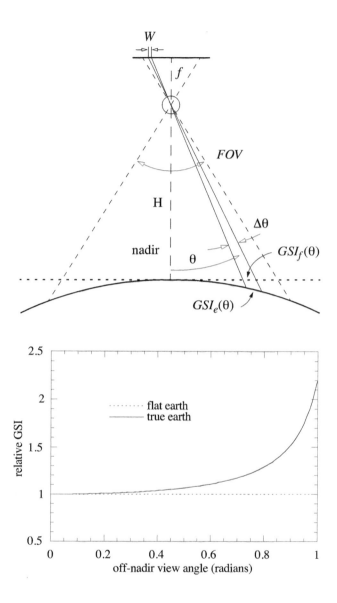

FIGURE 3-32. *Cross-track pushbroom scanner geometry for Eq. (3-42) and Eq. (3-44). The cross-track GSI is constant under the flat earth approximation, but increases with increasing view angle θ for the true spherical earth. The altitude used is 832km, corresponding to the SPOT satellites.*

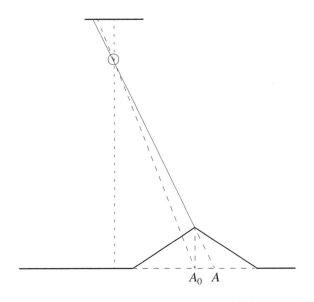

A_0 A

FIGURE 3-33. *Geometry for a pushbroom scanner imaging a topographic feature off-nadir in the cross-track direction. The in-track vector is pointing out of the page. The image of the peak appears to come from A because of optical parallax. If the image were orthographic, the image of the peak would appear to come from A_0.*

Given only one image, it is impossible to infer pixel-to-pixel elevation differences. However, with two images, it is possible to determine elevation differences between two points from their *parallax* (Fig. 3-34). It is easy to show that the image parallaxes of ground points A and C are,

$$p_a = a_1 - a_2 = \frac{fB}{H - Z_A} \tag{3-45}$$

and

$$p_c = c_1 - c_2 = \frac{fB}{H - Z_C}, \tag{3-46}$$

where H and Z are measured relative to the datum plane, and the image coordinates, $a_1, a_2, c_1,$ and c_2, are all measured relative to the optical center (*principal point*) of the respective image (Wolf, 1983). Combining Eq. (3-45) and Eq. (3-46), we have,

$$\Delta Z = Z_A - Z_C = \frac{\Delta p(H - Z_C)}{p_a}, \tag{3-47}$$

where Δp is the parallax difference, $p_a - p_c$. This equation can be simplified for high sensor altitude and moderate relief ($H \gg Z_C$) to,

$$\Delta Z \cong \Delta p \frac{H^2}{fB} \;=\; \Delta p \times \frac{H}{f} \times \frac{H}{B} \;=\; \frac{\Delta p/m}{B/H}\,, \qquad\qquad (3\text{-}48)$$

where B is the *base*, the ground distance between the centers of the two images (Ehlers and Welch, 1987). The quantity B/H is called the *base-to-height ratio*. For a given sensor focal length and altitude, the base determines the sensitivity of elevation measurements made from image parallax. The term $\Delta p/m$ is the image parallax difference scaled by the sensor magnification to the ground. From an image analysis viewpoint, the primary significance of Eq. (3-48) is that *the elevation difference between two ground points is proportional to the parallax difference between their image points*. This simple relationship is used to derive elevation maps from stereo imagery (Chapter 8).

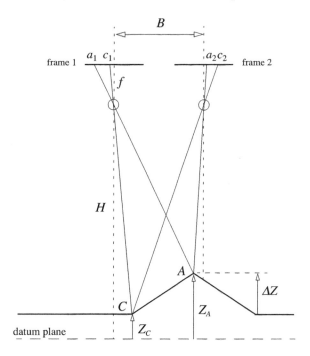

FIGURE 3-34. *Geometry of stereo imaging. The distance between the image points a_1 and c_1 in frame 1 is not equal to the distance between the image points a_2 and c_2 in frame 2 because of the elevation difference between ground points A and C and the different view points of the two frames. The distance between the two view points (called "camera stations" in aerial photography) is called the "base" B of the stereopair.*

We assumed a rather simple situation here, namely a pushbroom scanner parallel to the datum. In reality, they may not be parallel, causing a *tilted* imaging plane. Such a situation occurs in off-nadir viewing, for example. The resulting geometry is considerably more complex in detail but follows the basic principles outlined here.

3.10   Summary

In this chapter, we have seen how the sensor modifies the signal of interest in remote sensing. The sensor affects the spatial and radiometric quality of the signal. The important aspects of this sensor influence are:

- Scene spatial features are weighted by the sensor spatial response, which includes blurring by the optics, image motion, detector, and sensor electronics.
- Scene spectral radiances received at the sensor are weighted by the sensor spectral response in each band.
- The overall imaging process, under reasonable assumptions, is a linear transform of the at-sensor radiance.
- Geometric distortion arises from internal sensor and external platform and topographic factors.

In the next chapter, we discuss data models, which provide the link between the physical remote sensing models of Chapter 2, the sensor models of this chapter, and image processing algorithms.

3.11   Exercises

Ex 3-1. Verify the pixel *DN* values in Fig. 3-5.

Ex 3-2. Calculate the convolution of a square sensor *GIFOV* and a square wave radiance target, either by integration using Eq. (3-2) or by a graphical approach. Allow the *GIFOV* to be 1/2, 1, and 2 times the width of a single bar in the pattern and discuss the results in terms of sample-scene phase.

Ex 3-3. One reasonable design goal for an imaging system is to match the Airy Disk diameter to the detector size. That insures that a significant fraction of energy in the *PSF* is collected by the detector. If operational constraints specify that the detector is 12 μm square and the optical focal length is 10 m, what is the necessary optical aperture diameter at a blue wavelength? At a red wavelength? At an NIR wavelength? If one optical system is to meet the design goal at all three wavelengths, what aperture diameter should be used?

Ex 3-4. With the angle of 13.08° between a line target and sensor scan direction, as in Fig. 3-18, how is the effective subpixel sample interval calculated for the extracted profiles?

Ex 3-5. What is the east-west displacement, in pixels, due to earth rotation from the top to the bottom of a Landsat TM scene, centered on Tucson, Arizona? What is it for a scene centered on Anchorage, Alaska?

Ex 3-6. Derive Eq. (3-44), starting with Eq. (3-40).

Ex 3-7. Plot Eq. (3-40) and Eq. (3-44) as a function of H (over a range of 10 to 1000 km) for a fixed scan/view angle of 20°. Is the flat earth approximation better or worse for aerial scanners compared to satellite scanners?

Ex 3-8. Derive Eq. (3-47) and Eq. (3-48).

CHAPTER 4

Data Models

4.1 Introduction

The remote-sensing analyst sometimes views an image only as "data," disconnected from the underlying physical processes that created the "data." This is often a mistake, because it can lead to less than optimal processing algorithms. The various radiation and sensor models discussed in Chapters 2 and 3 are often implicit in *data models*. Since specific data models are often assumed in data processing, they can provide a link between the physics of remote sensing and the design of image processing algorithms (Fig. 4-1). In this chapter, we discuss the common data models used in remote sensing and relate them to the underlying physical process models.

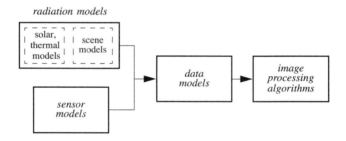

FIGURE 4-1. *Data models form the link between physical radiation and sensor models and image processing algorithms.*

4.2   A Word on Notation

The assortment of notations used to denote pixels and their locations in 2-D images is confusing, at best. Engineers use different notation than do earth scientists, who in turn use different notation than do statisticians. The strongly multidisciplinary nature of remote sensing brings these different disciplines together, making the issues of terminology and notation particularly evident.

Part of the problem is the inappropriate use of continuous notation for discrete functions. The real world of the scene is continuous in space, wavelength, and time (Chapter 2), and all functions expressed in that context should use continuous notation. After the sensor converts the signal to sampled, quantized *DN* values in the A/D converter (Chapter 3), mathematical descriptions should be discrete. All digital image processing equations, for example, should use discrete notation.

How do we designate a particular pixel in a digital image? We could think of the image as an array of numbers, with indexes i and j. The array values are the pixel *DN*s. A pixel value at row i and column j would be denoted DN_{ij} (or simply DN_p as we have done in previous chapters). Rows and columns (or lines and samples) are conventionally numbered from (1,1) at the upper left to (N,M) at the lower right of the image array, as viewed (Fig. 4-2).[1] This notation is natural for computer programs because of sequential data storage formats. One must be careful, however, in associating the row index with y and the column index with x, because that results in a left-handed coordinate system. If rows are associated with x and columns with y, as if the array were rotated -90°, the only difference between them (other than discrete versus continuous) is an offset in the origin from (0,0) to (1,1). This notation is used, for example, in Gonzalez and Woods (1992); it leads to a right-handed (x,y) system, but y is horizontal and x is vertical, contrary to a "natural" Cartesian system. In this book, we will use both the (row, column) and the "natural" right-handed (x, y) coordinate systems, as appropriate in a particular context.

If the spatial order of the pixels is unimportant in a particular calculation, it is convenient to replace the double subscript by a single subscript p and any accompanying two-dimensional spatial operators by a single operator, indexed over p. We will use this briefer notation occasionally.

4.3   Univariate Image Statistics

We start with definitions for basic *univariate* image statistics, which apply to single band images, and then extend the discussion to *multivariate* statistics to include multispectral and multisensor images.

---

1. The reader should be aware that a coordinate system beginning at (0,0) and extending to $(N$-1$, M$-1$)$ is also used by some computer programs. If one is switching between two programs that use different conventions, mistakes can arise!

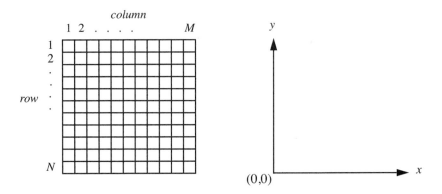

FIGURE 4-2. A discrete matrix (row, column) notation compared to a continuous, Cartesian (x,y) notation.

4.3.1 Histogram

The image histogram describes the statistical distribution of image pixels in terms of the number of pixels at each *DN*. It is calculated simply by counting the number of pixels in each *DN* "bin," and dividing by the total number of pixels in the image, *N*,

$$hist_{DN} = \text{count}(DN)/N. \qquad (4\text{-}1)$$

The histogram is often associated with the *Probability Density Function (PDF)* of statistics,

$$hist_{DN} \approx PDF(DN), \qquad (4\text{-}2)$$

but this association is mathematically problematic because (1) the PDF is defined for continuous variables, and (2) it is only properly used for statistical distributions from a random process. Aside from the issue of discrete versus continuous variables, images are seldom treated as instances of a random process, but rather as individual data arrays.[2] The histogram is, therefore, a more appropriate description for digital images.

The histograms of large images of land areas are typically unimodal (i.e., they have a single "peak"), with an extended tail toward higher *DN*s, i.e., higher scene radiances (Fig. 4-3). It is important to remember that an image histogram only specifies the number of pixels at each *DN*; it contains no information about the *spatial distribution* of those pixels. Sometimes, however, spatial information can be inferred from the histogram. For example, a strongly bimodal histogram usually indicates two dominant materials in the scene, such as land and water. What cannot be inferred is the extent to which pixels in each category are spatially connected. For example, NIR images of a scene with many small lakes and a scene of an ocean coastal area could have similar, bimodal histograms.

---

2. A notable exception may be the treatment of certain noise processes.

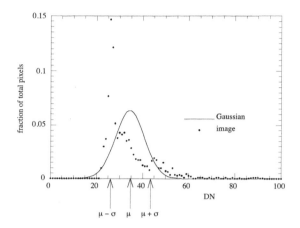

FIGURE 4-3. An example image histogram compared to a Gaussian distribution with the same mean and variance. Note the asymmetry in the histogram, which is not atypical, and that it appears to be multimodal.

The image histogram is a useful tool for *contrast enhancement*. For example, a common contrast enhancement technique "stretches" the range of *DN*s and "clips" or thresholds it at one or both ends, resulting in a certain percentage of saturated pixels. The appropriate *DN* thresholds can be obtained from the histogram as percentages of the total number of pixels in the image.

Normal distribution

In remote sensing, as in most science and engineering fields, it is often mathematically convenient to assume a *normal* (Gaussian) distribution for independent, identically-distributed samples from a random process. In one dimension, this continuous distribution has the form,

$$N(f;\mu, \sigma) = \frac{1}{\sigma\sqrt{2\pi}} e^{-\left[\frac{(f-\mu)^2}{2\sigma^2}\right]} \tag{4-3}$$

which has the familiar shape shown in Fig. 4-3. The centroid of the distribution is given by μ and its width is proportional to σ. While the normal distribution is not appropriate as a global model for most images, it is widely used to model the distribution of subsets of pixels having similar characteristics within an image for classification purposes (Chapter 9).

4.3.2 Cumulative Histogram

Some image processing algorithms, notably histogram equalization, histogram matching, and destriping (Richards and Jia, 1999), require a function, the *cumulative histogram (chist)*, derived from the histogram as follows,

$$chist_{DN} = \sum_{DN = DN_{min}}^{DN} hist_{DN} \ . \tag{4-4}$$

The cumulative histogram is the fraction of pixels in the image with a *DN* less than or equal to the specified *DN*. It is a monotonic function of *DN*, since it can only increase as each histogram value is accumulated. Because the histogram as defined in Eq. (4-1) has unit area, the asymptotic maximum for the cumulative histogram is one (Fig. 4-4). In this normalized form, the cumulative histogram is also called the *Cumulative Distribution Function (CDF)* (Castleman, 1996), although that association has the same theoretical problems as Eq. (4-2).

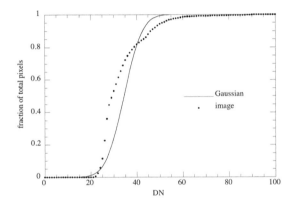

FIGURE 4-4. *Comparison of an example image cumulative histogram to a Gaussian cumulative histogram using the same data as in Fig. 4-3.*

4.3.3 Statistical Parameters

The *DN mean* can be calculated in two ways,

$$\mu = \frac{1}{N} \sum_{p = 1}^{N} DN_p = \sum_{DN = DN_{min}}^{DN_{max}} DN \times hist_{DN} \ . \tag{4-5}$$

The first approach adds the *DN*s of all the pixels in the image and divides by the total number of pixels to yield the average *DN*. The second approach weights each *DN* by the corresponding histogram value (the fraction of the image that has that *DN*) and sums the weighted *DN*s. If the histogram is already available, the latter form, known as the *first moment* of the histogram, is a more efficient calculation. The *DN* variance is given by,

$$\sigma^2 = \frac{1}{N-1} \sum_{p=1}^{N} (DN_p - \mu)^2 = \frac{N}{N-1} \sum_{DN=DN_{min}}^{DN_{max}} (DN - \mu)^2 \times hist_{DN} \quad , \tag{4-6}$$

where either calculation may be used, as in Eq. (4-5). The second form is the *second moment* of the histogram. In the sum over pixels, $N-1$ is used rather than N when the mean must be estimated from the data, rather than being known *a priori* from a distribution (Press *et al.*, 1992). The *DN standard deviation*, σ, is the square root of the variance.

The mean and variance are sufficient to specify a normal, or Gaussian, distribution (Eq. (4-3)). If the histogram is unimodal and symmetric, a Gaussian distribution may not be a bad model for the actual data. However, as noted earlier, global image histograms tend to be asymmetric and sometimes multimodal. Whether the distribution is normal or not, the *DN* mean and variance are useful. As discussed below, the image standard deviation can be used as a measure of image contrast, since it is a measure of the histogram width, i.e., the spread in *DN*s.

Other statistical parameters are sometimes useful, including the *mode* (*DN* at which the histogram is maximum), *median* (*DN* which divides the histogram area in half, with 50% of the pixels below the median and 50% above), and the higher order measures of *skewness* (asymmetry),

$$skewness = \frac{1}{N} \sum_{p=1}^{N} \left(\frac{DN_p - \mu}{\sigma}\right)^3 = \sum_{DN=DN_{min}}^{DN_{max}} \left(\frac{DN - \mu}{\sigma}\right)^3 \times hist_{DN} \tag{4-7}$$

and *kurtosis* (sharpness of the peak relative to a normal distribution) (Press *et al.*, 1986; Pratt, 1991),

$$kurtosis = \left[\frac{1}{N} \sum_{p=1}^{N} \left(\frac{DN_p - \mu}{\sigma}\right)^4\right] - 3 = \left[\sum_{DN=DN_{min}}^{DN_{max}} \left(\frac{DN - \mu}{\sigma}\right)^4 \times hist_{DN}\right] - 3 . \tag{4-8}$$

Skewness is zero for any symmetric histogram. A histogram with a long tail toward larger *DN*s has a positive skewness, and this is typical of remote-sensing images. Kurtosis is zero for a normal distribution. If a histogram has a positive kurtosis, then the peak is sharper than that of a Gaussian; a negative kurtosis means the peak is less sharp than that of a Gaussian. Note that both skewness and kurtosis are normalized by σ and are unitless, unlike the mean and standard deviation. For the image data represented in Fig. 4-3, we have $\mu = 34.1$, $\sigma = 9.12$, skewness = 1.78, and kurtosis = 4.32, indicating a degree of asymmetry and sharpness in the peak of the histogram. Skewness and kurtosis are quite sensitive to *outliers*, pixels with *DN*s far removed from the majority distribution, because of their high order.

4.4   Multivariate Image Statistics

The extension of the image statistical measures of the previous section to K dimensions is straight-forward. The data measurement variable, DN, becomes a measurement vector, DN, having K components (Fig. 4-5),

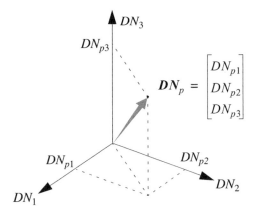

FIGURE 4-5.   *Visualization of a three-band multispectral image pixel* DN_p *as a vector in three-dimensional space.*

$$DN_{ij} = \begin{bmatrix} DN_{ij1} & DN_{ij2} & \cdots & DN_{ijK} \end{bmatrix}^T = \begin{bmatrix} DN_{ij1} \\ DN_{ij2} \\ \vdots \\ DN_{ijK} \end{bmatrix}. \qquad (4\text{-}9)$$

As we suggested before, a simpler notation can be used, where the subscript p denotes a particular pixel,

$$DN_p = \begin{bmatrix} DN_{p1} & DN_{p2} & \cdots & DN_{pK} \end{bmatrix}^T = \begin{bmatrix} DN_{p1} \\ DN_{p2} \\ \vdots \\ DN_{pK} \end{bmatrix}. \qquad (4\text{-}10)$$

One way to visualize two- or three-dimensional data is the *scatterplot*. An example for bands 2, 3, and 4 of a Landsat TM image is shown in Fig. 4-6. This is a binary plot which shows a dot if a particular multispectral vector has a histogram count of at least one. Therefore, the number of pixels with a particular multispectral vector is not shown. Since the histogram is three-dimensional, different viewing angles reveal different features in the data; some software programs allow interactive rotation to assist interpretation. The 3-D plot can be reduced to three 2-D plots by projecting every point onto one of the bounding planes (Fig. 4-7). The projection removes some of the spectral information, because all points along a given projection line are represented by only a single point in 2-D. Further information is lost if the 2-D scatterplot is projected to a 1-D histogram, as shown.

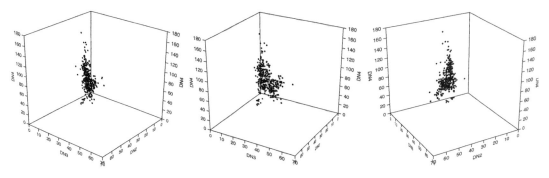

FIGURE 4-6.   *Three-band scatterplots of bands 2, 3, and 4 of a TM image, viewed from three different directions. Only every 20th sample and line of the image are used to calculate these scatterplots, so that they are not too dense. Every dot in the scatterplot represents one or more pixels with a particular spectral vector. Note that the image data occupy a small fraction of the total DN volume.*

Even with the large amount of data reduction achieved with scatterplots, visualization of multi-spectral data can be challenging. There are 21 possible unique scatterplots of band pairs for a seven-band Landsat TM image (Fig. 4-8). Some additional information about the *density* of the distribution along a projected line can be preserved by representing the integrated value as a grey level and displaying the scatterplot as a greyscale image (we call this a *scattergram*). Some examples from various sensor band combinations are shown in Fig. 4-9. It is possible to see the distribution of pixel counts over the multispectral space in these representations.

In K dimensions, we write the histogram as,

$$hist_{DN} = \text{count}(\boldsymbol{DN})/N. \tag{4-11}$$

Note that this is a *scalar* function of a K-D *vector*. In one dimension, the normal distribution is given by Eq. (4-3) and requires only two parameters, the data mean and variance, to specify the function completely. Similarly, the only parameters of a K-dimensional normal distribution are the *mean vector*, the expected value of the $\boldsymbol{DN}$,[3]

---

3. The expected value of a random variable q is sometimes denoted $E\{q\}$. For simplicity, we are assuming the sample mean and covariance of a finite number of samples from a random process equals the true mean and covariance of an infinite number of samples. In practice, only the limited sample statistics are available. See Press *et al.* (1992) and Fukunaga (1990) for further discussion.

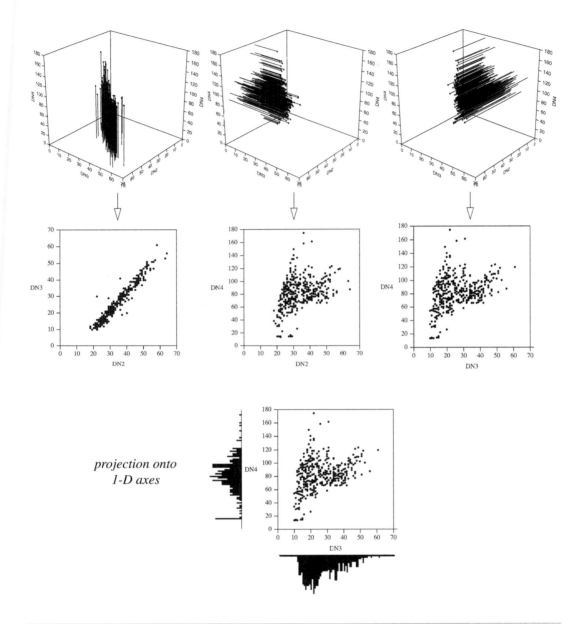

FIGURE 4-7. *Reduction of 3-D scatterplots to 2-D scatterplots by projections onto the three bounding planes. The 2-D scatterplots provide multiple views of the data, but do not contain all the information that exists in 3-D. If the 2-D scatterplot is projected onto either axis the marginal distribution of the 2-D data, i.e. the histogram of each band, results.*

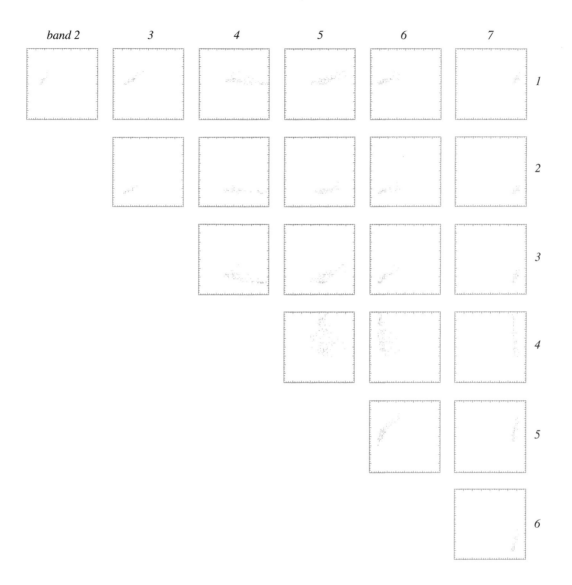

FIGURE 4-8. A full set of band-pair scatterplots for a seven-band TM image. The row of numbers across
the top indicates the TM band whose DN is plotted along the abscissa of each scatterplot, and the column
of numbers on the right indicates the TM band whose DN is plotted along the ordinate.

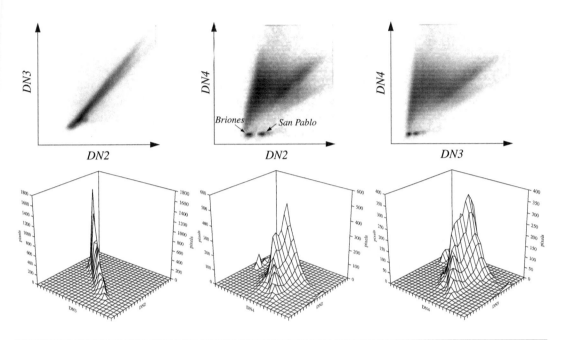

FIGURE 4-9. *Two-dimensional scattergrams with density coded as grey levels (inverted for better visibility) and displayed as surfaces. The data are from the TM image of Fig. 2-15; note the two "clusters" of pixels from the Briones and San Pablo Reservoirs. The units of the surface plots are number of pixels per* **DN**.

$$\mu = \begin{bmatrix} \mu_1 & \dots & \mu_K \end{bmatrix}^T = \langle DN \rangle \qquad (4\text{-}12)$$

where the mean in band k is,

$$\mu_k = \langle DN_k \rangle = \frac{1}{N} \sum_{p=1}^{N} DN_{pk} \qquad (4\text{-}13)$$

and the *covariance matrix*,

$$C = \begin{bmatrix} c_{11} & \dots & c_{1K} \\ \vdots & & \vdots \\ c_{K1} & \dots & c_{KK} \end{bmatrix} = \langle (DN - \mu)(DN - \mu)^T \rangle \qquad (4\text{-}14)$$

where the *covariance* between bands m and n is,

$$c_{mn} = \langle (DN_m - \mu_m)(DN_n - \mu_n) \rangle = \frac{1}{N-1} \sum_{p=1}^{N} (DN_{pm} - \mu_m)(DN_{pn} - \mu_n) . \qquad (4\text{-}15)$$

The covariance matrix is *symmetric*, i.e., c_{mn} equals c_{nm}. Also, because the diagonal elements, c_{kk}, are the variances of the distribution along each dimension (refer to Eq. (4-6)), they are always positive; however, *the off-diagonal elements may be negative or positive*. In other words, two bands of a multispectral image may have a negative covariance. That means that pixels with relatively low *DN*s in one band have relatively high *DN*s in the other, and vice versa.

The significance of the off-diagonal terms of the covariance matrix may be better appreciated by defining the *correlation matrix*,

$$\mathbf{R} = \begin{bmatrix} 1 & \cdots & \rho_{1K} \\ \vdots & & \vdots \\ \rho_{K1} & \cdots & 1 \end{bmatrix}, \quad -1 \le \rho_{mn} \le 1 \text{ or } |\rho_{mn}| \le 1 \qquad (4\text{-}16)$$

where the *correlation coefficient* between two bands m and n is defined as,

$$\rho_{mn} = c_{mn} / (c_{mm} \cdot c_{nn})^{1/2} , \qquad (4\text{-}17)$$

i.e., the covariance between two bands, divided by the product of their standard deviations. The value of ρ_{mn} must be between minus one and plus one; the diagonal terms, for which m equals n, are each normalized to one. Examples of the shape of a two-dimensional normal distribution for different correlation coefficients are shown in Fig. 4-10. Note that values of ρ_{mn} close to plus or minus one imply a strong linear dependence between the data in the two dimensions, whereas if ρ_{mn} is near zero there is little dependence between the two dimensions. We will see later (Chapter 5) how the off-diagonal elements in the covariance and correlation matrices may be changed to zero by appropriate transformation of the K-dimensional image. The K features of the transformed image are then *uncorrelated*, a useful property for data analysis.

A K-dimensional, discrete, normal (Gaussian) distribution may be defined similarly to a continuous, normal distribution as,

$$N(\mathbf{DN}; \mathbf{\mu}, \mathbf{C}) = \frac{1}{|\mathbf{C}|^{1/2}(2\pi)^{K/2}} e^{-(\mathbf{DN} - \mathbf{\mu})^T \mathbf{C}^{-1}(\mathbf{DN} - \mathbf{\mu})/2} \qquad (4\text{-}18)$$

which is a scalar function of the **DN** vector. The normal distribution is displayed in 2-D as a grey-scale image and with overlaying equal probability contours in Fig. 4-11. Note that the contours are elliptical (see Ex. 4-2). One of the normal distribution's important characteristics is that the probability never goes identically to zero, no matter how far away the **DN** vector is from the mean. A consequence of this is that a **DN** vector always has a nonzero probability under the normal distribution assumption. This fact impacts certain classification algorithms (Chapter 9).

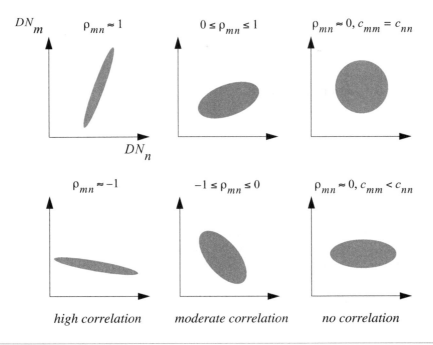

FIGURE 4-10. *The correlation coefficient indicates the shape of the scatterplot (or scattergram) of a multispectral image.*

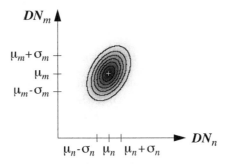

FIGURE 4-11. *The normal probability density function in 2-D, displayed as an inverted greyscale image with overlaying contours of equal probability.*

Examples of spectral scattergrams for different band pairs from a TM image are shown in Fig. 4-12. Real data obviously can deviate considerably from the idealized diagram of Fig. 4-10. Scatterplots can also be a useful visualization of the data distribution in color images (Plate 4-1).

4.4.1 Reduction to Univariate Statistics

In general, the full vector-matrix form of multivariate statistics is necessary. The mathematics reduces to a 1-D description *only if*,

$$PDF(\mathbf{DN}) = PDF(DN_1) \cdot PDF(DN_2) \cdot \ldots \cdot PDF(DN_K), \tag{4-19}$$

i.e., the *probability density functions are independent of each other* in all dimensions. This is an example of a *separable* function, used in signal processing (Chapter 3 and Appendix B). The 1-D distributions of each image band are called the *marginal distributions*. An important consequence of independence between bands is that the covariance matrix is then *diagonal*, with zeros in all off-diagonal elements. The correlation coefficient is also zero in this case. In practice, Eq. (4-19) can be approximated with Eq. (4-1), Eq. (4-2) and Eq. (4-11).

4.5   Noise Models

Noise is introduced into the data by the sensor. It is a variation in the sensor output that interferes with our ability to extract scene information from an image. Image noise occurs in a wide variety of forms (Fig. 4-13 to Fig. 4-16) and is often difficult to model; for these reasons, many noise reduction techniques are *ad hoc*. It is beneficial to categorize noise types and generalize their descriptive models, which we pursue in the following.

The simplest model for noise is an *additive, signal-independent* component at each pixel, p,

$$\mathbf{DN}_p = int[\mathbf{a}_p + \mathbf{n}_p] \tag{4-20}$$

where all three quantities are vectors in multispectral space. The noise can originate at several places along the imaging chain described in Chapter 3 and should be modeled accordingly when a specific source is known. We are treating it as being added to the amplified detector signal, $\mathbf{a}_p$ (see Eq. (3-16)).

The function $\mathbf{n}_p$ in Eq. (4-20) is unspecified at this point and can be tailored to describe several common types of noise. In most cases, it is reasonable to assume the noise has zero mean over a large area of the image and is therefore manifested as fluctuations above and below the noiseless signal, $\mathbf{a_p}$.

A *signal-dependent*, but still additive, noise model is useful in some situations,

$$\mathbf{DN}_p = int[\mathbf{a}_p + \mathbf{n}_p(\mathbf{a}_p)]. \tag{4-21}$$

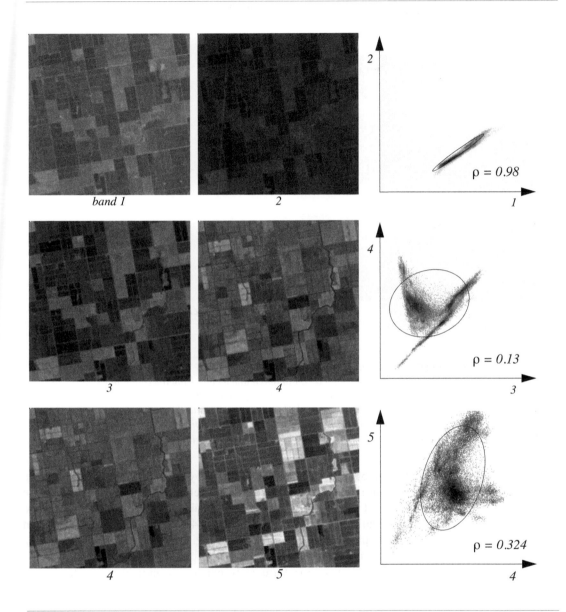

FIGURE 4-12. Band-to-band scattergrams for a TM image of an agricultural area (Plate 4-1). These are uncalibrated DN data. The visible bands 1 and 2 are highly correlated with dark vegetation and light soil. The red band 3 and NIR band 4 have low correlation because of the high reflectance of vegetation in the NIR and low reflectance in the red. The NIR band 4 and SWIR band 5 are moderately correlated. The ellipses are the two-sigma contours for a Gaussian model distribution.

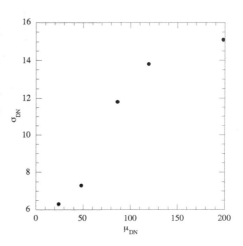

noise dependence on signal level

FIGURE 4-13.   *An example of photographic granularity. The image is an aerial photograph of Dulles Airport near Washington, D.C., digitized with a GSI of about one meter. Note the randomness of the noise, but that there also seems to be some spatial "clumping" of neighboring pixels, corresponding to film grains. The plot of DN standard deviation versus DN mean for five featureless areas in the image clearly shows the signal dependence of the noise. The exact form of this relationship is the combined result of granularity and film scanner characteristics, which are not known for this image.*

For example, the noise in photographic film (*granularity*), caused by randomly-located silver halide grains, is signal-dependent because the probability of a grain developing at any given point in the image depends on the exposure at that point. To illustrate, we take some sample areas, with no apparent scene detail, from the Dulles Airport photograph in Fig. 4-13 and calculate the *DN* mean and standard deviation, which are plotted against each other in Fig. 4-13. The noise, given by σ_{DN}, is clearly dependent on the signal, μ_{DN}, in this case.

Global, random noise like that in Fig. 4-13 is, fortunately, not prominant in electro-optical scanner systems, because the quantization interval (Fig. 3-26) is typically twice as large as the detector noise standard deviation. *Periodic* noise related to scanning is more likely. Whiskbroom scanners with multiple detectors per scan, like TM and MODIS, are subject to cross-track *striping*, caused by differences in calibration and response of individual detectors. For the same reason, pushbroom sensors can show in-track striping. TM also exhibits *banding* noise, in which *all* of the detectors

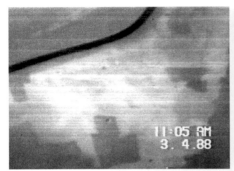

aerial video aperiodic scanline noise

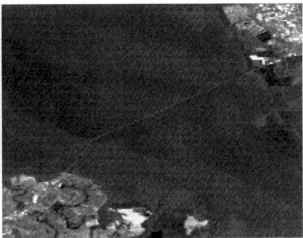

Landsat-4 MSS coherent noise

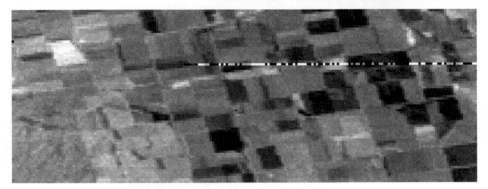

Landsat-1 MSS bad scanline noise

FIGURE 4-14. A gallery of scanner noise. The image contrast has been increased in all cases to emphasize the noise pattern.

can show a response change from scan to scan. Since the TM scans in both cross-track directions, the banding often appears in alternate scans and is usually associated with high contrast radiance transitions (Fig. 4-14).

Detector scan noise is modeled easily with either Eq. (4-20) or Eq. (4-21). As an example, we will model different (linear) gains and offsets for each of the 16 detectors in each band of a TM or ETM+ image. This scan noise is represented by modifying Eq. (3-17) for each band as follows,

$$DN_{ij}^{det} = int[gain^{det} \times e_{ij}^{det} + offset^{det}]$$ (4-22)

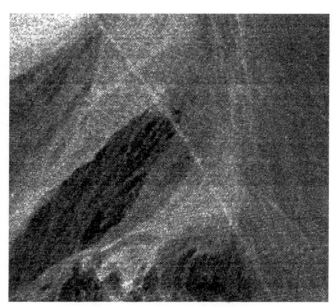

AVIRIS scanline and random pixel noise

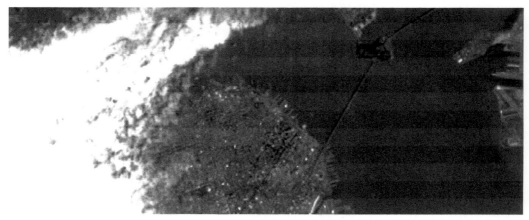

Landsat-4 TM banding noise

FIGURE 4-15. *More examples of scanner noise. TM banding noise is the result of detector saturation at one end of the scan, usually caused by bright clouds. In the case above, it affects mainly the eastward scans (left-to-right); the detectors slowly recover from the saturation and operate normally during the westward scans. An algorithm to reduce TM banding noise is described in Chapter 7.*

where the superscript, *det*, refers to the detector index $1, 2, 3, \ldots, 16$. The vector symbol for *DN* is dropped here, since Eq. (4-22) is band-specific, and the row-column index, *ij*, is made explicit, since we need to model the noise along the in-track direction only. The row (image line) index for a given detector is given by,

$$i = (scan - 1) \times 16 + det \qquad (4\text{-}23)$$

as a function of the scan index, *scan*, and detector index, *det*. Noise correction techniques based on this model are described in Chapter 7.

Image noise has traditionally been thought of as only *spatial* noise. Increasingly, however, it also has a *spectral* characteristic. One example is *spectral cross-talk*, which has been found in ASTER (Fig. 4-16), MODIS Terra (corrected by sensor modifications in MODIS Aqua) and Hyperion data. The complex focal planes of modern remote sensors (Chapter 1), with many spectral bands in close spatial proximity to each other, can lead to optical or electrical cross-talk between bands.

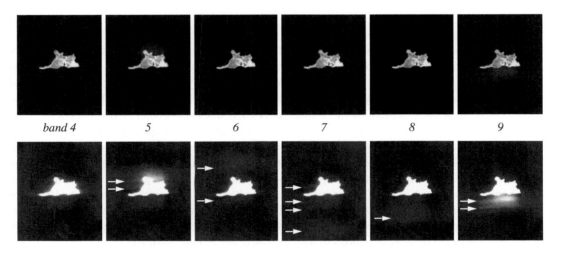

FIGURE 4-16. *Example of spectral crosstalk in ASTER data. The image is of Taraku Island northeast of Japan collected on April 28, 2001. The top row shows the SWIR bands 4 to 9. In the bottom row, each is contrast stretched to the same maximum value of 13. The multiple "ghost" images of the island (arrows) are where light from one or more bands has been recorded in the band of interest. There is a time difference between each band of 360ms, corresponding to 81 image lines (ASTER is a pushbroom sensor); thus the offset "ghost" effect. It has been deduced that light from band 4 is causing the "ghosts" in bands 5 and 9, and a correction algorithm has been developed (Iwasaki and Tonooka, 2005). Although the "ghosts" have a contrast of only a few DN out of 256, they could cause problems in derived geophysical products. (Imagery provided by Wit Wisniewski and the Remote Sensing Group, University of Arizona.)*

4.5.1 Statistical Measures of Image Quality

The quality of a digital or displayed image is influenced by many factors and is difficult to express in a single numerical *metric*. In this section, we present some of the common metrics for image quality.

Contrast

Numerical contrast may be defined in several ways, e.g.,

$$C_{ratio} = \frac{DN_{max}}{DN_{min}}, \tag{4-24}$$

$$C_{range} = DN_{max} - DN_{min} \tag{4-25}$$

or,

$$C_{std} = \sigma_{DN} \tag{4-26}$$

where DN_{max} and DN_{min} are the maximum and minimum DNs in the image, and σ_{DN} is the DN standard deviation. One definition or another may be appropriate in particular applications. For example, C_{ratio} and C_{range} may be inappropriate for some noisy images because one or two bad pixels could result in deceptively high contrast values; C_{std} would be much less affected by outliers. Note that, of these three definitions for contrast, only C_{ratio} does not depend on the units of the data.

The *visual* contrast of a displayed image is another indicator of the quality of the image. In this context, the quantitites in Eq. (4-24) to Eq. (4-26) should not be the DNs of the digital image, but rather the displayed GLs (Chapter 1) or preferably, direct measurements of the radiance from the display. However, the visual contrast depends not only on the digital image's DN range, but also on psychophysical factors such as the spatial structure within the image (see Cornsweet (1970) for examples), the ambient viewing light level, and the display monitor characteristics. Furthermore, visual and numerical contrast are both spatially-varying quantities, in that an image which has high global contrast may have local regions of low, intermediate, or high contrast.

Modulation

Another easily measured image property is modulation, M, defined as,

$$M = \frac{DN_{max} - DN_{min}}{DN_{max} + DN_{min}}. \tag{4-27}$$

Because DNs are always positive, this definition insures that modulation is always between zero and one and unitless.

Modulation is most appropriate for periodic (spatially repetitive) signals. An example, a periodic *sinusoidal* function (Fig. 4-17), is described as,

$$f(x) = mean + amplitude \times \sin\left(2\pi\frac{x}{period} - phase\right).$$ (4-28)

The *spatial frequency* of such a function is related to its period by

$$frequency = \frac{1}{period}$$ (4-29)

and has units of reciprocal distance.[4] If, for the periodic signal in Fig. 4-17, we divide the numerator and denominator in Eq. (4-27) by two, we obtain,

$$M = \frac{(DN_{max} - DN_{min})/2}{(DN_{max} + DN_{min})/2} = \frac{amplitude_{DN}}{mean_{DN}}$$ (4-30)

which expresses modulation as one-half the ratio of the data range to the data mean. For aperiodic signals, such as most images, Eq. (4-30) is still meaningful, but must be interpreted more carefully.

Note that modulation is related to one measure of contrast, C_{ratio}, as follows,

$$M = \frac{C_{ratio} - 1}{C_{ratio} + 1}.$$ (4-31)

In general, modulation suffers from the same sensitivity to outliers as C_{ratio} or C_{range}.

Signal-to-Noise Ratio (SNR)

A *signal* is the noiseless part of a measurement, i.e., the component that carries information of interest. The specific definition of a signal therefore changes with the application. In optical remote sensing, we might call reflectance the signal if our goal is to measure reflectance. Or, we might define the signal as the at-sensor radiance if our purpose is to analyze the sensor's influence on the radiation it receives. Since an image consists of spatial variations, we need to think of the signal as spatially variant, as well.

As just described, numerous types of *noise* can corrupt the signal and make information extraction more difficult. Some measure of the relative amounts of signal and noise is necessary for engineering design, data quality assessment, noise reduction algorithms, and certain information extraction algorithms. The *Signal-to-Noise Ratio (SNR)* is such a measure; it is unitless and therefore independent of the data units. The difficult problem is how to meaningfully define signal and noise. For an image contaminated by random noise at every pixel, an "amplitude" *SNR* can be defined as the ratio of the noise-free image contrast to the noise contrast,

$$SNR_{amplitude} = \frac{C_{signal}}{C_{noise}}$$ (4-32)

---

4. This type of function plays an essential role in Fourier transforms, as described in Chapter 6.

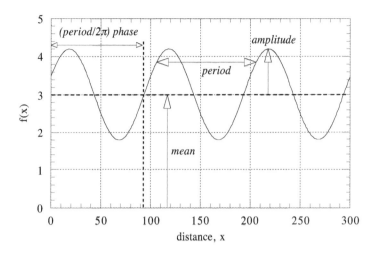

FIGURE 4-17. *A 1-D periodic function of spatial coordinates, with a period of 100 spatial units, a mean of 3 units, an amplitude of 1.2 units, and a phase of 5.8 radians, corresponding to an offset of 92.3 spatial units. The modulation of this function is 0.4.*

where C is any of the definitions in Eq. (4-24) to Eq. (4-26). Because of the problem of outliers, the measure C_{std} is generally the most reliable. We then have *SNR* defined in terms of the ratio of the signal standard deviation to the noise standard deviation,

$$SNR_{std} = \frac{\sigma_{signal}}{\sigma_{noise}} .$$

(4-33)

A "power" *SNR* is given by,

$$SNR_{power} = (SNR_{amplitude})^2 = \left(\frac{C_{signal}}{C_{noise}}\right)^2$$

(4-34)

and, if we again choose to use C_{std} as the measure of contrast, this yields

$$SNR_{var} = \frac{\sigma_{signal}^2}{\sigma_{noise}^2}$$

(4-35)

which is probably the most commonly-used definition of *SNR*. The use of variance in the *SNR* measure is compatible with other statistical descriptors of images, such as the *semivariogram* described later in this chapter, and with statistically-based image transforms, such as the *Maximum Noise Fraction (MNF)* transform described in Chapter 5.

The *SNR* expressed in decibels (dB) is given by,

$$SNR_{dB} = 10\log(SNR)$$

(4-36)

where *SNR* can be any of the previous definitions. This is not another definition of *SNR*, but simply a nonlinear transform of a given *SNR* measure; it is particularly useful for compressing the large dynamic range of SNR_{power} and also has some computational benefits in systems analysis.

By now it should be clear that there is no single definition of *SNR*. The definition should be chosen to be meaningful to the problem at hand. The reader should likewise always question the definition of *SNR* used in any publication; unfortunately it is not uncommon to simply specify *SNR* values without being specific about the definition. In some cases, it may be possible to infer the definition being used from the context of the discussion.

The overriding practical problem in trying to estimate *SNR* from data is, of course, that one does not usually have access to the pure signal. If the noise level is not high, it may suffice to use the noisy image as an approximation of the signal. The noise level (for uniform, random noise) can be estimated from uniform areas of the image which are assumed to have no signal content (see the discussion about Fig. 4-13). Simulations of an image with different amounts of global, additive random noise at each pixel are shown in Fig. 4-18.

Any type of periodic noise is more visible than random noise (Fig. 4-19), but is generally easier to correct (Chapter 7). Robust measures of *SNR* for striping or isolated (local) random noise have not been developed, however (see Ex 4-5.).

National Imagery Interpretability Scale (NIIRS)

Spatial characteristics, such as *GIFOV* or *GSI*, and radiometric characteristics, such as quantization level or detector noise, are required for numerical design, comparison, and evaluation of imaging systems. However, they do not relate the system parameters to the tasks expected to be achieved from the imagery. A summary metric that attempts to make just such a connection is the *National Imagery Interpretability Scale (NIIRS)*. NIIRS was developed for military applications where imagery is interpreted visually by experienced and certified analysts. It is primarily dependent on spatial resolution capability, i.e., *GSI*, but also includes image *SNR*- and *PSF*-related influences. A 10-level NIIRS scale has been developed for military application; an abbreviated description of the panchromatic imagery NIIRS is given in Table 4-1; a more detailed description can be found in Leachtenauer *et al.* (1997) and IRARS (1996). A multispectral imagery NIIRS has also been developed (IRARS, 1995). These scales are applied to unrated imagery by presenting those images to trained ("NIIRS certified") interpreters and asking them to rate each according to the level of detail that can be discerned in the image. An example of applying the process to IKONOS 1m panchromatic imagery, for which an average NIIRS rating of 4.5 was obtained, is described in Ryan *et al.* (2003).

At first glance, the NIIRS does not seem particularly relevant to civilian remote sensing as emphasized in this book. However, with the trend to higher resolution multispectral systems, such as IKONOS, QuickBird, and OrbView, there is increasing crossover between the civilian and military communities in the use of the same sensors for different applications. Also, the overall goal expressed by the NIIRS of relating system characteristics to task performance is a useful and potentially valuable approach for quantitative remote sensing system analysis. For example, it is possible to relate sensor parameters mathematically to NIIRS and thereby predict whether a particular

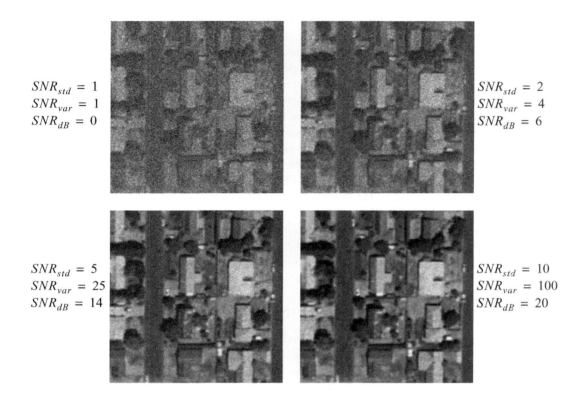

$SNR_{std} = 1$
$SNR_{var} = 1$
$SNR_{dB} = 0$

$SNR_{std} = 2$
$SNR_{var} = 4$
$SNR_{dB} = 6$

$SNR_{std} = 5$
$SNR_{var} = 25$
$SNR_{dB} = 14$

$SNR_{std} = 10$
$SNR_{var} = 100$
$SNR_{dB} = 20$

FIGURE 4-18. The effect of random noise on image quality is simulated in this aerial photograph of a residential neighborhood in Portland, Oregon, by adding a normally-distributed random number to each pixel. Even for an SNR of unity, there is discernable image structure because of the spatial correlation between neighboring pixels. This correlation means that the mean DN is fairly constant over local neighborhoods, but changes globally across the image. Therefore, one is still able to detect the roofs, trees, and streets even at very low SNR, but smaller objects, such as cars, are obscured. (The original image is courtesy of the U.S. Geological Survey.)

imaging system will perform adequately for specified tasks using the *General Image-Quality Equation (GIQE)* (Leachtenauer *et al.*, 1997). To apply this concept to civilian earth remote sensing systems, the set of tasks should be defined appropriately, e.g. mapping of the Anderson land cover and land use categories (Chapter 9). Another major change is that the tasks are not defined by visual analysis but by computer classification or other information extraction processes, whose performance needs to be described in a numerical way.

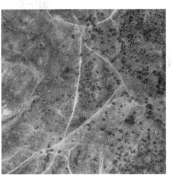

noiseless

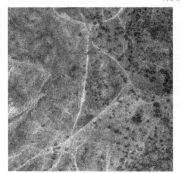

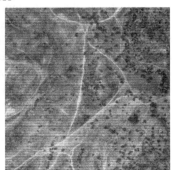

global random noise   *detector striping noise*

FIGURE 4-19. *Comparison of the visual effect of equal amounts of random and striping noise. The aerial photograph is used in a simulation of global, additive random noise and unequal detector offsets. The random noise is normally distributed, and the striping noise has a period of 16 lines (as in TM imagery). Both noise distributions have a mean of zero and a standard deviation of eight; the standard deviation of the original image is 18.*

TABLE 4-1. *Example of the National Image Interpretability Scale (NIIRS) .*

rating level	example criterion
0	cannot be interpreted due to clouds or poor quality
1	distinguish airport taxiways and runways
2	detect large buildings
3	identify large ship type
4	identify individual tracks in railroad yard
5	identify individual railcars by type

TABLE 4-1. *Example of the National Image Interpretability Scale (NIIRS) (continued).*

rating level	example criterion
6	identify automobiles as sedans or station wagons
7	indentify individual railroad ties
8	indentify vehicle windshield wipers
9	detect individual railroad tie spikes

4.5.2 Noise Equivalent Signal

Sometimes one is interested in the input signal level to an optical sensor that results in an output signal equal to the noise level. This is the *noise equivalent signal*; it can also be interpreted as the input signal level that results in an output *SNR* of one. Examples in remote-sensing usage are *Noise Equivalent Radiance* (*NER*), the input radiance to an optical sensor that produces an output *SNR* of one, and *Noise Equivalent Reflectance Difference* (*NEΔρ*), a differential measure that indicates the ground reflectance difference that produces an image contrast (C_{range} in Eq. (4-25)) of one (Schott, 1997). It is therefore implied that the value $\Delta\rho$ is just detectable, but that is an arguable point.

4.6   *Spatial Statistics*

The statistical properties of spatial data have been studied and modeled for years in applications such as mining and nonrenewable resource (oil, natural gas) exploration (Journel and Huijbregts, 1978). The techniques have been generalized to other spatial data in recent years (Isaaks and Srivastava, 1989; Carr, 1995), including remote-sensing data (Curran, 1988). The field of spatial statistics is called *geostatistics* by its practitioners.

A key concept in geostatistics is the notion of *spatial continuity*, which expresses the likelihood of a particular data value at a particular location (a pixel in an image in our case) given neighboring or regional data values. Techniques have been devised to use second-order statistics to measure data continuity (*covariance* and *semivariogram*) and to optimally estimate missing values by interpolation (*kriging*).

In the world of image processing, many spatial descriptors have been proposed to address the recognition of different spatial *textures*. If one thinks of an image as a tangible, physical surface defined by $DN(x,y)$, then texture describes the visual "roughness" of that surface, analogous to the sensation of moving a finger across a real surface. Texture is manifested in an image as local, quasi-periodic variations in *DN* and is caused by the spatial distribution of scene reflectance, and shading and shadows from topography. One texture descriptor is based on properties of the *spatial co-occurrence matrix* (Haralick *et al.*, 1973). Others are based on *fractal geometry* (Mandelbrot, 1983; Pentland, 1984; Feder, 1988). We will look at these as well as the geostatistical measures in the following sections.

4.6.1 Visualization of Spatial Covariance

The joint probability of a pair of *DN* values at two pixels, separated by an amount h in a given direction, may be approximated by the joint histogram,

$$hist_h = count(DN_1, DN_2)/N_h \qquad (4\text{-}37)$$

which is a special case of Eq. (4-11); N_h is the number of pixel pairs used in the count operation for the given h. We can visualize the histogram as a 2-D scattergram with one axis corresponding to the *DN* of a pixel and the other corresponding to the *DN* of another pixel which is h pixels away in the given direction. An example is shown in Fig. 4-20. At short separations (small h), the scattergram shows high correlation, i.e., close neighbors are similar. As the separation increases (larger h), the scattergram becomes less correlated and eventually nearly isotropic, indicating the dissimilarity of pairs of pixels. We will see later that such scattergrams are identical to the spatial co-occurrence matrix.

4.6.2 Covariance and Semivariogram

In geostatistics, the *covariance function* of a 1-D data set is defined as (Isaaks and Srivastava, 1989),

$$C(h) = \frac{1}{N(h)} \sum_{(i,j)|h_{ij}=h} DN_i DN_j - \mu_{-h}\mu_{+h} . \qquad (4\text{-}38)$$

The parameter, h_{ij}, is called the *lag* and is the distance between the pairs of data points located at i and j. The quantities, μ_{-h} and μ_{+h} are the mean *DN*s of all data points at a distance of $-h$ or $+h$ from some other data point; these two values are seldom equal in practice. $N(h)$ is the number of data point pairs separated by h.

The *semivariogram* is given by (Isaaks and Srivastava, 1989),

$$\gamma(h) = \frac{1}{2N(h)} \sum_{(i,j)|h_{ij}=h} (DN_i - DN_j)^2 \qquad (4\text{-}39)$$

or, as expressed in the equivalent notation of (Woodcock *et al.*, 1988a),

$$\gamma(h) = \frac{1}{2N(h)} \sum_{i=1}^{N} (DN_i - DN_{i+h})^2 . \qquad (4\text{-}40)$$

Equation (4-40) measures the moment of inertia about the 45° line in the spatial scattergrams of Fig. 4-20. It therefore is a measure of the spread in the scattergram and the degree of spatial correlation between different pixels.

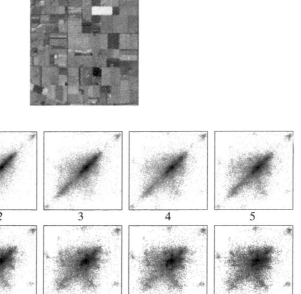

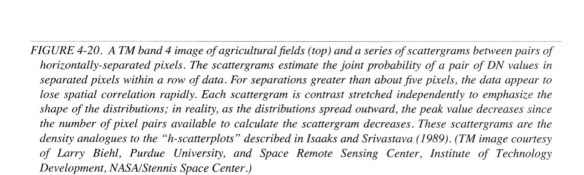

FIGURE 4-20. A TM band 4 image of agricultural fields (top) and a series of scattergrams between pairs of horizontally-separated pixels. The scattergrams estimate the joint probability of a pair of DN values in separated pixels within a row of data. For separations greater than about five pixels, the data appear to lose spatial correlation rapidly. Each scattergram is contrast stretched independently to emphasize the shape of the distributions; in reality, as the distributions spread outward, the peak value decreases since the number of pixel pairs available to calculate the scattergram decreases. These scattergrams are the density analogues to the "h-scatterplots" described in Isaaks and Srivastava (1989). (TM image courtesy of Larry Biehl, Purdue University, and Space Remote Sensing Center, Institute of Technology Development, NASA/Stennis Space Center.)

The covariance function and semivariogram are related by,

$$\gamma(h) = C(0) - C(h) \tag{4-41}$$

if a sufficiently large number of data samples are used so that the $C(0)$ is the same over any set of samples (this is the *stationarity* assumption from statistics). Thus, the covariance function and the semivariogram actually contain the same information about the image.

The semivariogram usually exhibits a characteristic shape, increasing from small lags to larger lags. The plateau where γ becomes more or less constant is the *sill*. The distance from zero lag to the onset of the sill is the *range*. The semivariogram is, by definition, zero for zero lag. However, for real data, noise may cause nonzero semivariogram values for lags smaller than the sample interval. This residual value at small lags is known as the *nugget* value. The nugget can be used to estimate spatially-uncorrelated image noise (Curran and Dungan, 1989). The sill, on the other hand, is the other extreme, i.e., the value of the semivariogram for very large lags. The sill is usually reached more or less asymptotically at the range, which is one measure of *correlation length* in the data.

The spatial characteristics of the data can be emphasized by normalizing the covariance functions by their value at the origin, namely the variance of the data. This normalization removes differences due to contrast among data sets, and more clearly shows the spatial correlation relative to that at zero lag. Examples are shown later.

Several continuous models for the 1-D covariance function and semivariogram are summarized in Table 4-2. The exponential *Markov covariance model*, widely used in image processing (Jain, 1989; Pratt, 1991), is equivalent (with a scale change on h) to the exponential model for semivariograms, in view of Eq. (4-41). The various models of Table 4-2 are plotted in Fig. 4-21.

TABLE 4-2. *Some 1-D continuous models for the discrete spatial covariance and semivariogram (Isaaks and Srivastava, 1989; Pratt, 1991; Carr, 1995). Note each of the semivariogram models is normalized to a sill value of one.*

spatial statistic	model	equation
covariance	exponential (Markov model)	$C(h) = C(0)e^{-\alpha h}$
semivariogram	Gaussian	$\gamma(h) = 1 - e^{-3(h/\alpha)^2}$, where $\alpha = \gamma(0.95 sill)$
	exponential	$\gamma(h) = 1 - e^{-3(h/\alpha)}$, where $\alpha = \gamma(0.95 sill)$
	spherical	$\gamma(h) = \begin{cases} 1.5h/\alpha - 0.5(h/\alpha)^3, & h \le \alpha \\ 1, & h > \alpha \end{cases}$, where $\alpha = \gamma(sill)$

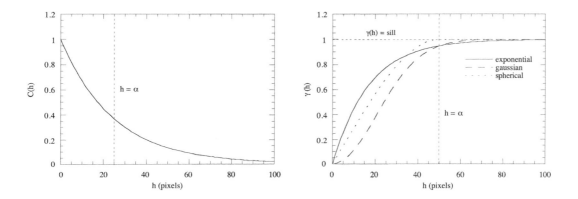

FIGURE 4-21. *Graphs of the exponential covariance model and three normalized semivariogram models.*

Geophysical data are often available only on an irregular sample grid, with large gaps between measured data points. Semivariograms are used in that context to estimate the missing values using a procedure known as *kriging* (Isaaks and Srivastava, 1989). Continuous models are needed to interpolate values of the semivariogram at locations other than the discrete values used in Eq. (4-40). Some interesting applications of the semivariogram in remote-sensing data analysis are summarized in Table 4-3.

TABLE 4-3. *Example applications of the semivariogram in remote sensing .*

application	data source	reference
agricultural fields, forestry	TM, aerial photography, TM airborne simulator	Woodcock *et al.*, 1988b
landcover categorization	airborne MSS	Curran, 1988
estimate *SNR*	AVIRIS	Curran and Dungan, 1989
sea surface temperature estimate random pixel noise	AVHRR AVHRR, CZCS	Wald, 1989
image misregistration	simulated MODIS	Townshend *et al.*, 1992
sea surface temperature	AVHRR	Gohin and Langlois, 1993
landcover categorization	field spectroradiometer, HCMM thermal, SPOT, TM	Lacaze *et al.*, 1994
landcover mapping in shadows	TM	Rossi *et al.*, 1994
kriging of stereo disparity maps	SPOT	Djamdji and Bijaoui, 1995

TABLE 4-3. *Example applications of the semivariogram in remote sensing (continued).*

application	data source	reference
forest canopy structure	MEIS-II	St-Onge and Cavayas, 1997
texture classification	IRS LISS-II, TM, SPOT, microwave	Carr and Miranda, 1998
flooded and upland vegetation	JERS-1 SAR	Miranda *et al.*, 1998

The aerial photograph of a desert area in Fig. 4-22 will be used to demonstrate calculation of semivariograms from an image. This image contains several distinct regions of different vegetation densities. Three transects were extracted in each of three characteristically different regions, the semivariogram was calculated for each transect and averaged for each region. The result clearly shows distinguishing features among the three vegetation density categories (Fig. 4-23). For example, the low density vegetation area shows the largest sill since it has the greatest contrast in the image. It also has a greater range than that for the high density vegetation area, indicating that the average plant size is larger in the low density area. These empirical data are consistent with the disks-on-a-uniform-background model described and analyzed in Woodcock *et al.* (1988a).

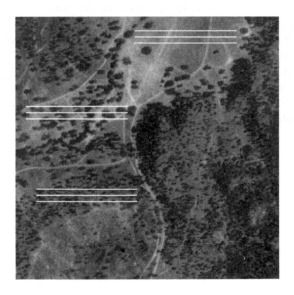

FIGURE 4-22. *Scanned aerial image used to illustrate spatial statistics. Three groups of three transects each are extracted from regions of distinctly different vegetation density and plant size: no vegetation (top), low density (middle), and high density (bottom). Each transect is 100 pixels long.*

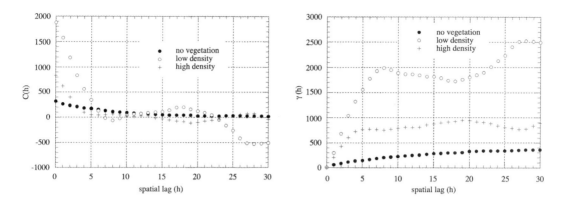

FIGURE 4-23.   *Covariance functions and semivariograms for the transects of Fig. 4-22. The average function for the three transects in each category is plotted.*

To isolate size variations from contrast, the covariances are normalized at zero lag in Fig. 4-24. From these spatial correlation curves, we can estimate the correlation length, or range, for each of the three vegetation density categories as the lag distance over which the data are correlated. We will use the exponential model of Table 4-2 and fit it to the covariance functions of Fig. 4-24. Only the covariances for smaller values of lag will be used; the exponential model clearly does not describe the behavior at large lags. The fitted model curves are shown in Fig. 4-25 and are reasonably good approximations to the data in each case. The correlation length may be defined as the reciprocal of the parameter α in the model, corresponding to the lag at which the covariance function falls to $1/e$ of its value at zero lag (Table 4-4).

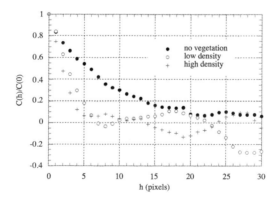

FIGURE 4-24. *Normalized covariance functions. In effect, the data variance is set to one for each class.*

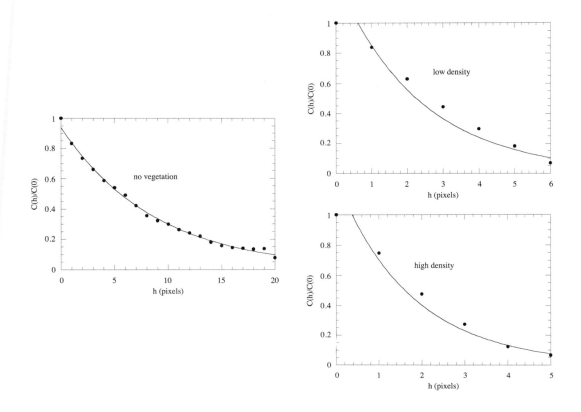

FIGURE 4-25. Exponential model fitted to covariance functions from Fig. 4-24. Note that the horizontal axis is scaled differently in each graph.

TABLE 4-4. Correlation lengths obtained from the exponential model fits in Fig. 4-25.

category	α (pixels^{-1})	correlation length (pixels)
no vegetation	0.114	8.78
low density	0.422	2.37
high density	0.560	1.79

These values can be used to quantify and characterize the spatial patterns of the different areas in the image. The slope of the covariance function or semivariogram near zero lag indicates the average plant size, because it measures how rapidly the pixels lose correlation as we move away from a given pixel. The slope for low-density vegetation is less than that for high-density vegetation because the plants are larger. The slope for the nonvegetated area is much lower than that for either of the areas with vegetation due to the lack of spatial structure.

Finally, note that the curves in Fig. 4-23 do not quite satisfy Eq. (4-41), because a finite (and fairly small) sample set was used. The relationship is satisfied, however, for all practical purposes.

Separability and anisotropy

Extension of the 1-D spatial statistics models of this section to 2-D images can be done in two ways. One is to simply replace the 1-D lag, h, by a 2-D radial distance, r. For example, the exponential covariance model (Table 4-2) becomes an *isotropic* function,

$$C(r) = C(0)e^{-\alpha r} \tag{4-42}$$

where

$$r = (h_x^2 + h_y^2)^{1/2} . \tag{4-43}$$

An isotropic function is the same in all directions from a given origin. Generally, however, remote-sensing images are not isotropic and, therefore, neither are their spatial statistics. Alternatively, we can create a *separable* model by multiplication of two 1-D functions, in orthogonal directions,

$$C(x, y) = C(0)e^{-\alpha|x|}e^{-\beta|y|} = C(0)e^{-(\alpha|x| + \beta|y|)} . \tag{4-44}$$

A more general *nonseparable, anisotropic* model has been suggested in Jain (1989) for images and in Isaaks and Srivastava (1989) for geostatistical quantities,

$$C(h_x, h_y) = C(0)e^{-\sqrt{\alpha h_x^2 + \beta h_y^2}} . \tag{4-45}$$

The isotropic, separable, and anisotropic 2-D forms of the exponential covariance function model are shown in Fig. 4-26. Which is more realistic? The answer, of course, depends on the image. To model images *in general*, the isotropic form should be used, since there is no *a priori* reason to believe the image statistics are different in different directions.[5] The anisotropic

---

5. If the sensor has an asymmetry between the in-track and cross-track directions due to an asymmetric *PSF* (Chapter 3), a consistent asymmetry can be expected also in the image spatial statistics. A broader *PSF* profile in one direction will cause a broader image covariance function (for example MODIS; see Fig. 3-12). Any asymmetry in *GSI* (for example MSS; see Fig. 1-12) will lead to a scaling of the covariance or semivariogram abscissa if plotted in pixel units. Such "asymmetry" should disappear, however, if the functions are plotted in absolute distance units.

exponential model is probably a good choice if one is modeling a *particular* image. The axes directions can be rotated if needed to fit the actual covariance. Combinations of these models to fit more complex data are discussed in Isaaks and Srivastava (1989).

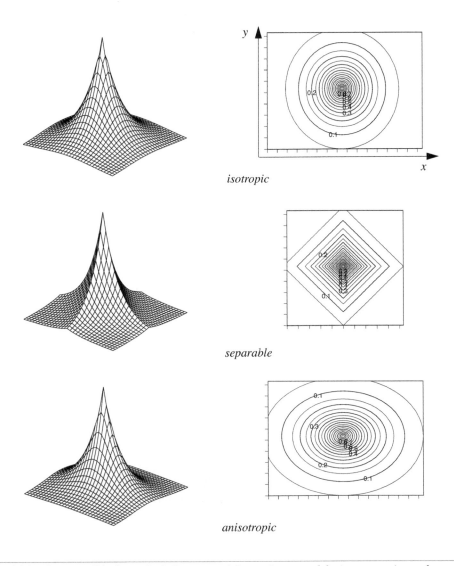

isotropic

separable

anisotropic

FIGURE 4-26. *Three possible forms for 2-D exponential covariance models, in perspective and as contour maps. For the anisotropic model,* $\alpha = \beta/4$.

4.6.3 Power Spectral Density

The Fourier transform of the covariance function is called the *Power Spectral Density (PSD)*, or *power spectrum.*[6] It characterizes the spatial frequency content in the covariance function. The width of the power spectrum is proportional to the inverse of the spatial correlation length. If the correlation length of the data is small, the covariance function is narrow, and the power spectrum is wide. Conversely, data that are correlated over a substantial distance will have a narrow power spectrum, with little power at high spatial frequencies.

Using the data of the previous section, we can calculate the 1-D spatial power spectrum for each of the cover types in Fig. 4-22 as the Fourier transform of the respective covariance function in Fig. 4-24. Figure 4-27 demonstrates that the widths of the power spectra are inversely proportional to the widths of the covariance functions.

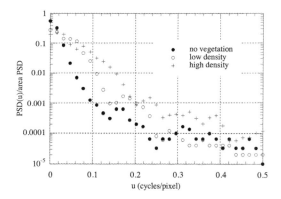

FIGURE 4-27. *Power spectra calculated from the data in Fig. 4-24. The normalization of the covariance functions corresponds to normalization of the areas under the power spectra. The latter, therefore, more clearly show the relative distribution of variance at each spatial frequency. Note the inverse ordering of the three curves compared to Fig. 4-24.*

Some example image blocks and their power spectra are shown in Fig. 4-28. The image was partitioned into adjacent 128-by-128 pixel blocks, and the power spectrum calculated for each block as the square of the Fourier transform amplitude of the image block.[7] The power spectrum has directional information about linear features (edges, lines) in the image. The highest power components are orthogonal to the direction of such objects in the image.

---

6. If the reader is unfamiliar with Fourier transforms, it may be useful to first read the appropriate sections in Chapter 6 before reading this section.

7. This alternate use of the term *power spectrum* is quite common in image processing. It is not a proper usage in a statistical context, but is so common, that we defer to the majority! The two quantities, the Fourier transform of the covariance function and the squared amplitude of the Fourier spectrum of a data block, are related and convey approximately the same information.

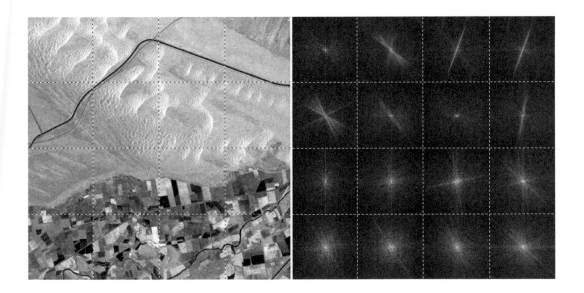

FIGURE 4-28. *Landsat TM band 4 image of sand dunes and irrigated agriculture near Yuma, Arizona, partitioned into 16 128-by-128 pixel blocks and the corresponding power spectrum of each block. See Chapter 6 for a description of the spatial frequency coordinate system used for power spectra.*

The exponential Markov covariance model (Table 4-2) leads to a corresponding model for the power spectrum,

$$PSD(u) \; = \; \frac{2\alpha}{\alpha^2 + (2\pi u)^2} \qquad\qquad (4\text{-}46)$$

where u is spatial frequency. This model is graphed in Fig. 4-29 for the three vegetation density categories. Just as the exponential model was reasonable for the covariance function, the model of Eq. (4-46) fits the PSD reasonably well.

Although we have gone into some detail to illustrate the modeling of image covariance and power spectral density functions, the reader is advised that most of the time *any given image* will not necessarily be modeled well by the functions we have used (note Fig. 4-28). They are best thought of as *average models* that can be used for analytical work.

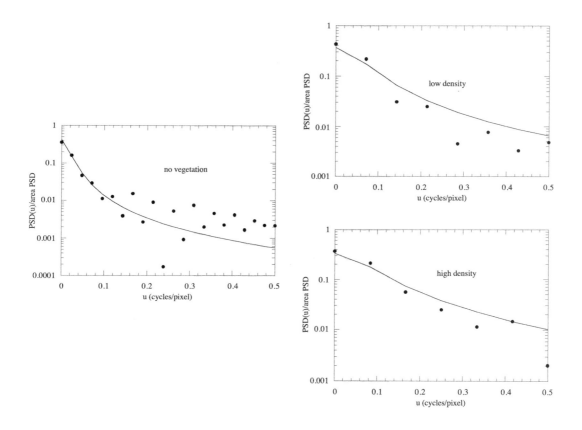

FIGURE 4-29. *Power spectra calculated by the Fourier transform of the covariance functions in Fig. 4-25, compared to the model of Eq. (4-46). The value of α used in each case is from Table 4-4.*

4.6.4 Co-Occurrence Matrix

The 2-D matrix of joint probabilities $p(DN_1,DN_2)$ between pairs of pixels, separated by a distance, h, in a given direction, is called the *Co-occurrence Matrix* (*CM*).[8] The scattergrams of Fig. 4-20 are examples. The *CM* contains a large amount of information about the local spatial information in an image; the major problem is how to extract that information. A set of texture features derived from the *CM* was suggested in Haralick *et al.* (1973) (Table 4-5). These features measure various properties of the *CM*, but all are not independent of each other.

---

8. The *CM* goes by several names in the literature, including *Grey-Tone Spatial-Dependence Matrix* (Haralick *et al.*, 1973), *Grey-Level Co-occurrence Matrix* (Gonzalez and Woods, 1992), *Grey Level Dependency Matrix* (Pratt, 1991), and *Concurrence Matrix* (Jain, 1989).

TABLE 4-5. *Some of the spatial texture features derivable from the CM, as originally proposed in Haralick et al. (1973). The quantities p_x and p_y are the marginal distributions of the CM, defined as $p_x = \sum_j p_{ij}$ and $p_y = \sum_i p_{ij}$.*

name	equation	related image property
Angular Second Moment (ASM)	$f_1 = \sum_i \sum_j p_{ij}^2$	homogeneity
Contrast	$f_2 = \sum_i \sum_j (i-j)^2 p_{ij}$	semivariogram
Correlation	$f_3 = \dfrac{\sum_i \sum_j ij p_{ij} - \mu_1 \mu_2}{\sigma_1 \sigma_2}$	covariance
Sum of Squares	$f_4 = \sum_i \sum_j (i-\mu)^2 p_{ij}$	variance
Inverse Difference Moment	$f_5 = \sum_i \sum_j \dfrac{p_{ij}}{1+(i-j)^2}$	—
Sum Average	$f_6 = \sum_{i=2}^{2^{Q+1}} i p_{x+y}$	—
Entropy	$f_9 = -\sum_i \sum_j p_{ij} \log(p_{ij})$	—

The *CM*, for a given lag h and direction, is 2^Q-by-2^Q, where Q is the number of bits/pixel. With a Q of 8 bits/pixel, the *CM* is therefore 256-by-256. For a complete description of the image, the *CM* should be calculated for all lags and in all directions (at least horizontal, vertical and diagonally). To reduce the computation burden, it is common practice to reduce the *DN* quantization (by averaging over adjacent *DN*s) to some smaller value, say 16 levels, and to average the *CM*s for different spatial directions.

To illustrate the calculation of texture features from the *CM*, three images with widely different spatial characteristics will be used (Fig. 4-30); one is the agriculture image used for Fig. 4-20. The entropy and contrast features were calculated for the original *DN* images, with requantization of the *DN* scale to 32 levels between the minimum and maximum *DN* of each image. The spatial parameter was varied in the horizontal direction from a shift of one pixel to 15 pixels.

In Fig. 4-31, the behavior of the *CM* contrast and entropy features with lag, h, are shown. There appear to be distinct differences among the three images, but the fact that their *DN* histogram characteristics are quite different confuses the comparison. To normalize this difference, a histogram equalization transform has been recommended (Haralick *et al.*, 1973); the resulting images are shown in Fig. 4-30 and the new contrast and entropy features in Fig. 4-32. Their shapes have not changed much, but they are now on a similar scale, at large *h*, for each image.

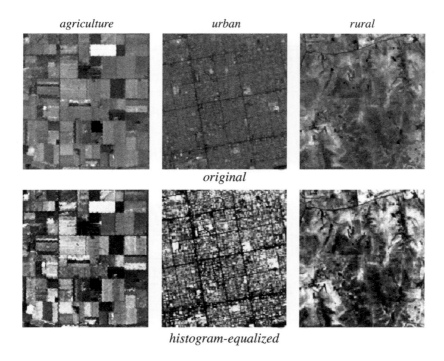

FIGURE 4-30.   *Three sample TM images used for co-occurrence matrix analysis. The histogram equalization process produces an image with an approximately equal pixel count-per-DN density in all parts of the DN range. It is described in Chapter 5.*

4.6.5 Fractal Geometry

Fractal geometry is a way to describe the "texture" of a surface.[9] There are four topological dimensions in traditional Euclidean geometry: 0-D for points, 1-D for straight lines, 2-D for planes, and 3-D for volumetric objects like cubes and spheres. An object that is "fractal" has an intermediate dimensionality, such as 1.6 for an irregular line or 2.4 for an image "surface". Generally, the higher the fractal dimension, the finer and "rougher" the texture.

---

9. We do not have space here to discuss the rich mathematical background of fractal geometry. Suffice it to say that there are many and varied examples in nature; the reader is recommended to Mandelbrot (1983) for the classic description and to Feder (1988) for a particularly readable discussion of applications.

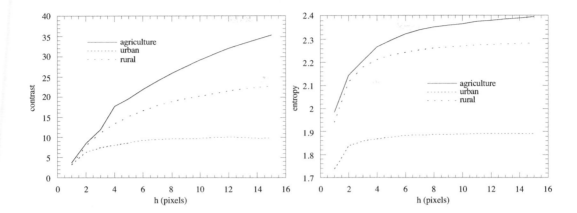

FIGURE 4-31. The contrast and entropy spatial features from the sample images of Fig. 4-30. Note the similarities to the semivariogram functions in Fig. 4-23 (which were calculated from an aerial image), namely the relatively rapid change as h increases from zero and the asymptotic behavior for large h.

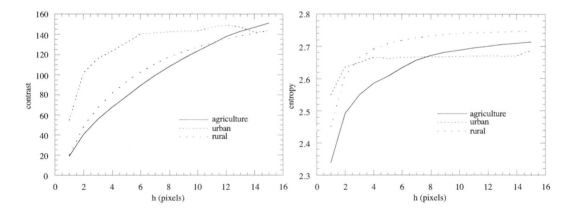

FIGURE 4-32. Contrast and entropy CM features after equalization of the image histograms. The shapes of the features as a function of h have not changed much from that in Fig. 4-31, but both features are now normalized to approximately the same value at large h for the three images.

Objects that are "fractal" also obey the notion that they are "self-similar" in their statistics. This means that the object is statistically the same, no matter what scale it is viewed at. A classic example is measurement of the length of the perimeter of the coastline of Britain (Mandelbrot, 1967). If the perimeter is measured with rulers of different lengths (scales), different values are obtained. Shorter rulers lead to longer values, and vice versa. The logarithm of the measured perimeter is found to obey a linear relationship to the logarithm of the ruler length (Feder, 1988); the slope of that log-log relation is related to the fractal dimension. Most natural objects and surfaces are found to be approximately fractal over a limited range of scales.

Several tools have been developed for estimating the fractal dimension, D, from an image. One approach uses 1-D data transects. The fractal dimension is related to either the semivariogram or power spectrum of the transect (Carr, 1995). For either the semivariogram or the power spectrum, a log-log plot is calculated, and a linear function is fit to the log-transformed data (equivalent to fitting a power function to the data before taking the logarithms). The fractal dimension, D, is then related to the slope of the fitted line, as given in Table 4-6. If one wishes to use data in multiple directions, but anisotropy is *not* of interest, then the 2-D power spectrum can be calculated over the image area of interest, averaged azimuthally to yield an average radial power spectrum, and the fractal dimension calculated in the 1-D manner. This technique was used in Pentland (1984) to produce fractal maps from images. The fractal dimension was measured over 8-by-8 pixel blocks across the image, and the result used to segment the image into regions of different texture.

An interesting technique for estimating the fractal dimension of imagery was tested by Pelig *et al.* (1984) and described in detail in Peli (1990). A "covering blanket" is calculated above and below the image surface, for different resolution scales. One visualizes the blanket as being draped over (and under) the image, so that it bounds the *DN* surface. As the resolution is reduced, the two "blankets" become smoother. The change in volume between the blankets at two successive scales is plotted versus the scale index in a log-log plot, and the slope used to estimate the fractal dimension as previously described. Although the processing is 2-D, the end result is an isotropic measure, since only the volume information is used.

TABLE 4-6. *Some ways to estimate fractal dimensions from images. The slope of the log-log straight line fit is denoted s.*

measurement	fractal dimension (D)	reference
semivariogram	$2 - s/2$	Carr, 1995
power spectrum	$2.5 - s/2$	Carr, 1995, Pentland, 1984
covering blanket volume	$2 + s$	Pelig *et al.*, 1984

An example of fractal dimension calculation from imagery is shown in Fig. 4-33. The TM image has two characteristic texture regions, the desert sand dunes in the upper portion and the agricultural area in the lower portion. The technique of Pentland was used, with 32-by-32 pixel blocks and the DC (zero) frequency component excluded from the power spectral analysis. The resulting fractal dimension map appears to correlate with the image texture regions. Now, if we look at the histogram of the original image, we do not see an obvious point at which a *DN* threshold might separate the two regions (Fig. 4-34). If we look at the fractal dimension histogram, however, a valley appears to signify two classes. In fact, if we threshold the fractal dimension image at the valley, the resulting map appears to reasonably map the two regions.

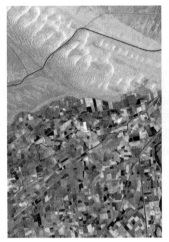 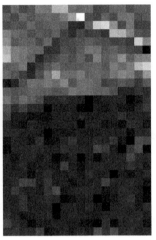

TM band 4 *fractal dimension map*

FIGURE 4-33. A TM band 4 image (the same area as Fig. 4-28) and the 32-by-32 pixel block fractal dimension map. (Fractal map calculation courtesy of Per Lysne, University of Arizona.)

4.7 Topographic and Sensor Effects

The most appropriate data models and associated parameters for given imagery are, in part, determined by terrain and sensor effects. We have seen in Chapters 2 and 3 how topography and sensor spectral and spatial resolution affect the reproduction of scene information in the recorded imagery. In this section, we will visualize some of these effects, with the aid of simulation and the spatial and spectral models discussed in this chapter.

4.7.1 Topography and Spectral Scattergrams

Assuming the surface has a Lambertian reflectance characteristic, its radiance is proportional to the cosine of the angle between the solar vector and the surface normal vector (Eq. (2-6)). Suppose we have such a scene consisting of two materials, soil and vegetation, each having a certain within-class variance but no band-to-band correlation. We can generate reflectance images in the red and

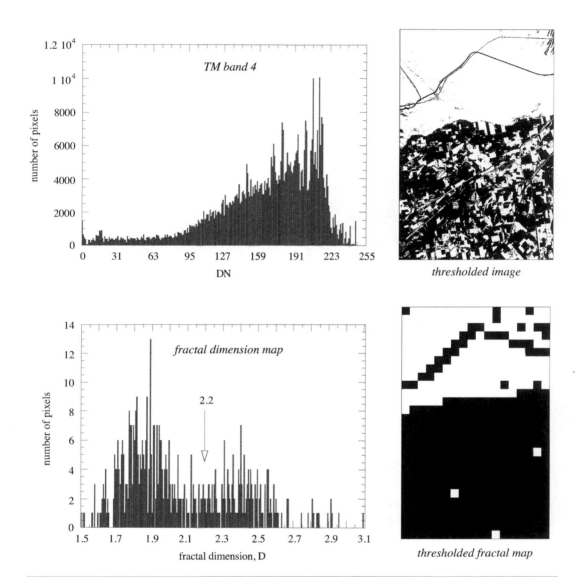

FIGURE 4-34. *The original image and fractal dimension map histograms. Threshold maps of the original image (DN = 170) and the fractal dimension map (D = 2.2) are also shown. Note, the two regions of the original image appear inseparable with simply a DN threshold, while they are separable in the fractal dimension map. The average fractal dimension of the sand area is 2.45 and that of the agriculture area is 1.88, which, being less than two, actually implies a nonfractal structure (Pentland, 1984). Pentland found that contrast boundaries generally were nonfractal; the irrigation canal in the upper half of the image and the dark-light field boundaries in the lower half are in that category.*

NIR spectral regions from random Gaussian distributions for each class in each band. A real topographic DEM will be used (Fig. 4-35), and we want to distribute the soil and vegetation pixels equally across the scene, assuming 50% coverage by each class.

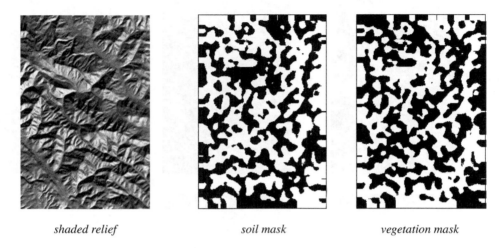

shaded relief   *soil mask*   *vegetation mask*

FIGURE 4-35. The shaded relief image generated from a DEM, part of the same area shown in Fig. 2-11, and the synthesized soil and vegetation masks. The algorithm to generate the relief image is simply a cosine calculation as in Eq. (2-6). The process used to generate the binary masks shown consisted of creating a random, Gaussian noise pattern with zero mean, smoothing that by convolution with a Gaussian low-pass filter (see Chapter 6), and thresholding at a DN of zero. The convolution step introduces spatial correlation in the noise, leading to the patterns above after thresholding.

To make the spatial characteristics more realistic, a soil mask and a complementary vegetation mask are created (Fig. 4-35).[10] Each mask is filled with the simulated reflectance pixels from the respective class, and the result multiplied by the topographic shaded relief image, to generate surface radiance, as given mathematically by Eq. (2-7). Finally, the soil and vegetation components are composited into a two-band, simulated scene (Fig. 4-36).

Looking at the NIR-red scattergram of the simulated images (Fig. 4-37), we see that the originally uncorrelated distributions are now strongly correlated along lines passing through the scattergram origin, or zero reflectance. This happens because the incident radiation is reduced by a cosine factor, scaled in this case to the range [0,1] by saturating negative values at zero. This *topographic modulation* has long been recognized as an influence on spectral signatures in remote sensing data, and detailed modeling of its characteristics has been done (Proy *et al.*, 1989). Although most work

10. The simulated spatial distribution of soil and vegetation approximates the actual land-cover distribution for this DEM, which is an area of hills near Oakland, California. They are uniformly covered by soil and grass, with patches of woodlands.

soil pixels   *vegetation pixels*   *final simulated images*

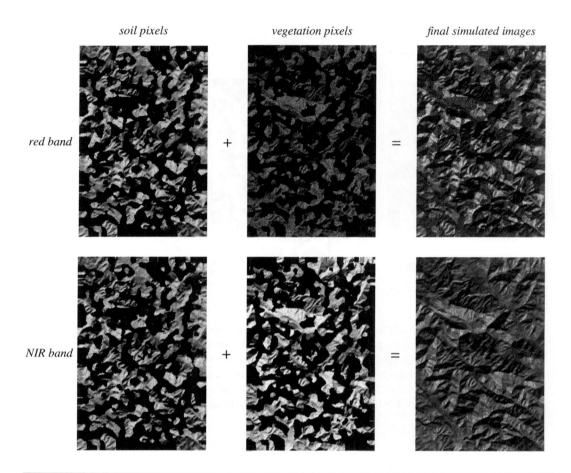

red band

NIR band

FIGURE 4-36. *The shaded relief image is multiplied by the pixels in each class mask to produce the soil and vegetation components on the left. These components are then composited, separately in the red and NIR bands, to produce the final simulated images on the right. The relative brightness of each class and band is preserved in this figure, e.g. vegetation is darker than soil in the red band and lighter than soil in the NIR band.*

has been directed at reducing topographic effects (Table 4-7), topography can be extracted from multispectral imagery using models (Eliason *et al.*, 1981), and physical models for topographic modulation can be used to decompose imaging spectrometer data into its spectral constituents (Gaddis *et al.*, 1996).

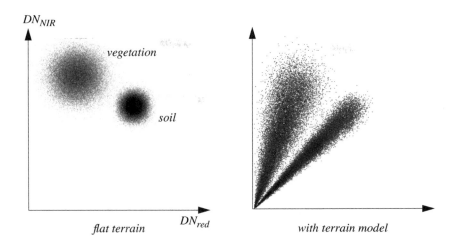

FIGURE 4-37. *NIR versus red scattergrams for the original soil and vegetation data, before and after merger with the terrain model. The signature of each class is now correlated between the two bands and their distributions extend through the zero reflectance origin of the scattergram.*

TABLE 4-7. *Examples of research on measurement and correction of topographic effects in analysis of remote sensing imagery.*

data	reference
field measurements	Holben and Justice, 1980
CASI	Feng *et al.*, 2003
DAIS, HyMap	Richter and Schlapfer, 2002
ETM+	Dymond and Shepherd, 2004
MSS	Holben and Justice, 1981; Eliason *et al.*, 1981; Justice *et al.*, 1981; Teillet *et al.*, 1982; Kawata *et al.*, 1988
TM	Civco, 1989; Conese *et al.*, 1993; Itten and Meyer, 1993; Richter, 1997; Gu and Gillespie, 1998; Gu *et al.*, 1999; Warner and Chen, 2001
SPOT-4	Shepherd and Dymond, 2003

4.7.2 Sensor Characteristics and Spatial Statistics

The sensor introduces many, but two particularly notable, effects into the imagery: noise and blurring. In this section, we want to see how these two factors affect the spatial statistics derived from the imagery. We will first look at image noise. The aerial photograph of Fig. 4-22 was modified by the addition of normally-distributed, spatially-independent noise at each pixel, with a *DN* standard deviation of 20. The noise level is set rather high to illustrate its effect (Fig. 4-38).

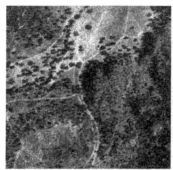 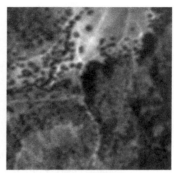

noisy sensor   *larger GIFOV*

FIGURE 4-38. *The two simulated images used to investigate the influence of the sensor on spatial statistics measured from its imagery. On the left, normally-distributed, uncorrelated random noise with a DN standard deviation of 20 was added to the original image of Fig. 4-22. This corresponds to the noise model of Eq. (4-20) and a C_{std} of about 2 by Eq. (4-26). On the right, the original noiseless image was convolved with a 5-by-5 uniform spatial response function, simulating a GIFOV of that size. The same transects of Fig. 4-22 were extracted for the analysis.*

The spatial covariances for the transects in Fig. 4-22 are shown in Fig. 4-39. The uncorrelated noise is seen in the spike at zero lag, which in each case is about 400 units higher than the original covariance, i.e., an amount equal to the added noise variance. At nonzero lags, the image noise introduces some covariance noise but no overall shift because the added noise is uncorrelated from pixel-to-pixel. The semivariograms shown in Fig. 4-39 contain the same information in a different form; the sills are about 400 *DN* units higher at all lags, but still zero at zero lag, by definition. The nugget value, obtained by extrapolating the lowest, nonzero values, to a lag of zero, now equals about 400. The fact that the semivariogram nugget value represents uncorrelated noise can be used to estimate image noise (Curran and Dungan, 1989).

To see the effect of the sensor *GIFOV* on spatial statistics, we return to the original aerial photograph, without added noise, and introduce a uniform, square *GIFOV* of 5-by-5 pixels by convolution (Fig. 4-38). The scene details are blurred, and we see a large reduction in covariance at all lags (Fig. 4-40). The shapes of the covariance functions are more easily compared by normalization to one for zero lag (Fig. 4-41). Now, the *GIFOV* is seen to make the covariance function broader because the spatial averaging within the *GIFOV* adds correlation to neighboring points. Also, the

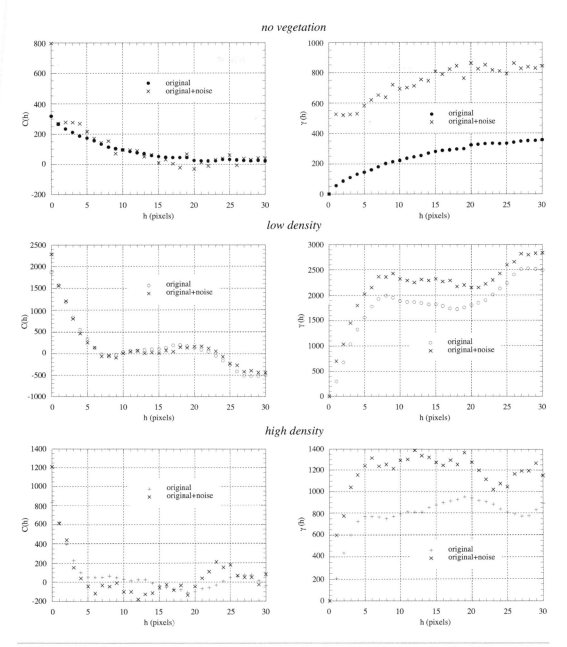

FIGURE 4-39. *Covariance and semivariogram functions for the original image and with added spatially-uncorrelated noise. Note that the vertical scale is different for each category (compare to Fig. 4-23).*

GIFOV has changed the shape of the covariance for small lags; it now is not as sharp and approaches zero slope at zero lag. This implies that composite data models may be a better fit to image data than any one of the three discussed earlier. The semivariograms exhibit a decrease in the sills and an increase in range, corresponding to the increase in width of the covariance functions (Fig. 4-41).

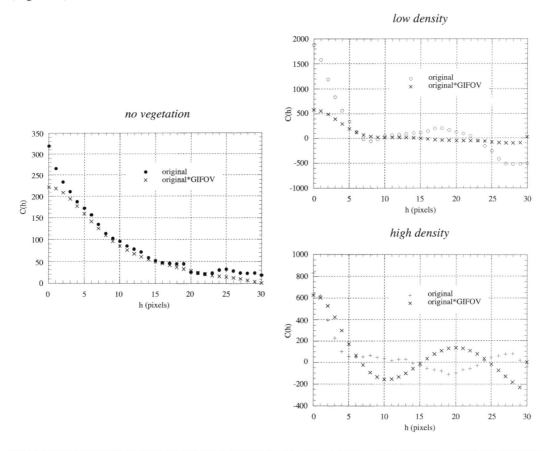

FIGURE 4-40. *Covariance functions for the original image with a simulated GIFOV of 5-by-5 pixels. Note that the vertical scale is different for each category.*

An experimental investigation of semivariogram dependence on *GIFOV*, in which an airborne sensor was flown at two different altitudes over the same test site, is reported in Atkinson (1993). The semivariogram data presented there exhibit the same behavior as in our simulation, even to the point of suggesting a composite model for the semivariogram of the lower resolution imagery.

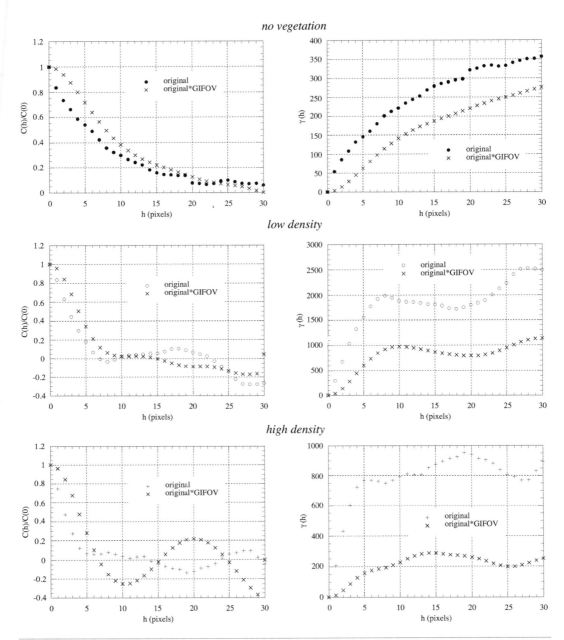

FIGURE 4-41.   *Normalized covariance and semivariogram functions for the original image with a simulated GIFOV of 5-by-5 pixels. Note that the vertical scale is different for each category (compare to Fig. 4-23).*

4.7.3 Sensor Characteristics and Spectral Scattergrams

Having seen how the sensor affects spatial statistics, we now turn to its effect on spectral statistics. In this case, we create a two-band, three-class scene with a computer paint program. The scene may be thought of as showing asphalt roads traversing a landscape of soil and vegetation in the form of grass, shrubs, and trees. The radiance values in the simulated scene approximate the relative values that might be recorded in red and near IR spectral bands (top row, Fig. 4-42).

Now, we want to add some spatial "texture" to each class. First, spatially- and spectrally-uncorrelated random variation is added, with a standard deviation of five in each band for the asphalt and soil classes, and ten for the vegetation class (middle row, Fig. 4-42). Then, spatial correlation over a distance of four scene points is introduced within each class (bottom row, Fig. 4-42).[11] Note that the added spatial correlation, while clearly changing the visual appearance in image space, is not particularly evident in the spectral domain.

Finally, spatial averaging by the sensor (Chapter 3) is simulated by convolution with a Gaussian system spatial response having a diameter of five scene elements. The effect on the data in the image and spectral domains, shown in Fig. 4-43, illustrates two significant points:

- The sensor's spatial response reduces the texture variance in the spatially-uncorrelated case by averaging neighboring scene points which are independent of each other. In the spatially-correlated case, the reduction of within-class variance is much less because the scene data are already correlated over several points before the sensor integration occurs.

- For both types of scene texture, the sensor's spatial response *mixes* the spectral signatures wherever two classes are spatially adjacent. Thus, the scattergram, which originally consisted of only three points, begins to fill in with mixed signatures. We will discuss this phenomenon in terms of image classification in Chapter 9.

In this simulation, we have not included the 3-D nature of the scene, specifically shading and shadowing. If topographic variation is also simulated, the net scattergrams exhibit characteristics of both Fig. 4-37 and Fig. 4-43. The two synthetic band images of Fig. 4-36 are subjected to the same 5×5 *GIFOV* simulation and the resulting scattergram between the two bands shows the elongation of the data clusters toward the origin and mixing of the two classes, soil and vegetation (Fig. 4-44). Even without topography, the final images, particularly the one with spatially-correlated scene texture, are visually quite believable. That is not to say, however, that they necessarily represent accurate models of the real world! More detailed and scene-specific modeling is presented in the literature (Woodcock *et al.*, 1988a; Woodcock *et al.*, 1988b; Jupp *et al.*, 1989a; Jupp *et al.*, 1989b). A detailed analysis of information about semivegetated landscapes contained in the red-NIR reflectance scattergram is given in Jasinski and Eagleson (1989 and Jasinski and Eagleson (1990).

---

11. The correlation is achieved by convolving the added noise with an exponential function, $e^{-r/2}$. This is not quite the same as the exponential autocovariance model discussed earlier, but is satisfactory for the present demonstration.

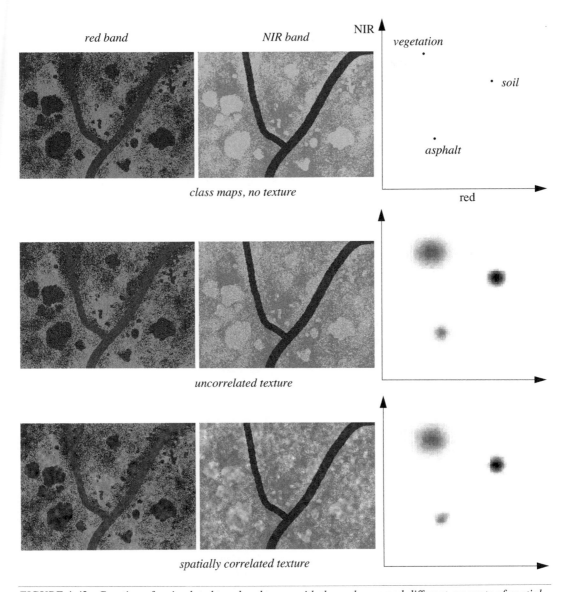

FIGURE 4-42. *Creation of a simulated two-band scene with three classes and different amounts of spatial "texture," and the associated scattergrams. Spatial correlation was introduced (bottom row) by convolution with a LPF, separately for each class. The results were then combined so there was no blurring introduced across class boundaries. The class variances were then renormalized to five and ten to correspond to the uncorrelated texture case. In these and the later scattergram figures the contrast has been greatly increased to make relatively less-populated scattergram bins visible.*

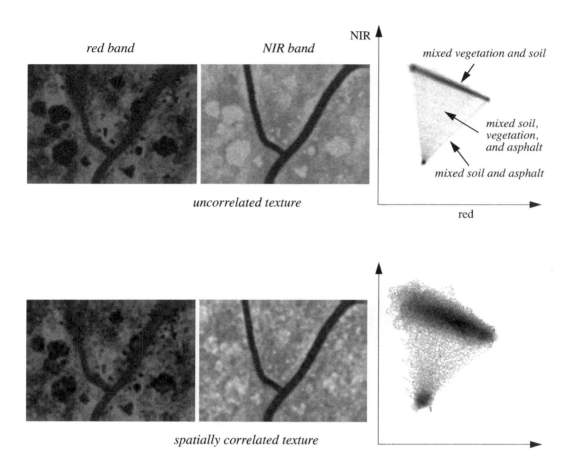

FIGURE 4-43. *The simulated images resulting from spatial averaging of the two textured-scene cases in Fig. 4-42 and the associated scattergrams. The spatial averaging used a 5 × 5 pixel Gaussian weighting function to simulate a sensor PSF. The sensor has reduced the spectral variance of each class for the spatially-uncorrelated scene texture case. The spectral variance is much less reduced for spatially-correlated scene texture. In both cases, new spectral vectors have been created by the mixing of the three classes within the spatial response function of the sensor. The new vectors are interior to the triangle defined by the original three signatures (Adams et al., 1993); these mixed pixels are significant in thematic classification, as discussed in Chapter 9.*

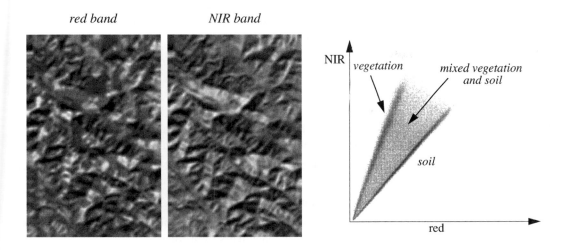

FIGURE 4-44. Simulation of a 5 × 5 pixel GIFOV on the synthetic images of Fig. 4-36 and the resulting scattergram, which shows both topographic-induced correlation and spectral mixing.

4.8   Summary

Some of the common models used in image data analysis have been described and illustrated with examples. The connection between radiation and sensor models and data models was also discussed and explored by simulation. The important points of this chapter are:

- The spectral statistics of remote-sensing image data are influenced by the topography in the scene. The topographic effect tends to correlate the data between spectral bands along a straight line through the origin of the reflectance scattergram.

- The spectral statistics of remote-sensing image data are also influenced by the sensor's spectral passband locations and widths, and noise characteristics. Random, spatially-uncorrelated sensor noise increases the within-class variance of all surface materials equally.

- The spatial *and* spectral statistics of remote-sensing image data are influenced by the sensor's spatial response function, which increases the spatial correlation length, reduces within-class variance, and creates mixed spectral vectors.

With the background provided by Chapters 2, 3, and 4, we are now ready to look at image processing techniques and algorithms in the physical context of remote sensing. Chapters 5 and 6 present tools for spectral and spatial transforms that are useful in the correction and calibration of images for atmospheric and sensor effects, described in Chapter 7, and in the fusion of images, described in Chapter 8. Finally, many of the data models discussed in this chapter will prove useful in our discussion of thematic classification in Chapter 9.

4.9   *Exercises*

Ex 4-1. If a multispectral image dataset is uncorrelated among spectral bands, are the data necessarily statistically independent? What if the data are normally distributed in each band?

Ex 4-2. Show mathematically that the equal probability contours of a 2-D Gaussian distribution are ellipses.

Ex 4-3. Verify the modulation and phase values for the function in Fig. 4-17.

Ex 4-4. For 100 data points in an image transect, how many pairs of points can be used to calculate the semivariogram for h equal to 2? How many pairs for h equal to 50?

Ex 4-5. *SNR*, as defined by Eq. (4-32), Eq. (4-34) or Eq. (4-36), is a useful metric for evaluating image quality in the presence of noise that is distributed more or less uniformly across an image. These definitions are not well-suited to isolated noise, such as dropped pixels or lines, and detector striping, however. Suggest *SNR* measures that would be meaningful in these three cases and describe how they could be measured.

Ex 4-6. Cover the TM image in Fig. 4-28 with a sheet of paper and examine the power spectra to identify the image blocks that have

   · few or no linear features

   · a relatively small correlation length

Confirm your selections for both cases by comparison to the image.

Ex 4-7. What value of horizontal spatial lag, h, provides the best discrimination among the three types of images, based on the contrast and entropy *CM* features of Fig. 4-32?

Ex 4-8. Sketch the scattergrams of Sect. 4.7.1 if the incident cosine factor is in the range [0.2,0.5].

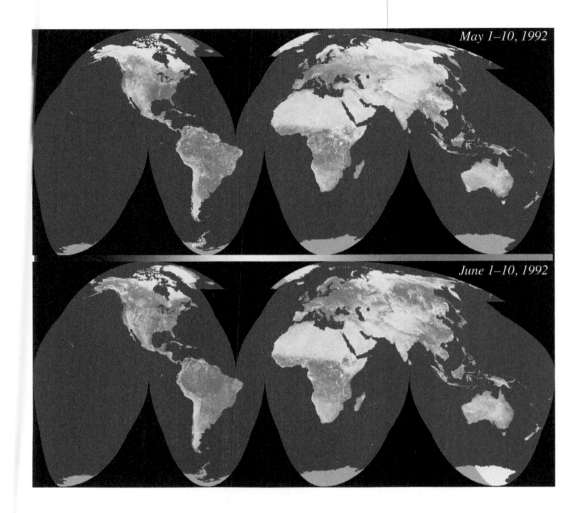

Plate 1-1

The potential for remote sensing in global applications is illustrated in these two images. They show the Normalized Difference Vegetation Index (NDVI), calculated from AVHRR satellite data and spatially averaged to about 16km. To increase the potential for cloud-free coverage at every pixel, each image is a composite of many individual AVHRR images acquired throughout a 10-day period. The compositing process uses a set of rules to choose the "best" cloud-free image pixel at each location for the global mosaic. The result is that neighboring pixels in the final composite image may actually be from different dates, but they are never more than 10 days apart. The degree of "greeness" indicates the amount of green vegetation at the time. Note the increase in vegetation between the two periods corresponding to spring growth in the Northern Hemisphere. Similar global vegetation index products are also produced from MODIS data. The geometric projection used here is known as the Interrupted Goode Homolosine equal-area projection (Goode, 1925; Steinwand, 1994). (NDVI data courtesy U.S. Geological Survey; image coloring by the author.)

Plate 1-2

This TM Color InfraRed (CIR) composite from June 9, 1984, shows the All-American Canal used to irrigate agriculture in the desert west of Yuma, Arizona. The U.S.-Mexico border is between the canal and the fields; crops such as lettuce and melons are typically grown in this area. The degree of "redness" indicates the growth stage of the crop in a particular field; grey fields are fallow and black fields are flooded with water. The Algodones Dunes at the top show periodic sand dune patterns. An image acquired at a different date would likely show changes in these patterns, as well as in the agricultural area.

AVIRIS 51:27:17

at-sensor spectral radiance

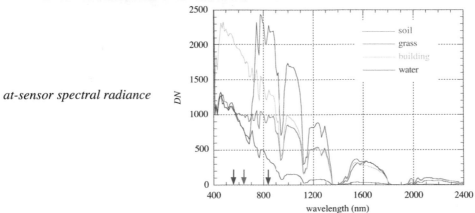

Plate 1-3

This July 23, 1990 AVIRIS image of Palo Alto, California, illustrates hyperspectral data. The CIR composite is made from bands 51, 27 and 17 (at the wavelengths marked by the arrows in the graph) displayed in red, green and blue, respectively, and represents only about 1.5% of the total data in the image. Five pixels (center of yellow squares) were selected for extraction of spectral profiles. They represent dry grassland on the Stanford University campus, the roof of a large commercial building, the Palo Alto golf course, and a San Francisco Bay salt evaporator pond. The AVIRIS data are in at-sensor radiance units and are not corrected for atmospheric effects. Atmospheric scattering increasing towards the blue spectral region is evident in all the spectra, as are the major H_2O and CO_2 absorption bands. (Image courtesy NASA/JPL.)

Hyperion 30:21:10

43:30:21

204:150:93

Plate 1-4

Three different color composites made from Hyperion hyperspectral data are shown. Hyperion was on the NASA testbed satellite Earth Observing-1 (EO-1), which also carried the Advanced Land Imager (ALI). The three bands used for each composite are assigned to red:green:blue respectively. The image on the left is a "natural color" composite that approximates a natural visual image. The one in the middle is a color IR composite, with vegetation appearing as red, and the image on the right uses the SWIR bands to display non-visual information. The image shows the east end of San Francisco Bay on July 31, 2002, and includes the Dumbarton Bridge at the bottom and the cities of Hayward, Union City and Fremont in the center. The large green features in the true color and color IR composites are salt evaporation ponds, many of which are in a plan of restoration to natural wetlands (http://www.southbayrestoration.org).

Plate 1-5

This image of the Burchardkai Container Terminal at the Port of Hamburg, Germany, was collected by QuickBird on May 10, 2002. The image was produced by fusing the 0.6m panchromatic band and three 2.4m multispectral bands into a "natural color" composite (see Chapter 8). The many colored shipping containers are clearly distinguishable from each other in this "pan-sharpened" product. (Image © 2002 DigitalGlobe.)

Plate 1-6

This example of high-resolution airborne digital imagery shows an overlapping pair of natural color images (contrast enhanced) collected by Photo Science Inc. over the University of Kentucky on January 12, 2006, with an Applanix DSS 322 digital camera. Stereo analysis of such overlapping images can produce elevation and object height information (Chapter 8). Collection of imagery in winter during "leaf-off" conditions is especially useful for mapping purposes in temperate climate regions. These images show part of the football stadium, tall light standards, and a parking lot. Note the change in perspective and in the location of moving vehicles in the two frames. The GSI is 0.18m, achieved with 9 μm detectors and a 60 mm focal length lens at an altitude above ground level of about 1200m. Each image above is only about 5.8% of a full camera frame, which is 4092 lines-by-5436 pixels. A frame can be recorded every 2.5 seconds, and the camera can produce either natural color or CIR images by a choice of filters. (Imagery and camera data courtesy Applanix Inc.)

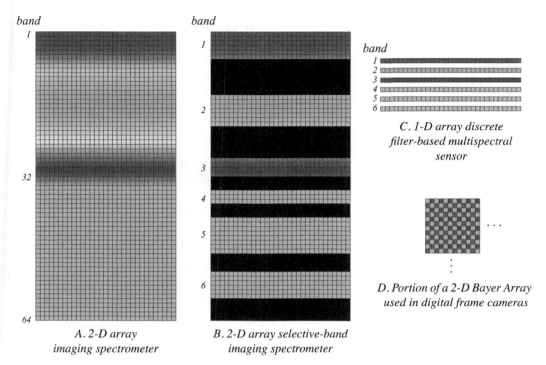

band 1

band 32

64

*A. 2-D array
imaging spectrometer*

band 1
band 2
band 3
band 4
band 5
band 6

*B. 2-D array selective-band
imaging spectrometer*

band 1
2
3
4
5
6

*C. 1-D array discrete
filter-based multispectral
sensor*

...

...

*D. Portion of a 2-D Bayer Array
used in digital frame cameras*

*HYDICE flocal plane array imaging spectrometer
assembly in a cold dewar. (photo courtesy of Bill
Rappoport, Goodrich Electro-Optical Systems.)*

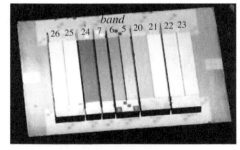

band
26 25 24 7 6 5 20 21 22 23

*MODIS detector bezel and filters for the
SWIR/MWIR focal plane. The spectral bands
are indicated. (Raytheon photograph courtesy
of Ken J. Ando.)*

Plate 1-7

*Four generic focal plane array (FPA) designs for multispectral and hyperspectral imaging systems (top):
A. An FPA of the type used in Hyperion. B. An FPA similar to that used in MERIS. Both of these designs have
continuous spectral dispersion across the detector array. C. An FPA similar to the arrays in ALI. D. A Bayer
Array used in some digital frame cameras, such as the DSS 322 that produced the images in Plate 1-6; the
pixels in each color must be interpolated 2 × to produce registered bands. In all cases, the grey detectors are
in the non-visible NIR and SWIR spectral regions. Photographs of the actual HYDICE 2-D array and MODIS
SWIR/MWIR FPA are shown below. The colors in those FPAs do not represent the spectral sensitivity of the
detectors, but result from anti-reflection coatings on the detectors in both cases.*

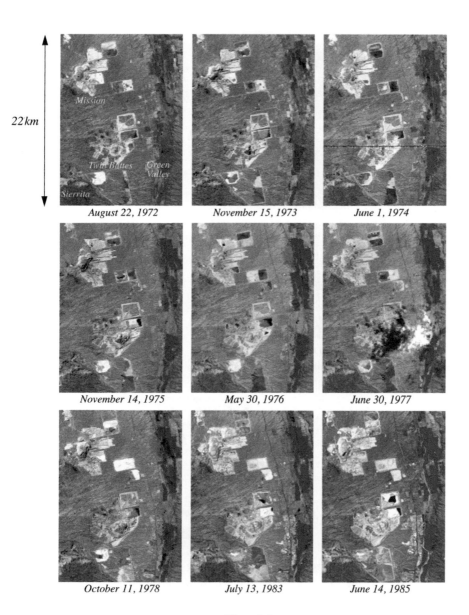

22 km

August 22, 1972   November 15, 1973   June 1, 1974

November 14, 1975   May 30, 1976   June 30, 1977

October 11, 1978   July 13, 1983   June 14, 1985

Plate 1-8

Remote sensing imagery can provide a long-term record of man's impact on the environment. This series of Landsat MSS CIR images of the large copper mining complex south of Tucson, Arizona, also includes irrigated pecan orchards along the right side, and the growing retirement community of Green Valley between the orchards and the mines. There are three large open pit mines in this area: Mission, Twin Buttes and Sierrita. The large polygonal features are tailings waste ponds from the open pit mining, and the dark blue–black areas are moist material and standing surface water. Note the noisy data in the 1974 image and the cloud and its shadow in the 1977 image. The expansion of mining activity and Green Valley development during this 13 year period are evident.

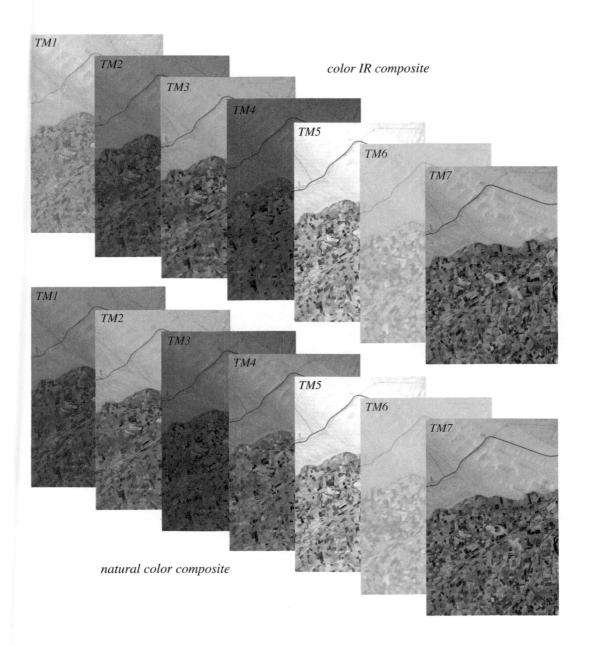

TM1 TM2 TM3 TM4 TM5 TM6 TM7

color IR composite

TM1 TM2 TM3 TM4 TM5 TM6 TM7

natural color composite

Plate 1-9

Two common color composite combinations for Landsat TM and ETM+ imagery are shown.

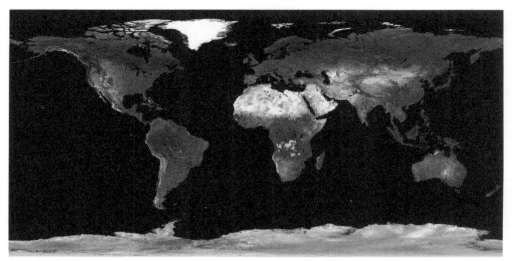

Fire Map May 11–20, 2005

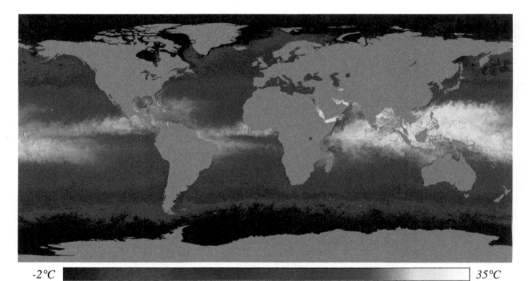

-2°C   35°C

Sea Surface Temperature June 2–9, 2001

Plate 1-10

Two regular MODIS products are global fire and sea surface temperature maps. In the fire map, red indicates at least one fire per pixel and yellow indicates a large number of detected fires per pixel during the 10-day compositing period. The details of the fire products and algorithms are given in Justice et al. (2002) and Giglio et al. (2003). The MODIS sensor measures sea surface temperature accurate to within about 0.25°C (as determined by comparison to sample data from ships and buoys), which is better than twice the accuracy of previous satellites. Daily global measurements of sea surface temperature (SST) accurate to within half a degree have been a goal of oceanographers for decades. (Images and descriptions courtesy NASA.)

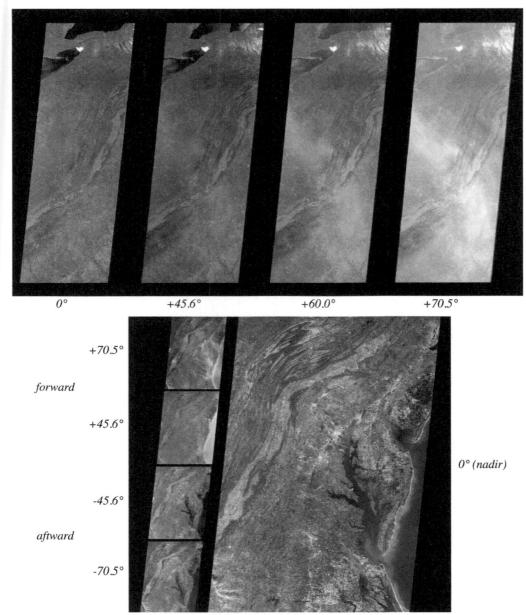

0°   +45.6°   +60.0°   +70.5°

+70.5°

forward

+45.6°

                                                              0° (nadir)

-45.6°

aftward

-70.5°

Plate 2-1

Natural color images from the Multi-Angle Imaging Spectro-Radiometer (MISR) instrument on NASA's Terra satellite. At the top is a view of the eastern United States on June 14, 2000, from Lake Ontario to northern Georgia, and spanning the Appalachian Mountains. As the slant angle increases, the line-of-sight through the atmosphere grows longer, and haze becomes progressively more apparent. The lower set of images are of the east coast of the U.S. centered on Chesapeake Bay on June 28, 2000. The images are about 400 km wide, and the GSI is 1.1 km. (Images and description courtesy NASA/GSFC/JPL, MISR Science Team.)

Plate 2-2

These digital photographs taken from a commercial airliner illustrate atmospheric effects on remote sensing imagery. The top image shows how atmospheric scattering increases as the view angle from nadir increases (compare with Plate 2-1). The area is in the American Midwest in the summer, and the atmosphere probably has high water vapor content. The lower left image shows the same effect, but from a much higher altitude (about 10km) which reveals the transition from the lower to the upper atmosphere. Both of these images are contrast enhanced. The lower right image is not contrast enhanced and shows the appearance of irrigated agriculture and an urban area in southern California as viewed from aircraft altitude. The "true" visual colors are muted and shifted towards blue by path radiance scattering.

daytime
(MODIS 1:4:3)

Terra

Aqua

nightime
(MODIS 31:29:20)

Terra

Aqua

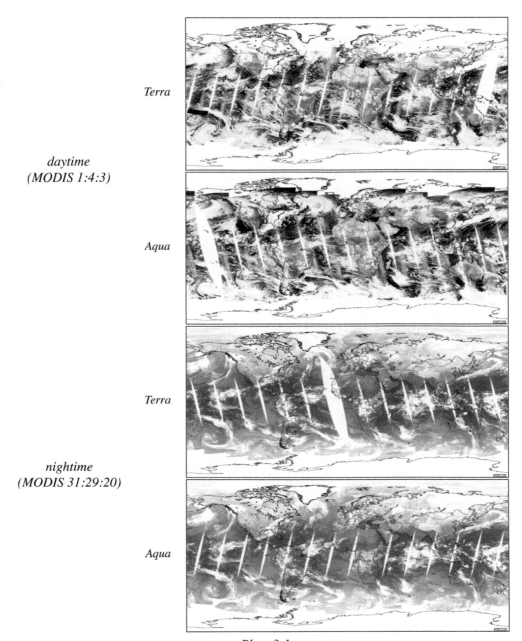

Plate 3-1

Terra and Aqua MODIS orbital coverage on November 16, 2002. Half of each orbit is in the sun-illuminated side of the earth (daytime) and the other half is in the non-illuminated side (nightime). Terra is in a descending orbit (southbound across the equator in daytime) and Aqua is in an ascending orbit (northbound across the equator in daytime). During daytime and at the poles, all bands collect data. However, during nightime the reflective bands (1–19) are turned off, and only the emissive bands (20–25, 27–36) continue collecting data. (Images and description courtesy NASA MODIS Data Support Team.)

TM 4:3:2

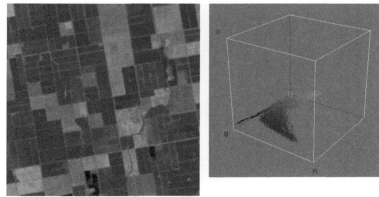

Hyperion 43:30:21

Hyperion 204:150:93

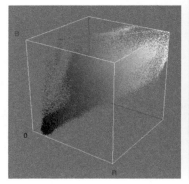

Plate 4-1

Three examples of color image composites and their 3-D color scatterplots. Note the clear separation of soil, water and vegetation spectral vectors in the TM CIR composite and the Hyperion CIR composite. The Hyperion SWIR composite has some saturated pixels (top and right faces of 3-D scatterplot) resulting from contrast enhancement for display. The software used to produce these color scatterplots allows interactive rotation of the color cube and scatterplot (Barthel, 2005).

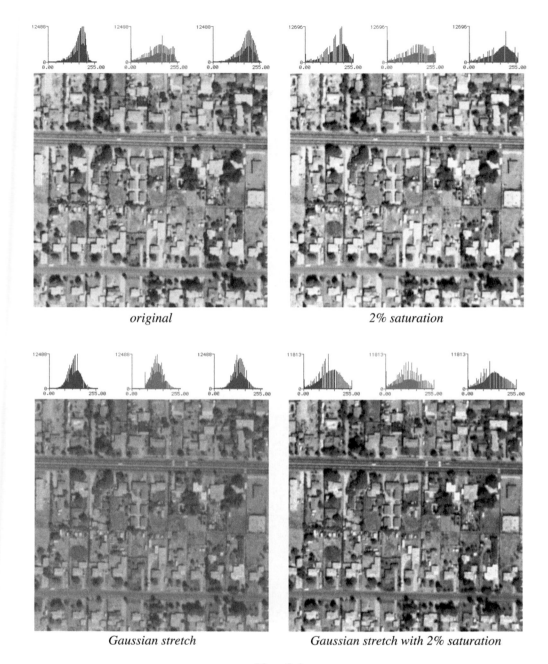

original   *2% saturation*

Gaussian stretch   *Gaussian stretch with 2% saturation*

Plate 5-1

A CIR aerial photograph is processed with a Gaussian stretch to normalize the color balance. The original image has an incorrect red hue (probably resulting from poor camera exposure or film processing), affecting even the color of asphalt roads which is corrected to grey in the normalized images. The linear stretch with 2% saturation does not normalize the color balance.

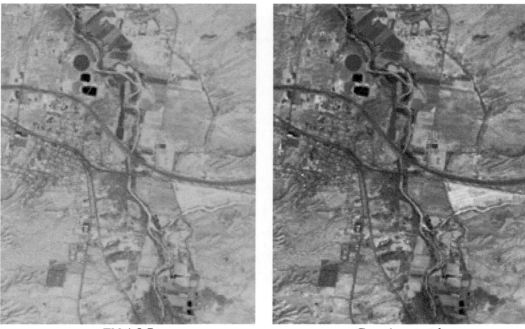

TM 4:5:7   *Gaussian stretch*

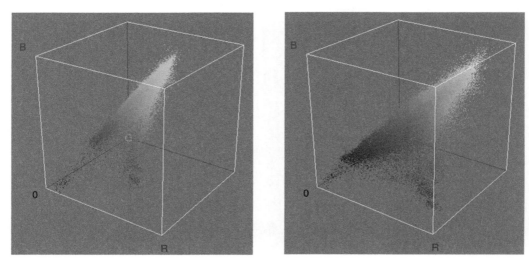

RGB scatterplots

Plate 5-2

A TM image of Benson, Arizona, Interstate 10, and the San Pedro River Valley is shown before and after a Gaussian stretch. Since the three bands in the TM color composite are not at visible wavelengths, there is no preconceived notion of "correct" colors. The scatterplot of the processed image has the soil pixels aligned with the color cube diagonal, and the vegetation pixels are stretched more toward red.

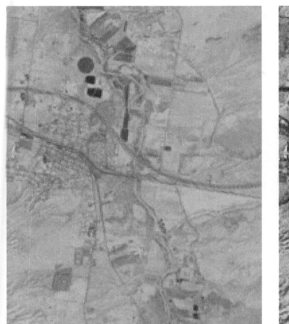

decorrelation stretch TM 4:5:7   *decorrelation stretch with 2% saturation*

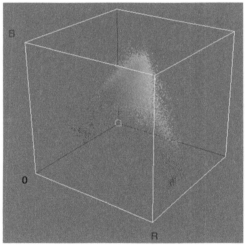
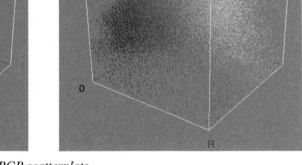

RGB scatterplots

Plate 5-3

A decorrelation stretch makes the data spectrally uncorrelated and greatly increases the color contrast.

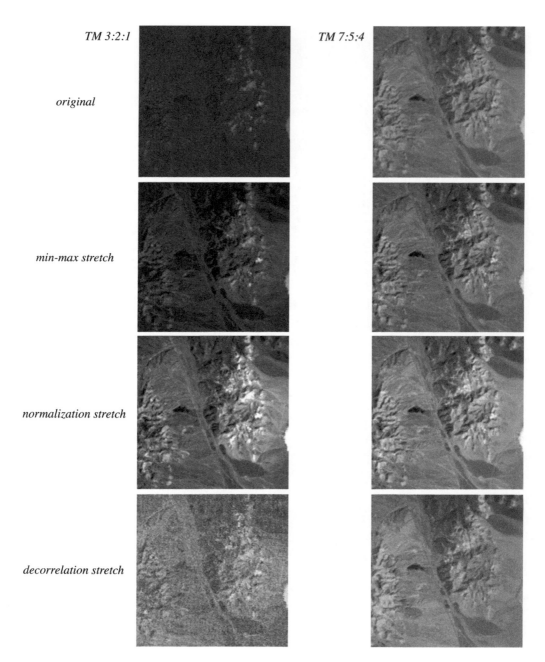

Plate 5-4

Examples of color contrast enhancement of TM imagery of Cuprite, Nevada. This desert area has relatively little spectral content in the visible (left column); the decorrelation stretch manages to enhance the small spectral variations present in the data, but results in a noisy image. In the NIR and SWIR (right column), however, the varied mineralization is clearly evident and easily enhanced.

Cylindrical color space coordinate system

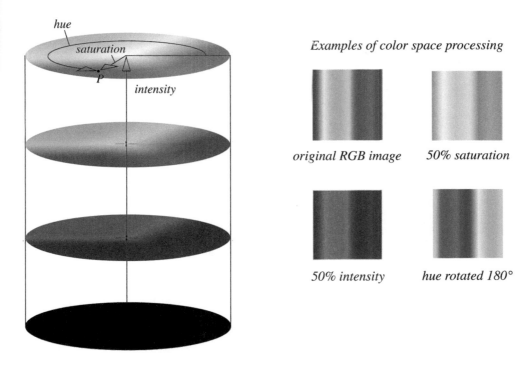

Examples of color space processing

original RGB image   *50% saturation*

50% intensity   *hue rotated 180°*

Hexcone color space transform

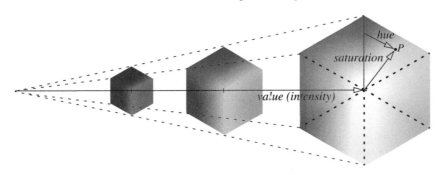

Plate 5-5

Two examples of color spaces, a cylindrical color space and the hexcone color space described in the text. They share the general properties of intensity, hue, and saturation components of a pixel vector at P. In the upper right, the color images resulting from an inverse CST of the individual color space components in Fig. 5-30.

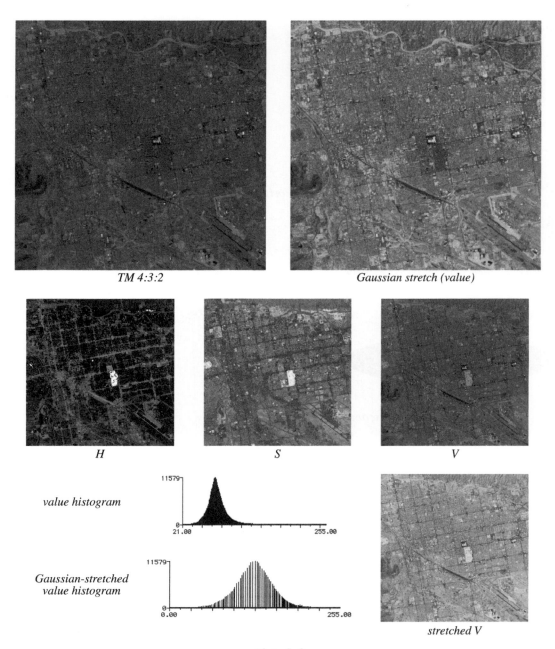

TM 4:3:2

Gaussian stretch (value)

H

S

V

value histogram

Gaussian-stretched value histogram

stretched V

Plate 5-6

Value (intensity) processing of a color image of Tucson, Arizona, using the approach of Fig. 5-29. The hexcone color space components are shown in the middle row. The value component is processed with a CDF reference stretch to match a Gaussian with a mean of 128 and a standard deviation of 32, and the inverse CST is then calculated to produce the color image at top right. The contrast of the image is improved with little or no change in hue or saturation.

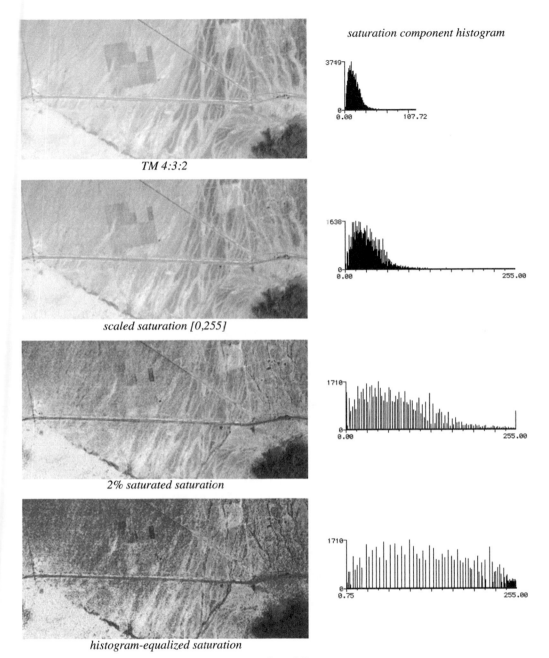

saturation component histogram

TM 4:3:2

scaled saturation [0,255]

2% saturated saturation

histogram-equalized saturation

Plate 5-7

Saturation processing of a color image using the approach of Fig. 5-29. The original TM composite of a desert area has subtle spectral content, which can be amplified by stretching the saturation component. The range for saturation is [0,255] because of the specifics of the hexcone CST alogrithm used here; other algorithms may have a different operating range, e.g. [0,1] or [0,100].

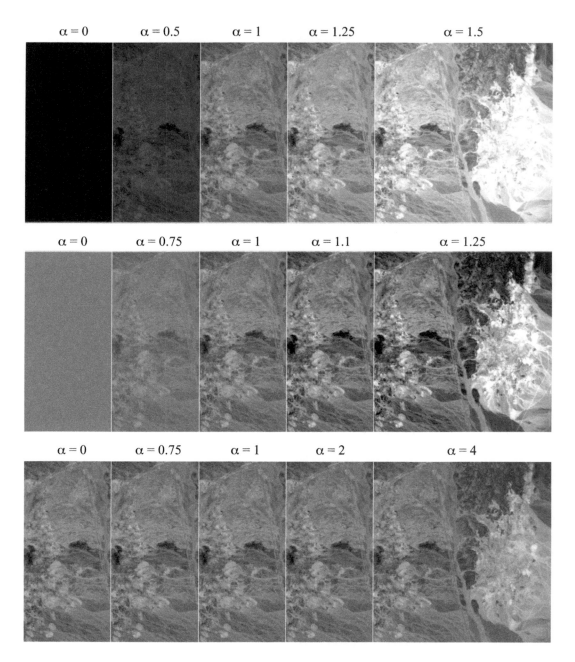

Plate 5-8

The image blending algorithm is used to vary intensity (top row), contrast (middle row) and saturation (bottom row) of an AVIRIS CIR image of Cuprite, Nevada. The case $\alpha = 0$ corresponds to image0, and the case $\alpha = 1$ corresponds to image1.

SPOT-4 3:4:2

corrected

Plate 7-1

The correction of a SPOT-4 multispectral image of a mountainous region of North Canterbury, New Zealand for topographic and reflectance variation (Shepherd and Dymond, 2003). Corrections were applied for the $\cos\theta$ variation in incident irradiance and for reflectance variation as a function of incident and view angles, obtained from a co-registered DEM. Note that the topographic shading is largely removed by the processing. (Images courtesy James Shepherd, Manaaki Whenua Landcare Research, New Zealand, and Taylor & Francis Ltd, http://www.tandf.co.uk/journals.)

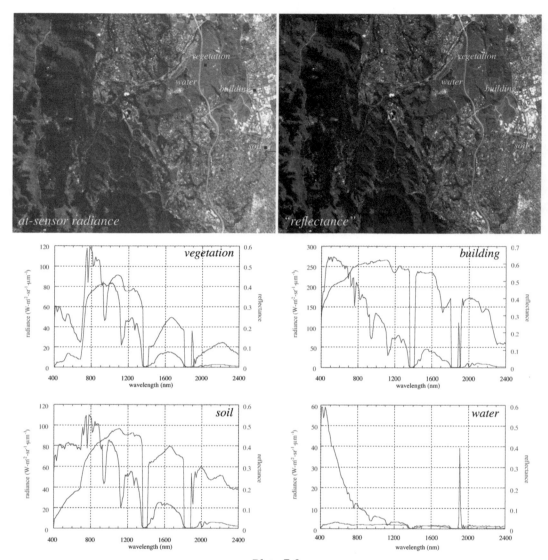

Plate 7-2

Atmospheric correction of an AVIRIS image of Jasper Ridge, California, acquired on April 3, 1997, with an ATREM-like procedure that included a single ground-measured soil reflectance spectrum in the correction process. The at-sensor radiances and "reflectances" are shown as natural color composites of bands 30, 20, and 9. The three bands in each image are displayed with equal minimum and maximum values to preserve their relative color balance. The blue haze due to view path radiance, solar spectral irradiance, and atmospheric transmittance are removed by the correction. There is no topographic correction in this case, however, so the "reflectances" are not fully calibrated at every pixel. The spectral plots of pixels at four sites in the image show at-sensor radiance (blue) and reflectance (red). The "spike" in reflectance near the edge of the water absorption band at 1900 nm is a common artifact of high spectral resolution atmospheric correction algorithms; some correction programs, for example HATCH, attempt to reduce such artifacts by enforcing a smoothness criterion on the retrieved spectrum (Qu et al., 2003). (Imagery and processing description courtesy NASA Jet Propulsion Laboratory.)

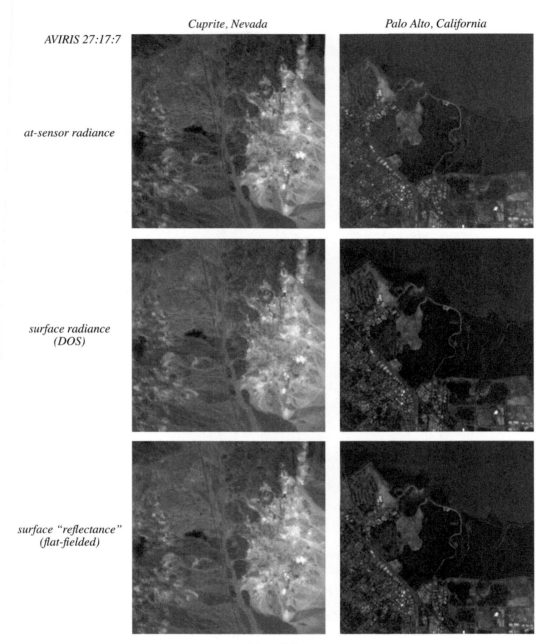

Cuprite, Nevada   Palo Alto, California

AVIRIS 27:17:7

at-sensor radiance

surface radiance
(DOS)

surface "reflectance"
(flat-fielded)

Plate 7-3

The effects of atmospheric path radiance correction by DOS, followed by in-scene flat-field calibration using a bright target are shown for natural color composites of two AVIRIS images, one of an arid desert area (Cuprite) and the other of a coastal area (Palo Alto). The same display LUT is used for all images to preserve their relative colors. The surface "reflectance" images have no correction for topographic shading. These normalization techniques do not require an atmospheric model like that used for the image in Plate 7-2.

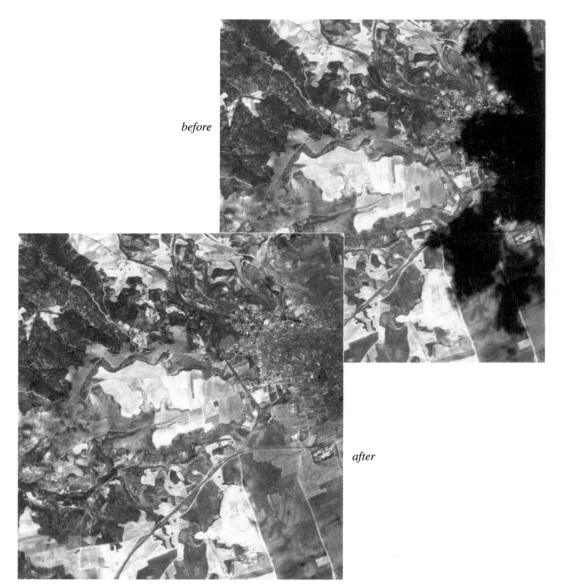

before

after

Plate 7-4

De-shadowing of an airborne HyMap image of Chinchon, Spain, acquired on July 12, 2003. The processing technique uses three bands of the hyperspectral image (at 0.85 μm, 1.6 μm and 2.2 μm) to find cloud shadows (Richter and Muller, 2005). In these NIR and SWIR bands the direct solar radiation is at least 80% of the total irradiance at the ground. A series of processing steps that includes masking of clouds and water, soft spectral classification for shadow versus non-shadow, thresholding to define the core (darkest) shadow areas, and spatial expansion of the core shadow areas is performed. The final correction is a spatially-variant modification of the direct solar component and is performed only within the shadowed areas. (Images courtesy Dr. Rudolf Richter, DLR/DFD, Germany, and Taylor & Francis Ltd, http://www.tandf.co.uk/ journals.)

bilinearly resampled to 10m

CST fusion with pan band

PCT fusion with pan band

Plate 8-1

Feature space fusion examples for an ALI image collected on July 27, 2001, showing the intersection of the Pima Freeway 101 and the Red Mountain Freeway 202 in Mesa, Arizona. The original ALI data are Level 1R, i.e. with radiometric corrections but not geometric corrections. Note how the color of vegetation is not retained in the CST fused CIR image. This is attributed to the fact that the ALI pan band does not include the NIR spectral region and hence is not well correlated with the replaced value component. The hexcone CST with intensity defined as value (see text) was used here; a CST algorithm that uses a different definition for the intensity component might be less or more sensitive to the degree of correlation.

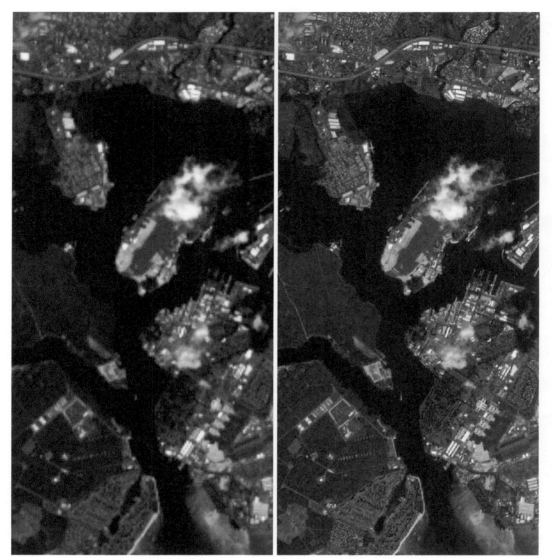

ALI 3:2:1 bilinearly resampled to 10m   *HFM fusion with pan band*

Plate 8-2

Demonstration of HFM fusion for an ALI image of Pearl Harbor, Hawaii, collected on November 23, 2002. The original Level 1R data were bilinearly resampled from the 30m GSI of the multispectral bands to the 10m GSI of the panchromatic band before fusion. (ALI image courtesy George Lemeshewsky, U.S. Geological Survey.)

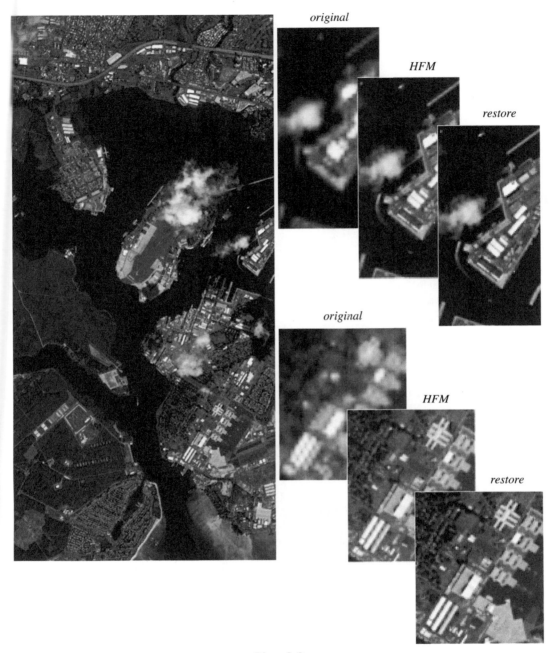

Plate 8-3

The result obtained with the iterative spatial domain restoration/fusion algorithm described in the text. Zoomed comparisons to the original (bilinearly resampled to 10m) and the HFM result (Plate 8-2) are also shown. The iterative algorithm appears to produce some additional edge sharpening compared to HFM. (Restored image courtesy George Lemeshewsky, U. S. Geological Survey.)

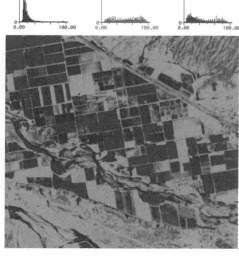

ANN output map

red: crop
green: light soil
blue: dark soil

TM 4:3:2

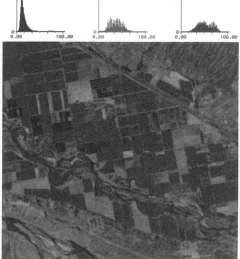

mixing fractions map

red: crop
green: light soil
blue: dark soil

Plate 9-1

Training sites for three classes are outlined in the Marana, Arizona, TM image. These data are used in the two band (TM3 and TM4) supervised classification examples in the text. On the right are the soft classification maps produced by the ANN classifier and by linear unmixing. The spatial variations in both of the soft maps mimic the spatial variations in the spectral band composite. Note how the light and dark soil ANN output values are more evenly distributed in the histogram than are their fractions, which explains the higher contrast of the ANN map. The two soft maps are strongly correlated, as explained in the text.

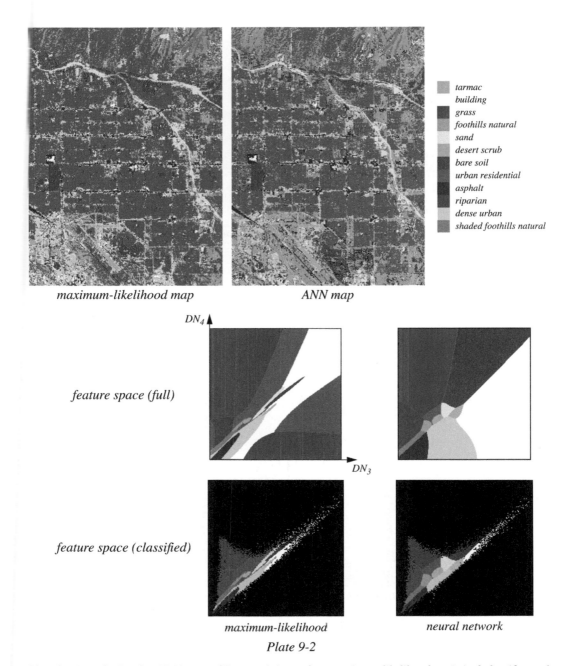

maximum-likelihood map   ANN map

tarmac
building
grass
foothills natural
sand
desert scrub
bare soil
urban residential
asphalt
riparian
dense urban
shaded foothills natural

feature space (full)

feature space (classified)

maximum-likelihood   neural network

Plate 9-2

Classification of a Landsat TM image of Tucson, Arizona, by a maximum-likelihood statistical classifier and a 3-layer ANN with 12 urban land-use classes. The band 4 (NIR) versus band 3 (red) feature space diagrams are projections of the 6 band feature space used in the classification (Paola and Schowengerdt, 1995b). The middle row shows the full extent of the feature space decision boundaries and the bottom row shows only the regions of feature space that are populated by the image pixels. (class maps and full feature space diagrams © 1995 IEEE.)

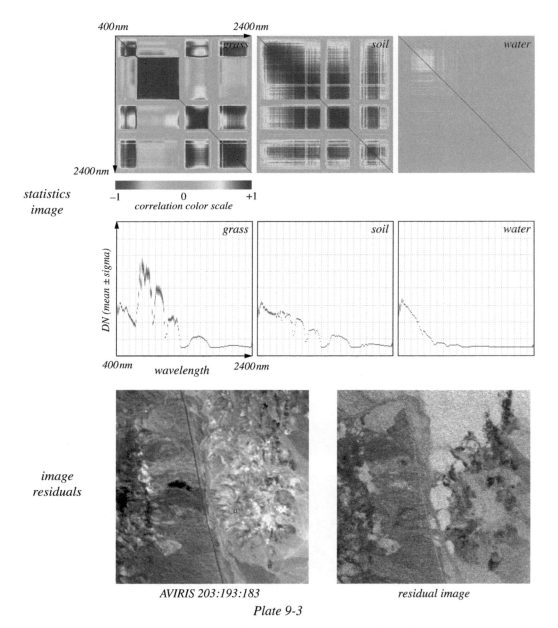

400nm → 2400nm

grass *soil* *water*

statistics image

−1   0   +1
correlation color scale

$DN\ (mean \pm sigma)$

grass *soil* *water*

400nm *wavelength* 2400nm

image residuals

AVIRIS 203:193:183   *residual image*

Plate 9-3

Techniques for hyperspectral image analysis and feature extraction. The statistics image created from training data in the AVIRIS image of Plate 1-3 provides a visual tool for class differentiation based on second-order statistics. The correlation matrices for three training classes are pseudo-colored in the top row and the class mean and standard deviation spectral data are plotted in the middle row in red and orange−yellow, respectively. Note the regions of negative and positive correlation in the grass statistics image. In the bottom row, image residual feature extraction from three hyperspectral SWIR bands provides color discrimination of different minerals, similar to that achieved in a classification map. In this AVIRIS residual image of Cuprite, Nevada, red corresponds to alunite, dark blue corresponds to kaolinite and yellow-orange corresponds to the rare mineral, buddingtonite (the two small sites near the center).

CHAPTER 5

Spectral Transforms

5.1 Introduction

Image processing and classification algorithms may be categorized according to the *space* in which they operate. The *image space* is $DN(x,y)$, where the spatial dependence is explicit. In Chapter 4, we presented the concept of a multidimensional *spectral space*, defined by the multispectral vector *DN*, where spatial dependence is not explicit. Spectral transformations, discussed in this chapter, alter the spectral space; and spatial transformations, discussed in the next chapter, alter the image space. Many of these transformed spaces are useful for thematic classification (Chapter 9), and are collectively called *feature spaces* in that context. In this chapter, we describe various feature spaces that can be derived from the spectral space. These derived spaces do not add new information to the image, but rather redistribute the original information into a more useful form. We will discuss various linear and nonlinear transformations of the *DN* vector, motivated by the possibility of finding a feature space that may have advantages over the original spectral space.

5.2   Feature Space

We're interested in tranformations f of the image spectral vector $\textbf{DN}$ to a feature space vector $\textbf{DN}'$,

$$\textbf{DN}' = f(\textbf{DN}) \tag{5-1}$$

If f is a *linear* transform, then,

$$\textbf{DN}' = \textbf{W} \cdot \textbf{DN} + \textbf{B}$$

$$= \begin{bmatrix} w_{11} & \cdots & w_{1K} \\ \cdots & \cdots & \cdots \\ w_{K1} & \cdots & w_{KK} \end{bmatrix} \begin{bmatrix} DN_1 \\ \cdots \\ DN_K \end{bmatrix} + \begin{bmatrix} B_1 \\ \cdots \\ B_K \end{bmatrix} \tag{5-2}$$

where $\textbf{W}$ is a weight matrix applied to the original spectral bands, and B is a bias vector. This transformation corresponds to a rotation, scaling, and offset of the $\textbf{DN}$ spectral space (Fig. 5-1).[1] Many important spectral transforms are linear, as we'll see later. If W is the identity matrix and B is zero, then we simply have the original spectral space,

$$\textbf{DN}' = \begin{bmatrix} 1 & \cdots & 0 \\ \cdots & \cdots & \cdots \\ 0 & \cdots & 1 \end{bmatrix} \begin{bmatrix} DN_1 \\ \cdots \\ DN_K \end{bmatrix} + \begin{bmatrix} 0 \\ \cdots \\ 0 \end{bmatrix} = \textbf{DN} \tag{5-3}$$

The linear transform in Eq. (5-2) is called an *affine* transform and is also useful in geometric processing of images (Chapter 7). In that application, K is two, corresponding to spatial coordinates (x, y) and (x', y'), the matrix W describes operations such as rotation, scale change, and shear, and the vector $\textbf{B}$ describes a spatial shift.

If f cannot be written in the form of Eq. (5-2), then it is a nonlinear transform. For example,

$$\textbf{DN}' = \textbf{DN}^2 \tag{5-4}$$

is a nonlinear transform of *DN*. Some particularly useful linear and nonlinear transformations can be interpreted in terms of the physical aspects of remote sensing imaging, which we will do in the following sections.

---

1. In some literature, for example (Fukunaga, 1990), the linear transformation is defined as $\textbf{DN}' = \textbf{U}^T \cdot \textbf{DN}$, where $\textbf{U} = \textbf{W}^T$. This is completely equivalent to our definition, since $(\textbf{W}^T)^T = \textbf{W}$.

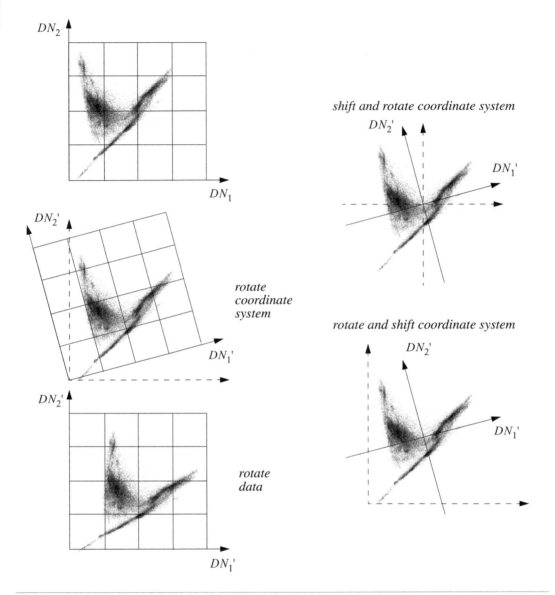

FIGURE 5-1. Some properties of a linear feature space transform in 2-D. A transform like the PCT described later in this chapter can be viewed as either a rotation of the coordinate system or a (opposite) rotation of the data. We'll use the former view in many of the diagrams later in this chapter. Also, a shift of the origin before rotation is equivalent to a shift after rotation. In all of these cases, the angle between the data and the new coordinate system is the same; the only difference may be a shift, which is inconsequential for the PCT (Richards and Jia, 1999).

5.3   *Multispectral Ratios*

One of the earliest feature spaces used for remote sensing images was calculated by the *ratio* of *DN*s in spectral band m to those in band n, pixel-by-pixel,

$$R_{mn}(x, y) = \frac{DN_m(x, y)}{DN_n(x, y)}. \tag{5-5}$$

This is a nonlinear transformation of the **DN** vector.

Based on the analyses of Chapters 2 and 3 (in particular, Eq. (2-10) and Eq. (3-30)), the *DN* in band b is approximately a linear function of earth surface reflectance,[2]

$$DN_b(x, y) \approx a_b \rho_b(x, y) \cos[\theta(x, y)] + b_b. \tag{5-6}$$

The ratio of Eq. (5-5) is then,

$$R_{mn}(x, y) \approx \frac{a_m \rho_m(x, y) \cos[\theta(x, y)] + b_m}{a_n \rho_n(x, y) \cos[\theta(x, y)] + b_n} \tag{5-7}$$

which is not simply related to the reflectances in the two bands. If, however, the *DN* bias b_b can be estimated and subtracted from the data in each band, for example by sensor offset and atmospheric haze correction (Chapter 7), then a simplification results,

$$\textit{bias-corrected:}\quad R_{mn}(x, y) \approx \frac{a_m \rho_m(x, y)}{a_n \rho_n(x, y)} = k_{mn} \frac{\rho_m(x, y)}{\rho_n(x, y)}. \tag{5-8}$$

The incident topographic irradiance factor does not appear in Eq. (5-8); in fact, this is one of the primary benefits of band ratios, namely that they can suppress topographic shading if a bias correction is done first. The ratio of Eq. (5-8) is directly proportional to the ratio of reflectances in the two bands, since k_{mn} is a constant for any two bands, and is more representative of surface properties than Eq. (5-5). The dynamic range of the ratio image is normally much less than that of the original image because the radiance extremes caused by topography have been removed. Thus, the *reflectance contrast* between surface materials can be enhanced by color composites of different band ratios (Chavez *et al.*, 1982).

Alternatively, the coefficients a_b and b_b can be fully calibrated, as described in Chapter 7, for solar irradiance, atmospheric transmittance and scatter, and sensor gain and offset. We then obtain the ratio,

$$\textit{fully-calibrated:}\quad R_{mn}(x, y) = \frac{[DN_m(x, y) - b_m]/a_m}{[DN_n(x, y) - b_n]/a_n} = \frac{\rho_m(x, y)}{\rho_n(x, y)}. \tag{5-9}$$

---

2. Only the solar reflective radiation component is considered here.

Correction for spatial irradiance variation due to topography is not necessary because the cosine term cancels. The various vegetation indexes described later and used to estimate biomass and vegetative land cover are defined in terms of surface reflectances, so remote sensing data must be calibrated and corrected for the atmosphere first.

Multispectral ratios of NIR to visible bands can enhance radiance differences between soils and vegetation. Soil and geology will exhibit similar ratio values near one, while vegetation will show a relatively larger ratio of two or more. A TM image of an agricultural area provides an illustration of this mapping in Fig. 5-2. A ratio value of about one corresponds to the concentration in the data arising from the large amount of bare soil in the image. In fact, a *soil line* has been defined in the NIR-red spectral space (Richardson and Wiegand, 1977). Spectral vectors along this line represent the continuum of dark to light soils in this image. As the ratio increases, the isolines rotate counterclockwise toward the vegetation region of the scattergram. Ratio values around three in this image represent the active crop fields, and values between one and three represent areas of partial vegetative cover, such as emergent crops and the desert area in the upper right of the image.

The *modulation ratio* is a useful variant of the simple ratio,

$$M_{mn} = \frac{DN_m - DN_n}{DN_m + DN_n} = \frac{R_{mn} - 1}{R_{mn} + 1}.$$ (5-10)

The relation between the two is shown in Fig. 5-2 and Fig. 5-3. Any gain factor that is common to both bands is removed by the modulation ratio, as it is by the simple ratio.

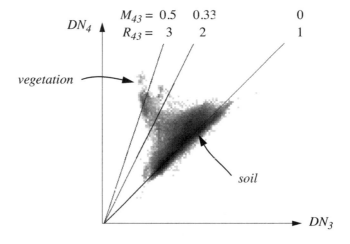

FIGURE 5-2. *The DN scattergram for band 4 versus band 3 of a TM image of an agricultural scene near Marana, Arizona (Plate 9-1). Some ratio isolines are superimposed on the scattergram; values of the modulation and simple ratios along each isoline are indicated at the top. These data are uncalibrated DNs, so M and R would be different after calibration to reflectance.*

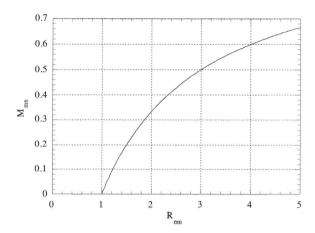

FIGURE 5-3. Plot of the modulation ratio as a function of the simple ratio. If we think of this as a contrast enhancement transform of a simple ratio image, the modulation ratio provides increased contrast of lower ratio values relative to higher values. The physical interpretation is that M is more sensitive to low ratio values than is R.

5.3.1 Vegetation Indexes

As shown in the previous section, the ratio of an NIR band to a red band can indicate vegetation. To define an index that is sensor-independent, the ratio must be specified in terms of a geophysical parameter, such as reflectance. The *Ratio Vegetation Index (RVI)* was one of the earliest such indices used in remote sensing applications,

$$RVI = \frac{\rho_{NIR}}{\rho_{red}}.$$ (5-11)

Similarly, the modulation ratio for NIR and red reflectances is called the *Normalized Difference Vegetation Index (NDVI)*,

$$NDVI = \frac{\rho_{NIR} - \rho_{red}}{\rho_{NIR} + \rho_{red}} = \frac{RVI - 1}{RVI + 1}.$$ (5-12)

It is used extensively to monitor vegetation on continental and global scales using AVHRR data (Townshend and Justice, 1986; Tucker and Sellers, 1986; Justice *et al.*, 1991), but appears to be a poor indicator of vegetation biomass if the ground cover is low, as in arid and semi-arid regions (Huete and Jackson, 1987). It also tends to saturate and lose sensitivity for dense vegetative cover, such as rain forest canopies (Huete *et al.*, 2002).

The *Soil-Adjusted Vegetation Index (SAVI)* is a superior vegetation index for low cover environments (Huete, A. R., 1988),

$$SAVI = \left(\frac{\rho_{NIR} - \rho_{red}}{\rho_{NIR} + \rho_{red} + L}\right)(1 + L)$$ (5-13)

where L is a constant that is empirically determined to minimize the vegetation index sensitivity to soil background reflectance variation. If L is zero, *SAVI* is the same as *NDVI*. For intermediate vegetation cover ranges, L is typically around 0.5. The factor $(1 + L)$ insures the range of *SAVI* is the same as *NDVI*, namely [–1,+1].

SAVI incorporates a *canopy background correction*, i.e., a correction for the underlying soil background. Another external factor not related to vegetation itself that influences vegetation index behavior is the atmosphere. Incorporation of an empirical term for atmospheric correction leads to the *Enhanced Vegetation Index (EVI)*,

$$EVI = G\left(\frac{\rho_{NIR} - \rho_{red}}{L + \rho_{NIR} + C_1\rho_{red} - C_2\rho_{blue}}\right)$$ (5-14)

where the empirical parameters are

$$G = 2.5, L = 1, C_1 = 6, \text{ and } C_2 = 7.5$$ (5-15)

for application to MODIS data (Huete *et al.*, 2002). *EVI* does not exhibit the saturation seen with *NDVI* when the vegetation canopy is dense and appears to be sensitive to canopy structural characteristics, such as *Leaf Area Index (LAI)* (Gao *et al.*, 2000).

A *Transformed Vegetation Index (TVI)* has been applied to biomass estimation for rangelands (Richardson and Wiegand, 1977). Two indices derived from Landsat MSS data are given by,

$$MSS: \quad \begin{aligned} TVI_1 &= \left(\frac{DN_4 - DN_2}{DN_4 + DN_2} + 0.5\right)^{1/2} = \left(\frac{R_{42} - 1}{R_{42} + 1} + 0.5\right)^{1/2} \\ TVI_2 &= \left(\frac{DN_3 - DN_2}{DN_3 + DN_2} + 0.5\right)^{1/2} = \left(\frac{R_{32} - 1}{R_{32} + 1} + 0.5\right)^{1/2} \end{aligned}$$ (5-16)

where the 0.5 bias term automatically prevents negative values under the square root for most images. As seen from Eq. (5-16), the *TVI* is a function of *RVI*, and consequently contains no additional information. The *TVI* shows an advantage over the *NDVI* in some situations, in that it tends to be more linearly related to biomass, thus simplifying regression calculations.

A *Perpendicular Vegetation Index (PVI)*, orthogonal to the soil line in two dimensions, was defined in Richardson and Wiegand (1977),

$$PVI = \left[(\rho_{red}^{soil} - \rho_{red}^{veg})^2 + (\rho_{NIR}^{soil} - \rho_{NIR}^{veg})^2 \right]^{1/2} . \qquad (5\text{-}17)$$

The *PVI* is interpreted geometrically as the perpendicular distance of a pixel vector from the soil line in an NIR versus red reflectance space (Fig. 5-4). Since the *PVI* is defined in terms of reflectance, without scaling constants, it can be applied to calibrated data from any sensor.

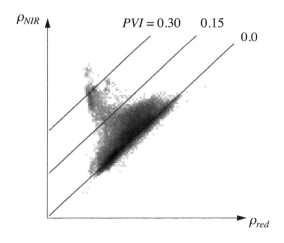

FIGURE 5-4. *Isolines for the PVI. The data are the same as that in Fig. 5-2 and are uncalibrated DNs. Because the PVI is defined in terms of a physical scene property, namely the soil line, it is only strictly valid in reflectance space. A reflectance range from zero to 0.5 is arbitrarily assumed here for illustration. The line where the PVI is zero corresponds to the soil line for this image.*

A general approach to defining vegetation indexes in *K* dimensions was described by Jackson (1983). The methodology provides a unified description of the various 2-D and *K*-D indices, such as the *Tasseled Cap Transform* described later. The above discussion of vegetation indices is necessarily limited to only a few of the more commonly used indices; a more complete survey and description of interrelationships can be found in Liang (2004). The discussion here also emphasizes *empirical* measures of vegetation coverage and chlorophyll content that can be derived directly from remotely-sensed data. There are a number of important *physical model-based* measures such as *LAI*, *Fraction of Photosynthetically Active Radiation (FPAR)*, and *Fraction of Absorbed Photosynthetically Active Radiation (FAPAR)*, that model the geometry, chemical composition, and radiation transfer of a leaf canopy and then integrate the individual components to obtain a total signal (Liang, 2004). Such measures can be linked to the remote sensing-derived measures described here by application of radiative transfer models (Myneni *et al.*, 1997), empirical relationships (Liang, 2004), and inversion methods applied to remote sensing data (Knyazikhin *et al.*, 1998).

5.3.2 Image Examples

The *RVI* and *NDVI* vegetation indexes are demonstrated with an agriculture scene (Fig. 5-5). Both indices clearly indicate the presence of vegetation and appear correlated with the amount of vegetation in particular fields (compare to Plate 4-1). The degree and nature of those correlations cannot be determined without careful ground measurements, however.

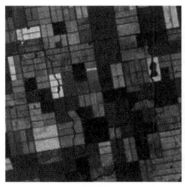
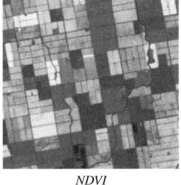

RVI   *NDVI*

FIGURE 5-5. The RVI and NDVI index images for the TM agriculture image of Plate 4-1. Bands 4 and 3 were used and the indices are calculated from uncalibrated DNs.

One of the benefits of the spectral ratio transform is suppression of the topographic shading factor. The sensor data must first be calibrated, however, for sensor offset and atmospheric path radiance. The importance of this is illustrated in Fig. 5-6. The original data were calibrated to at-sensor radiance using the sensor gain and offset coefficients, and then to surface radiance by subtracting a dark object value from each band; the process is described in Chapter 7. Notice how the topographic shading remains in the *DN* ratios of band 2/1 and band 3/2, while it is not evident in the ratios of bias-corrected data.

The Cuprite area is highly mineralized, with little vegetative cover. The enhancement of different mineral signatures is particularly apparent in the 5/4 and 7/5 ratios. The dark areas in the latter ratio image indicate clay-rich minerals in general, which absorb strongly in band 7 and have high reflectance in band 5 (Avery and Berlin, 1992); in this case they correspond to the clay mineral, kaolinite, as indicated by analysis of imaging spectrometer data of Cuprite (Kruse *et al.*, 1990). Another property of band ratios can be seen in these data. When the reflectances in the two bands are similar (as for the TM 2/1 and 3/2 ratios of geologic materials), and particularly for low reflectances, any image noise is amplified by the ratio calculation.

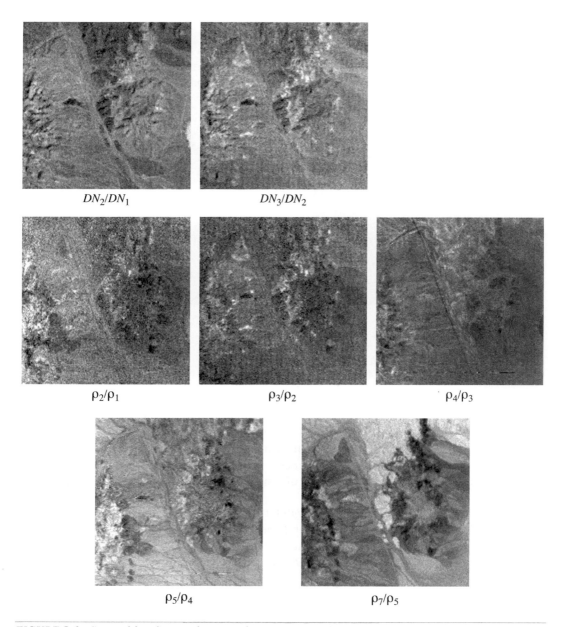

DN_2/DN_1

DN_3/DN_2

ρ_2/ρ_1

ρ_3/ρ_2

ρ_4/ρ_3

ρ_5/ρ_4

ρ_7/ρ_5

FIGURE 5-6. Spectral band ratios for a Landsat TM image of Cuprite, Nevada, acquired on October 4, 1984. The bands used in the top row have had no atmospheric or detector offset correction; the bands used in the lower two rows have had a bias correction and are contrast stretched for display; the result is proportional to the ratio of reflectances in the two bands (Eq. (5-8)).

5.4 Principal Components

Multispectral image bands are often highly correlated, i.e., they are visually and numerically similar (Fig. 5-7). The correlation between spectral bands arises from a combination of factors:

- *Material spectral correlation.* This correlated component is caused by, for example, the relatively low reflectance of vegetation across the visible spectrum, yielding a similar signature in all visible bands. The wavelength range of high correlation is determined by the material spectral reflectance.

- *Topography.* For all practical purposes, topographic shading is the same in all solar reflectance bands and can even be the dominant image component in mountainous areas and at low sun angles. It therefore leads to a band-to-band correlation in the solar reflective region, which is independent of surface material type (Fig. 4-44). The effect is different in the thermal region (see Fig. 2-23).

- *Sensor band overlap.* Ideally, this factor is minimized in the sensor design stage, but can seldom be avoided completely. The amount of overlap is typically small (Fig. 3-22), but is nevertheless important for precise calibrations.

Analysis of all of the original spectral bands is inefficient because of this redundancy. The *Principal Component Transformation (PCT)*[3] is a feature space transformation designed to remove this spectral redundancy (Ready and Wintz, 1973). It is a linear transform of the type in Eq. (5-2), with an image-specific matrix W_{PC} and zero bias B,

$$PC = W_{PC} \cdot DN .\qquad(5\text{-}18)$$

This transformation alters the covariance matrix as follows,

$$C_{PC} = W_{PC}CW_{PC}{}^T\qquad(5\text{-}19)$$

and the PCT is optimum in the sense that, of all possible transformations, W_{PC} is the *only* one that diagonalizes the covariance matrix of the original multispectral image, so that,

$$C_{PC} = \begin{bmatrix} \lambda_1 & \dots & 0 \\ \vdots & & \vdots \\ 0 & \dots & \lambda_K \end{bmatrix} .\qquad(5\text{-}20)$$

The *K eigenvalues* λ_k are found as the K roots of the *characteristic equation*,

$$|C - \lambda I| = 0 ,\qquad(5\text{-}21)$$

---

3. Also known as the *Karhunen-Loeve (KL)* or *Hotelling* transformation.

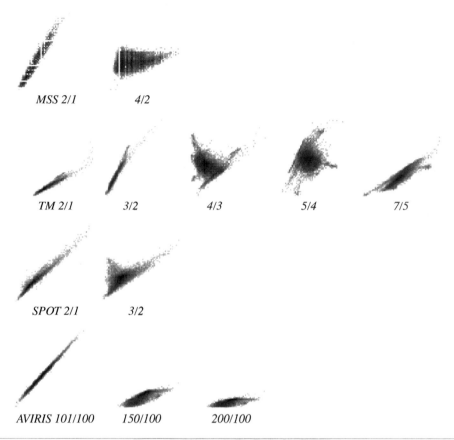

FIGURE 5-7. *Two-band scattergrams of different images from different sensors. Note the similarities between the NIR/visible scattergrams for different sensors (implying vegetation in the respective images) and the relatively high degree of correlation between visible bands for all sensors.*

where C is the original data covariance matrix and I is the (diagonal) identity matrix. Each eigenvalue is equal to the variance of the respective *PC* image along the new coordinate axes, and the sum of all the eigenvalues must equal the sum of all the band variances of the original image, thus preserving the total variance in the data. Since C_{PC} is diagonal, the principal component images are uncorrelated and, by convention, are ordered by decreasing variance, such that PC_1 has the largest variance and PC_K has the lowest. The result is removal of any correlation present in the original K-dimensional data, with a simultaneous compression of most of the total image variance into fewer dimensions.

The *PC* coordinate axes are defined by the *K eigenvectors*, e_k, obtained from the vector-matrix equation for each eigenvalue λ_k,

$$(C - \lambda_k I)e_k = 0 , \qquad (5-22)$$

which form the rows of the transformation matrix W_{PC},

$$W_{PC} = \begin{bmatrix} e_1^t \\ : \\ e_K^t \end{bmatrix} = \begin{bmatrix} e_{11} & \cdots & e_{1K} \\ : & & : \\ e_{K1} & \cdots & e_{KK} \end{bmatrix} , \qquad (5-23)$$

where e_{ij} is the jth element of the ith eigenvector. The eigenvector components are the direction cosines of the new axes relative to the original axes. This completes the eigenanalysis, and the transformation of Eq. (5-18) is applied; the eigenvector component, e_{ij}, becomes the weight on band j of the original multispectral image in the calculation for the principal component i.[4] For a particularly clear description of how to find the eigenvalues and eigenvectors of a given data covariance matrix, see Richards and Jia (1999).

The PCT has several interesting properties:

- *It is a rigid rotation in K-D of the original coordinate axes to coincide with the major axes of the data* (Fig. 5-8). The data in the PC images result from projecting the original data onto the new axes. In this example, the PC_2 component will have negative *DN*s. That is not a problem because the origin of the PC space can be arbitrarily shifted to make all PC values positive (routinely done for display), with no other change in the PCT properties (Richards and Jia, 1999).

- *Although the PC axes are orthogonal to each other in K-D, they are generally not orthogonal when projected to the original multispectral space.* The depiction in Fig. 5-8 is for a 2-D PCT; these orthogonal PC axes do not align with the projection of a higher dimensional transformation as shown in Fig. 5-9. This makes it difficult to interpret the eigenvectors of a high dimensional PCT in terms of 2-D scattergrams.

- *It optimally redistributes the total image variance in the transformed data.* The first PC image contains the maximum possible variance for any linear combination of the original bands, the second PC image contains the maximum possible variance for any axis orthogonal to the first PC, and so forth. The *total* image variance is preserved by the PCT.[5] This property, illustrated in Fig. 5-10, is why principal components are important as a data compression tool. If the low variance (low contrast) information in the higher order components can be ignored, significant savings in data storage, transmission, and processing time can result. Also, any uncorrelated noise in the original image will

---

4. The eigenvector weighting is called a *loading* or *factor loading* in statistics circles.

5. The *trace*, or sum of diagonal elements, of the original covariance matrix, and of the new covariance matrix are equal.

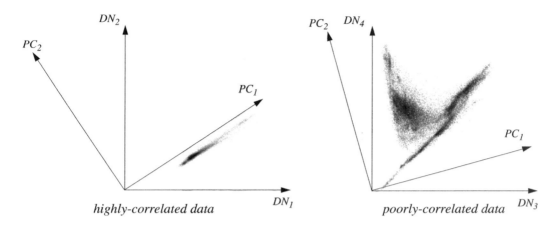

FIGURE 5-8. *2-D PCTs of two highly-correlated ($\rho = 0.97$) and two nearly-uncorrelated ($\rho = 0.13$) TM bands from a nonvegetated desert scene and a partially-vegetated agriculture scene, respectively. In the latter case, there are approximately equal amounts of vegetative cover and bare soil, which explains the orientation of the PC axes.*

usually appear only in the higher order components (see Fig. 7-27), and can therefore be removed by setting those PC images to a constant value. Caution is advised, however, because small, but significant band differences may also appear only in the higher-order components.

· *The transform, W_{PC}, is data-dependent.* Some of these dependencies are demonstrated in Fig. 5-11. The eigenvalues of a PC calculation for parts of two TM scenes, one largely vegetated (north of San Francisco, California) and the other almost without vegetation (a desert area near Cuprite, Nevada), shows that the vegetated scene has a relatively higher contribution in the second eigenvalue, reflecting the higher dimensionality of that data. If the TIR band is included in the analysis of the vegetated scene, we see that it contributes some small amount of additional variance, which appears primarily in PC_4 for this particular image. The small contribution of the TIR band is expected because it usually has much lower variance (contrast) than the non-TIR bands. A comparison is also made between the TM and the MSS eigenvalues for the nonvegetated scene of Fig. 5-12. The TM appears to have somewhat higher dimensionality than MSS, probably because of spectral contrast between different minerals that TM senses in the SWIR (see Fig. 5-14 and related discussion).

An example of principal components transformation of Landsat TM images is shown in Fig. 5-12 and Fig. 5-13. These are the six non-TIR bands of the nonvegetated scene used in the preceding discussion. The positive-negative relationship of the first two components, a common characteristic of this transformation, expresses the uncorrelated nature of the new coordinates. The

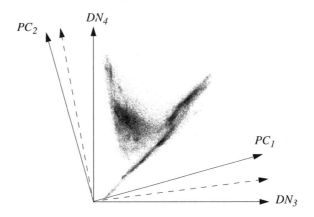

FIGURE 5-9. *Comparison of the 2-D PC axes from Fig. 5-8 (dashed arrows) to the 6-D PC_1 and PC_2 axes (solid arrows) projected to the band 4-band 3 data plane. They are not the same, and the axes projected from 6-D are not orthogonal.*

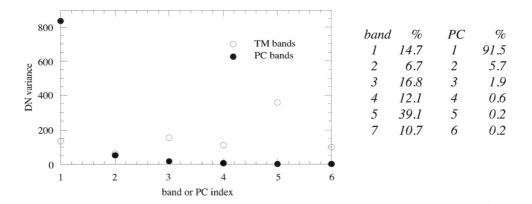

FIGURE 5-10. *Distribution of total image variance across the original spectral bands and across the principal components. The image is the nonvegetated scene of Fig. 5-12. The TIR band 6 is excluded; band index 6 above is actually TM band 7. The percent of the total variance is shown for the TM bands and the PCs.*

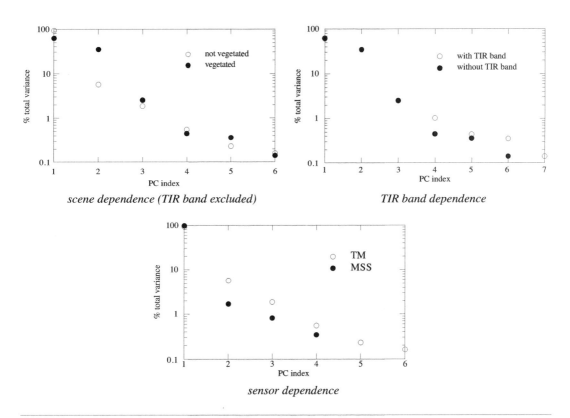

FIGURE 5-11. *The dependence of the percentage of total image variance captured by each eigenvalue.*

higher order components typically contain less image structure and more noise, indicating the data compression that has occurred. The contrast of each of the images in Fig. 5-12 and Fig. 5-13 has been stretched for display, but their variances reveal the redistribution of image contrast achieved by the PCT (Fig. 5-10). The eigenvectors for the various cases discussed are plotted in Fig. 5-14.

The PCT can also be applied to multitemporal datasets (Table 5-1). If the multispectral images from multiple dates are simply "stacked," as in a single, multispectral image, the PCT performs a *spectral-temporal* transformation and can be difficult to interpret because of the coupling of the two dimensions. If it can be assumed that the percentage area of change is small relative to areas of no change, then the latter will dominate the lower-order PCs, and the areas of change will be emphasized in the higher-order PCs (Richards, 1984; Ingebritsen and Lyon, 1985). The analysis of temporal changes may be facilitated if a feature such as the difference between the two multispectral images is first calculated, followed by a PCT on the multispectral difference image.

TABLE 5-1. Example applications of the PCT to multitemporal imagery.

data	description	PCT	reference
MSS	landcover on two dates	8-D	Byrne *et al.*, 1980
	fire burn, two dates	8-D	Richards, 1984
	surface mining, two dates	8-D	Ingebritsen and Lyon, 1985
	landcover, two dates	8-D	Fung and LeDrew, 1987
AVHRR-derived *NDVI*	36 monthly *NDVI* datsets	36-D	Eastman and Fulk, 1993
TM	temporal normalization for *NDVI*	2-D	Du *et al.*, 2002
review article	ecosystem change detection	–	Coppin *et al.*, 2004
review article	change detection	–	Lu *et al.*, 2004

5.4.1 Standardized Principal Components (SPC)

Some researchers suggest that a PCT based on the *correlation*, rather than covariance, matrix has advantages for remote sensing analysis (Singh and Harrison, 1985). In using the correlation matrix, we are effectively normalizing the original bands to equal and unit variance (the covariance matrix is then identically the correlation matrix). This may have advantages when data with different dynamic ranges are combined in a PCT. The SPCT does not exhibit the optimum compression characteristic of the PCT, of course. Two examples of the SPCT, using the same TM data used for Fig. 5-12 and Fig. 5-13, are shown in Fig. 5-15.

5.4.2 Maximum Noise Fraction (MNF)[6]

Also known as the *Noise-Adjusted Principal Components (NAPC)* transform (Lee *et al.*, 1990), this technique was introduced in Green *et al.* (1988). The MNF is a modification of the conventional PCT to improve the isolation of image noise that may occur in one or only a few of the original bands. It's particularly useful for hyperspectral data, where the *SNR* may vary considerably in different bands due to different signal levels. Unfortunately, the noise spectral covariance matrix must be available (or estimated) for the MNF transform.

---

6. The MNF is also known as the *Minimum Noise Fraction* transform in some circles (RSI, 2005). However, the original paper by Green *et al.* (1988) calls it the *Maximum Noise Fraction*.

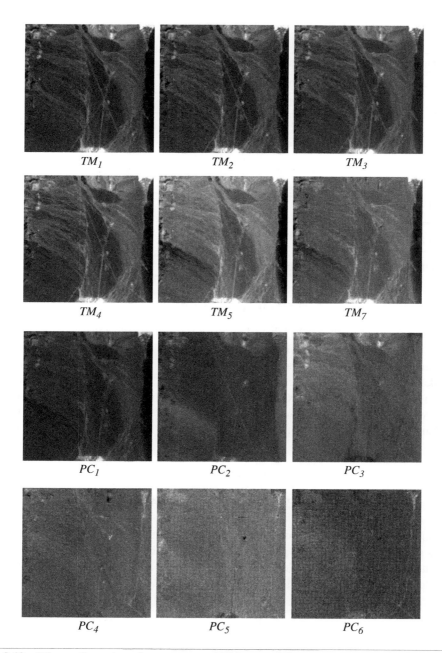

FIGURE 5-12. PC transformation of a nonvegetated TM scene. PC components are individually contrast stretched.

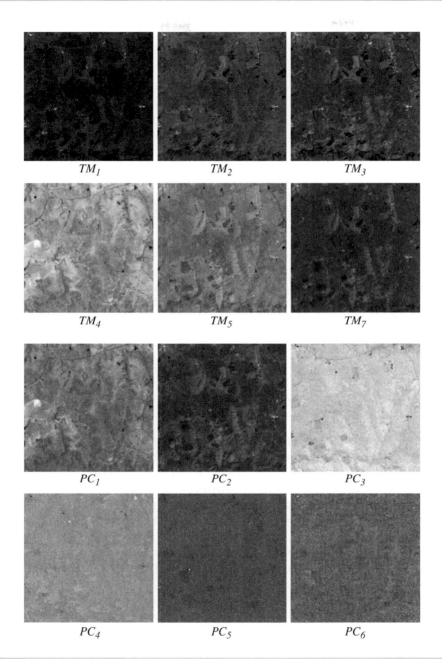

FIGURE 5-13. PC transformation of a vegetated TM scene. PC components are individually contrast stretched.

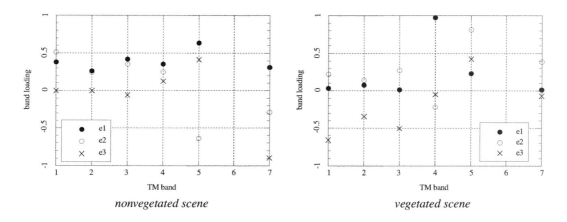

FIGURE 5-14. The first three eigenvectors for the nonvegetated and vegetated scenes. These are for a six band PCT, excluding the thermal band 6. In both cases, the first eigenvector has all positive band components. For the nonvegetated scene, the major spectral contrast appears to be in bands 5 and 7 (where mineral signatures are found; see Fig. 5-6), while for the vegetated scene, bands 3, 4, and 5 contain most of the spectral contrast because of the vegetation reflectance "edge" at 700nm.

5.5   Tasseled-Cap Components

The PCT is data dependent; the coefficients of the transformation are a function of the data spectral covariance matrix. While this feature allows the PCT to adapt to a given dataset and produce the best transformation from a compression standpoint, it makes the comparison of PCTs from different images difficult. The PCs can usually be interpreted in terms of physical characteristics of a given scene, but the interpretation must change from scene to scene. Early in the Landsat era, it was recognized that a *fixed* transformation, based on physical features, would be useful for data analysis.

   A fixed-feature space, designed specifically for agricultural monitoring, was first proposed for Landsat MSS data by Kauth and Thomas (1976). They noted that the *DN* scattergrams of Landsat MSS agricultural scenes exhibit certain consistent properties, for example, a triangular-shaped distribution between band 4 and band 2 (Fig. 5-7). Visualization of such distributions in K dimensions as crops grow yields a shape described as a "tasseled cap," with a base called the "plane of soils." Crop pixels move up the tasseled cap as the crops grow and back down to the plane of soils as the crop progresses to a senescent stage. Kauth and Thomas first derived a linear transformation of the four MSS bands that would yield a new axis called "soil brightness," defined by the signature of nonvegetated areas. A second coordinate axis, orthogonal to the first and called "greenness," was

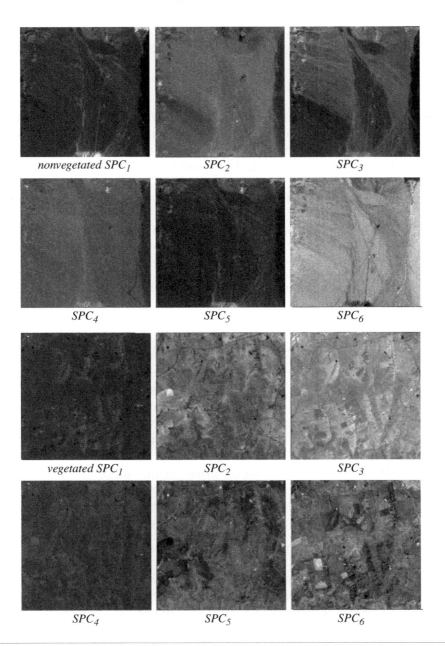

FIGURE 5-15. Standardized principal components; compare to the conventional PCs for the two scenes in Fig. 5-12 and Fig. 5-13.

then derived in the direction of vegetation signatures. The third and fourth transformed axes, called "yellow stuff" and "non-such," respectively, were derived to be orthogonal to the first two axes. The *Tasseled-Cap Transform (TCT)* is a special case of Eq. (5-2), with a specific transformation matrix W_{TC},

$$TC = W_{TC} \cdot DN + B . \tag{5-24}$$

Unlike the PCT, the weights in the TCT matrix are *fixed for a given sensor and independent of the scene*. The matrix transformation coefficients are given for the Landsat-1 MSS (Kauth and Thomas, 1976), Landsat-2 MSS (Thompson and Whemanen, 1980), Landsat-4 TM (Crist and Cicone, 1984), Landsat-5 TM (Crist et al., 1986), and Landsat-7 ETM+ (Huang et al., 2002) in Table 5-2. Like the PCT, the new axes produced by the TCT are orthogonal in the *K*-dimensional space, but they are *not* orthogonal when projected onto a lower-dimension data space (Fig. 5-16).

TABLE 5-2. *Tasseled-cap coefficients for several sensors. The MSS coefficients are for a 0-63 DN scale in band 4 and a 0-127 DN scale in the other bands, i.e., the same as data supplied to users. The Landsat-4 and -5 coefficients are for DN data; the Landsat-7 coefficients are for reflectance data, after atmospheric correction. The Landsat-5 TM coefficients were derived by reference to the Landsat-4 coefficients; they include extra additive terms that must be used to compare the TC features from the two sensors (Crist et al., 1986) .*

sensor	axis name	W_{TC}						bias
		band 1	2	3	4			
L-1 MSS	soil brightness	+0.433	+0.632	+0.586	+0.264			
	greenness	−0.290	−0.562	+0.600	+0.491			
	yellow stuff	−0.829	+0.522	−0.039	+0.194			
	non-such	+0.223	+0.120	−0.543	+0.810			
L-2 MSS	soil brightness	+0.332	+0.603	+0.676	+0.263			
	greenness	+0.283	−0.660	+0.577	+0.388			
	yellow stuff	+0.900	+0.428	+0.0759	−0.041			
	non-such	+0.016	+0.428	−0.452	+0.882			
	band	1	2	3	4	5	7	
L-4 TM	soil brightness	+0.3037	+0.2793	+0.4743	+0.5585	+0.5082	+0.1863	
	greeness	−0.2848	−0.2435	−0.5436	+0.7243	+0.0840	−0.1800	
	wetness	+0.1509	+0.1973	+0.3279	+0.3406	−0.7112	−0.4572	
	haze	+0.8242	−0.0849	−0.4392	−0.0580	+0.2012	−0.2768	
	TC5	−0.3280	+0.0549	+0.1075	+0.1855	−0.4357	+0.8085	
	TC6	+0.1084	−0.9022	+0.4120	+0.0573	−0.0251	+0.0238	

TABLE 5-2. *Tasseled-cap coefficients for several sensors. The MSS coefficients are for a 0-63 DN scale in band 4 and a 0-127 DN scale in the other bands, i.e., the same as data supplied to users. The Landsat-4 and -5 coefficients are for DN data; the Landsat-7 coefficients are for reflectance data, after atmospheric correction. The Landsat-5 TM coefficients were derived by reference to the Landsat-4 coefficients; they include extra additive terms that must be used to compare the TC features from the two sensors (Crist et al., 1986) (continued).*

sensor	axis name	$\mathbf{W}_{TC}$						bias
L-5 TM	soil brightness	+0.2909	+0.2493	+0.4806	+0.5568	+0.4438	+0.1706	+10.3695
	greeness	−0.2728	−0.2174	−0.5508	+0.7221	+0.0733	−0.1648	−0.7310
	wetness	+0.1446	+0.1761	+0.3322	+0.3396	−0.6210	−0.4186	−3.3828
	haze	+0.8461	−0.0731	−0.4640	−0.0032	−0.0492	+0.0119	+0.7879
	TC5	+0.0549	−0.0232	+0.0339	−0.1937	+0.4162	−0.7823	−2.4750
	TC6	+0.1186	−0.8069	+0.4094	+0.0571	−0.0228	+0.0220	−0.0336
L-7 ETM+	soil brightness	+0.3561	+0.3972	+0.3904	+0.6966	+0.2286	+0.1596	
	greeness	−0.3344	−0.3544	−0.4556	+0.6966	−0.0242	−0.2630	
	wetness	+0.2626	+0.2141	+0.0926	+0.0656	−0.7629	−0.5388	
	haze	+0.0805	−0.0498	−0.1950	−0.1327	+0.5752	−0.7775	
	TC5	−0.7252	−0.0202	+0.6683	+0.0631	−0.1494	−0.0274	
	TC6	+0.4000	−0.8172	+0.3832	+0.0602	−0.1095	+0.0985	

The motivation behind the development of the TCT was to find a fixed set of axes in feature space that represented the physical properties of growing crops in the American midwest. The soil axes represents the pre-emergence time regime and as crops grow, their feature space signature moves away from this axis in the direction of greenness. Then, as crops mature and become senescent, their signature increases in the yellow stuff dimension and decreases in greenness. If this transformation, derived for midwestern agriculture, is applied to other types of agriculture and vegetation in different climate zones, the transformation coefficients can still be interpreted in a similar fashion, but some new scene components may not be represented, such as non-green vegetation and different soil types (Crist, 1996). The yellow stuff and non-such dimensions for MSS data have been shown to indicate changes in atmospheric haze conditions and, therefore, are useful for relative atmospheric calibration between images (Lavreau, 1991). In the TM TCT space, the fourth component is called "haze" and has a similar interpretation.

In Fig. 5-17, we provide a comparison of the PCT and TCT for a TM agricultural scene (see the CIR composite in Plate 4-1 for a visual reference). The most highly-correlated components of the two transformations are PC_1 and TC_1 (correlation of 0.971) or TC6 (0.868), and PC_4 and TC_5 (0.801). It is not possible to draw general conclusions from these observations, however, because the PCT will change from scene to scene while the TCT will not. In general, it appears that the TCT

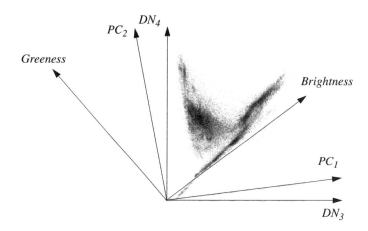

FIGURE 5-16. Projection of two of the component axes from 6-D PC and TC transformations onto the TM band 4 versus band 3 data plane. The Brightness-Greenness projection is fixed and independent of the data. Notice how the Brightness-Greenness space "captures" the physical components of soil and vegetation better than does the PC₁-PC₂ space.

leads to superior separation of scene components, particularly soil and vegetation. A significant drawback is that it is sensor-dependent (primarily on the spectral bands), and the coefficients must be rederived for every sensor.

5.6   Contrast Enhancement

The spectral transforms discussed so far are motivated by quantitative analysis of multispectral imagery. Contrast enhancement, on the other hand, is only intended to improve the visual quality of a displayed image. The transformations used for enhancing contrast are generally not desirable for quantitative analysis.

The need for contrast enhancement derives from the requirements on sensor design described in Chapter 3. A remote-sensing system, particularly a satellite sensor with global coverage, must image a wide range of scenes, from very low radiance (oceans, low solar elevation angles, high latitudes) to very high radiance (snow, sand, high solar elevation angles, low latitudes). Therefore, the sensor's dynamic range must be set *at the design stage* to accomodate a large range of scene radiances, and it is desirable to have as many bits/pixel as possible over this range for precise measurements. However, any particular scene will generally have a radiance range much less than the full range. When imaged and converted to *DN*s, it therefore uses less than the full quantization range, typically eight bits/pixel or more. Since most display systems use eight bits/pixel in each color, with the extremes of that range adjusted to yield black and white on the monitor, the displayed image will have low contrast because it is not using the full range available in the display.

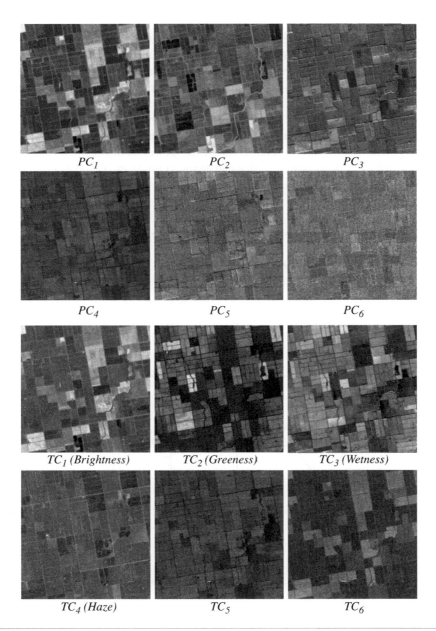

FIGURE 5-17. *The PC and TC components for the Yuma agricultural scene. The PCT achieves more data compression (PC$_5$ and PC$_6$ contain almost no scene content), but the TCT better represents soil and vegetation in TC$_1$ and TC$_2$. Note PC$_2$ is inverted relative to TC$_2$.*

5.6.1 Global Transforms

Contrast enhancement is a mapping from the original *DN* data space to a *GL* display space. Each *DN* in the input image is transformed through the mapping function to a *GL* in the output (display) image. In the simplest case, the mapping function is determined by the global (full image) statistical properties and is the same for every pixel. Since the primary goal is usually to expand the *DN* range to fill the available display *GL* range, the transform is called a *contrast stretch*; typical transformations are shown in Fig. 5-18. The slope of the transformation, $d(GL)/d(DN)$, can be considered the *gain* in contrast, much like the gain parameter in an electronic amplifier. For the linear stretch, the gain is the same at all *DN*s. In the piecewise-linear and histogram equalization stretches, the gain varies with *DN*, thus preferentially enhancing contrast in certain *DN* ranges (low water radiances, for example).

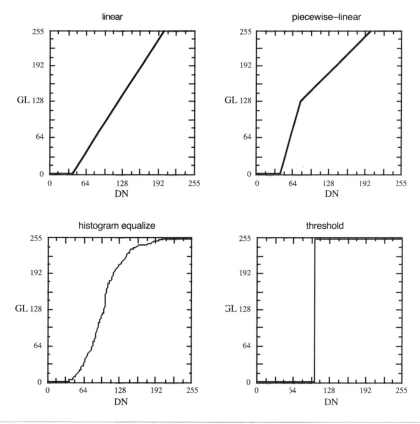

FIGURE 5-18. *Some types of DN-to-GL transformations for contrast enhancement. The local slope of the transformation represents the gain applied at each DN.*

Linear stretch

Examples of linear contrast stretching and the effect on the image histogram are depicted in Fig. 5-19. The *min-max stretch* expands the image *DN* range to fill the dynamic range of the display device, e.g., [0,255]. Since it is based on the image minimum and maximum *DN*, it is sensitive to outliers, i.e., single pixels that may be atypical and outside the normal *DN* range. The resulting contrast will then be less than that possible if the outliers were excluded in determining the minimum and maximum *DN* in the image.

To achieve a greater contrast increase, a *saturation stretch* may be used with a linear stretch for all pixels within a *DN* range smaller than the min-max range. Pixels with a *DN* outside that range are then transformed to a *GL* of either zero or 255. This saturation ("clipping") at the extremes of the *DN* range is acceptable unless important image structure is lost in the saturated areas of the image. Typically, saturation of one or two percent of the image pixels is a safe level, but it depends, of course, on the image and intended application. If one is primarily interested in viewing cloud structure, for example, there should be little saturation at high *DN* values.

A linear transformation also can be used to *decrease* image contrast if the image *DN* range exceeds that of the display. This situation occurs for radar imagery, some multispectral sensors such as AVHRR (10 bits/pixel), MODIS (12 bits/pixel), and most hyperspectral sensors (12 bits/pixel). The displayed image will not contain all the radiometric information in the raw image, but on the other hand, the human eye cannot usually distinguish more than about 50 *GL*s at once anyway.

Nonlinear stretch

If the image histogram is asymmetric, as it often is, it is impossible to simultaneously control the average display *GL* and the amount of saturation at the ends of the histogram with a simple linear transformation. With a *piecewise-linear transformation*, more control is gained over the image contrast, and the histogram asymmetry can be reduced, thus making better use of the available display range (Fig. 5-19). This example is a two segment stretch, with the left segment having a higher gain than the right segment. The transformation parameters are selected to move the input minimum and maximum *DN*s to the extremes of the display *GL* range and to move the mode of the histogram to the center of the display range (128). More than two linear segments may be used in the transformation for better control over the image contrast.

Histogram equalization is a widely-used nonlinear transformation (Fig. 5-19). It is achieved by using the *Cumulative Distribution Function (CDF)* of the image as the transformation function, after appropriate scaling of the ordinate axis to correspond to output *GL*s. Equalization refers to the fact that the histogram of the processed image is approximately uniform in density (number of pixels/*GL*) (Gonzalez and Woods, 2002). Because of the unimodal shape of most image histograms, equalization tends to automatically reduce the contrast in very light or dark areas and to expand the middle *DN*s toward the low and high ends of the *GL* scale. Where the *CDF* increases rapidly, the contrast gain also increases. The highest gain therefore occurs at *DN*s with the most pixels. This effect is seen in Fig. 5-19 as the variable spacing of *GL*s in the enhanced image histogram. The contrast of an equalized image is often rather harsh, so equalization is not recommended as a general purpose stretch. However, no parameters are required from the analyst to implement the transformation, making it easy to apply.

Normalization stretch

The normalization stretch is a robust contrast enhancement algorithm. The image is transformed with a linear stretch such that it has a specified *GL* mean and standard deviation, and then it is clipped at the extremes of the display *GL* range. It is, therefore, a variation on the saturation stretch. Two parameters are required to achieve the desired second-order statistics, given the input *DN* mean and standard deviation (see Table 5-3). The clipping is necessary because the resulting *GL*s will usually extend beyond the limits of the display range. This two-parameter algorithm provides a convenient way to explicitly control the mean of the resulting image and, at the same time, vary the contrast by controlling the standard deviation. Examples are shown in Fig. 5-20.

Reference stretch

Differences between atmospheric conditions at different times, differences between sensors, and scene changes over time make it difficult to compare multisensor and multitemporal imagery. Even adjacent scenes from two orbits of the same sensor can differ significantly in their radiometric characteristics because they normally are acquired one or more days apart, and the sensor view angle and atmospheric conditions are different. There is a need, therefore, to match the global radiometric characteristics of one image to another. It must, of course, be presumed and reasonable that the histograms of the two images *should in fact* be similar. That is, if the presumption of similarity is not justified, for example, if there have been substantial surface changes between the two image dates, then reference matching is not justified.

    One way to match the contrast and brightness characteristics of two images is with the linear normalization stretch of the previous section. However, that can only match the means and standard deviations of the two images. A natural generalization of the normalization stretch is to match histogram *shapes*. The *CDF reference stretch* does this by matching the *CDF* of an image to a reference *CDF* (Richards and Jia, 1999). This is achieved by forward mapping the image *DN*s through the image *CDF*, and then backward mapping through the reference CDF_{ref} to DN_{ref} (Fig. 5-21).[7] CDF_{ref} can be from another image, or a hypothetical image with a specified *CDF*, for example, one with a Gaussian histogram. The two-stage mapping creates a nonlinear look-up table for $DN_{ref}(DN)$. One complication is that the matching of the *CDF* levels does not necessarily lead to an integer DN_{ref}. The CDF_{ref} must therefore be *interpolated* between *DN*s to obtain the output DN_{ref} to be used. Linear interpolation is sufficient for this purpose since the *CDF* is normally a smooth function of *DN*.

---

7. This technique is sometimes called *histogram matching* or *histogram specification*.

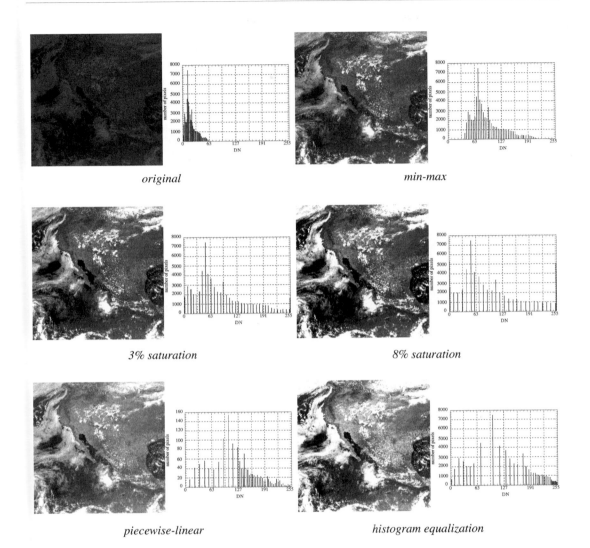

FIGURE 5-19. *Examples of contrast enhancement using point transformations and global statistics. The test image is from the GOES visible wavelength sensor and shows cloud patterns over North America. As the gain in the stretch transformation increases, the occupied DN bins spread further apart. This is generally not noticeable in the image, unless only a few DNs are spread across the whole range. Also note the piecewise-linear transformed histogram resembles the equalized histogram in the sense of larger gaps between DNs for the lower DNs relative to the higher DNs.*

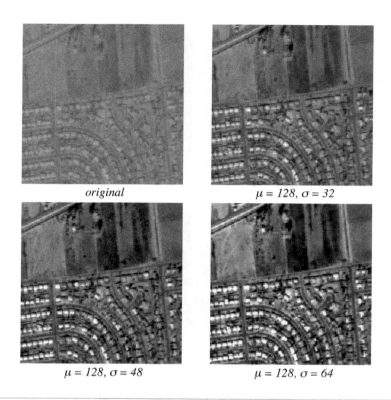

original   *μ = 128, σ = 32*

μ = 128, σ = 48   *μ = 128, σ = 64*

FIGURE 5-20. Examples of the normalization stretch. Note how the GL mean can be held constant, while the contrast is controlled by a single parameter, the GL standard deviation.

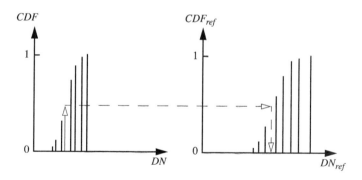

FIGURE 5-21. The procedure to match the CDFs of two images. Note the calculated DN_{ref} must be estimated by interpolation of the CDF_{ref} between existing values.

The first transformation, through the *CDF* of the image to be stretched, is the same as a histogram equalization. The second transformation represents an *inverse* histogram equalization, to an image having a *CDF* the same as that of the reference. The concept here is that the *CDF* of *any* histogram-equalized image is the same (i.e., a linear ramp), except for possibly a horizontal shift along the *DN* axis. An example of this contrast matching technique is shown in Fig. 5-22. The two images are eight months apart and show significant differences, in addition to that due to the change in solar irradiance. The *CDF* reference stretch result is compared to that from a linear stretch, with parameters determined from two targets in both images, one dark and one light. The latter is an empirical calibration transform for two images from different dates (see Chapter 7) and is preferred for radiometric comparison. The *CDF* stretch achieves a better visual match of the two images, but is a nonlinear transform that may not be appropriate because of the apparent changes between the two dates. For example, notice how the airport pavement is dark in December and bright in August. This indicates either a non-Lambertian reflectance characteristic or perhaps an actual change in pavement material. The image histograms are compared in Fig. 5-23.

Note that histogram matching can be applied with any specified *CDF* as the reference. In particular, a Gaussian *CDF* is used for a *Gaussian stretch*. For an image range of [0,255], the mean of the Gaussian is set to 128 and the standard deviation to a value such as 32, which allows ±2 standard deviations of data within the *DN* range. The *CDF* transformation of Fig. 5-21 is applied, and the resulting image histogram is approximately Gaussian with the given parameters. This is like the normalization stretch, but with the addition of histogram shaping, and is particularly useful for normalizing images with color imbalance. A small amount of saturation is helpful to increase overall contrast. Some color examples are shown in Plates 5-1 and 5-2.

A second technique for reference stretching, *statistical pixel matching*, was originally developed for medical images (Dallas and Mauser, 1980). It requires that the two images are registered and cover the same area on the ground (not a requirement for the *CDF* reference stretch). Therefore, in remote sensing, it is only useful for either multitemporal or multisensor images of the same scene. For mosaicing applications, the adjacent images would have to overlap, and only the region of overlap can be used in the algorithm. The procedure is,

- Create an accumulator table with an entry for each *DN* of the image to be stretched and initialize its elements to zero.

- For every pixel in the image to be stretched, accumulate the DN_{ref} of the corresponding pixel of the reference image, in the appropriate *DN* bin of the table.

- After all pixels have been counted, find the average DN_{ref} at each *DN* of the table. After a linear scaling of the DN_{ref} to *GL*, this becomes the desired *GL(DN)* transformation function.

- Do the transform.

This algorithm works best for images which are already quite similar everywhere except in small areas.

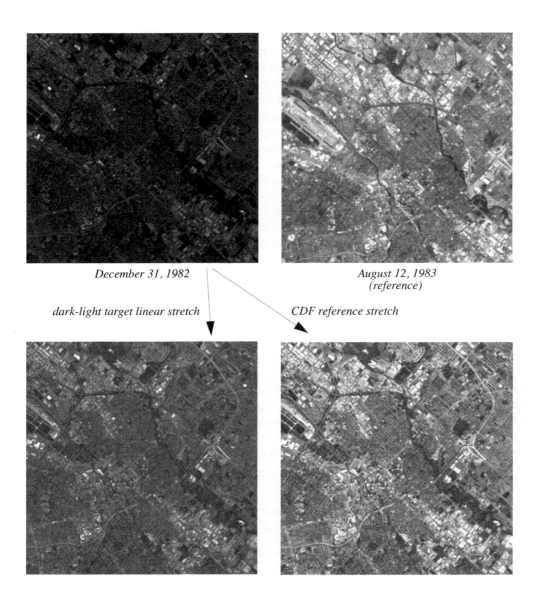

December 31, 1982   August 12, 1983
                                                          (reference)

dark-light target linear stretch   CDF reference stretch

FIGURE 5-22. *Contrast matching of two TM band 3 images of San Jose, California. The December image is radiometrically adjusted to match the August image using two techniques described in the text.*

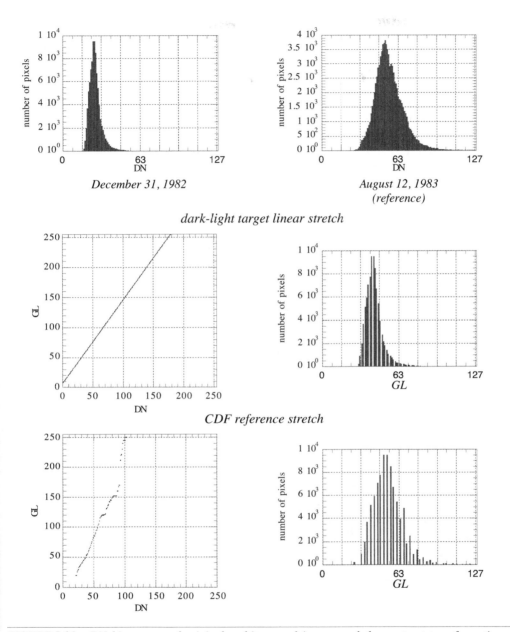

FIGURE 5-23. *DN histograms of original multitemporal images and the contrast transformations and resulting histograms for the December image after matching to the August image. The CDF reference stretch adjusts the entire histogram to match the histogram of the reference image, while the dark-light target stretch relies on only two points on the DN scale and is sensitive to target selection.*

Thresholding

Thresholding is a type of contrast manipulation that *segments* an image into two categories defined by a single *DN* threshold. The use of a binary threshold on certain types of images, such as those with land and water or snow, results in sharply-defined spatial boundaries that may be used for masking portions of the image. Separate processing may then be applied to each portion and the results spatially recombined. Thresholding is a simple classification technique, as we will see in Chapter 9. An example of thresholding of a GOES weather satellite image is shown in Fig. 5-24. The lowest threshold separates areas of water from clouds and land, the middle threshold isolates most clouds, and the highest threshold selects only the brightest and most dense clouds.

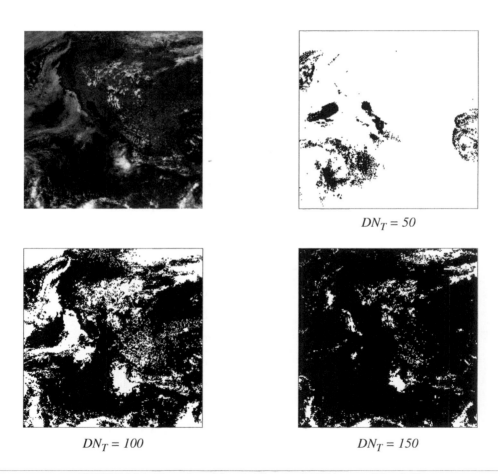

$DN_T = 50$

$DN_T = 100$ $DN_T = 150$

FIGURE 5-24. Examples of thresholding the GOES image of Fig. 5-19.

DN thresholding can be easily implemented in an interactive mode with a display and operator-controlled cursor, but selection of the "best" threshold level is a difficult task and requires prior knowledge about the scene to be meaningful. There are techniques for the quantitative determination of optimum threshold levels using only information from the image histogram (Pun, 1981).

5.6.2 Local Transforms

The contrast enhancement examples discussed thus far are point transformations using global image statistics; the same contrast stretch is applied to all image pixels (Table 5-3). Obviously, contrast can vary locally within an image. A more optimal enhancement may therefore be achieved by using an *adaptive* algorithm whose parameters change from pixel-to-pixel according to the local image contrast. We will describe a robust algorithm, *Local Range Modification (LRM)*, which illustrates the nature of adaptive processing (Fahnestock and Schowengerdt, 1983).

TABLE 5-3. *Summary of contrast enhancement algorithms. A display GL range of [0,255] is assumed.*

algorithm	equation	remarks
min-max	$GL = \dfrac{255}{DN_{max} - DN_{min}}(DN - DN_{min})$	sensitive to outliers
histogram equalization	$GL = 255\,CDF(DN)$	produces uniform histogram
normalization	1. $GL = \dfrac{\sigma_{ref}}{\sigma}(DN - \mu) + \mu_{ref}$ 2. $GL = 255, GL > 255$ $GL = 0, GL < 0$	matches means and variances
threshold	$GL = 255, DN \geq DN_T$ $GL = 0, DN < DN_T$	binary output
reference	$GL = CDF_{ref}^{-1}[CDF(DN)]$	matches histograms

The essential idea is to partition the image into adjoining blocks (designated *A*, *B*, etc.) and derive a contrast stretch, different at each pixel, which is dependent on the local contrast within the corresponding block and surrounding blocks (Fig. 5-25). The stretch must change smoothly from pixel-to-pixel; otherwise brightness discontinuities occur at the boundaries between blocks (Fig. 5-26). Also, one premise in LRM is that the final *GL* range of the enhanced image should be predictable and not exceed specified minimum and maximum *GL*s.

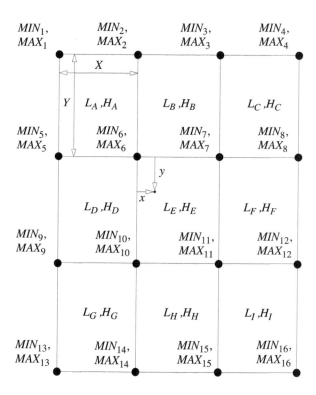

FIGURE 5-25. Blocking parameters for the LRM adaptive contrast enhancement.

The first step is to find the lowest and highest *DN*s, *L* and *H*, within each block. The overall minimum and maximum *DN*s of four adjoining blocks are then found by (using node 6 as the example),

$$MIN_6 = minimum(L_A, L_B, L_D, L_E)$$
$$MAX_6 = maximum(H_A, H_B, H_D, H_E)$$

$$(5\text{-}25)$$

and assigned to their shared node. This is equivalent to using overlapping blocks of twice the width and height in the initial image partitioning. At the corner of the image there is only one block, so in the upper left, for example,

$$MIN_1 = L_A$$
$$MAX_1 = H_A$$

$$(5\text{-}26)$$

and along the left image edge,

$$MIN_5 = minimum(L_A, L_D)$$
$$MAX_5 = maximum(H_A, H_D).$$

(5-27)

The values *MIN* and *MAX* set the *GL* range of enhanced pixels within the corresponding set of blocks. They are then linearly interpolated in two dimensions at each pixel location (x,y) within a block,

$$GL_{min} = \left[\frac{x}{X}MIN_7 + \left(\frac{X-x}{X}\right)MIN_6\right]\left(\frac{Y-y}{Y}\right) + \left[\frac{x}{X}MIN_{11} + \left(\frac{X-x}{X}\right)MIN_{10}\right]\frac{y}{Y}$$

$$GL_{max} = \left[\frac{x}{X}MAX_7 + \left(\frac{X-x}{X}\right)MAX_6\right]\left(\frac{Y-y}{Y}\right) + \left[\frac{x}{X}MAX_{11} + \left(\frac{X-x}{X}\right)MAX_{10}\right]\frac{y}{Y}$$

(5-28)

to obtain *estimated* range extremes, GL_{min} and GL_{max}, at each pixel. Using linear interpolation insures that these values are within the range of the surrounding *MIN* and *MAX*, i.e.,

$$MIN \le GL_{min} \le MAX$$
$$MIN \le GL_{max} \le MAX$$

(5-29)

The estimated range values found from Eq. (5-6) are then used to transform the pixel at (x,y) by a linear stretch,

$$GL' = \frac{255}{GL_{max} - GL_{min}}(DN - GL_{min}) .$$

(5-30)

Figure 5-26 is an example of adaptive contrast enhancement with the LRM algorithm. Note how the LRM algorithm provides a smooth contrast transition across blocks without obvious artifacts. Although the global, relative radiometry of the image is altered by the processing, it is much easier to distinguish previously low-contrast features in the darker and brighter portions of the enhanced image, and there are no transition artifacts between blocks.

Other artifacts can appear in adaptive enhancements, however. If the area used in the calculation for each pixel is smaller than some uniform objects in the scene (such as a lake), the algorithm will stretch that very low-contrast region excessively, resulting in high-contrast noise at those pixels.

Any global stretch can be used within an adaptive framework, for example histogram equalization (Pizer *et al.*, 1987). For adaptive histogram equalization, the nonlinear *CDF* is calculated for each block and interpolated between blocks to yield a smooth transition across the image, as is done in LRM.

5.6.3 Color Images

The use of color in display and enhancement of remote-sensing images is an important aspect of image processing. Color may be used simply for display of multispectral images or may be manipulated directly by processing techniques to enhance visual information extraction from the images.

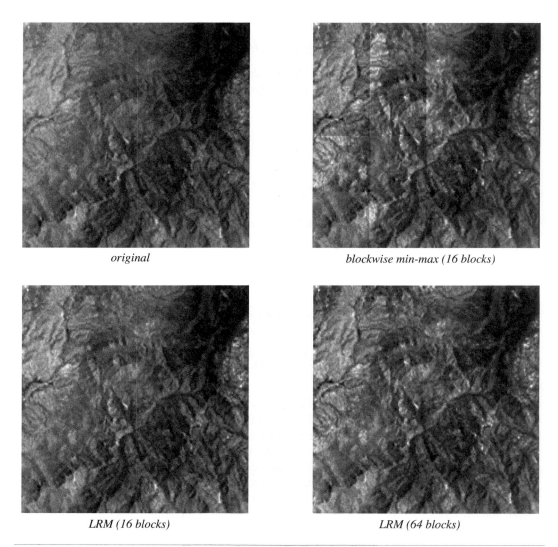

original   *blockwise min-max (16 blocks)*

LRM (16 blocks)   *LRM (64 blocks)*

FIGURE 5-26. Adaptive contrast enhancement with blockwise stretching and the LRM algorithm, which does not introduce block discontinuities.

At first glance, it would seem that one could simply extend the techniques used for black and white image contrast enhancement to color images. This is true to some extent, but it is wise to specifically consider the color aspects of the processing to achieve the best results. In this section we describe some relatively simple heuristic techniques for numerical color manipulation; theories that model the visual perception of color can be quite complex and will not be discussed here.

Min-max stretch

The most obvious way to enhance a color image is to simply stretch the histogram of each band according to its minimum and maximum *DN*. This will indeed improve the contrast of the color composite, but the color balance is likely to change in an unpredictable way from that of the original image. The problem is that the min-max stretch is controlled by the image minimum and maximum *DN*, which can be sensitive to outliers, as noted earlier.

Normalization stretch

The normalization stretch discussed previously is a good, robust tool for consistent color contrast enhancement. It overcomes the shortcomings of the min-max stretch by incorporating the image mean and standard deviation. For color images, the normalization stretch is applied to each band independently, setting their means and standard deviations equal across the bands (Fig. 5-27). The *average* color of the resulting color composite is therefore grey, and other spectral information in the image appears as deviations about that mean color. For some images, dominated by a single cover type such as vegetation or water, this may not be good for interpretation because of the bias toward a grey color. For most images that contain a mixture of cover types, it does appear to be a broadly useful tool.

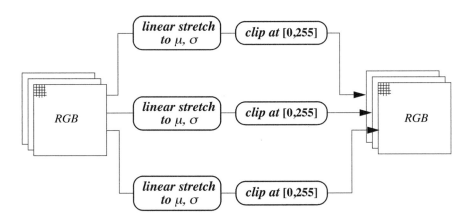

FIGURE 5-27. *The normalization algorithm for color images.*

Reference stretch

The technique of matching an image histogram to that of another image, or to a specified histogram, such as a Gaussian, works well with color images. The *DN* transformation should be applied separately to each band of the color image. Some examples are shown in Plates 5-1 and 5-2.

Decorrelation stretch

One of the difficulties in color contrast enhancement is spectral band correlation. If three correlated bands are displayed, their data distribution lies nearly along a line in the color cube, from the darkest pixels to the brightest ones, and very little of the available color space is utilized. If we decorrelate the bands, stretch the PCs to better fill the color space, and then inverse transform to the RGB color space, we will enhance whatever spectral information is present in the data. This idea is the basis of the PCT decorrelation stretch (Gillespie *et al.*, 1986; Durand and Kerr, 1989; Rothery and Hunt, 1990). The process is indicated in Fig. 5-28. The contrast stretch applied to the PCs shifts the data to all positive values and equalizes the variances, thereby insuring that the data occupy a spherical volume in the PC space. The inverse PCT then calculates transformed bands that are decorrelated. Since the eigenvector transformation matrix is orthogonal, its inverse is equal to its transpose, and is therefore easily calculated by swapping rows and columns. Example applications to TM images are shown in Plates 5-3 and 5-4.

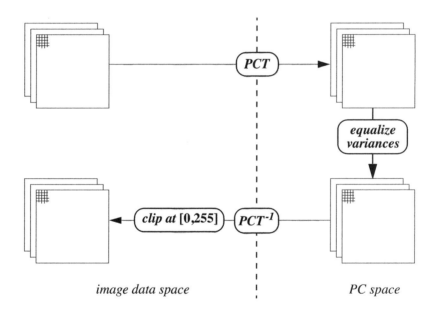

FIGURE 5-28. *The PCT decorrelation contrast stretch algorithm. The image produced by the inverse PCT is not constrained to an 8-bit range and may need to be clipped for display purposes. The importance of controlling the range of values in and out of a color transform algorithm is discussed in Schetselaar (2001).*

Color-space transforms

To describe the visually-perceived color properties of an image, we do not naturally use the proportions of red, green, and blue components, but rather terms such as "brightness," "color," and "color purity," which have the equivalent computer graphics terms of "intensity," "hue," and "saturation," respectively. Similarly, it is often easier to anticipate the visual results of an intensity or hue manipulation on an image than it is for the results of red, green, and blue manipulation. A transformation of the RGB components into hue, saturation, and intensity (HSI) components before processing may therefore provide more control over color enhancement. Whatever algorithm is used, we will call the conversion of RGB to HSI a *Color-Space Transform (CST)*. In the perceptual color space of HSI, we can modify any or all of the components somewhat predictably. For example, the intensity component can be stretched as for any single band image, or the saturation might be increased at all pixels by an appropriate transformation. The processed images are then converted back to RGB for display using the inverse of the CST (Fig. 5-29).[8]

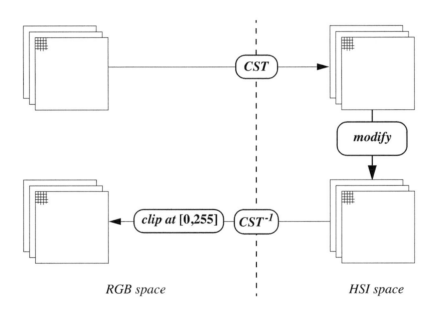

FIGURE 5-29. *The use of a Color-Space Transform (CST) to modify the perceptual color characteristics of an image.*

Many HSI coordinate systems have been defined; a cylindrical coordinate system illustrates the general properties of most HSI spaces in Plate 5-5. For image processing, we're interested in the forward and reverse coordinate transformation algorithm. One particular RGB-to-HSI

8. The acronym HSI is sometimes permutated to IHS in the literature.

tranformation will be used to illustrate the concepts involved. This transformation, the *hexcone CST*, is heuristic and not based on any particular color theory, but it is representative of most algorithms used in color image processing and color graphics (Smith, 1978; Schowengerdt, 1983).

The hexcone CST algorithm is based on the RGB color cube introduced in Chapter 1. We consider internal RGB subcubes defined by their vertex location along the greyline (Fig. 5-30). Imagine the projection of each subcube onto a plane perpendicular to the greyline at the subcube vertex. Moving the vertex from black to white, the projection onto a plane results in a series of hexagons of increasing size, as shown in Fig. 5-30 and Plate 5-5. The hexagon at black degenerates to a point; the hexagon at white is the largest. This series of hexagons define a *hexcone*. The distance along the grey line defines the intensity of each hexagonal projection.[9] For a pixel with a given intensity, the color components, hue and saturation, are defined geometrically in the appropriate hexagon. The hue of a point is determined by the angle around the hexagon, and the saturation is determined by the distance of the point from the center, i.e., the grey point. Points farther from the center represent purer colors than those closer to the grey point. The use of simple linear distances for defining hue and saturation make the hexcone algorithm more efficient than similar transformations involving trigonometric functions.

Examples of processing in the HSI space and its effect on the RGB space are shown in Fig. 5-31 and Plates 5-5, 5-6, and 5-7. By reducing the saturation component, for example, we obtain a "pastel" RGB image. By reducing the intensity component, we obtain a darker RGB image. The double ramp transform for hue in Fig. 5-31 rotates the colors in RGB space in a cyclic fashion, since the hue dimension is periodic.

Spatial domain blending

A simple blending algorithm for color manipulation was described in (Haeberli and Voorhies, 1994). The operation is a linear combination of two images,

$$output = (1 - \alpha) \times image0 + \alpha \times image1 \tag{5-31}$$

where *image1* is the image of interest and *image0* is a base image to be mixed with *image1*. The parameter α can take any value. If α is between zero and one, the two images are linearly interpolated to create a third; if α is outside the range [0, 1], one image is extrapolated "away" from the other. If α is greater than one, a portion of *image0* will be subtracted from a scaled *image1*; if α is less than zero, a portion of *image1* will be subtracted from a scaled *image0*. The color enhancement possibilities are given in Table 5-4 and image examples are shown in Plate 5-8. This algorithm is easy to implement either in interactive software or in hardware, since it requires only a weighted average of two images and avoids the computational cost of the HSI type of algorithm. It can also be used to control sharpening of an image, as we'll see in Chapter 6.

9. Smith used an alternate quantity, *value*, given by the maximum of R, G, and B (Smith, 1978). Value is more closely related to artist's terminology for describing color. The difference between value and intensity (e.g., given by the average of R, G, and B) can affect operations such as multispectral-panchromatic fusion (Chapter 8).

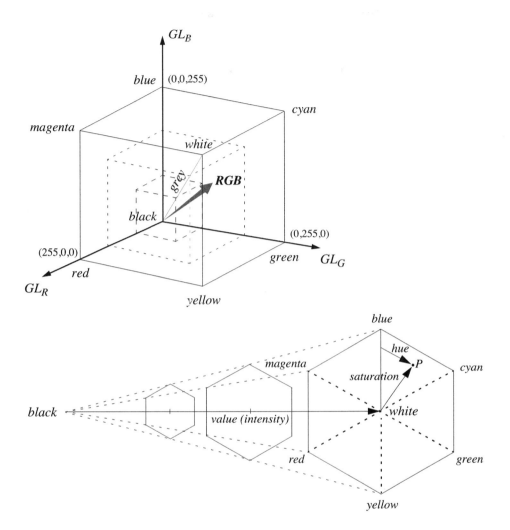

FIGURE 5-30. *Generation of the hexcone CST. Three possible RGB cubes are shown at the top and the resulting projections onto the plane perpendicular to the greyline are shown below. Such projections for all possible subcubes define the hexcone. The projection of a particular RGB point, P, into one of the hexagons is shown. At P, the intensity is the center grey value, the hue is the angle around the hexagon, and the saturation is the fractional distance from the center to the perimeter. The hue and saturation were defined as simple linear, rather than trigonometric, relations to make the original algorithm efficient (Smith, 1978). A color version of this diagram is shown in Plate 5-5.*

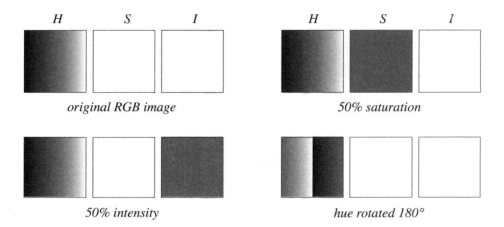

FIGURE 5-31. *Four examples of the HSI space components for a test image. The inverse hexcone model is applied to these components and the resulting RGB images are shown in Plate 5-5.*

TABLE 5-4. *Base image to be used to manipulate different color image properties with the interpolation/extrapolation blending algorithm. The magnitude of the effect for all properties is increased for $\alpha \geq 1$ and decreased for $0 \leq \alpha \leq 1$.*

property to be modified	base *image0*
intensity	black
contrast	grey
saturation	greyscale version of *image1*

Color transformations can be useful for displaying diverse, spatially-registered images. For example, a high-resolution visible band image may be assigned to the intensity component, a lower-resolution thermal band image may be assigned to the hue component, and a uniform value may be assigned to the saturation component (Haydn *et al.*, 1982). In geology, co-registered datasets, such as airborne gamma ray maps, can be combined with Landsat imagery to enhance interpretation (Schetselaar, 2001). After the inverse HSI transform, the resulting color image contains the detail structure of the remote sensing image expressed as intensity, with the gamma ray variation superimposed as color variation. This is a type of *image fusion*, discussed in detail in Chapter 8.

5.7   Summary

A variety of spectral transformations were examined, ranging from nonlinear spectral band ratios to linear transformations of various types. Some are designed to improve quantitative analysis of remote-sensing images, while others simply enhance subtle information so that it is visible. The following important points can be made:

- Spectral band ratios can help isolate spectral signatures from imagery by reducing topographic shading. Ratios involving NIR and red spectral bands are useful for vegetation measurements.

- The principal components transform is optimal for data compression, but because it is data dependent, the resulting features have different interpretations for different images.

- The tasseled-cap transform provides a fixed, but sensor-specific, transform based on soil and vegetation signatures.

- Color-image composites can be enhanced for visual interpretation by a variety of spectral transforms. These transforms, however, are not to be used for quantitative data analysis.

In the next chapter, we will look at spatial transforms, which can be combined with spectral transforms for certain applications such as image fusion and feature extraction for classification.

5.8   Exercises

Ex 5-1. What is the required W in Eq. (5-1) to extract the NIR and red bands from 7-band TM data, i.e., subset the bands? What is the required W for a color min-max stretch of three TM bands, given the following DN ranges?

$$
\begin{array}{ll}
\text{TM4:} & 53\text{--}178 \\
\text{TM3:} & 33\text{--}120 \\
\text{TM2:} & 30\text{--}107
\end{array}
$$

Ex 5-2. Suppose a *GIFOV* contains soil and vegetation and that the net reflectance measured at the pixel is a weighted sum of the reflectances of each material, where the weights are simply the fractional coverages of each (i.e., the linear mixing model; see Chapter 9). For the two cases below, calculate and plot the *NDVI* and *SAVI*, with an L value of 0.5, versus the fraction of vegetation within the *GIFOV*. Which index has a more linear relation to vegetation coverage?

	Case I: dry soil		Case II: wet soil	
	ρ_{red}	ρ_{NIR}	ρ_{red}	ρ_{NIR}
soil	0.35	0.40	0.20	0.20
vegetation	0.10	0.50	0.10	0.50

Ex 5-3. Fill in the details on the facts and approximations that result in Eq. (5-6). What underlying physical constants or quantities do the two parameters, a_b and b_b, contain?

Ex 5-4. Given a multispectral image with a *DN* covariance matrix,

$$C = \begin{bmatrix} 1900 & 1200 & 700 \\ 1200 & 800 & 500 \\ 700 & 500 & 300 \end{bmatrix} .$$

What is the correlation matrix? Now, suppose you do a calibration of the data to at-sensor radiance *L* as follows,

$$L_1 = 2 \times DN_1 + 11$$
$$L_2 = 3 \times DN_2 + 4$$
$$L_3 = 5 \times DN_3 + 2$$

What are the covariance and correlation matrices of the calibrated data? Will the PCT of the calibrated data be the same as the PCT of the *DN* data? Will the standardized PCT be the same? Provide a mathematical and graphical explanation.

Ex 5-5. Explain with a hypothetical 2-D scatterplot why small areas of change between two images will be "captured" in the multitemporal PC_2 component.

Ex 5-6. Suppose you do a PC decorrelation stretch of a color image and compare the histograms of the result in HSI space to the histograms of the original image in HSI space. What differences do you expect and why?

Ex 5-7. Given the following histograms of the (actual) HSI components of a three-band image, specify three *DN* transformations that will make the saturation of every pixel equal to 200, linearly stretch the intensity to increase the contrast, and leave the hue unchanged.

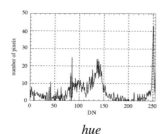

hue

saturation

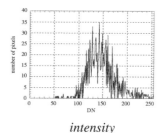

intensity

CHAPTER 6

Spatial Transforms

6.1 Introduction

Spatial transforms provide tools to extract or modify the spatial information in remote-sensing images. Some transforms, such as convolution, use only local image information, i.e., within relatively small neighborhoods of a given pixel. Others, for example the Fourier transform, use global spatial content. Between these two extremes, the increasingly important category of scale-space filters, including Gaussian and Laplacian pyramids and the wavelet transform, provide data representations that allow access to spatial information over a wide range of scales, from local to global.

6.2   An Image Model for Spatial Filtering

A useful concept for understanding spatial filtering is that any image is made from spatial components at different scales. Suppose we process an image such that the value at each output pixel is the average of a small neighborhood of input pixels (say 3 × 3), as illustrated in Fig. 6-1. The result is a blurred version of the original image. Now we subtract this result from the original, yielding the image on the right in Fig. 6-1, which represents the difference between each original pixel and the average of its neighborhood. To look at spatial information at a different scale, we repeat the process with a larger neighborhood, such as 7 × 7 pixels. Anticipating our later discussion, we call the blurred image a *Low-Pass (LP)* version of the image and the difference between it and the original image a *High-Pass (HP)* version, and write the mathematical relation,

$$\text{image}(x, y) \;=\; LP(x, y) + HP(x, y) \tag{6-1}$$

which is valid for any size (scale) neighborhood. As the neighborhood size is increased, the LP image isolates successively larger and larger structures, while the HP image picks up the smaller structures lost in the LP image, to maintain the relation in Eq. (6-1).

This *decomposition* of an image into a sum of components at different scales is the basis of all spatial filtering. The inverse process, namely adding the components together to synthesize the image, is *superposition*.[1] In Eq. (6-1), each of the two components actually contains a range of scales. In a similar fashion, scale-space filtering decomposes an image into several components, each containing a range of scales. We will see later that it is also possible to decompose an image into a large set of components, each representing a single scale, using the Fourier transform.

6.3   Convolution Filters

The underlying operation in a convolution filter is the use of a moving window on the image. An operation is performed on the input image pixels within the window, a calculated value is put in the output image, usually at the same location as the center of the window in the input image, and the window is then moved one pixel along the same line to process the next neighborhood of input image pixels, which are unchanged for subsequent calculations within the window. When a line of pixels is finished, the window is then moved down one row, and the process is repeated (Fig. 6-2). Almost any function can be programmed within the moving window; some examples are listed in Table 6-1 and discussed below in the following sections.

---

1. These dual transformations are sometimes called *analysis* and *synthesis*, particularly in the wavelet literature.

image(x,y)   —   LP(x,y)   =   HP(x,y)

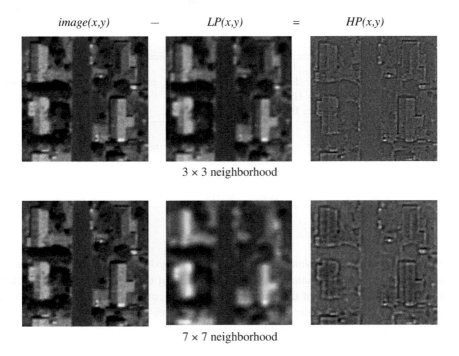

3 × 3 neighborhood

7 × 7 neighborhood

FIGURE 6-1. *Examples of the global spatial frequency image model at two scales. The difference image has both positive and negative DNs but is scaled to all positive GLs for display.*

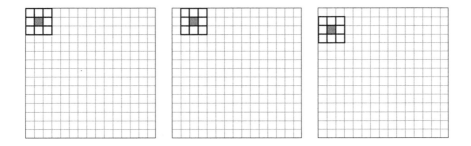

FIGURE 6-2. *A moving window for spatial filtering. The first output pixel is calculated (left), the next output pixel is calculated for the same row (center), and after the row is completed the process is repeated for the next row (right). The output pixel is located at the coordinate of the shaded pixel in the output image.*

TABLE 6-1. Catalog of local filter types.

type	output	examples	applications
linear	weighted sum	Low-Pass Filter (LPF) High-Pass Filter (HPF) High-Boost Filter (HBF) Band-Pass FIlter (BPF)	enhancement, sensor simulation, noise removal
statistical	given statistic	minimum, maximum median standard deviation mode	noise removal, feature extraction, *SNR* measurement
gradient	vector gradient	Sobel, Roberts	edge detection

6.3.1 Linear Filters

In Chapter 3 we introduced *convolution* as a fundamental physical process behind instrument measurements. Convolution in that context described the effect of the system response function (spectral or spatial) on the resolution of data provided by the instrument. The same type of operator is also quite useful in processing digital images. In this case, however, we *specify* the response function and have nearly complete freedom to use *any* weighting function suited to the application at hand. That flexibility makes convolution one of the most useful tools in image processing. In effect, we are creating a *virtual instrument* with a given response, and applying it to an input image. The output image represents the output of the virtual instrument.

Convolution

A linear filter is calculated in the spatial domain as a weighted sum of pixels within the moving window. This *discrete convolution* between the input image f and the window response function w, both of size $N_x \times N_y$, is written mathematically for the output pixel g_{ij},

$$g_{ij} = \sum_{m=0}^{N_x-1} \sum_{n=0}^{N_y-1} f_{mn} w_{i-m,j-n} \tag{6-2}$$

and expressed symbolically in the convenient form,

$$g = f * w. \tag{6-3}$$

Since the nonzero extent of the window is typically much smaller than the image, the sum in Eq. (6-2) does not have to be over every pixel. If the window is $W_x \times W_y$ pixels, we can write an alternate expression,

$$g_{ij} = \sum_{m = i - W_y/2}^{i + W_y/2} \sum_{n = j - W_x/2}^{j + W_x/2} f_{mn} w_{i-m, j-n}, \tag{6-4}$$

where w is centered at (0,0) and is nonzero over $\pm W_x/2$ and $\pm W_y/2$.[2] In this form, we can clearly see that the output pixel is a weighted sum of pixels within a neighborhood of the input pixel.

The distinguishing characteristic of a linear filter is the *principle of superposition*, which states that the output of the filter for a sum of two or more inputs is equal to the sum of the individual outputs that would be produced by each input separately. This is achieved with a convolution because Eq. (6-2) is a linear weighted sum of the input pixels. Furthermore, the filter is *shift-invariant* if the weights do not change as the window moves across the image. The reader may compare Eq. (6-3) and Eq. (6-4) with Eq. (3-1) and Eq. (3-2) to better appreciate the relationship between the continuous convolution of physical functions and the discrete convolution of data arrays.

The implementation of Eq. (6-2) involves the following steps:

1. Flip the rows and columns of the window function (equivalent to a 180° rotation).
2. Shift the window such that it is centered on the pixel being processed.
3. Multiply the window weights and the corresponding original image pixels.
4. Add the weighted pixels and save as the output pixel.
5. Repeat steps 2 through 4 until all pixels have been processed.

The first step, flipping of the window, is often forgotten because many window functions are symmetric. However, for asymmetric window functions, it is important in determining the result.

Low-pass and high-pass filters (LPF, HPF)

Equation (6-1) defines a complementary pair of images, the sum of which equals the original image. Using the notation for convolution, we can rewrite it as[3]

$$\begin{aligned} \text{image}(x, y) &= LPF * \text{image(x,y)} + HPF * \text{image(x,y)} \\ &= (LPF + HPF) * \text{image(x,y)} \\ &= IF * \text{image(x,y)} \end{aligned} \tag{6-5}$$

For Eq. (6-5) to hold, the *LPF* and *HPF* must sum to an *identity filter*, *IF*, i.e., a single one surrounded by zeros.[4] This relationship then defines two *complementary* convolution filters. For example, a 1×3 *LPF* has the weights [+1/3 +1/3 +1/3], and its complementary *HPF* is [-1/3 +2/3 -1/3]; their sum is the *IF* with the weights [+0 +1 +0].

---

2. If W is an odd number, then $W/2$ is rounded down, e.g., for W equal 5, $W/2$ equals 2.
3. We will use a notation where, e.g., *LPF*[image(x,y)] or $LPF * \text{image(x,y)}$ denotes a Low-Pass Filter operating on an image and LP(x,y) denotes the low-pass filtered image that results.
4. Known as a *delta function* in science and engineering fields.

Examples of 1×3 and 1×7 filters applied to a one-dimensional signal are shown in Fig. 6-3. The *LPF*s preserve the local mean (the sum of their weights is one) and smooth the input signal; the larger the window, the more the smoothing. The *HPF*s remove the local mean (the sum of their weights is zero) and produce an output which is a measure of the deviation of the input signal from the local mean. These same characteristics also pertain to two-dimensional LP and HP filters (Table 6-2).

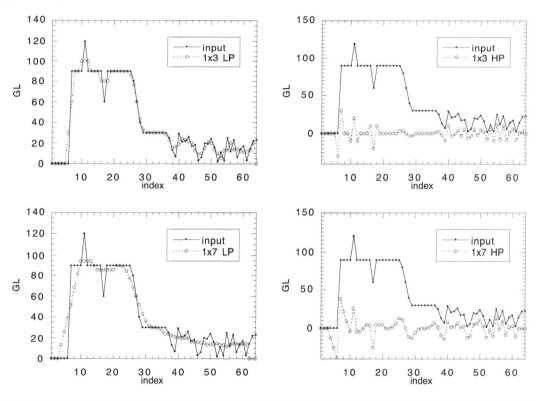

FIGURE 6-3. 1-D signal processed with 1×3 and 1×7 LPFs and HPFs. The LP result approximates the original signal, but is somewhat smoother. The HP result has zero mean and zero-crossings wherever there is step or pulse in the original signal. The 1×7 LP result is smoother than that for the 1×3 LPFs, and the HP result has a broader response around the zero-crossings.

High-boost filters (HBF)

An image and its HP component can be additively combined to form a *high-boost* image,

$$HB(x, y;K) = \text{image(x,y)} + K \cdot HP(x, y) , \ K \geq 0 \tag{6-6}$$

TABLE 6-2. *Examples of simple box filters, which have uniform weights in the LPF and the complementary weights in the HPF. The normalizing factor is required to preserve the relation in Eq. (6-1).*

size	LPF	HPF
3×3	$1/9 \cdot \begin{bmatrix} +1 & +1 & +1 \\ +1 & +1 & +1 \\ +1 & +1 & +1 \end{bmatrix}$	$1/9 \cdot \begin{bmatrix} -1 & -1 & -1 \\ -1 & +8 & -1 \\ -1 & -1 & -1 \end{bmatrix}$
5×5	$1/25 \cdot \begin{bmatrix} +1 & +1 & +1 & +1 & +1 \\ +1 & +1 & +1 & +1 & +1 \\ +1 & +1 & +1 & +1 & +1 \\ +1 & +1 & +1 & +1 & +1 \\ +1 & +1 & +1 & +1 & +1 \end{bmatrix}$	$1/25 \cdot \begin{bmatrix} -1 & -1 & -1 & -1 & -1 \\ -1 & -1 & -1 & -1 & -1 \\ -1 & -1 & +24 & -1 & -1 \\ -1 & -1 & -1 & -1 & -1 \\ -1 & -1 & -1 & -1 & -1 \end{bmatrix}$

which is a sharpened version of the original image; the degree of sharpening is proportional to the parameter K.

Example HB box filters are given in Table 6-3 (see Ex 6-2 to derive an *HBF*); the sum of their weights is one, which means that the output image will have the same mean *DN* as the input image. Image examples are shown in Fig. 6-4 for different values of K. In HB processing, we are reinforcing ("boosting") the high-frequency components of the original image, relative to the low-frequency components (Wallis, 1976; Lee, 1980). The parametric nature of the filter allows "tuning" of K to achieve the degree of enhancement desired and would be ideally implemented in an interactive mode.

TABLE 6-3. *Example 3×3 HB box filters for different values of K.*

$K = 1$	$K = 2$	$K = 3$
$1/9 \cdot \begin{bmatrix} -1 & -1 & -1 \\ -1 & +17 & -1 \\ -1 & -1 & -1 \end{bmatrix}$	$1/9 \cdot \begin{bmatrix} -2 & -2 & -2 \\ -2 & +25 & -2 \\ -2 & -2 & -2 \end{bmatrix}$	$1/9 \cdot \begin{bmatrix} -3 & -3 & -3 \\ -3 & +33 & -3 \\ -3 & -3 & -3 \end{bmatrix}$

Band-pass filters (BPF)

A band-pass version of an image can be constructed as a sequence of a *LPF* followed by a *HPF*,

$$BP(x, y) = HPF[LPF[\text{image}(x,y)]] = HPF[LP(x, y)]. \qquad (6\text{-}7)$$

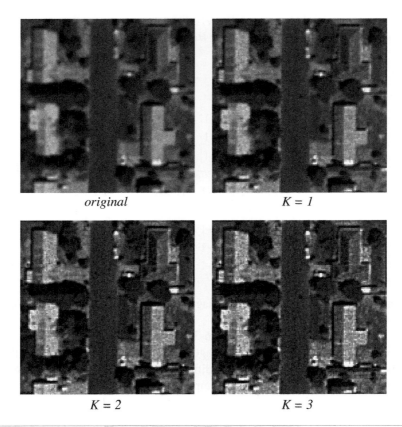

FIGURE 6-4. Example application of 3 × 3 HB filters from Table 6-3. Each image is stretched or saturated to [0,255], as necessary.

BPFs are primarily useful for periodic noise isolation and removal. Some examples will be shown in Chapter 7.

Directional filters

It is possible to design a convolution filter to process images in a particular direction; some directional filters are shown in Table 6-4. These are all variations on the notion of a discrete derivative, which is a type of HP filter. Fig. 6-5 depicts the results of preferential processing for oriented-features. Note how the diagonal filter accentuates horizontal *and* vertical edges, in addition to those at −45°. Considerable care obviously must be exercised when interpreting spatially-filtered images, particularly directional enhancements, because of their abstract nature.

TABLE 6-4. Example directional filters. For the aximuthal filter, the angle α is measured counterclockwise from the horizontal axis; features in the directions α ± 90° will be enhanced.

type	direction of enhanced features			
	vertical	horizontal	diagonal	azimuthal
1st derivative	$\begin{bmatrix} -1 & +1 \end{bmatrix}$	$\begin{bmatrix} -1 \\ +1 \end{bmatrix}$	$\begin{bmatrix} -1 & 0 \\ 0 & +1 \end{bmatrix}$, $\begin{bmatrix} 0 & -1 \\ +1 & 0 \end{bmatrix}$	$\begin{bmatrix} \sin\alpha & 0 \\ -\sin\alpha-\cos\alpha & \cos\alpha \end{bmatrix}$
2nd derivative	$\begin{bmatrix} -1 & +2 & -1 \end{bmatrix}$	$\begin{bmatrix} -1 \\ +2 \\ -1 \end{bmatrix}$	$\begin{bmatrix} -1 & 0 & 0 \\ 0 & +2 & 0 \\ 0 & 0 & -1 \end{bmatrix}$, $\begin{bmatrix} 0 & 0 & -1 \\ 0 & +2 & 0 \\ -1 & 0 & 0 \end{bmatrix}$	

The border region

In spatial filtering, it is desirable to have the size of the output and input images equal, because that allows further algebraic operations such as discussed at the beginning of this chapter. If a $W \times W$ window (W being an odd integer) is used for a convolution filter, the *border region* includes the first and last $W/2$ (truncated to an integer) rows and columns of the input image (Fig. 6-6). Output pixels within the border region cannot be directly calculated, and since the window cannot extend beyond the borders of the original image, some "trick" must be used to maintain the size of the image from input to output. A number of techniques can be used to fill this area and make the output image the same size as the input image:

- Repeat the nearest valid output pixel in each border pixel.
- Reflect the input pixels in the border area outward to effectively increase the size of the input image. The convolution is then done on the larger image, and the result is trimmed to the same size as the input image.
- Reduce the width and/or height of the window to one pixel in the border area.
- Set the border pixels to zero or to the mean *DN* of the output image.
- Wrap the window around to the opposite side of the image and include those pixels in the calculation. This is not an obvious thing to do, but is motivated by the fact that the resulting *circular convolution* is equivalent to convolution performed by Fourier transforms.

Each approach causes a different type of artifact around the border. In general, the first and second are the most effective in maintaining image size without introducing severe artifacts.

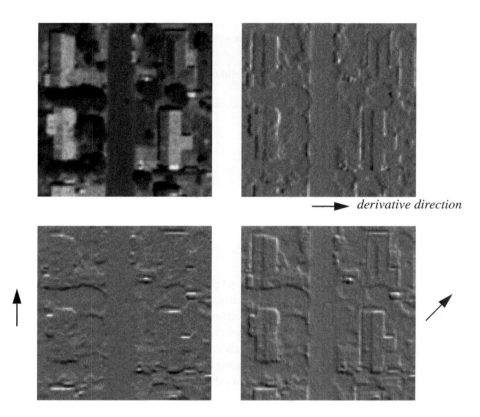

derivative direction

FIGURE 6-5. *Examples of directional enhancement using derivative filters. The direction of the derivative is indicated by the arrows. In some consumer image processing programs, this effect is called "embossing" or "relief" because of the 3-D impression conveyed by the filtering.*

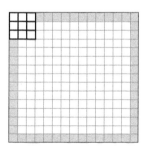

FIGURE 6-6. *The border region for a 3 × 3 filter. If the image size is to be maintained from input to output, some "trick" must be used to calculate the output pixels at the shaded locations.*

Characteristics of filtered images

The LP component of an image is statistically *nonstationary*, i.e., its properties (local mean and variance) change from point to point, while the HP component can be modeled as having a statistically *stationary mean* (zero), with a variance that scales depending on local image contrast (Hunt and Cannon, 1976). One sees this empirically in that the histograms of HP images universally exhibit a Gaussian-like shape with zero mean. The variance of an HP image is typically much less than that of the original image. The histogram of an LP image resembles that of the original image, with a slightly reduced *DN* range (Fig. 6-7).

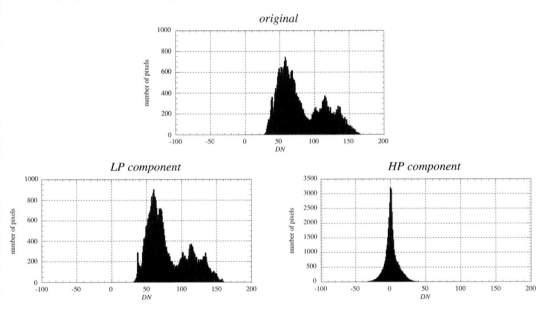

FIGURE 6-7. *Histograms of a TM image, and its LP and HP components. A 5 × 5 convolution window was used to generate the two components. Note the larger amplitudes of the LP and HP histograms because of their narrower widths and that the HP histogram has a near-zero mean with both positive and negative DNs.*

Application of the blending algorithm to spatial filtering

The "blending" algorithm for color processing described in Chapter 5 can also be applied to spatial filtering. The blending operation is repeated here for convenience,

$$output = (1 - \alpha) \times image0 + \alpha \times image1 \qquad (6\text{-}8)$$

The image of interest is *image1*, which is blended with the base image, *image0*. The type of filtering is determined by the nature of the base image. For example, if *image0* is a *LPF* version of *image1*, the result can be a LP, HB or HP version of *image1*, depending on the value of α. If α is between zero and one, the output image is a LP version of *image1*. If α is greater than one, the output image is a HB version of *image1*, which approaches a pure HP image as α is increased to large values (Fig. 6-8).

$\alpha = 0$ $\qquad$ $\alpha = 0.5$ $\qquad$ $\alpha = 1$ $\qquad$ $\alpha = 5$ $\qquad$ $\alpha = 100$

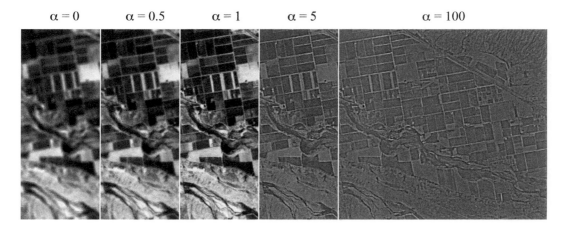

FIGURE 6-8. *Application of the blending algorithm to variable spatial filtering. The base image ($\alpha = 0$) is a 5×5 box-filter LP version of the image of interest ($\alpha = 1$).*

The box-filter algorithm

The simplest LP box filter can be programmed in a very efficient *recursive* form. Consider a 3×3 linear convolution filter, as shown in Fig. 6-9. A straightforward calculation of the average at each pixel would require eight additions.[5] However, if we save and update the sum of the input pixels in each of the three columns of the window (C_1, C_2, C_3), the eight additions need to be computed only once at the beginning of an image line. The next output pixel is calculated as,

$$
\begin{aligned}
\text{output pixel} &= C_2 + C_3 + C_4 \\
&= C_1 + C_2 + C_3 - C_1 + C_4 \qquad , \\
&= \text{previous output pixel} - C_1 + C_4
\end{aligned}
\tag{6-9}
$$

---

5. We are ignoring the multiplication by the filter weights. For the LP box filter, this multiplication can be done after summing the pixels in the window and is one additional operation.

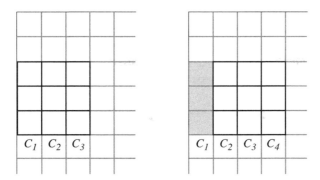

FIGURE 6-9. *Depiction of the box-filter algorithm applied to neighboring regions along an image row. The weighted sum of pixels in the window column j is C_j.*

an algorithm that can obviously be extended to all pixels in the line. Thus, except for the first output pixel of each line, only three additions and one subtraction are needed to calculate each output pixel for a 3 × 3 filter, yielding a computational advantage of two for the recursive algorithm over the direct calculation. The advantage obviously increases with the size of the window, and the recursion algorithm also can be applied in the vertical direction by maintaining three full lines of data in memory for the previously calculated output image line, the oldest input image line, and the newest input image line (McDonnell, 1981). With a full bidirectional implementation of the recursion algorithm, the number of computations required for each output pixel is *independent of the size of the filter window.*

The box-filter algorithm can be applied along image rows only if the filter weights are constant within each row of the window. In that case, it is unquestionably the fastest algorithm for convolution. Otherwise, Fourier domain implementation of the convolution can be a competitive alternative (see Sect. 6.4.4), particularly for large windows.

Cascaded linear filters

A series of filters applied sequentially to an image can be replaced by a single, net filter that is equal to the convolution of the individual filters,

$$g = (f * w_1) * w_2 = f * (w_1 * w_2) = f * w_{net} \qquad (6\text{-}10)$$

where,

$$w_{net} = w_1 * w_2 . \qquad (6\text{-}11)$$

Suppose that w_1 and w_2 are each $W \times W$ pixels in size. The net window function w_{net} is then $(2W - 1) \times (2W - 1)$; examples are given in Table 6-5. Any number of cascaded filters may be replaced by a single filter, as long as there are no nonlinear operations (such as *DN* thresholding) in the processing chain. This property of linear filters was used in Chapter 3 to construct the net sensor spatial response from the individual component responses.

TABLE 6-5. Example cascaded filters and their equivalent net filter.

filter 1	filter 2	net filter
$1/9 \cdot \begin{bmatrix} +1 & +1 & +1 \\ +1 & +1 & +1 \\ +1 & +1 & +1 \end{bmatrix}$	$1/9 \cdot \begin{bmatrix} +1 & +1 & +1 \\ +1 & +1 & +1 \\ +1 & +1 & +1 \end{bmatrix}$	$1/81 \cdot \begin{bmatrix} +1 & +2 & +3 & +2 & +1 \\ +2 & +4 & +6 & +4 & +2 \\ +3 & +6 & +9 & +6 & +3 \\ +2 & +4 & +6 & +4 & +2 \\ +1 & +2 & +3 & +2 & +1 \end{bmatrix}$
$1/3 \cdot \begin{bmatrix} +1 & +1 & +1 \end{bmatrix}$	$1/3 \cdot \begin{bmatrix} +1 \\ +1 \\ +1 \end{bmatrix}$	$1/9 \cdot \begin{bmatrix} +1 & +1 & +1 \\ +1 & +1 & +1 \\ +1 & +1 & +1 \end{bmatrix}$

6.3.2 Statistical Filters

Statistical filters output a local statistical property of an image. Statistical measures calculated over small neighborhoods have low statistical significance because of the small sample size, but are nevertheless useful for tasks such as noise reduction (local median), edge detection (local variance), or texture feature extraction (local variance). Some examples are shown in Fig. 6-10.

The *median filter* is a particularly useful statistical filter. If the pixel *DN*s within the window are sorted into decreasing or increasing order, the output of the median filter is the *DN* of the pixel at the middle of the list (the number of pixels must be odd). The median operation has the effect of excluding pixels that do not fit the "typical" statistics of the local neighborhood, i.e., outliers. Isolated noise pixels can therefore be removed with a median filter. One can see this by generating a sequence of numbers with one or more outliers, for example,

$$10, 12, 9, 11, 21, 12, 10, \text{ and } 10.$$

If the median of five points is calculated in a moving window, the result is (excluding the two border points on either end),

$$. \quad . \quad , 11, 12, 11, 11, \quad . \quad .$$

and the outlier value, 21, has magically disappeared! If we apply a median filter to the data of Fig. 6-3, we see that it very neatly removes the two spikes in the data, with minimal smoothing of other features (Fig. 6-11). Efficient data sorting algorithms, such as HEAPSORT and QUICKSORT (Press *et al*., 1992), can speed median filter processing for large windows.

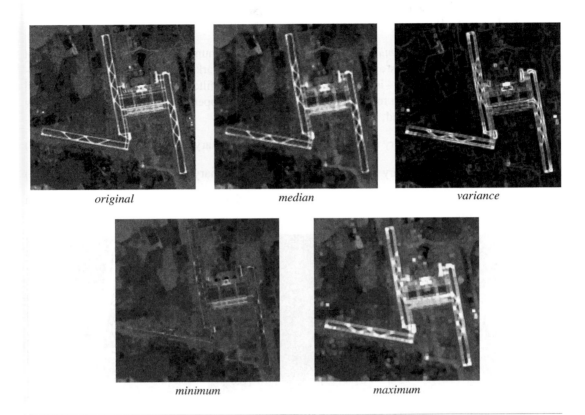

original   *median*   *variance*

minimum   *maximum*

FIGURE 6-10. Example processing by 3 × 3 statistical filters.

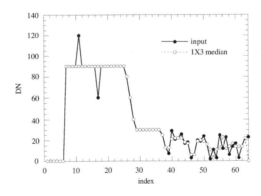

FIGURE 6-11. Application of the median filter to the 1-D signal used earlier.

Morphological filters

The *minimum filter* and *maximum filter* output the local minimum or maximum *DN* of the input image. If a minimum filter is applied to a binary image of dark and light objects, the result— expansion of the dark objects—is the same as that of a *dilation* filter, and a maximum filter applied to a binary image is equivalent to an *erosion* filter. If these two operations are cascaded, an *opening* or *closing* operation is obtained,

$$\text{opening[binary image]} = \text{dilation[erosion[binary image]]} \qquad (6\text{-}12)$$

$$\text{closing[binary image]} = \text{erosion[dilation[binary image]]}. \qquad (6\text{-}13)$$

This type of image processing is illustrated in Fig. 6-12 and has application to spatial segmentation and noise removal.

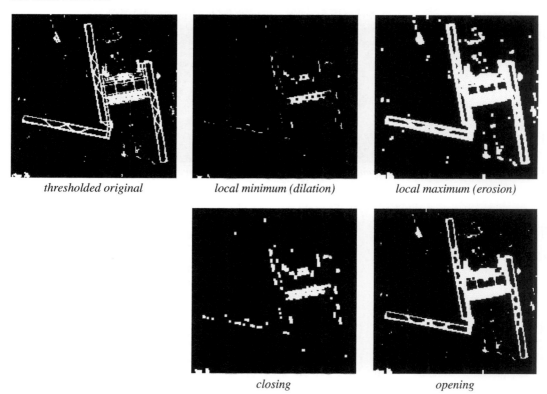

thresholded original	*local minimum (dilation)*	*local maximum (erosion)*

closing	*opening*

FIGURE 6-12. *Examples of 3 × 3 morphological filter processing of the image in Fig. 6-10. The terms "erosion" and "dilation" always refer to the dark objects in a binary image, independent of its content. In our example, the dilation operation "erodes" the airport, and the erosion operation "dilates" the airport!*

Dilation and erosion filters are examples of *morphological image processing* (Serra, 1982; Giardina and Dougherty, 1988; Dougherty and Lotufo, 2003; Soille, 2002). The window shape is important because it affects the changes induced in binary objects; in addition to being simply square or rectangular, it can be a plus shape, a diagonal shape, or in fact, any desired pattern. In morphological image processing, the shaped window is called a *structuring element* and can be designed to perform pattern matching or modification of particular shapes. A good set of examples is presented in (Schalkoff, 1989).

6.3.3 Gradient Filters

The detection of significant *DN* changes from one pixel to another is a common problem in image processing. Such changes usually indicate a physical boundary in the scene, such as a coastline, a paved road, or the edge of a shadow. Although many different approaches to this problem have been proposed over many years (Davis, 1975), a combination of high-pass spatial filtering and *DN* thresholding provides a simple and effective technique that is widely used.

The directional high-pass filters of Table 6-4 produce images whose *DNs* are proportional to the difference between neighboring pixel *DNs* in a given direction, i.e., they calculate the *directional gradient*. An isotropic gradient may be calculated by filtering the image in two orthogonal directions, e.g., horizontally and vertically, and combining the results in a vector calculation at every pixel (Fig. 6-13). The *magnitude* of the local image gradient is given by the length of the composite vector, and the *direction* of the local gradient is given by the angle between the composite vector and the coordinate axis, as shown in Fig. 6-13.

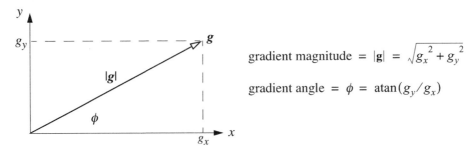

$$\text{gradient magnitude} = |\mathbf{g}| = \sqrt{g_x^2 + g_y^2}$$

$$\text{gradient angle} = \phi = \text{atan}(g_y/g_x)$$

FIGURE 6-13. Vector geometry for calculating image gradients. The x-derivative component of the image at the given pixel is g_x and the y-derivative component is g_y; their vector sum is $\mathbf{g}$.

Numerous local gradient filters have been proposed (Table 6-6), but there often is little visual difference in their results, however (Fig. 6-14). The edge boundaries produced by the 3 × 3 Sobel and Prewitt filters are not as sharp as those produced by the 2 × 2 Roberts filter. On the other hand, the 3 × 3 filters produce less noise enhancement. Another difference is that the output of the Roberts filter is shifted by one-half pixel relative to the filters with odd sizes. No matter what filter is used, the overall gradient calculation is nonlinear because of the vector magnitude calculation.

TABLE 6-6. *Example local gradient filters. See also Robinson (1977).*

filter	horizontal component	vertical component
Roberts	$\begin{bmatrix} 0 & +1 \\ -1 & 0 \end{bmatrix}$	$\begin{bmatrix} +1 & 0 \\ 0 & -1 \end{bmatrix}$
Sobel	$\begin{bmatrix} +1 & +2 & +1 \\ 0 & 0 & 0 \\ -1 & -2 & -1 \end{bmatrix}$	$\begin{bmatrix} -1 & 0 & +1 \\ -2 & 0 & +2 \\ -1 & 0 & +1 \end{bmatrix}$
Prewitt	$\begin{bmatrix} +1 & +1 & +1 \\ +1 & -2 & +1 \\ -1 & -1 & -1 \end{bmatrix}$	$\begin{bmatrix} -1 & +1 & +1 \\ -1 & -2 & +1 \\ -1 & +1 & +1 \end{bmatrix}$

The *detection* of edges is a binary classification problem that may be addressed with a *DN* threshold applied to the gradient magnitude. A threshold that is too low results in many isolated pixels classified as edge pixels and thick, poorly defined edge boundaries, while a threshold that is too high results in thin, broken segments (Fig. 6-14). These problems arise because we are using only local information in the gradient filters. The use of scale-space filtering and zero-crossing mapping, as described later in this chapter, greatly increases the robustness of edge detection and the spatial continuity of edge maps.

6.4   Fourier Transforms

Fourier theory dates from the eighteenth century and has applications in almost all areas of science and engineering and even the fine arts, such as stock market analysis! It is a general framework for analysis of signals in one or more dimensions as linear combinations of basic sinusoidal functions. We will introduce the topic in the context of synthesis of 1-D signals and 2-D images.

6.4.1 Fourier Analysis and Synthesis

In Sect. 6.2 we introduced the idea that an image could be represented by a sum of two components with different spatial scales. The Fourier transform is an extension of that idea to *many* scales. We will use a 1-D signal example to illustrate and then extend to two dimensions. In Fig. 6-15 we have a 1-D periodic, infinite square wave signal with a spatial period of 100 units. It can be shown that this signal is composed of an infinite set of sine wave signals of different frequencies, amplitudes, and phases (see Fig. 4-17). The first component actually has zero frequency because it represents

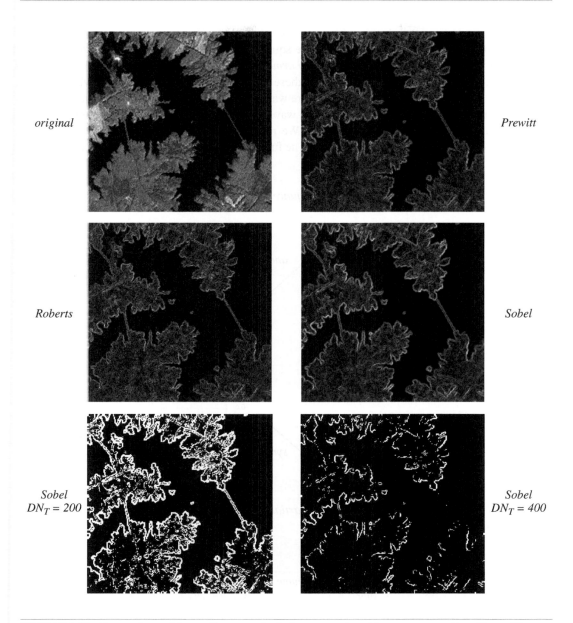

FIGURE 6-14. Comparison of the gradient magnitude images produced by common gradient filters. Binary edge maps produced by thresholding the Sobel gradient magnitude at two levels are in the bottom row.

the mean amplitude of the signal and is sometimes called the *DC* term.[6] The lowest nonzero frequency component has the same period as the square wave and is known as the *fundamental*; the next highest frequency component, the *third harmonic*, is three times the frequency of the fundamental, and so on. The relative strengths of these components are 1, 1/3, and 1/5 when they are added together to synthesize the original square wave. As each additional higher frequency component is included, the sum approaches a square wave, with the edges of the pulses becoming sharper and the tops and bottoms becoming flatter. We need to add an infinite number of sine waves, however, to retrieve the square wave exactly. The full sum is termed the *Fourier Series* of the square wave.

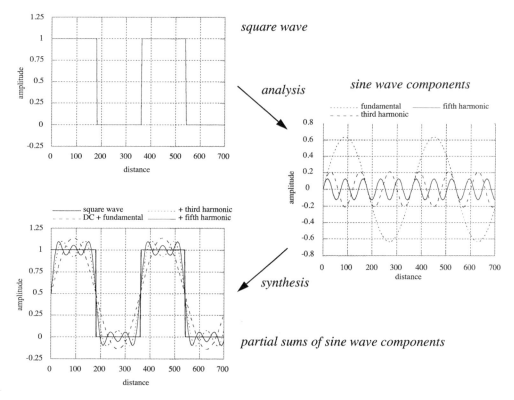

FIGURE 6-15. *Fourier analysis of a 1-D continuous square wave into its sine wave components and synthesis of the square wave by superposition of sine waves. Even with only the DC term and three sine waves, the synthesized signal is a fair approximation of the square wave. The residual oscillations near the transitions in the square wave are known as Gibbs phenomenon and disappear completely only when an infinite number of sine waves are included in synthesis (Gaskill, 1978).*

6. DC is an acronym for Direct Current and is used in physics and electrical engineering to describe a constant current from a stable source such as a battery. Its use has persisted to describe the constant mean level of any signal.

In Fig. 6-16 and Fig. 6-17 we extend this idea to synthesis of 2-D digital images. Because the images are discrete (arrays of digital numbers), the Fourier series is a finite sum of sines and cosines. If all the terms are included, the original image is obtained. The examples show how larger partial sums provide better approximations to the original image. Each partial sum amounts to an LP version of the image, and the error maps are the complementary HP versions. They satisfy the relation of Eq. (6-1), namely,

$$\text{image} = \text{partial sum} + \text{error}$$
$$= LP + HP. \tag{6-14}$$

6.4.2 Discrete Fourier Transforms in 2-D

The 2-D discrete Fourier series is defined mathematically as,

$$f_{mn} = \frac{1}{N_x N_y} \sum_{k=0}^{N_x - 1} \sum_{l=0}^{N_y - 1} F_{kl} \, e^{j2\pi\left(\frac{mk}{N_x} + \frac{nl}{N_y}\right)} \tag{6-15}$$

where F_{kl} is the complex amplitude of component kl. The complex exponential term can be written as,[7]

$$e^{j2\pi\left(\frac{mk}{N_x} + \frac{nl}{N_y}\right)} = \cos 2\pi\left(\frac{mk}{N_x} + \frac{nl}{N_y}\right) + j \sin 2\pi\left(\frac{mk}{N_x} + \frac{nl}{N_y}\right) \tag{6-16}$$

The coefficient j is the imaginary number $\sqrt{-1}$.

Equation (6-15) thus represents the *superposition of $N_x N_y$ cosine and sine waves to synthesize the original image f_{mn}*. The inverse of Eq. (6-15) is defined as,

$$F_{kl} = \sum_{m=0}^{N_x - 1} \sum_{n=0}^{N_y - 1} f_{mn} \, e^{-j2\pi\left(\frac{mk}{N_x} + \frac{nl}{N_y}\right)} \tag{6-17}$$

which yields the *Fourier coefficients* of the image. Equation (6-17) is the *Discrete Fourier Transform (DFT)* of f_{mn} and Eq. (6-15) is the *inverse* DFT of F_{kl}.

When we do the 2–D Fourier transform of a digital image, we implicitly assume that the image is replicated infinitely in all directions (Fig. 6-18), by virtue of the periodicity of the component cosines and sines.[8] Likewise, its Fourier transform is infinitely periodic. It is this periodic

---

7. This fundamental relation is known as *Euler's Theorem*.

8. This is the fundamental connection between the *continuous* Fourier transform and the *discrete* Fourier transform.

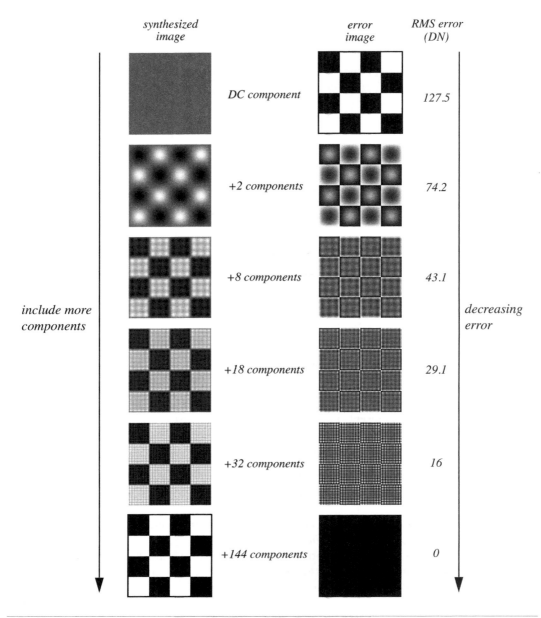

FIGURE 6-16. *Fourier synthesis of a 2-D square wave. The error images are stretched to [0,255]. Only the nonzero Fourier components are counted in the partial sums. As more components are included in the approximation of the original square wave, the residual error decreases. Unlike the residual synthesis error in the continuous signal of Fig. 6-15, the error here will go to zero with a finite number of components because the signal is discrete (sampled).*

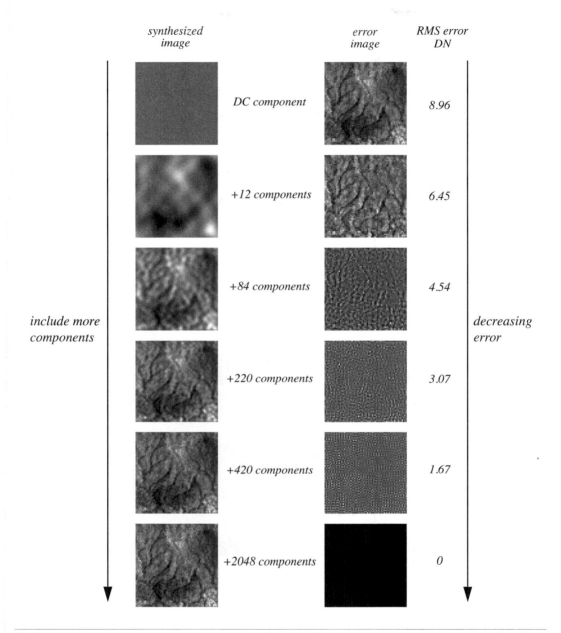

FIGURE 6-17. *Fourier synthesis of a portion of a TM image. The error images are stretched to [0,255]. The 64 × 64 image has 4096 complex Fourier coefficients, but because of symmetry, only 2048 are unique.*

representation that leads to the wraparound effect (circular convolution) when Fourier transforms are used to implement spatial convolution. The Fourier array, as produced by Eq. (6-17), is often reordered to the so-called *natural order*, which arises from optical Fourier transforms (Fig. 6-19).

The spatial frequency units for image processing, when we are not concerned with absolute quantities, are *cycles/pixel*. The spatial frequency *intervals* along each axis are given by,

$$\Delta u = 1/N_x , \quad \Delta v = 1/N_y \quad \text{(cycles/pixel)} \tag{6-18}$$

and the spatial frequency *coordinates* are given by,

$$u = k\Delta u = k/N_x , \quad v = l\Delta v = l/N_y \text{ (cycles/pixel).} \tag{6-19}$$

If absolute units are necessary, the spatial frequency intervals are divided by the pixel intervals Δx and Δy, in units of length, resulting in

$$\Delta u = \frac{1}{N_x \Delta x} , \quad \Delta v = \frac{1}{N_y \Delta y} \quad \text{(cycles/unit length).} \tag{6-20}$$

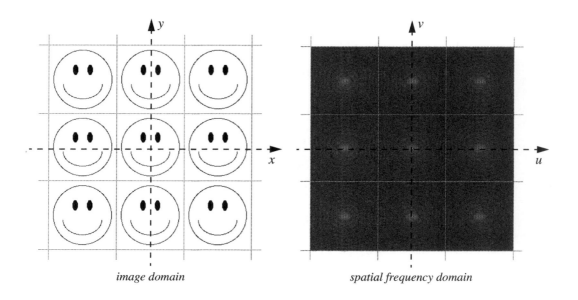

image domain   *spatial frequency domain*

FIGURE 6-18.   *The implied periodicity of the discrete Fourier transform extends infinitely in both directions. Each square on the left is a copy of the original image and each square on the right is a copy of the Fourier transform of the image. The replication is known as the "postage stamp" representation!*

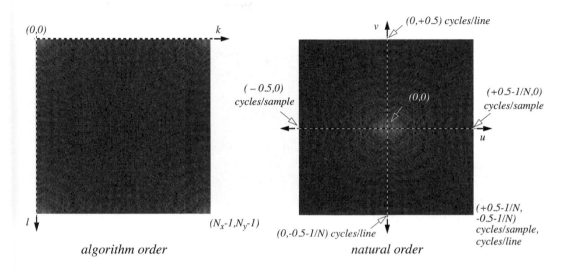

FIGURE 6-19. Coordinate geometry in the Fourier plane for an N × N array. The upper picture shows the spectrum as produced by the FFT algorithm. Because of the periodicity of the array, we can shift the origin, as shown below. The directions of k and l, and u and v are consistent with the discrete and natural coordinate systems described in Chapter 4.

6.4.3 The Fourier Components

The Fourier transform (Eq. (6-17)) produces *real* and *imaginary* components. These are related by the complex number relationship,

$$F_{kl} = Re(F_{kl}) + jIm(F_{kl}) \tag{6-21}$$

where j is $\sqrt{-1}$. A complex number can equivalently be written in terms of *amplitude*, A_{kl}, and *phase*, ϕ_{kl},

$$F_{kl} = A_{kl}e^{-j\phi_{kl}} \tag{6-22}$$

where,

$$A_{kl} = |F_{kl}| = \sqrt{[Re(F_{kl})]^2 + [Im(F_{kl})]^2} \tag{6-23}$$

and

$$\phi_{kl} = \text{atan}[Im(F_{kl})/Re(F_{kl})]. \tag{6-24}$$

The various components of the Fourier transform are illustrated in Fig. 6-20.

real and imaginary components

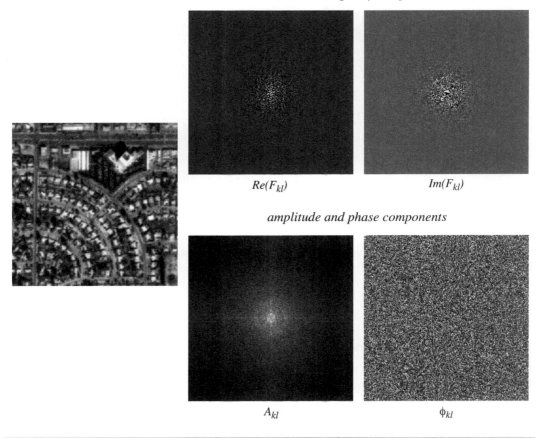

$Re(F_{kl})$ $Im(F_{kl})$

amplitude and phase components

A_{kl} ϕ_{kl}

FIGURE 6-20. The Fourier components of an image. To completely specify the transform, only the real and imaginary, or the amplitude and phase components are needed, not both. The natural coordinate system is used in this figure. The amplitude component is difficult to display as a greyscale image because it has a large dynamic range, dominated by the DC value. One way to address this is to take the logarithm of the spectrum amplitude and then set the DC component to zero, which greatly reduces the dynamic range of the displayed data.

The phase component is critical to the spatial structure of the image (Oppenheim and Lim, 1981). An impressive demonstration of this fact can be made by setting the amplitude component to a constant value, and doing an inverse Fourier transform of the modified spectrum. The result shows that the phase component carries information about the relative positions of features in the image. Conversely, if the phase part is set to a constant value (zero), with the original amplitude component retained, and the inverse Fourier transform performed, the result is unintelligible (Fig. 6-21). Thus, while the phase component is often ignored in Fourier filter design, it should be treated with respect!

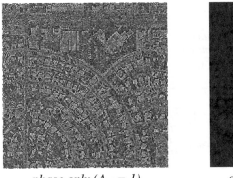 

phase only ($A_{kl} = 1$)   *amplitude only ($\phi_{kl} = 0$)*

FIGURE 6-21. *Evidence for the importance of spatial phase information. The spectrum of the image in Fig. 6-20 was modified by retaining only one component, as indicated above, and inverse Fourier transformed to the spatial domain. The resulting image is shown here.*

6.4.4 Filtering with the Fourier Transform

The major application of the Fourier transform to filtering is in implementation of Eq. (6-2). Taking the Fourier transform of both sides of that equation, we obtain the relatively simple product relation in the Fourier domain,[9]

$$\mathcal{F}[g_{ij}] = G_{kl} = \mathcal{F}[f * w]$$
$$= F_{kl}W_{kl} .$$

(6-25)

The arrays *F* and *G* are the *spatial frequency spectra* of the input and output images, respectively. *W* is the *transfer function* of the filter. In general, all three functions, *G*, *F*, and *W*, are complex. To use Fourier transforms for a spatial domain convolution, we must take the Fourier transform of the original image and the window weighting function, multiply the image spectra and the transfer function, which yields the filtered image spectrum, and then take the inverse of that product to obtain the filtered image in the spatial domain. This sequence is depicted in Fig. 6-22. Since the spatial window is normally much smaller than the image, it must be "padded" to the same size before taking its Fourier transform to insure that the Fourier components are at the same locations for both *F*

---

9. This transform relationship, i.e., that a convolution in one domain is a product of Fourier transforms in the Fourier domain, is known as the *Convolution Theorem* (Castleman, 1996; Jain, 1989).

and W (Eq. (6-18)). This padding is accomplished by surrounding the window function with zeros. If one filter is to be applied to many different images, its transfer function can be calculated once and used directly in the multiplication step, thus avoiding one of the Fourier transform operations.

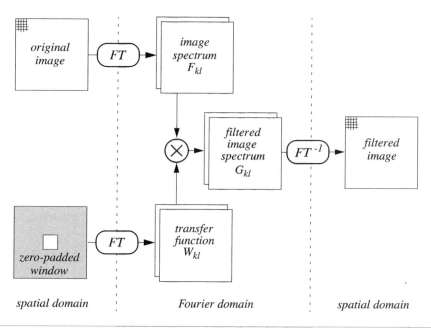

FIGURE 6-22.   *A filtering algorithm that uses the Fourier transform to compute a spatial domain convolution. The Fourier domain arrays are shown as doubles because the data are complex in that domain.*

The amplitude and phase components of Eq. (6-25) are given by,

$$|G_{kl}| = |F_{kl}||W_{kl}|$$
$$\phi(G_{kl}) = \phi(F_{kl}) + \phi(W_{kl})$$

$$(6\text{-}26)$$

These equations describe how the filter W affects the modulation and spatial phase, respectively, of the input digital image spectrum F.

A general guide for choosing spatial or Fourier domain processing is to use spatial domain convolution if the window is 7×7 or smaller and Fourier domain filtering otherwise (Pratt, 1991). Of course this "rule" depends also on the size of the image, the speed of the particular algorithms used, and the availability of special purpose hardware for performing convolutions or FFTs. The most efficient software implementation of Eq. (6-15) is the *Fast Fourier Transform (FFT)* (Brigham, 1988). Common FFT algorithms require that the size of the input array is a power of two; less efficient algorithms can transform images of other sizes. The choice in remote sensing is usually in favor of spatial domain convolution because of the large size of remote-sensing images, although the steadily increasing computation speed of computers makes the Fourier domain approach worth considering in some cases. As we will see in Chapter 7, the Fourier approach provides a unique view of the data that is quite useful for analysis and removal of periodic noise components.

Transfer functions

One of the most significant advantages of Eq. (6-25) over Eq. (6-2) is that it allows us to view filters as multiplicative "masks" in the spatial frequency domain. For example, an "ideal" amplitude LP filter has the characteristics,

$$
\begin{aligned}
|W_{kl}| &= 1, |k| \le k_c, |l| \le l_c \\
|W_{kl}| &= 0, |k| > k_c, |l| > l_c
\end{aligned}
\tag{6-27}
$$

where k_c and l_c are the effective "cutoff" frequencies for the *LPF*. The ideal *LPF* is a binary mask in the Fourier domain that transmits frequency components below the cutoff frequency unchanged, and does not transmit frequency components above the cutoff frequency. The amplitude filter, $|W_{kl}|$, is called the *Modulation Transfer Function (MTF)*. The *MTF*s of some of the box filters described earlier are shown in Fig. 6-23. A useful relationship to remember is that the value of the *MTF* at zero frequency is the sum of the weights in the corresponding spatial domain convolution filter,

$$
|W_{kl}(0, 0)| = \sum_{m=0}^{N_x - 1} \sum_{n=0}^{N_y - 1} w_{mn}.
\tag{6-28}
$$

The mean value of a filtered image is equal to the mean value of the input image times the *MTF* at zero frequency. Pure LP and HB filters do not change the mean, while pure HP filters set it to zero. Combination filters have an intermediate effect on the mean (see Ex 6-8).

When filters are viewed in the frequency domain, two potential problems with the simple box filters described earlier are seen—their lack of rotational symmetry and inconsistent behavior at higher frequencies. The latter is an indication of spatial *phase reversal*, which can cause artifacts for small, periodic targets (Castleman, 1996). For simple visual enhancement, these problems may not matter, but for modeling and more precise work, it is desirable to use filters that have an equal response in all directions (unless one is specifically modeling an asymmetric response). If a Gaussian function is used within the square window, the situation is improved. For example, a 3×3 Gaussian *LPF*, with a 1/e radius of 1.5 pixels has the weights,

$$
w_g = \begin{bmatrix} +0.079 & +0.123 & +0.079 \\ +0.123 & +0.192 & +0.123 \\ +0.079 & +0.123 & +0.079 \end{bmatrix},
\tag{6-29}
$$

and the *MTF* shown in Fig. 6-24. The Gaussian is still spatially truncated by the 3 × 3 window, but the effect in the frequency domain is much less than for a box filter. A larger window would reduce the remaining rotational asymmetry. HP and HB filters corresponding to Eq. (6-29) can be easily derived.

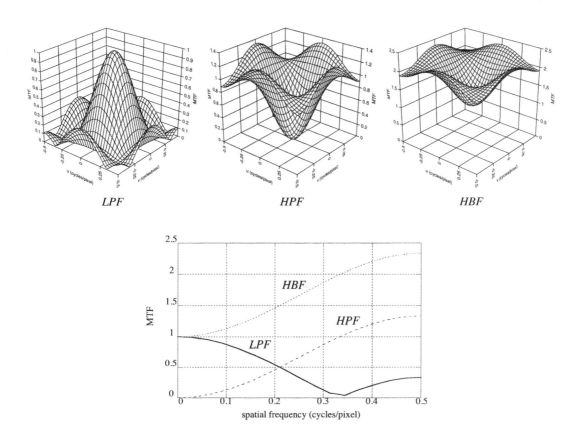

FIGURE 6-23. *The 2-D MTFs for the 3 × 3 box filters of Table 6-2 and Table 6-3. The HBF is for a K value of one. Note how each of these filters shows significantly different responses in different directions, i.e., they are not rotationally symmetric. The graph below shows profiles of the three MTFs along either the k or l axis. Since the MTFs are symmetric about (k,l) equal (0,0), it is convention to show only half the function. An MTF of one passes that frequency component unchanged in modulation through the filter; values greater than one increase the output modulation and values less than one decrease the output modulation, relative to that of the input signal. The terminology "low-pass," "high-pass," and "high-boost" refers to the effect on spatial frequency components, which is clearly shown in this graph.*

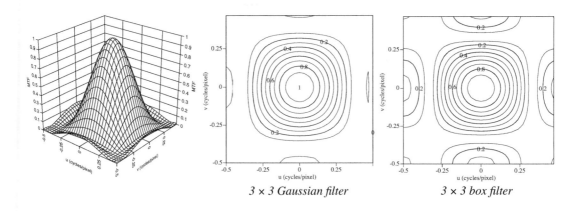

3 × 3 Gaussian filter   *3 × 3 box filter*

FIGURE 6-24. *The MTF of a Gaussian filter with a spatial radius of 1.5 pixels and truncated to 3 × 3 pixels. The contour map is compared to that for a uniform box filter.*

6.4.5 System Modeling Using the Fourier Transform

The discussion in this chapter so far has been in terms of digital image processing, where the functions involved are all sampled. The linear systems tools of convolution and the Fourier transform were used for image processing; however, they also apply to the physical systems of image formation (Gaskill, 1978). An expansion of Fig. 3-1 shows the connection between these two applications (Fig. 6-25). In this section we describe how convolution and Fourier transforms are used for sensor system modeling.

In Chapter 3, we described the components and total spatial response for sensors. The description was in terms of the sensor *Point Spread Function (PSF) in the spatial domain*. We now have the Fourier transform tools to describe sensor spatial response in terms of the *Transfer Function (TF)*, the 2-D Fourier transform of the *PSF*, in the spatial frequency domain. Just as with digital filters, the *Modulation Transfer Function (MTF)*, the amplitude of the *TF*, is usually of most interest. For completeness, the optical spatial response diagram in Chapter 3 can now be expanded to include the Fourier component (Fig. 6-26).

A key concept for sensor modeling is that the sensor *TF* filters the spatial frequency components of the analog continuous image before it's sampled (digitized). The Convolution Theorem applied in this case is (compare to Eq. (6-25)),

$$\mathcal{F}\left[i(x,y)\right] = I(u,v) = \mathcal{F}\left[i_{ideal} * PSF_{net}\right]$$
$$= I_{ideal}TF_{net} \tag{6-30}$$

where i_{ideal} is an idealized image formed by an optical system with no degradations. It can be thought of as simply a scaled (by sensor magnification) version of the scene radiance distribution.

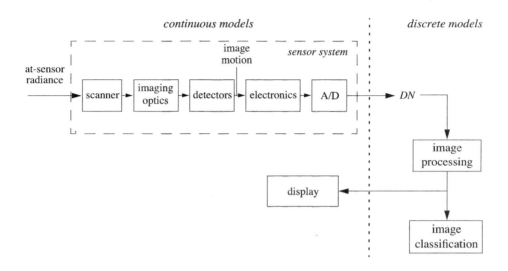

FIGURE 6-25. *The continuous-discrete-continuous model for image acquisition, processing, and display is shown in this simplification and expansion of Fig. 3-1. In the continuous model space, convolution and Fourier transforms can be used to describe how continuous physical functions are related. In the discrete model space, convolution and Fourier transforms are applied to discrete arrays, representing sampled continuous functions, to produce new discrete arrays. One may think of the continuous domain as describing the physics of remote sensing imaging, while the discrete domain describes the engineering of digital images produced by remote sensing systems.*

For reference, we repeat Eq. (3-18) here,

$$PSF_{net}(x, y) = PSF_{opt} * PSF_{det} * PSF_{IM} * PSF_{el}. \qquad (6\text{-}31)$$

This equation defines the relationship between an ideal, undegraded image, i_{ideal}, and the actual blurred image, i, formed by the imaging system. Taking the Fourier transform, we have by the Convolution Theorem a multiplication of each component's complex filter (see Eq. (6-21)),

$$TF_{net}(u, v) = TF_{opt}TF_{det}TF_{IM}TF_{el}. \qquad (6\text{-}32)$$

and similarly for the amplitude part (see Eq. (6-22)),

$$MTF_{net}(u, v) = |TF_{net}| = MTF_{opt}MTF_{det}MTF_{IM}MTF_{el} \qquad (6\text{-}33)$$

Equation (6-32) is completely equivalent to Eq. (6-31), but in the Fourier domain. Note that all the functions in this discussion are functions of *continuous* variables, either space (x,y) or frequency (u,v). That's because we are ignoring the effect of pixel sampling in the overall imaging process. It is possible to include pixel sampling in system *TF* analysis; the net effect is an additional blurring in the average spatial response due to sampling. Unfortunately, we do not have the space to treat sampling and its effect on spatial response properly here; the reader is referred to Park *et al.* (1984) for a full analysis.

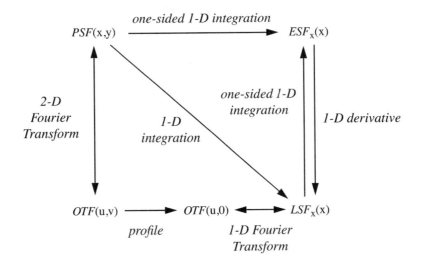

FIGURE 6-26. *Expansion of Fig. 3-17 to include the Fourier component, the Optical Transfer Function (OTF, or TF_{opt} in our notation). The TF_{opt} is the 2-D Fourier transform of the optical PSF, and the PSF is the inverse Fourier transform of TF_{opt}. The 1-D Fourier transform of the LSF results in a 1-D profile (cross-section) of the 2-D TF_{opt}.*

There are several examples in the literature of modeling the sensor *TF* using pre-launch measurements on the various components in the system (Table 6-7). Modeling of the Advanced Land Imager (ALI), a multispectral push-broom sensor, serves to illustrate the process. The in-track and cross-track components are detailed in Table 6-8 and the modeled *MTF*s are shown in Fig. 6-27. The ALI detector, image motion, and electronics model parameters are specified in Hearn (2000). The optics model in that reference is a more detailed one than used here; the in-track and cross-track Gaussian optics model parameters used here were set to make our TF_{net} match the TF_{net} for band 5 (NIR) in Hearn (2000). The in-track and cross track *LSF*s can be calculated from the model *TF*s by inverse Fourier transformation. They are shown in Fig. 6-28 to be comparable to the measured *LSF*s from Chapter 3.

TABLE 6-7. *Examples of sensor PSF and MTF measurement and modeling.*

sensor	reference
ALI	Hearn, 2000
AVHRR	Reichenbach *et al.*, 1995
ETM+	Storey, 2001
MODIS	Barnes *et al.*, 1998; Rojas, 2001
MSS	Slater, 1979; Friedmann, 1980; Park *et al.*, 1984
TM	Markham, 1985

TABLE 6-8. *The MTF modeling parameters used for ALI. The units of spatial frequency (u,v) are cycles-*μm^{-1}, *as measured in the focal plane of the system, or cycles-pixel*$^{-1}$, *which are normalized by the detector width w.*

in-track			
component	**TF model**	**parameter value**	**remarks**
optics	$MTF_{opt} = e^{-b^2 v^2}$	$b = 38\mu m = 0.95$ pixel	Gaussian PSF
detector	$MTF_{det} = |sinc(wv)|$	$w = 40\mu m = 1$ pixel	square pulse LSF
image motion	$MTF_{IM} = |sinc(sv)|$	$s = 36\mu m = 0.9$ pixel	square pulse LSF
electronics	$MTF_{el} = e^{-|v/v_0|^g}$	$v_0 = 0.2$ cycles-μm^{-1} $= 8$ cycles-pixel^{-1} $g = 1$	charge diffusion
cross-track			
optics	$MTF_{opt} = e^{-b^2 u^2}$	$b = 36\mu m = 0.9$ pixel	Gaussian PSF
detector	$MTF_{det} = |sinc(wu)|$	$w = 40\mu m = 1$ pixel	square pulse LSF
electronics	$MTF_{el} = e^{-|u/v_0|^g}$	$v_0 = 0.2$ cycles-μm^{-1} $= 8$ cycles-pixel^{-1} $g = 1$	charge diffusion

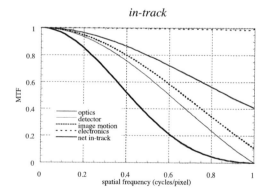 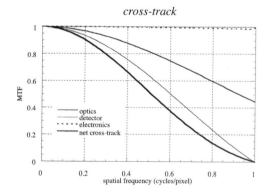

FIGURE 6-27. *The ALI in-track and cross-track model MTFs from Table 6-8.*

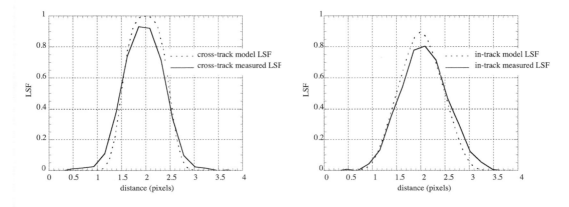

FIGURE 6-28. *The model and measured ALI LSFs. The model LSFs are calculated by the inverse Fourier transform of the transfer functions in Fig. 6-27. The measured LSFs are from the analysis discussed in Chapter 3 (see Fig. 3-18).*

6.4.6 The Power Spectrum

Applications of the Fourier transform also exist in image pattern analysis and recognition. In particular, the *power spectrum* (the square of the Fourier amplitude function) is a useful tool. Application of the power spectrum for fractal analysis of a TM image was presented in Chapter 4, including some example spectra (Fig. 4-28). Additional example power spectra for 256×256 aerial photographic images are shown in Fig. 6-29; more examples can be found in Jensen (2004) and Schott (1996).

From these examples and knowledge of the Fourier transform we can make the correspondances of Table 6-9. Because the spatial frequency power spectrum *localizes* information about *global* patterns in the spatial domain, it is useful as a diagnostic tool for global periodic noise or as a pattern recognition tool for global spatial patterns. In Chapter 7, we will illustrate its use in designing noise filters.

6.5   Scale-Space Transforms

In many cases, we would like to extract the spatial information from an image over a range of scales, from fine details in local areas to large features that extend across the image. The human vision system does a remarkable job at this task, without even thinking about it! Implementation of a similar capability in computer algorithms is a challenging task. A number of approaches are promising, however, and we have grouped them within the category *scale-space filtering*. The algorithms generally behave as filters, but are applied repeatedly on scaled versions of the image

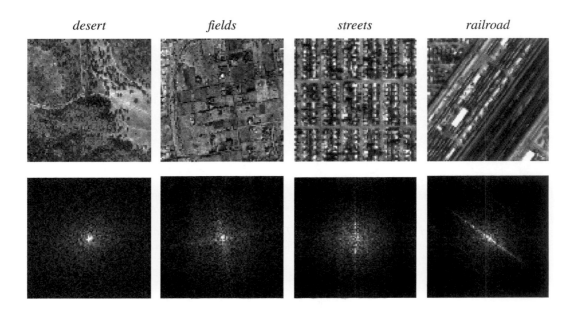

FIGURE 6-29. *The dependence of power spectra on image spatial structure. The "desert" image is relatively isotropic, with no directional features; the others have directional content of various degrees. Note how the direction of rays in the power spectra are related to the direction of image features and how the subtle periodic structure in "streets" results in faint periodic bright spots in its spectrum. More examples of these characteristics are in Fig. 4-28.*

TABLE 6-9. *Descriptive relationships between the spatial and spatial frequency domains.*

spatial description	spatial frequency description
periodic patterns	high-amplitude "spikes," localized at the frequencies of the patterns
linear, quasi-periodic features	high-amplitude "line" through zero frequency, oriented orthogonal to the spatial patterns
nonlinear, aperiodic features	high-amplitude "cloud," primarily at lower frequencies

(resolution pyramids), or the filter itself is scaled (zero-crossing filters). Some of the ideas, such as the scale-space LoG filter and zero-crossings, originated in human vision system models (Marr and Hildreth, 1980; Marr, 1982; Levine, 1985).

6.5.1 Image Resolution Pyramids

Image resolution pyramids are a way to efficiently include global, intermediate, and local scales in an analysis. A pyramid can be defined as,

$$i_L = \text{REDUCE}(i_{L-1}) \qquad (6\text{-}34)$$

where i_L is the image at level L and REDUCE is any operation on the image at level *L-1* that reduces its size for level L. For example, REDUCE can be simply an average over 2×2 pixel neighborhoods, with an equal down-sampling[10] along rows and columns, producing a *box pyramid* (Fig. 6-30). The linear size of the image at level L is related to the size at level *L-1* by,

$$N_L = \frac{N_{L-1}}{2}. \qquad (6\text{-}35)$$

One popular REDUCE operator was proposed in Burt (1981) and Burt and Adelson (1983). The weighting function in this case is separable,

$$w_{mn} = w1_m \cdot w2_n \qquad (6\text{-}36)$$

where both 1-D functions are parametric,

$$w1_m = \left[0.25 - a/2,\ 0.25,\ a,\ 0.25,\ 0.25 - a/2\right]^T. \qquad (6\text{-}37)$$

$$w2_n = \left[0.25 - a/2,\ 0.25,\ a,\ 0.25,\ 0.25 - a/2\right] \qquad (6\text{-}38)$$

A Gaussian-like function is generated with a equal to 0.4 in both directions. The 2-D weighting function is then,

$$w_{mn} = \begin{bmatrix} 0.0025 & 0.0125 & 0.0200 & 0.0125 & 0.0025 \\ 0.0125 & 0.0625 & 0.1000 & 0.0625 & 0.0125 \\ 0.0200 & 0.1000 & 0.1600 & 0.1000 & 0.0200 \\ 0.0125 & 0.0625 & 0.1000 & 0.0625 & 0.0125 \\ 0.0025 & 0.0125 & 0.0200 & 0.0125 & 0.0025 \end{bmatrix}. \qquad (6\text{-}39)$$

In the REDUCE operation the image is convolved with the weighting function, but only every other calculated pixel and line are used in the next level. Thus, a full convolution is not necessary, making the algorithm particularly efficient. If the weights in Eq. (6-39) are used, a *Gaussian pyramid* results (Burt, 1981), as shown in Fig. 6-31.

---

10. *Down-sampling* means sampling of every other row and column of an image, discarding the intermediate pixels. Conversely, *up-sampling* means insertion of rows and columns of zero *DN* between existing rows and columns, which is usually followed by interpolation to replace the zeros.

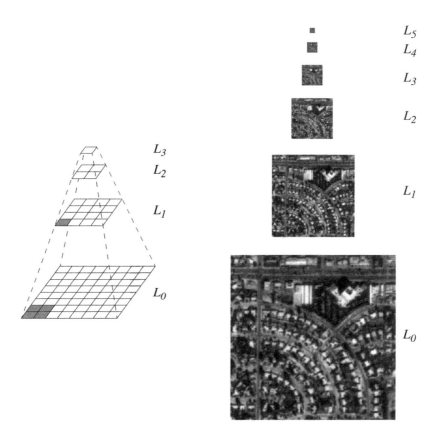

FIGURE 6-30. Construction of a 2 × 2 box pyramid and an example of six levels, starting with a 256 × 256 image. The grey area in level 0 indicates the four pixels that are averaged to calculate the lower left pixel in level 1.

The REDUCE operation with this filter is depicted in Fig. 6-32. The corresponding EXPAND operation consists of expansion of the image at level L by insertion of one row and column of zeros between existing rows and columns, and convolution of the resulting array with the filter in Eq. (6-39), to yield the image at level L-1.

The process used to create and reconstruct one level of the Gaussian pyramid from another level is shown in Fig. 6-33. Also shown is level 0 of the *Laplacian pyramid*, constructed by subtracting the up-sampled, reconstructed Gaussian level 1 image from the level 0 image. The Laplacian pyramid is useful for image coding and compression (Burt and Adelson, 1983) and for finding contrast

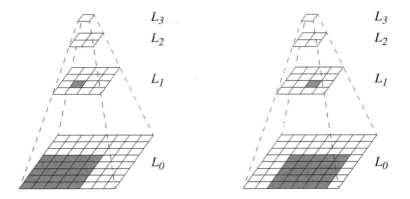

FIGURE 6-31. *Gaussian pyramid construction with the 5 × 5 pixel weighting function of Eq. (6-39). On the left, the weighted average of 25 pixels in level 0 gives the shaded pixel in level 1. On the right, the weighting function is moved two pixels along a row and the weighted average of the corresponding pixels gives the next pixel in level 1. This process is repeated across the image in level 0, and then performed on level 1 to calculate level 2, and so forth. In this way, the linear size of the image is reduced by two from level to level. This is equivalent to a full convolution of level 0 with the weighting function, followed by a down-sample, but avoids unnecessary calculations. The border pixels require special attention as discussed earlier.*

edges at different scales as discussed in the next section. The box and Gaussian pyramids are compared in Fig. 6-34. The Gaussian algorithm avoids the discontinuities characteristic of aliasing due to undersampling (Chapter 3) and produces a smoother image at each level.

Because of the down-sampling operation, the Gaussian pyramid algorithm is equivalent to successive convolutions of the original image with a weight window that expands by a factor of two at each pyramid level (Fig. 6-35). In Fig. 6-36, the filtered images at the first three levels are shown, without down-sampling. The down-sampling used in construction of the pyramid allows the use of a single weight function (Eq. (6-39)) *directly on the image at each level*, thereby avoiding unnecessary convolution calculations.

6.5.2 Zero-Crossing Filters

A significant limitation of the local edge gradient filters described earlier in this chapter is that they only use information from a local neighborhood about each pixel. Large-scale edges, extending over many pixels, are not found explicitly, but only by connecting individual "edge pixels" produced by the gradient operation. The pyramid representation provides a way to access multiple image scales with a single-size filter.

In the original descriptions of the Gaussian pyramid (Burt and Adelson, 1983), a Laplacian pyramid was also calculated as the difference between level k and $k+1$ of the Gaussian pyramid (Fig. 6-33). The name comes from the Laplacian second derivative operator (Castleman, 1996), a

REDUCE   EXPAND

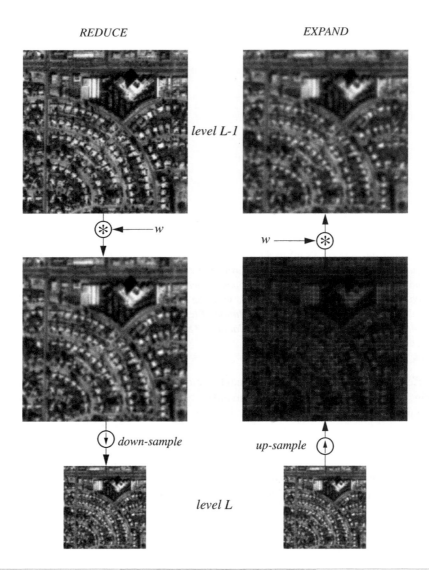

FIGURE 6-32. The REDUCE and EXPAND procedures as defined in Burt and Adelson (1983). Any spatial filter can be used, but these examples use the Gaussian window of Eq. (6-39).

connection that will be made in the following. Just as the Gaussian pyramid represents a series of LPFs with different bandpasses, the Laplacian pyramid is a series of *Band-Pass Filters (BPFs)* over different frequency regions.

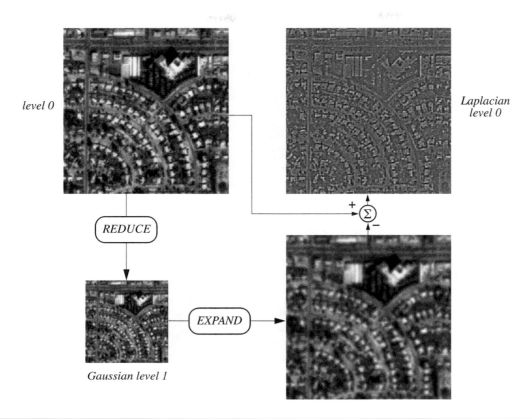

level 0

Laplacian level 0

REDUCE

EXPAND

Gaussian level 1

FIGURE 6-33. *The construction of level 1 of the Gaussian pyramid and level 0 of the Laplacian pyramid. Level 2 would be constructed by the same process, starting at level 1. The Laplacian at level 1 would be calculated as the difference between the level 1 image and the EXPANDed Gaussian level 2 image.*

Laplacian-of-Gaussian (LoG) filters

Suppose we have a 1-D function and want to find the location of "significant" changes in that function. If we calculate its second derivative (Fig. 6-37), we see that it crosses the zero ordinate at the locations of such changes. These locations indicate changes in the second derivative, from either negative to positive, or positive to negative, and are therefore called *zero-crossings*. Now, suppose we convolve the original function with a smoothing Gaussian five points wide and again calculate the second derivative. Notice that the zero-crossings from the more significant changes remain in approximately the same locations as before, and some from the less significant changes have disappeared.

Starting with a unit area, Gaussian function with zero mean and standard deviation σ,

$$g(x) = \frac{1}{\sigma\sqrt{2\pi}}e^{-x^2/2\sigma^2}, \qquad (6\text{-}40)$$

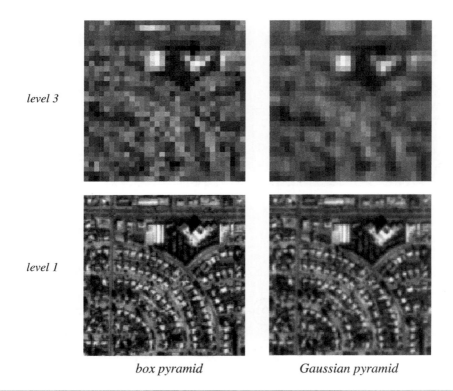

level 3

level 1

box pyramid   Gaussian pyramid

FIGURE 6-34. Level 3 and level 1 images compared for the box and Gaussian pyramids. The level 3 images are magnified to the level 1 scale by pixel replication.

the first derivative is,

$$g'(x) = -\left(\frac{x}{\sigma^2}\right)g(x) \tag{6-41}$$

and the second derivative, or the *Laplacian-of-Gaussian (LoG)*, is,

$$g''(x) = \left(\frac{x^2 - \sigma^2}{\sigma^4}\right)g(x) . \tag{6-42}$$

These functions are plotted in Fig. 6-38. Now, using the principles of linear systems theory, it is not difficult to show that, if a function f is convolved with g,

$$s(x) = f(x) * g(x) , \tag{6-43}$$

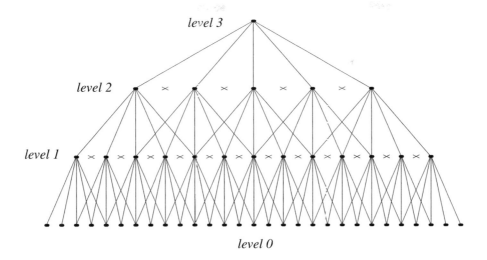

FIGURE 6-35. The links between a pixel in level 3 of the Gaussian pyramid and pixels at lower levels. The x-marks indicate pixels that are not included at each level because of down-sampling. The effective convolution window size at the original level 0 for a pyramid that reduces by two at each level L is $4(2^L - 1) + 1$ (Burt, 1981).

then the second derivative of s is given by,

$$s''(x) = f'(x) * g(x) = f(x) * g''(x).$$ (6-44)

This means that the second derivative of a function convolved with a Gaussian is equal to the function convolved with the second derivative of the Gaussian. The implication is that we can generate the LoG filter (Eq. (6-42)) once, and use it on different functions to find their second derivatives.

The Laplacian of a 2-D, rotationally-symmetric Gaussian is nearly the same form as Eq. (6-42) (Castleman, 1996). An additional factor of two occurs in the second term because of the 2-D context,

$$g''(r) = \left(\frac{r^2 - 2\sigma^2}{\sigma^4}\right) g(r)$$ (6-45)

where,

$$g(r) = \frac{1}{\sigma^2 2\pi} e^{-r^2/2\sigma^2}.$$ (6-46)

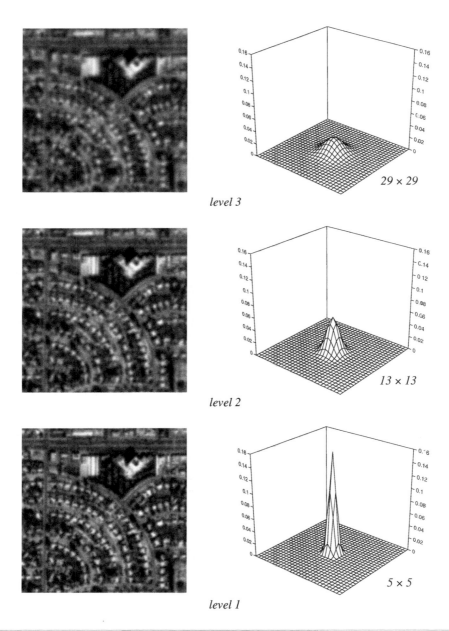

level 3

level 2

level 1

FIGURE 6-36.   *Levels 1 through 3 of the Gaussian pyramid, without down-sampling. The right-hand column shows the effective weight function at level 0. Note the weight amplitudes decrease as the spatial size increases, in order to maintain image normalization.*

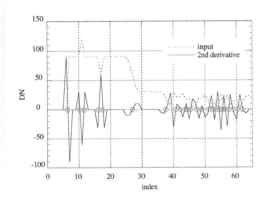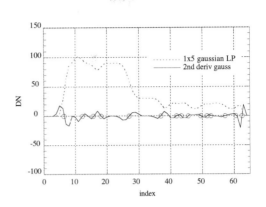

FIGURE 6-37. *Example of the relation between slope changes in a function and zero-crossings in its second derivative (marked with small circles). The more significant zero-crossings are retained even if the function is smoothed with a low-pass filter, but some of the less significant zero-crossings are lost. Note the similarities to Fig. 6-3.*

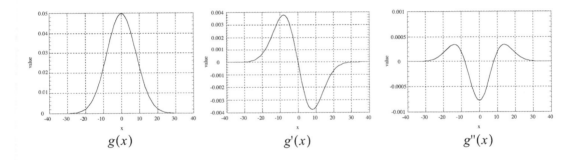

$$g(x) \qquad\qquad g'(x) \qquad\qquad g''(x)$$

FIGURE 6-38. *First and second derivatives of the 1-D Gaussian function g(x). The first derivative g'(x) has a zero-crossing at the location of the Gaussian's maximum, and the second derivative has zero-crossings at the locations of the Gaussian's maximum gradient, which are at x equal ± σ (eight in this example).*

A special algorithm can be used to find zero-crossings in filtered images. However, a simple series of standard algorithms works as well. To produce a zero-crossing map, we first threshold the filtered image at a *DN* of zero (which is the mean). This segments the image into negative and positive regions. Then a Roberts gradient filter (Table 6-6) is applied, and the gradient magnitude is thresholded to find the transition pixels between negative and positive regions, i.e., the zero-crossings. The process is depicted in Fig. 6-39.

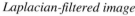

Laplacian-filtered image   *thresholded at zero*   *thresholded Roberts gradient*

FIGURE 6-39. The process used here to find zero-crossings. The Laplacian image is thresholded at zero DN and a Roberts gradient filter is applied. The gradient magnitude is then thresholded at a DN of one to form the binary zero-crossing map.

Difference-of-Gaussians (DoG) filters

The DoG filter is created by subtracting two Gaussian functions of different widths. The volume under each Gaussian is first normalized to one, so that their difference has a mean of zero,

$$DoG_{mn}(\sigma_1, \sigma_2) = \frac{1}{2\pi}\left[\frac{1}{\sigma_1^2}e^{-\left(\frac{m^2+n^2}{2\sigma_1^2}\right)} - \frac{1}{\sigma_2^2}e^{-\left(\frac{m^2+n^2}{2\sigma_2^2}\right)}\right] , \quad \sigma_2 > \sigma_1 \qquad (6\text{-}47)$$

The shape of the resulting filter is similar to the LoG filter, and zero-crossings are found in the same way.

Assuming that each Gaussian is normalized as mentioned, there are two ways to alter the DoG filter characteristics. One is to hold the smaller Gaussian fixed and to vary the size of the larger (subtracted) Gaussian. The resulting filters maintain high spatial resolution in the zero-crossings, but progressively ignore smaller features as the ratio, R_g, increases,

$$R_g = \sigma_2/\sigma_1 . \qquad (6\text{-}48)$$

Cross-sections of the DoG filter for different values of R_g are shown in Fig. 6-40. Example zero-crossing maps are shown in Fig. 6-41 for the same aerial photograph used previously.

The second way to change the DoG filter characteristics is to hold R_g constant, while allowing the overall size of the filter to vary (Fig. 6-42). In this case, the resulting zero-crossing maps lose resolution as the filter size increases. Larger filters tend to map larger features with their zero-

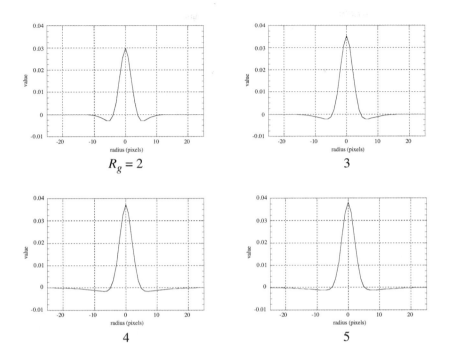

$$R_g = 2$$

FIGURE 6-40. *Profiles of the DoG filter for various size ratios between the two Gaussian functions in Eq. (6-47). These filters were used to produce the zero-crossing maps in Fig. 6-41. Note the locations of the zero-crossings are a weak function of R_g.*

crossings. Three DoG filters with a range of parameters were used to find the zero-crossings for our test image. The results illustrate the general parametric behavior of this class of scale-space filter (Fig. 6-43).

Finally, a comparison of the zero-crossings produced by a DoG(1,2) filter and a Roberts gradient edge map is made in Fig. 6-44. The zero-crossing map is fully connected and only one pixel wide, but does not indicate the *slope* of the filtered image at the zero-crossing, i.e., the *strength* of the local gradient. Therefore, low-contrast features, such as outside the lake boundary, and high-contrast features are both mapped by the zero-crossings. It is possible to combine the gradient information at the zero-crossings with the thresholding step (see Fig. 6-39) to produce a more selective zero-crossing map.

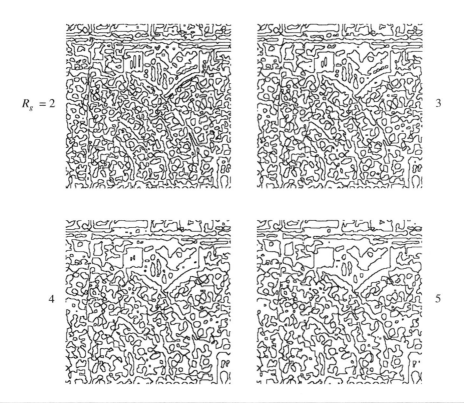

$R_g = 2$ 3

4   5

FIGURE 6-41. Zero-crossing maps for different size ratios in the DoG filter. The smaller Gaussian has a sigma of two. The contrast is reversed from that in Fig. 6-39.

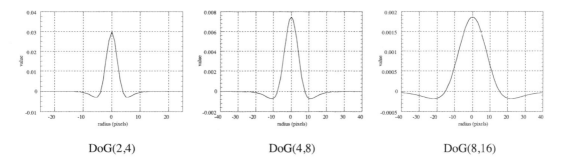

DoG(2,4)   DoG(4,8)   DoG(8,16)

FIGURE 6-42. Profiles of the DoG filter for different overall sizes, but the same size ratio between the two Gaussians. These filters were used to produce the zero-crossing maps in Fig. 6-43. The locations of the zero-crossings here are a strong function of overall size.

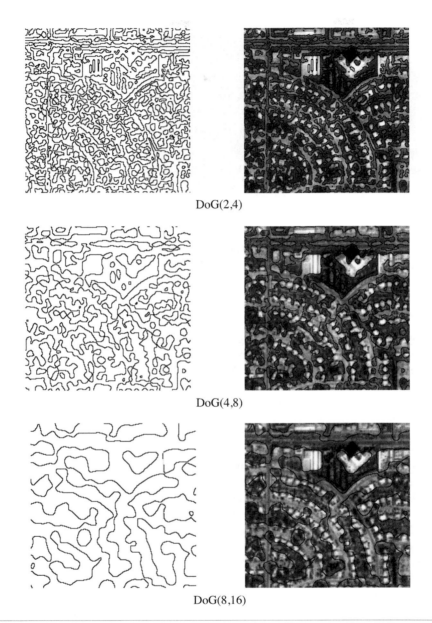

DoG(2,4)

DoG(4,8)

DoG(8,16)

FIGURE 6-43. Zero-crossing maps for constant size ratio, but different overall DoG filter size. The size ratio of the two Gaussians is two in all cases. Note that some edges are mapped at more than one scale; these are "significant" extended edges with high contrast.

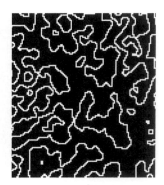

thresholded Roberts gradient   *zero-crossing map*

FIGURE 6-44. *Comparison of the Roberts thresholded gradient edge map with a zero-crossing map for the image on the left.*

6.5.3 Wavelet Transforms

Resolution pyramids and scale-space filtering provide a foundation for the relatively newer *wavelet transform*, which has attracted a great deal of attention in the last few years. Their introduction in the literature was in 1984 (Grossman and Morlet, 1984); other background papers of interest include Daubechies, 1988; Cohen, 1989; Mallat, 1989; Mallat, 1991; and Shensa 1992. The primer by Burrus *et al.* (1998) is also recommended. We want to put our discussion in the context of image resolution pyramid representations, which are only a subset of the entire subject of wavelets. The clear presentation in Castleman (1996), which is oriented to image processing, will be largely followed here.

Wavelet theory provides a general mathematical framework for decomposition of an image into components at different scales and with different resolutions. Just as the 2-D discrete Fourier transform expands an image into a weighted sum of global cosine and sine functions, the 2-D discrete wavelet transform expands the image into a sum of four components at each resolution level (Fig. 6-45). The wavelet transform operation is separable, consisting of two 1-D operations along rows and columns (Fig. 6-46).

A multiscale wavelet pyramid is constructed in a series of steps similar to those for the Gaussian and Laplacian pyramids. Each step is a convolutional filtering operation, followed by a down-sampling by two. It is therefore a type of REDUCE operation as described for the Gaussian and Laplacian pyramids. A major difference in wavelet decompostion is, however, that four components are calculated from the different possible combinations of row and column filtering. These four components may be most simply thought of as isotropic low-pass and high-pass components, plus horizontal and vertical high-pass components.

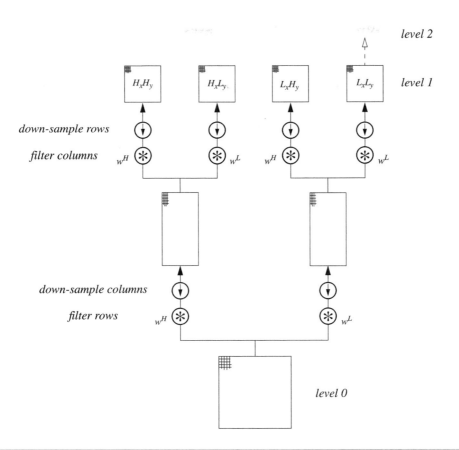

FIGURE 6-45. *Wavelet decomposition from one pyramid level to the next. The letter L means low-pass and H means high-pass. Thus, L_xL_y means a low-pass filtered version that has been filtered in x (along rows) and y (along columns) and down-sampled by two. Likewise, H_xL_y means the image from the previous level is first high-pass filtered and down-sampled in x and then low-pass filtered and down-sampled in y. The L_xL_y component at each level is used as input to the calculations for the next level. This processing architecture is known as a filter bank.*

Much of the important content of wavelet theory is related to determining the particular window functions to use in the convolutions. Many functions are possible, some looking quite unusual, with multiple peaks and asymmetry (Castleman, 1996). The *biorthogonal filters* are either even or odd symmetric; they also may be any length as long as the low-pass w^L and high-pass w^H filters are consistent. The constraint between them is,

$$w_m^H = (-1)^{1-m} w_{1-m}^L \qquad (6\text{-}49)$$

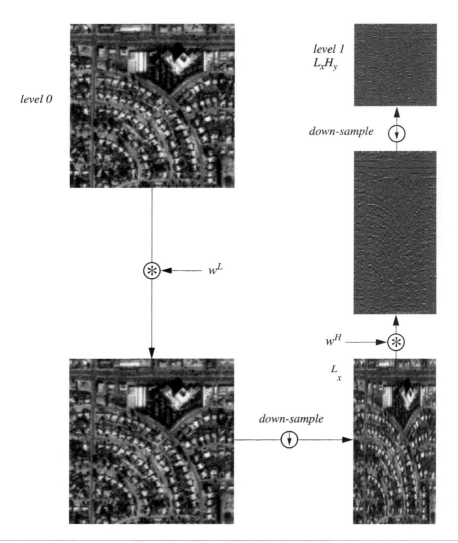

FIGURE 6-46. *Calculation of one of the four wavelet components in level 1. The filtering and down-sampling combinations are similar to the REDUCE operation described earlier, except that here they are done one direction at a time.*

where m is the index of the weight location in the high-pass filter w^H. In signal processing, these types of dual filters are called *quadrature mirror filters*.

Example wavelet components obtained with the following 1-D convolution filters,

$$w^L = \begin{bmatrix} -0.05 & +0.25 & +0.6 & +0.25 & -0.05 \end{bmatrix}$$

$$w^H = \begin{bmatrix} +0.05 & +0.25 & -0.6 & +0.25 & +0.05 \end{bmatrix} \tag{6-50}$$

are shown in Fig. 6-47; these filters are collectively called a "Laplacian analysis filter" in Castleman (1996), and are analogous to Burt's Gaussian window function (Eq. (6-39)). Each level in the wavelet pyramid contains a low-pass filtered version and three high-pass filtered versions of the original image. The original image can be reconstructed, just as in the Gaussian and Laplacian pyramids, by an inverse wavelet transform consisting of up-sampling and convolution.

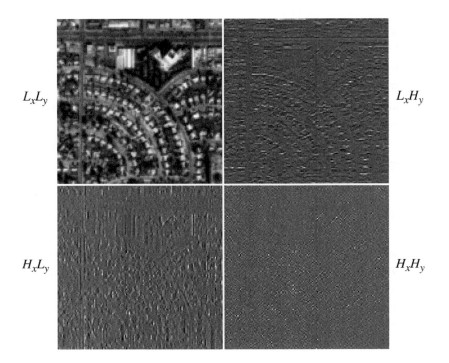

FIGURE 6-47. *The four wavelet transform components of level 1 produced by the filter bank in Fig. 6-45. To create level 2, the upper left image, L_xL_y, is processed by the wavelet transform to produce a similar set of four images that are smaller by a factor of two in each direction.*

Wavelet representations have been used for mapping high-frequency features (points, lines, and edges) for automated registration of two images (Chapter 7) and as a scale-space framework for fusion of images from sensors with different *GIFOV*s (Chapter 8). Although the mathematics of wavelet transforms can be quite complex, the transform's effect is the same as that of the Gaussian and Laplacian pyramid generation scheme; a multiresolution, scale-space representation of the image is created in each case.

6.6   Summary

A variety of spatial transforms were discussed, including convolution, Fourier transforms, various types of filters, and scale-space filtering. The important points of this chapter are:

- Convolution and Fourier transform filtering are equivalent global processing techniques, except in the border region.
- A wide range of processing can be performed with small neighborhood windows, including noise removal and edge detection.
- Scale-space filters allow access to image features according to their size, which is not possible with linear convolution or Fourier filters.
- The resolution pyramid provides a unified description of scale-space filters. Gaussian, Laplacian, and wavelet pyramids are examples.

In Chapters 7, 8, and 9 we will see some specific applications of spatial transforms to solve image processing problems in remote sensing.

6.7   Exercises

Ex 6-1.   Find the complementary to satisfy Eq. (6-1) for the Gaussian *LPF* in Eq. (6-29).

Ex 6-2.   Eq. (6-6) says a high-boost image can be calculated by adding two images. Rewrite it in terms of convolutions of two filters, *IF* and *HPF*, with the original image. For the 3×3 *HPF* in Table 6-2, determine a single filter that achieves Eq. (6-6) for an arbitrary value of K and show that it equals the specific filters in Table 6-3 for $K = 1, 2$, and 3.

Ex 6-3.   Rewrite Eq. (6-7) in terms of convolutions of an and an *LPF* with the image. Apply the relation in Eq. (6-10) and Eq. (6-11) to find a net BPF for the 3×3 and *LPF* in Table 6-2.

Ex 6-4.   Determine the 2-D weighting functions for the Burt pyramid for values of *a* equal to 0.6, 0.5, and 0.3.

Ex 6-5.   Suppose you calculate the local mean and standard deviation in a 3×3 pixel window, moved across an image. What is the statistical uncertainty in the mean values? What if the window is 5×5 pixels?

Ex 6-6.   Show that for large α, the output image in Eq. (6-8) approaches the pure HP image in Eq. (6-1), except for a multiplication factor.

Ex 6-7.   Show that the first step of a convolution, the flipping of the window function, is included in Eq. (6-4).

Ex 6-8.   What is the amplitude at zero frequency of the Fourier domain *MTF*s corresponding to each of the spatial filters in Fig. 6-8? (hint: use Eq. (6-28).)

Ex 6-9. Explain Table 6-5 in terms of what each filter does to the image *DN*s.

Ex 6-10. Explain why the right and bottom spatial frequencies in Fig. 6-19 are less than 0.5 by 1/N cycles/sample or cycles/line, respectively, when N is some power of 2 (2^2, 2^3, 2^9, etc).

Ex 6-11. If the original Landsat ETM+ image in Fig. 6-14 has a *DN* range of [0, 255], how are *DN* gradient thresholds of 200 and 400 possible for the gradient magnitude image of Fig. 6-14? How would you modify the definitions in Table 6-6 to insure that the gradient magnitude range would not exceed [0, 255]?

CHAPTER 7

Correction and Calibration

7.1 Introduction

Remote-sensing imagery sometimes must be corrected for systematic defects or undesirable sensor characteristics in order to extract reliable information. Users who obtain level 1 data for cost or other reasons must do corrections and calibration as needed for their application. Typical corrections needed include reduction of detector striping or other noise, geometric rectification, and various levels of radiometric calibration. These topics rely on the models and image processing techniques described in previous chapters.

7.2   Distortion Correction

Remote-sensing imagery requires spatial distortion corrections to maximize its usefulness for information extraction. The distortion arises from scanner characteristics and their interaction with the airborne platform or satellite orbital geometry and figure of the earth, as discussed in Chapter 3.

In correcting for distortion, we must essentially reposition pixels from their original locations in the data array into a specified reference grid. There are three components to the process:

- selection of suitable mathematical distortion model(s)
- coordinate transformation
- resampling (interpolation)

These three components are collectively known as *warping* (Wolberg, 1990). With the increasing complexity of remote-sensing systems, some having significant off-nadir viewing capability, the interest in multisensor image registration, and the increasing demands of earth scientists for precision, temporal resolution, and repeatability in remote-sensing measurements, there is a continuing need for geometric processing algorithms that are more accurate, autonomous, and efficient.

Various terms are used to describe geometric correction of imagery, and it is worthwhile defining them before proceeding:

- *Registration*: The alignment of one image to another image of the same area. Any two pixels at the same location in both images are then "in register" and represent two samples at the same point on the earth (Chapter 8).
- *Rectification*: The alignment of an image to a map so that the image is planimetric, just like the map (Jensen, 2004). Also known as *georeferencing* (Swann *et al.*, 1988). An example rectification of Landsat TM data to map coordinates is shown in Fig. 7-1.
- *Geocoding*: A special case of rectification that includes scaling to a uniform, standard pixel *GSI*. The use of standard pixel sizes and coordinates permits convenient "layering" of images from different sensors and maps in a GIS.
- *Orthorectification*: Correction of the image, pixel-by-pixel, for topographic distortion. The result, in effect, is that every pixel appears to view the earth from directly above, i.e., the image is in an orthographic projection.

Two approaches to rectification of satellite imagery are shown in Fig. 7-2. In Chapter 3, we described the pure satellite modeling approach, in which information on satellite position, attitude, orbit, and scan geometry are used with an earth figure model to produce system-corrected products. Further correction to rectify the system-corrected imagery can be done with polynomial distortion functions and *Ground Control Points (GCPs)*. This second stage, polynomial correction, is the subject of this section. It results in satisfactory rectification in many applications, but has the following disadvantages:

- two resamplings are performed on the data, resulting in unnecessary degradation
- polynomials of any reasonable degree cannot correct for local topography

 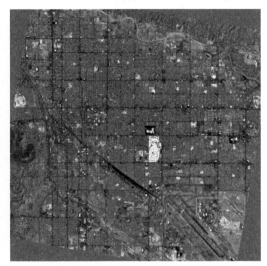

original   *rectified*

FIGURE 7-1. *Rectification of a Landsat TM band 4 image of Tucson, Arizona. The primary distortion in the original is due to the non-polar orbit; thus, rotation is the dominant correction (the major streets in Tucson are oriented north and south, and east and west). Note the corners of the rectified image are clipped by the image format; a larger output format would avoid this. Also, there are some areas in the rectified frame for which no input pixels exist in the original image. These pixels are filled with a uniform DN in the output.*

- a large number of GCPs are sometimes necessary for low residual error

It is possible to produce the scene-corrected, or precision level, product with a hybrid, single-stage correction and resampling (Westin, 1990), which presumably results in a superior product. Such correction, however, requires access to the raw imagery and detailed satellite ephemeris and attitude data, plus the necessary software algorithms.

Orthorectification requires a *Digital Elevation Model (DEM)*, since every pixel's location must be adjusted for topographic relief displacement. We will discuss this subject in the next chapter, where algorithms for automated registration of images and DEM extraction are also discussed.

7.2.1 Polynomial Distortion Models

The selection of an appropriate model (or models) is critical to the accuracy of any distortion correction. At least for satellite sensor or platform-induced distortions, there exist specific and accurate mathematical expressions for the distortion (Chapter 3). These types of models, however, require precise parameter input on sensor position, attitude, and scan angle as a function of time.

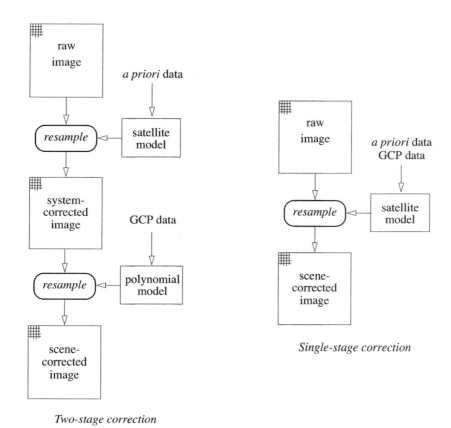

Two-stage correction

Single-stage correction

FIGURE 7-2. *Geometric processing data flow for the common two-stage process and a single-stage process for rectification. In the two-stage process, a generic polynomial model is used to remove residual distortions in the system-corrected imagery. The polynomial correction approach is widely used and available in all major software systems. A more sophisticated and efficient approach is to use physical models and include all corrections in a single resampling. (Adapted from Westin (1990).)*

Because of the difficulties in supplying the parameters for a precise sensor distortion model, data received by the user may contain residual geometric errors. Moreover, if the image is a lower level, system-corrected product (Chapter 1), it will not have undergone controlled rectification to a map projection. Therefore, it has become common practice in the user community to use a generic polynomial model for registering images to each other and to maps.

The polynomial model relates the global coordinates in the distorted image to those in the reference image or map,

$$x = \sum_{i=0}^{N}\sum_{j=0}^{N-i} a_{ij}x_{ref}^{i}y_{ref}^{j}, \quad y = \sum_{i=0}^{N}\sum_{j=0}^{N-i} b_{ij}x_{ref}^{i}y_{ref}^{j}. \tag{7-1}$$

These are bivariate polynomials, in which each variable x and y depends on both x_{ref} and y_{ref}. There is no physical basis for the choice of polynomial functions in this task, unlike the models described in Chapter 3. Polynomials are widely used as an approximating function in all types of data analysis, however. The level of detail in the data that can be approximated depends directly on the order of the polynomial, N.

A quadratic polynomial (N equals 2) is sufficient for most problems in satellite remote sensing, where the terrain relief is small and the *FOV* is not large. The terms in the quadratic can be interpreted as components in the total warp (Fig. 7-3); when used in combination, they can produce special warps, as shown in Fig. 7-4.

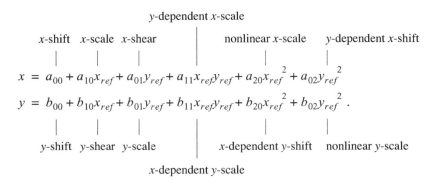

FIGURE 7-3. *Contributions of each quadratic polynomial term to the total warp.*

If the input imagery has been processed accurately for systematic distortions, a linear polynomial may suffice for further correction. In these cases, the higher-order quadratic terms may be dropped, leading to the linear polynomial,

$$x = a_{00} + a_{10}x_{ref} + a_{01}y_{ref}$$
$$y = b_{00} + b_{10}x_{ref} + b_{01}y_{ref} \tag{7-2}$$

also known as an *affine* transformation (Wolf, 1983). It can simultaneously accommodate shift, scale and rotation. Moreover, Eq. (7-2) can be written in a compact vector-matrix form,

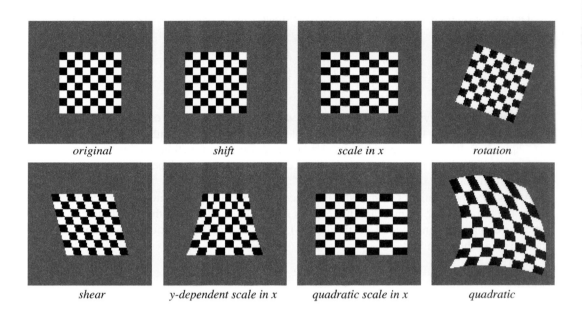

original	shift	scale in x	rotation

shear	y-dependent scale in x	quadratic scale in x	quadratic

FIGURE 7-4. Polynomial geometric warps. Nearest-neighbor resampling is used in these examples.

$$\begin{bmatrix} x \\ y \end{bmatrix} = \begin{bmatrix} a_{10} & a_{01} \\ b_{10} & b_{01} \end{bmatrix} \begin{bmatrix} x_{ref} \\ y_{ref} \end{bmatrix} + \begin{bmatrix} a_{00} \\ b_{00} \end{bmatrix} \tag{7-3}$$

or, equivalently,

$$p = Tp_{ref} + T_0 . \tag{7-4}$$

The affine transformation can be used for approximate correction of satellite sensor distortions (Anuta, 1973; Steiner and Kirby, 1976; Richards and Jia, 1999). The matrix T takes on specific values for a particular sensor. For example, several affine transformations can be applied to correct scanner- and orbit-related distortions in Landsat MSS imagery and produce an approximately rectified product (Table 7-1). These individual matrices can be combined into a single matrix transformation,

$$T_{total} = T_1 T_2 T_3 \tag{7-5}$$

for processing efficiency and to avoid multiple resamplings. This type of processing was common in the early years of the Landsat MSS because only the raw, uncorrected imagery was available. Much of the archived "legacy" imagery, dating to 1972, is in that raw format, so the need will remain for such simple system geometric correction techniques in the future.

TABLE 7-1. *Specific affine transformations for Landsat MSS data (Anuta, 1973; Richards and Jia, 1999). The various angles used are:*
   i : orbit inclination at the equator (degrees)
   ϕ: *geocentric latitude at frame center*
   θ: *orbit inclination at latitude* ϕ, $\theta = 90 - \text{acos}[\sin i / \cos\phi]$ *(degrees)*
   γ: *skew angle between the top and bottom of the image due to earth rotation*

distortion	transformation matrix	remarks
aspect ratio	$T_1 = \begin{bmatrix} 1 & 0 \\ 0 & 0.709 \end{bmatrix}$	$GSI_x = 0.709\,GSI_y$
earth rotation	$T_2 = \begin{bmatrix} 1 & -\alpha \\ 0 & 1 \end{bmatrix}, \ \alpha = \tan\gamma\cos\phi\cos\theta$	x-shear
rotation to north	$T_3 = \begin{bmatrix} \cos\theta & -\sin\theta \\ \sin\theta & \cos\theta \end{bmatrix}$	map orientation

Ground Control Points (GCPs)

The coefficients in the polynomial model Eq. (7-1) must be determined. The procedure is to specify *Ground Control Points (GCPs)* to constrain the polynomial coefficients. GCPs should have the following characteristics:

   · high contrast in all images of interest

   · small feature size

   · unchanging over time

   · all are at the same elevation (unless topographic relief is being specifically addressed)

Examples include road intersections, corners of agricultural fields, small islands, and river features. It is advisable to confirm that the GCPs have not changed over the time interval between the images being registered. This is particularly important for GCPs that rely on water levels for their location. Finding corresponding GCPs in an image and a map is generally more difficult than in two images because the map is an abstraction of the area's features and is often considerably older than the image.

   Visual location of GCPs is a widely-used paradigm in remote-sensing applications. Wherever available, man-made features prove the most reliable for GCPs. The primary drawbacks are the difficulty in finding satisfactory control points that are well distributed across the image area and the intensive manual labor involved. It can also be difficult, if not impossible, to locate GCPs in low-resolution imagery such as from the AVHRR. That is one reason why there has been considerable interest in accurate orbit and platform modeling for the AVHRR, as described in Chapter 3.

Techniques have also been developed to automatically find GCPs using a library of image "chips" of selected natural features, such as coastal areas (Parada *et al.*, 2000). These techniques are discussed in Chapter 8.

To see how GCPs are used to find the polynomial coefficients, suppose we have located M pairs of GCPs in the distorted image and reference (image or map) coordinate systems. Assuming the global polynomial distortion model, we can write, for *each* pair m of GCPs, a polynomial equation of degree N in each variable, x and y,

$$x_m = a_{00} + a_{10}x_{refm} + a_{01}y_{refm} + a_{11}x_{refm}y_{refm} + a_{20}x_{refm}^2 + a_{02}y_{refm}^2 ,$$
$$y_m = b_{00} + b_{10}x_{refm} + b_{01}y_{refm} + b_{11}x_{refm}y_{refm} + b_{20}x_{refm}^2 + b_{02}y_{refm}^2 \tag{7-6}$$

leading to M pairs of equations. This set of equations can be written in vector-matrix form for the x coordinates of the image as,

$$\begin{bmatrix} x_1 \\ x_2 \\ \vdots \\ x_M \end{bmatrix} = \begin{bmatrix} 1 & x_{ref1} & y_{ref1} & x_{ref1}y_{ref1} & x_{ref1}^2 & y_{ref1}^2 \\ 1 & x_{ref2} & y_{ref2} & x_{ref2}y_{ref2} & x_{ref2}^2 & y_{ref2}^2 \\ \vdots & \vdots & \vdots & \vdots & \vdots & \vdots \\ 1 & x_{refM} & y_{refM} & x_{refM}y_{refM} & x_{refM}^2 & y_{refM}^2 \end{bmatrix} \begin{bmatrix} a_{00} \\ a_{10} \\ a_{01} \\ a_{11} \\ a_{20} \\ a_{02} \end{bmatrix} \tag{7-7}$$

or,

$$X = WA \tag{7-8}$$

and similarly for the y coordinates of the image,

$$Y = WB . \tag{7-9}$$

Now, if M equals the number of polynomial coefficients K,[1] we have an exact solution by inverting the $M \times M$ matrix W,

$$M = K : \quad \begin{matrix} A = W^{-1}X \\ B = W^{-1}Y \end{matrix} , \tag{7-10}$$

and the error in the polynomial fit will be exactly zero at every GCP. It is usually desirable to have more GCPs because some may be in error. In this case we write,

---

1. $K = (N+1)(N+2)/2$

$$M \geq K: \quad \begin{matrix} X = WA + \varepsilon_X \\ Y = WB + \varepsilon_Y \end{matrix} \qquad (7\text{-}11)$$

where the added terms represent anticipated error vectors in the GCP locations. But when M is greater than K, the $M \times K$ matrix W cannot be inverted. It is possible to calculate a so-called *pseudoinverse* solution (Wolberg, 1990),

$$M \geq K: \quad \begin{matrix} \hat{A} = (W^T W)^{-1} W^T X \\ \hat{B} = (W^T W)^{-1} W^T Y. \end{matrix} \qquad (7\text{-}12)$$

which is equivalent to a *least-squares* solution for A and B that minimizes the total squared error in the polynomial fit to the GCPs,

$$min[\varepsilon_X{}^T \varepsilon_X] = (X - W\hat{A})^T (X - W\hat{A})$$
$$min[\varepsilon_Y{}^T \varepsilon_Y] = (Y - W\hat{B})^T (Y - W\hat{B}). \qquad (7\text{-}13)$$

The numerical stability of the solution for W can be improved by using matrix decomposition techniques instead of the pseudoinverse solution (Wolberg, 1990).

It is important to realize that the goodness-of-fit exhibited by the polynomial is *only* that. It does *not* indicate the validity of the polynomial as a model for the physical distortion. The fit may be quite good at the GCPs (zero error, in fact, if M equals K), while simultaneously there may be large error at other points. Although not always done in practice, any use of GCPs for distortion correction should include a GCP subset for control of the polynomial, i.e., determination of its coefficients, and another, distinct GCP subset for evaluation of the residual errors at other points. The latter subset may be called just *Ground Points (GPs)*, since they are not used for control of the model. Such a procedure is analogous to the use of independent training and testing sites in multi-spectral classification (Chapter 9).

An example of GCP selection and polynomial fitting can be generated from the aerial photograph and map in Fig. 1-2. A set of six GCPs and four test GPs are shown on the photograph and the scanned map in Fig. 7-5. The direct mapping of image points to map points is shown in Fig. 7-6. The dominate distortion between the two seems to be a rotation (the image needs to be rotated clockwise), and the GPs appear to be consistent with the GCPs. The GCP pairs were used to find the polynomial distortion function for K values of three (affine) to six (quadratic). The RMS deviation at the GCPs and GPs for each polynomial is plotted in Fig. 7-7. As the number of terms in the polynomial is increased, the RMS deviation at the GCPs decreases, eventually to zero when K equals M. At the same time the polynomial is providing a better fit to the GCPs, the RMS deviation at the GPs increases. This implies that the higher-order polynomial is not necessarily a better model for whatever actual distortion exists between the image and the map.

The selection of GCPs for a registration or rectification task is typically an iterative process in which an initial set of GCPs are used to find the polynomial. The GCP coordinates in the reference coordinate system are subjected to the polynomial transformation, and the goodness-of-fit to the

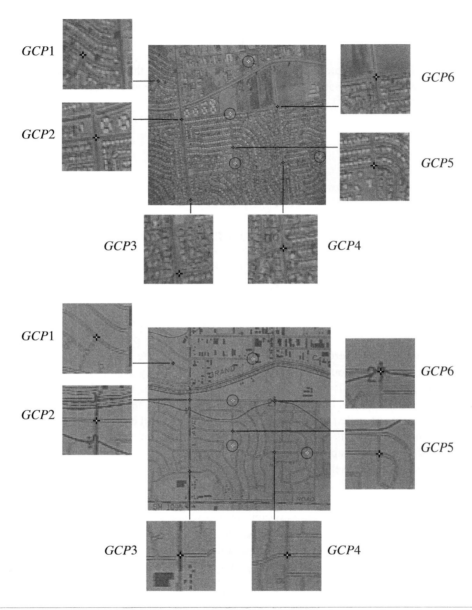

FIGURE 7-5. GCP location for rectifying the aerial photograph (top) to the scanned map (bottom). Six GCPs are selected for control (black cross with white center) and four GPs for testing (white cross in black circle). The contrast of the image and map are purposely made low to emphasize the GCPs and GPs in this figure.

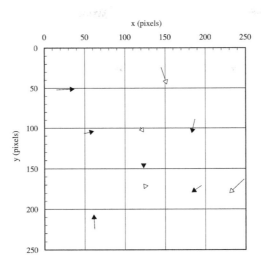

FIGURE 7-6. *Direct mapping of the image GCPs (black arrowheads) and GPs (white arrowheads) to those in the map. The coordinate system is line- and sample-based and has the origin at the upper left. Note the clockwise rotation, with also some scale differential around the borders indicated by the outer arrows pointing inward to the center.*

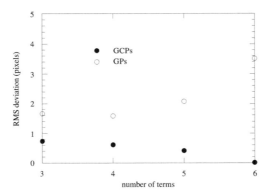

FIGURE 7-7. *RMS deviations between the GCP and GP locations as predicted by the fitted-polynomial and their actual locations for different numbers of terms in the polynomial. Note the deviation at the GPs is larger than that at the GCPs by a pixel or more.*

actual GCP locations examined in the image coordinate system. If the transformed GCP coordinates for any GCP are found to exceed a specified deviation from the image GCP, that point is assumed to be in error and either remeasured or dropped from the list of candidate GCPs. An example is shown in Fig. 7-8, where initially six GCPs were selected. After a trial fit with a bilinear polynomial, one point was considered to be an outlier since it deviated from the polynomial approximation by more than a pixel. After dropping that point from the analysis, the polynomial fit was repeated and the remaining GCPs showed smaller deviations from the polynomial model. The process is stopped when the total error at the GCPs falls within an acceptable level.

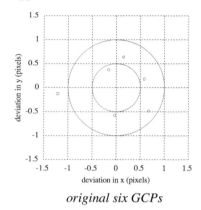
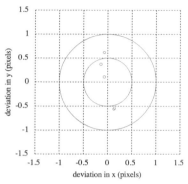

original six GCPs   *five GCPs, with outlier deleted*

FIGURE 7-8.   *Refinement of GCPs. The plot on the left shows that one GCP deviates from the affine polynomial model by more than one pixel in the x direction. The RMS deviation for all GCPs is 0.76 pixels. After the offending point is removed from the GCP list, and the polynomial is fit to the remaining five GCPs, the plot on the right results, and the RMS deviation is reduced to 0.49 pixels.*

Care should be exercised in application of global polynomials to uncorrected images with a large *FOV*. For example, cross-track panoramic distortion in whiskbroom sensor images is caused by uniform time sampling during the scan sweep (Chapter 3). The image scale (e.g., in m/pixel) off-nadir is related to that at-nadir by Eq. (3-40). This equation is graphed in Fig. 7-9 along with the quadratic polynomial fit to Eq. (3-40) over ±50° (approximately the *FOV* of the AVHRR and MODIS). An error of up to 15% at the extremes of the scan will occur if a global quadratic polynomial is used to approximate the actual distortion function. The error can be reduced to less than 5% with a quartic polynomial. Additional error may result when only a few GCPs are used to calculate the fit, as would be the case in practice. A quadratic polynomial model for panoramic distortion is quite good, however, over the much smaller ±5° *FOV* (approximately that of the Landsat MSS, TM, and ETM+).

The global polynomial model can be modified to better match distortion over large areas, or rapidly changing distortions such as those caused by topographic relief or platform attitude, by subdividing the image into small, contiguous sections. The distortion within each section is then modeled by a separate polynomial, and the results pieced together. This technique is known as a *piecewise polynomial model*. For example, the coordinate transformation within a quadrilateral defined by four control points may be modeled by a power series polynomial with the first four

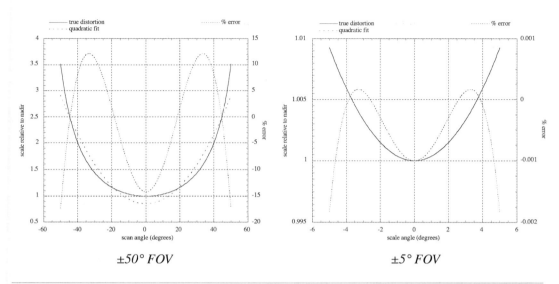

±50° FOV   ±5° FOV

FIGURE 7-9. Comparison of actual scanner panoramic distortion and a polynomial model over the FOVs of the AVHRR and Landsat sensors.

terms of Eq. (7-6). By defining a net of contiguous quadrilaterals across the image, each with a different polynomial transformation, a complex transformation can be accomplished by piecewise approximation (Fig. 7-10). A dense net of control points will generally yield higher accuracy, but a large number of GCPs, distributed across the image, may be difficult to achieve in some images. Also, a four term polynomial is discontinuous across the boundary between two quadrilaterals, so ground features that cross the boundary may be discontinuous after processing. The use of contiguous triangles between three control points, with a linear polynomial (affine) transformation within each triangle, eliminates the boundary discontinuities.[2] The global distortion is then modeled by a piecewise set of planes, much like a faceted surface.

The technique of piecewise distortion modeling was originally used by the Jet Propulsion Laboratory (Castleman, 1996) to correct severely-distorted planetary spacecraft images. Applications to airborne scanner imagery are presented in Craig and Green (1987) and Devereux *et al.* (1990), and application to SPOT imagery is reported in Chen and Lee (1992).

The image of Fig. 3-28 provides a good example of the limitations of the global polynomial model and the use of the piecewise model. A rectangular grid was created approximating the corrected geometry of the fields (a map was not available), six GCPs were selected in the image, and global affine and quadratic warps performed. The results clearly show that higher frequency distortions are present (Fig. 7-11). It is possible that a fifth, or higher-order polynomial would improve the situation, but not eliminate all distortion. Even if platform attitude data were available, it would

---

2. However, the *direction* of features can still change across boundaries.

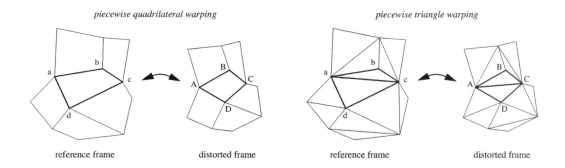

FIGURE 7-10. *Piecewise polynomial mapping. On the left, a net of quadrilaterals is defined by groups of four GCPs in both coordinate systems. The mapping function within each quadrilateral is a four-term polynomial. On the right, the same GCPs are used to define a net of triangles; the mapping function within each triangle is a three-term, affine polynomial. The coefficients for the polynomial within each polygon, for example, triangles a-b-c and a-c-d, will in general be different. The triangular network can be generated by Delaunay triangulation (Devereux et al., 1990; Chen and Lee, 1992), which is the dual of Dirichlet tessellation (also known as the Voronoi diagram) used in pattern recognition (Jain and Dubes, 1988; Schürmann, 1996) and in modeling of spatial statistics (Ripley, 1981).*

have to be sampled very frequently (say every 10 or 20 image lines) to account for the rapid changes evident in the image. This example also serves to emphasize the importance of stabilized platforms in airborne remote sensing.

7.2.2 Coordinate Transformation

Once the coordinate transformation f is found, it is used to define a mapping from the reference frame coordinates (x_{ref}, y_{ref}) to the distorted image frame coordinates (x,y),

$$(x, y) = f(x_{ref}, y_{ref}) .$$ (7-14)

The transformation of Eq. (7-14) is implemented by stepping through the integer coordinates (x_{ref}, y_{ref}) one-by-one, and calculating the transformed (x,y) values. Since the (x,y) coordinates will not be integer values in general, a new pixel must be estimated between existing pixels by an interpolation process known as *resampling*. The corrected output image is then created by "filling" each location (x_{ref}, y_{ref}) of a new, initially empty array, with the pixel calculated at (x,y) in the distorted image, as depicted in Fig. 7-12. This procedure, which may seem backwards at first, is preferable to a mapping from (x,y) to (x_{ref}, y_{ref}) because it avoids "overlapping" pixels and "holes" that have no assigned pixels in the output image.

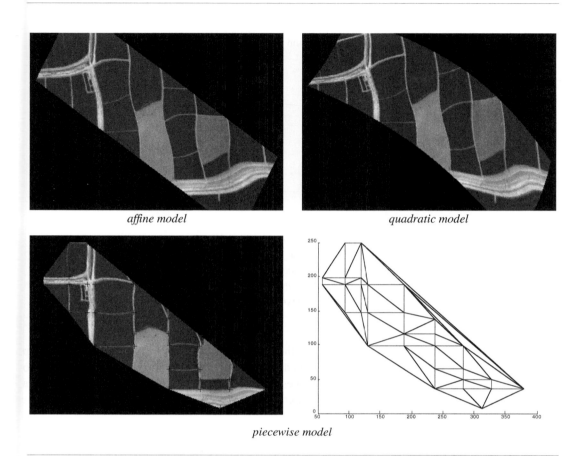

affine model *quadratic model*

piecewise model

FIGURE 7-11. *Rectification of the airborne ASAS image of Fig. 3-28 using six GCPs from a synthesized orthogonal grid and using global polynomial models is shown above. The affine transformation accounts for most of the global distortion; the quadratic transformation does not improve local distortion. A piecewise polynomial correction, as described in the text, is shown below with the 30 GCPs marked and a diagram of the triangular segments used for the correction. Note that regions around the perimeter that have no control points show severe residual distortion in the piecewise correction.*

Map projections

Maps are the spatial framework for analysis of remote-sensing images. However, every map has spatial distortion since it is a projection of the 3-D spherical earth onto a 2-D plane. A wide variety of projection schemes have been developed by trading off one type of distortion for another (Gilbert, 1974; Bugayevskiy and Snyder, 1995). For example, the relative areas of objects on the ground may be preserved at the expense of increased distortion in distances or shapes. The mathematical definitions of several common map projections are given in Table 7-2.

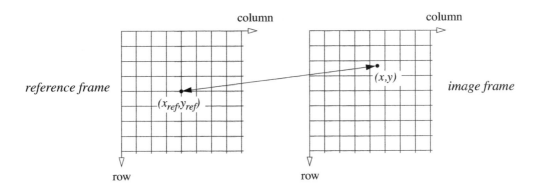

FIGURE 7-12. *The two-way relationship between the reference and distorted coordinate systems, expressed as rows and columns in the digital data. The arrow from (x_{ref}, y_{ref}) to (x,y) indicates a pointer, given by Eq. (7-14), from an integer coordinate in the reference frame to a (generally) non-integer coordinate in the image frame. The arrow from (x,y) to (x_{ref}, y_{ref}) indicates the transfer of an estimated pixel value, resampled at the non-integer location (x,y) in the original, distorted image, to the output location (x_{ref}, y_{ref}) in the reference frame.*

In general, the smaller the area of interest, the less important are the differences among map projections. Global datasets are being assembled as mosaics of many remote-sensing images to support the increased interest in long-term changes of the earth's environment. Global coverage in a single-map coordinate system demands special consideration (Steinwand, 1994; Steinwand *et al.*, 1995).

7.2.3 Resampling

The interpolation of new pixels between existing pixels is required to implement the geometric transformation as previously explained. The geometry of resampling is illustrated in Fig. 7-13, where a new pixel is to be resampled at the location R. Resampling can be thought of as convolution of the distorted image with a moving window function, as in spatial filtering. The resampled output values are calculated, however, *between* original pixels. Therefore, the resampling weighting function must be defined as a continuous function, rather than the discrete array used in convolution filtering.

The fastest scheme for calculating a resampled pixel is *nearest-neighbor* assignment (sometimes called *zero-order interpolation*). For the value of each new pixel at (x_{ref}, y_{ref}) in the output image, the value of the original pixel nearest to (x,y) is selected. In Fig. 7-13, this would result in pixel C at the output location. Because of this "round-off" property, geometric discontinuities on the order of plus or minus one-half pixel are introduced into the processed image. This effect is often negligible in displayed images, but may be important for subsequent numerical analyses. The significant advantage of nearest-neighbor assignment over other algorithms is that no calculations

TABLE 7-2. *Projection plane equations for several common map projections (Moik, 1980). The latitude of a point on the earth is* φ *and its longitude is* λ*. The projected map coordinates, x and y, are called "easting" and "northing," respectively. R is the equatorial radius of the Earth and* ε *is the Earth's eccentricity (Chapter 3). The subscripted values of latitude and longitude pertain to the definition of a particular projection.*

projection	x	y
polar stereographic	$2R\tan\left(\dfrac{\pi}{2}-\varphi\right)\sin\lambda$	$2R\tan\left(\dfrac{\pi}{2}-\varphi\right)\cos\lambda$
Mercator	$R\lambda$	$R\lambda\ln\left[\tan\left(\dfrac{\pi}{4}+\dfrac{\varphi}{2}\right)\left(\dfrac{1-\varepsilon\sin\varphi}{1+\varepsilon\sin\varphi}\right)^{\varepsilon/2}\right]$
oblique Mercator	$R\,\mathrm{atan}\left[\dfrac{\cos\varphi\sin(\lambda-\lambda_p)}{\sin\varphi\cos\varphi_p-\beta\sin\varphi_p}\right]$ $\beta = \cos\varphi\cos(\lambda-\lambda_p)$	$\dfrac{R}{2}\ln\left[\dfrac{1+\alpha+\beta\cos\varphi_p}{1-\alpha+\beta\cos\varphi_p}\right]$ $\alpha = \sin\varphi\sin\varphi_p$
transverse Mercator	$R\,\mathrm{atan}\left[\cos\varphi\sin(\lambda-\lambda_p)\right]$	$\dfrac{R}{2}\ln\left[\dfrac{1+\beta}{1-\beta}\right]$
Lambert normal conic	$\rho\sin\theta$	$\rho_0 - \rho\cos\theta$ $\rho = \dfrac{R\cos\varphi_1}{\sin\varphi_0}\left[\dfrac{\tan\left(\dfrac{\pi}{4}-\dfrac{\varphi}{2}\right)}{\tan\left(\dfrac{\pi}{4}-\dfrac{\varphi_1}{2}\right)}\right]^{\sin\varphi_0}$ $\theta = \lambda\sin\varphi_0$ $\sin\varphi_0 = \dfrac{\ln\left(\dfrac{\cos\varphi_1}{\cos\varphi_2}\right)}{\ln\left[\dfrac{\tan\left(\dfrac{\pi}{4}-\dfrac{\varphi_1}{2}\right)}{\tan\left(\dfrac{\pi}{4}-\dfrac{\varphi_2}{2}\right)}\right]}$

are required to derive the output pixel value, once the location (x,y) has been calculated, which must be done in any case. Nearest-neighbor assignment is equivalent to convolving the input image with a uniform spatial weighting function that is one sample interval wide, as shown in Fig. 7-14.

A smoother interpolated image is obtained with *bilinear* (first-order) resampling. This algorithm uses the four input pixels surrounding the point (x,y) to estimate the output pixel (Fig. 7-13). Bilinear resampling is usually implemented by first convolving the input image along its rows, creating new, resampled columns of pixels, and then along the new columns to create new, resampled rows of pixels, with a triangle weighting function in both directions (Fig. 7-14).

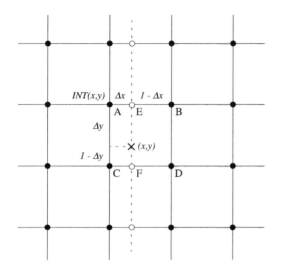

bilinear resampling:

$$DN_E = \Delta x DN_B + (1 - \Delta x) DN_A$$

$$DN_F = \Delta x DN_D + (1 - \Delta x) DN_C$$

$$DN(x, y) = \Delta y DN_F + (1 - \Delta y) DN_E$$

FIGURE 7-13. Geometry for resampling a new pixel at (x,y). Pixels in the distorted image (solid circles) are located at integer row and column coordinates in the image frame (Fig. 7-12). The coordinates of the local origin at pixel A are the integer parts and the offsets (Δx,Δy) are the fractional parts of (x,y) as calculated in Eq. (7-14). For the two-step, separable implementation of bilinear resampling, the intermediate pixels E and F are first calculated by interpolation between A and B, and C and D, respectively. Then the DN at (x,y) is estimated by interpolation between E and F. If cubic resampling is used, the four intermediate pixels (open circles) are first interpolated from the nearest four pixels in each row and then used in a fifth interpolation to estimate the DN at (x,y).

The difference between the nearest-neighbor and bilinear algorithms may be appreciated by the comparison in Fig. 7-15. The image has been enlarged digitally by the two methods to illustrate the continuous nature of bilinear resampling versus the discrete nature of nearest-neighbor resampling. The difference between the two resampling functions is especially apparent in a surface plot of the resampled image (Fig. 7-16). Bilinear resampling produces a smoother surface, at the expense of being considerably slower than nearest-neighbor resampling, because Eq. (7-14) must be calculated for every output pixel.

The smoothing incurred with bilinear interpolation may be avoided with *cubic* (second-order) interpolation, at the expense of more computation. The cubic interpolating function is a piecewise cubic polynomial that approximates the theoretically ideal interpolation function for imagery, the *sinc* function (Pratt, 1991).[3] The sinc function is not used for image interpolation because a large pixel neighborhood is required for accurate results. The cubic interpolator yields results

---

3. Defined as $sinc(x) = \sin(\pi x)/\pi x$.

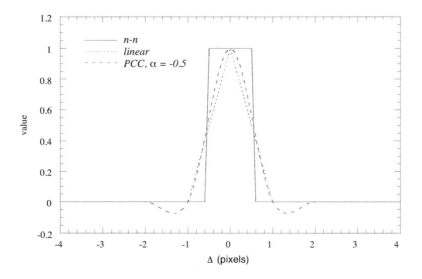

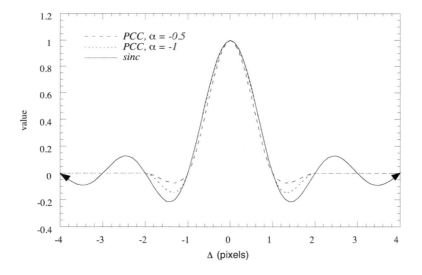

FIGURE 7-14. *Nearest-neighbor, linear, and PCC resampling spatial-weighting functions are compared at the top. The distance, Δ, is measured from the location, (x,y), of the resampled pixel (Fig. 7-13). Note the range of the nearest-neighbor function is ±0.5 pixels, the range of the linear resampling function is ±1 pixels, and the range of PCC is ±2 pixels. In the bottom graph, two PCC functions are compared to the sinc resampling function, which has an infinite range.*

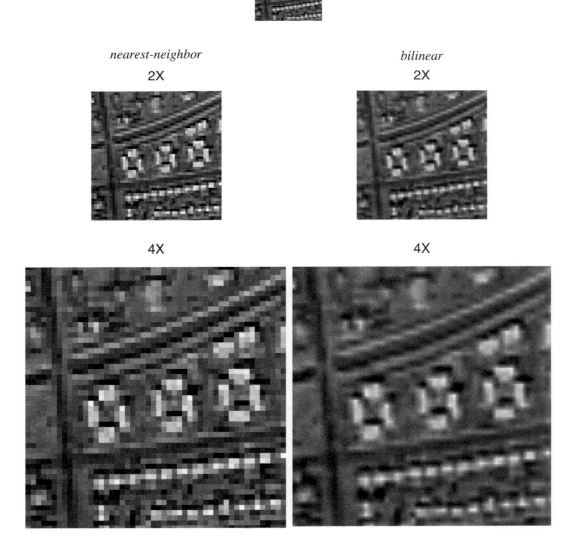

FIGURE 7-15. *Image magnification using nearest-neighbor and bilinear resampling. Although the bilinear resampling is smooth, there remains a jagged "stairstep" pattern along the curved road at the top. This is aliasing that originated in the digitization of the photograph.*

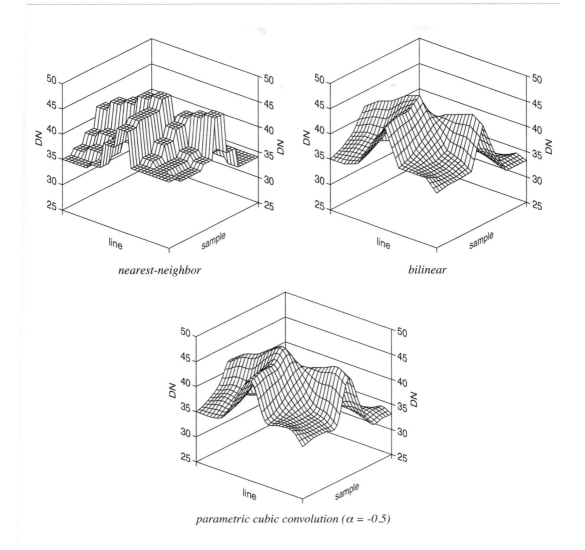

nearest-neighbor   *bilinear*

parametric cubic convolution (α = -0.5)

FIGURE 7-16. *Surface plots of DN(x,y) from 4 × -zoomed images for three resampling functions. The 4 × replication of pixels by nearest-neighbor resampling is evident. Bilinear resampling produces a function that is continuous between the original samples, but whose first derivative is not continuous, while the cubic resampling function produces a surface that is continuous and also has a continuous first derivative (Park and Schowengerdt, 1983).*

approximating those from a sinc function and requires only a 4-by-4 neighborhood in the input image. The cubic resampling function is actually a *parametric cubic convolution (PCC)* family of functions, defined by a single parameter α,

$$
\begin{aligned}
r(\Delta;\alpha) &= (\alpha + 2)|\Delta|^3 - (\alpha + 3)|\Delta|^2 + 1 & |\Delta| \le 1 \\
&= \alpha(|\Delta|^3 - 5|\Delta|^2 + 8|\Delta| - 4) & 1 \le |\Delta| \le 2 \\
&= 0 & |\Delta| \ge 2
\end{aligned}
\tag{7-15}
$$

where Δ is the local offset in either x or y, as appropriate (Park and Schowengerdt, 1983).

A drawback of cubic resampling is that it produces *DN* overshoot on either side of sharp edges. The magnitude of the overshoot is directly proportional to the magnitude of α. Although this characteristic contributes to the visual sharpness of the processed image, it is undesirable for further numerical analyses where radiometric accuracy is of prime importance. It has been shown (Keys, 1981; Park and Schowengerdt, 1983) that -0.5 is an optimal value for α (rather than the -1 commonly used), and the value of -0.5 was implemented early in the production of TM data (Fischel, 1984). The two interpolation functions are compared to the sinc function in Fig. 7-14. The effect of the nearest-neighbor, linear, and PCC resampling functions on Landsat TM imagery can be seen in Fig. 7-17. PCC generally does not yield a pronounced visual improvement over bilinear interpolation and requires more computation time. For applications in which the image sharpness is not critical, nearest-neighbor resampling may produce a satisfactory product with a large savings in computer processing time compared to the other interpolation functions.

Nearest-neighbor resampling does not introduce new pixel *DN* vectors into the image statistical distribution, whereas bilinear and bicubic resampling create new vectors. This is sometimes used as an argument for using only nearest-neighbor resampling if the image will later be classified by statistical techniques (Chapter 9). The effect of resampling on the quality of TM system-corrected imagery, relative to uncorrected imagery, has been reported in Wrigley *et al.* (1984). A comparison of bilinear and bicubic resampling on subsequent co-occurrence texture calculations (see Chapter 4) is presented in Roy and Dikshit (1994), on multispectral classification in Dikshit and Roy (1996), and on the NDVI calculated from resampled AVHRR imagery in Khan *et al.* (1995). An illustration of the effect of resampling on the spectral scattergram is shown in Fig. 7-18. The impact of resampling on the image spectral content will obviously depend on the relative proportion of changed pixel vectors. For general geometric corrections, involving rotation, scale change, and higher order factors, resampling can be expected to alter the spectrum as shown in Fig. 7-18, where up to 94% of the pixel vectors are created by resampling.

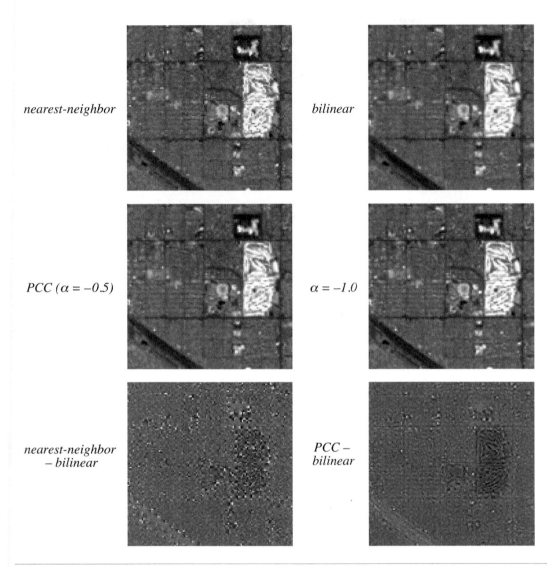

FIGURE 7-17. A portion of the rectified TM image of Fig. 7-1, as obtained with four different resampling functions. Nearest-neighbor resampling results in a blockiness, particularly for diagonal features. Bilinear resampling is continuous, while PCC results in a slightly sharper image. The visual difference between PCC with α equal to –0.5 and –1.0 is small; a value of –1.0 produces slightly more edge sharpening and overshoot. The difference maps show negative differences as dark and positive differences as light values. They illustrate the blockiness of nearest-neighbor relative to bilinear, and the fact that PCC(α = –0.5) has a high-boost filtering effect relative to bilinear. The latter is expected from a transfer function analysis of resampling (Park and Schowengerdt, 1983).

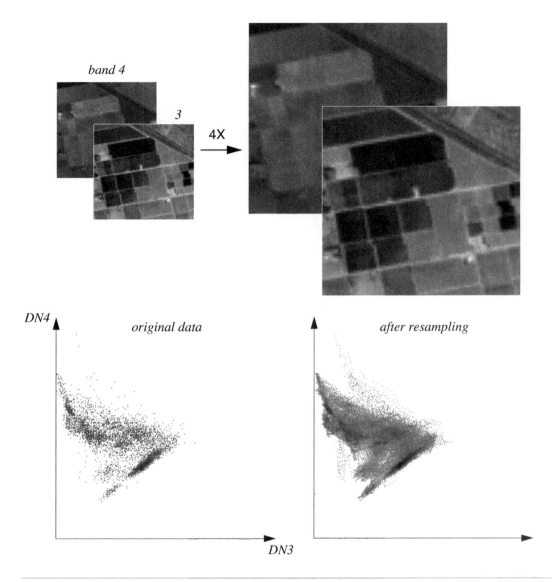

FIGURE 7-18. *Demonstration of the effect of resampling on the spectral scattergram. The two TM band images are enlarged 4 × using bilinear resampling above, and the before and after scattergrams are shown below. Note how the resampling has filled previously empty **DN** vector cells in the scattergram because of the pixel averaging caused by resampling. Cubic resampling would produce similar results, but may create some pixel vectors outside the range of the original data because of its "overshoot" behavior near DN edges mentioned in the text.*

7.3 Sensor MTF Compensation

Knowing that the sensor *PSF* (or equivalently, the *TF*) blurs the image recorded by the sensor, it is natural to consider how to remove, or at least reduce, its effect. This is an old topic in image processing, known as *image restoration* (Andrews and Hunt, 1977), with a large research literature. Its application in remote sensing dates to the mid-1980s (Wood *et al.*, 1986), but was not commonly available as a production processing option until about 2000. Some examples of published applications of restoration to remote-sensing imagery are given in Table 7-3.

TABLE 7-3. *Some of the research published on MTF compensation of remote-sensing imagery.*

sensor	reference
AVHRR	Reichenbach *et al.*, 1995
MODIS	Rojas *et al.*, 2002
SPOT	Ruiz and Lopez, 2002
SSM/I	Sethmann *et al.*, 1994
TM	Wood *et al.*, 1986; Wu and Schowengerdt, 1993
TM, simulated MODIS	Huang *et al.*, 2002

In sensor *MTF* correction, an attempt is made to undo the blurring effects of the sensor optics, detector, image motion, and any electronic filters to find a best estimate of i_{ideal}. The obvious approach is the *inverse filter* applied in the Fourier spatial frequency domain (see Eq. (6-30)),

$$I_{ideal}(u, v) = \frac{1}{TF_{net}}I = W_{inv}I \qquad (7\text{-}16)$$

or by the inverse Fourier transform and the Convolution Theorem,

$$\begin{aligned} i_{ideal}(x, y) &= \mathcal{F}^{-1}[W_{inv}I] \\ &= \mathcal{F}^{-1}[W_{inv}] * i \end{aligned} \qquad (7\text{-}17)$$

In this form, restoration is also known as *deconvolution*.

This direct approach in either the frequency or spatial domain has several pitfalls, however. First, if the TF_{net} is zero at any frequency, the inverse filter is not defined. A more likely situation is that division by very small values of TF_{net} will amplify noise at those generally higher frequencies. To avoid the latter problem, a threshold can be applied to limit the modulation boosting of the inverse filter at higher frequencies (Fig. 7-19).

inverse filter

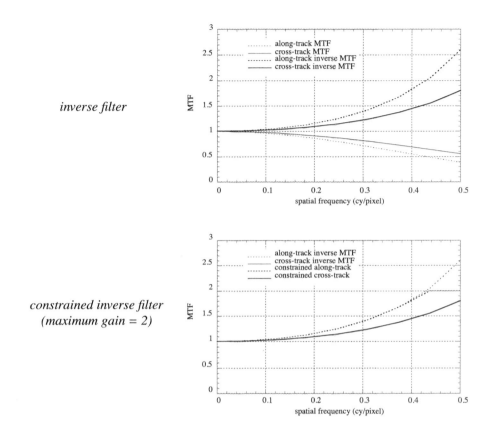

*constrained inverse filter
(maximum gain = 2)*

FIGURE 7-19. *The MTF, inverse MTF and a constrained inverse MTF are shown, using the ALI as an example system (Fig. 6-27). The upper limit on amplitude gain for the constrained inverse MTF (arbitrarily set to 2 in this example) should be determined by the SNR of the imagery; a lower SNR will generally require a lower limit on the inverse MTF to avoid excessive noise amplification.*

If there is noise in the image (which there always is), there are optimized filters, such as the *Wiener Filter*, that include a modification of the inverse filter by an LP term that depends on the SNR of the data. The Wiener Filter minimizes the error in estimating the ideal image from the noisy image by a linear filter operation. Other filters, in particular, non-linear filters, can be devised which may perform in practice better than the Wiener Filter (Andrews and Hunt, 1977; Bates and McDonnell, 1986). They do not however have the statistically optimal behavior of the Wiener Filter. We will not derive the Wiener Filter here; it is commonly discussed in signal and image processing books (Castleman, 1996; Gonzalez and Woods, 2002; Jain, 1989; Pratt, 1991). In the frequency domain, the Wiener Filter is,

$$W_{Wiener}(u, v) = \left[\frac{PSD_{i_{ideal}}|TF_{net}|^2}{PSD_{i_{ideal}}|TF_{net}|^2 + PSD_{noise}}\right] W_{inv}$$

$$= \left[\frac{1}{1 + \dfrac{PSD_{noise}}{PSD_{i_{ideal}}|TF_{net}|^2}}\right] W_{inv} \qquad (7\text{-}18)$$

$$= \left[\frac{1}{1 + 1/SNR}\right] W_{inv}$$

$$= LP_{Wiener} W_{inv}$$

where *PSD* is the *Power Spectral Density* described in Chapter 4. All quantities in Eq. (7-18) are functions of spatial frequency (u,v). At higher frequencies, W_{inv} behaves like an HB filter as seen in Fig. 7-19; the term LP_{Wiener} is a LP filter since the SNR of a noisy image will decrease at higher frequencies. It therefore prevents excessive boosting from W_{inv} at higher frequencies.

7.3.1 Examples of MTF compensation

MTF compensation, or *MTFC* as it is commonly referred to, is available as a processing option for IKONOS imagery. The restoration kernel is not published, as it is considered a proprietary "value-added" tool for commerical products. An example of MTFC applied to an IKONOS image is shown in Fig. 7-20, clearly demonstrating the visual "sharpening" obtained.

An example of restoration of Landsat TM imagery using the sampled Wiener Filter described in Wu and Schowengerdt (1993) is shown in Fig. 7-21. This technique was implemented in Fourier space, so a convolution kernel was not explicitly calculated. The correction in this case was a *partial restoration*, including compensation for all *PSF* terms *except PSF*$_{det}$. The rationale was that since the data were to be used later for unmixing (Chapter 9), it was desirable to retain the spatial integration of the detector in the restored image. The sharpening in the restored image is therefore generally less than produced by algorithms intended solely for visual enhancement.

A last example is the optional MTFC processing available in production ETM+ data from the USGS National Center for EROS shown in Fig. 7-22.[4] In this case, the MTFC is implemented as a deconvolution *simultaneous with image resampling* for geometric correction. The advantage of this approach is that only one convolution operation is performed on the imagery, thus minimizing degradation and computational cost. The MTFC kernels currently used for ETM+ data are compared in Fig. 7-22 to a PCC resampling function. Note that both MTFC kernels have larger negative lobes

---

4. The MTFC algorithm coefficients (not the kernel directly) are included with the image data in every ETM+ image's Calibration Parameter File (CPF). See Reichenbach and Shi (2004) for details on the underlying model used for the kernel; the specific kernel currently used for ETM+ data is slightly modified (Storey, 2006).

MTFC-off

MTFC-on

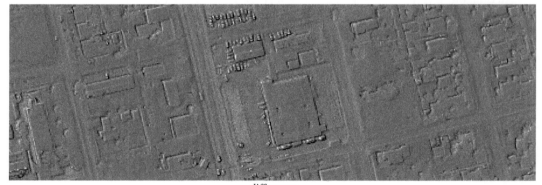

difference

FIGURE 7-20. An IKONOS panchromatic band image of Big Spring, Texas, collected on August 5, 2001, is shown. Note the high-pass filtered appearance of the difference between the MTFC-on and MTFC-off versions, as expected. (Imagery provided by NASA Science Data Program and Space Imaging, LLC.)

no MTFC

MTFC by sampled
Wiener Filter

difference

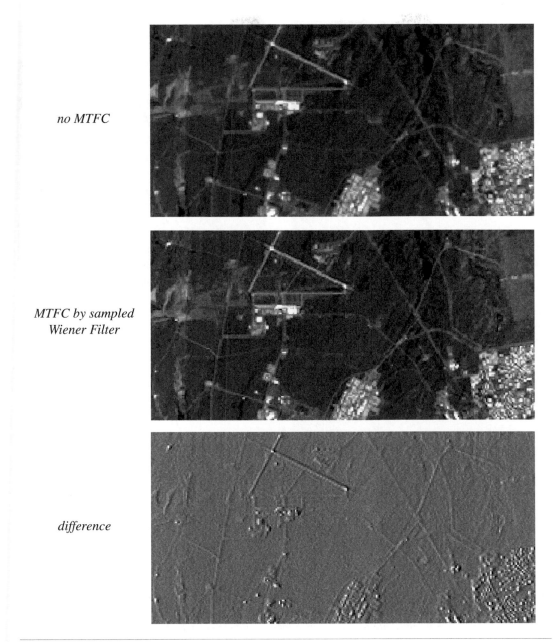

FIGURE 7-21. A TM image of Sierra Vista, Arizona, acquired in 1987, is shown. The difference image has a DN range of 238 but a standard deviation of only 9DN. (Image processing by Prof. Hsien-Huang P. Wu, National Yunlin University of Science and Technology, Taiwan, R.O.C.)

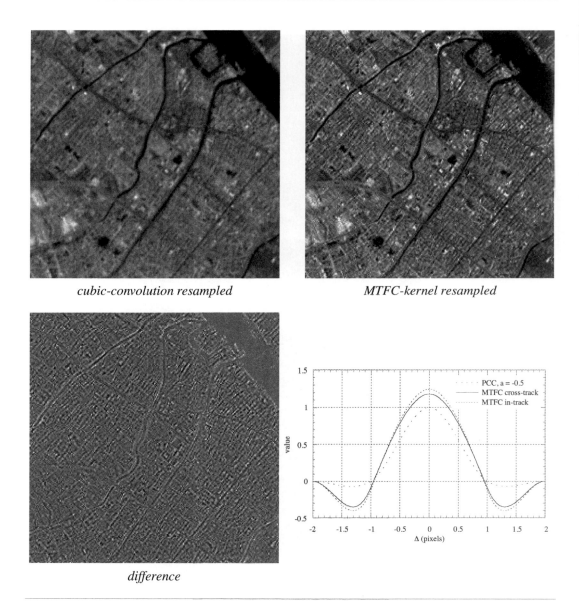

cubic-convolution resampled   *MTFC-kernel resampled*

difference

FIGURE 7-22. *Landsat-7 ETM+ processing with MTFC is shown for a band 8 pan image of Basra, Iraq, acquired on February 23, 2000. The MTFC is implemented in the resampling kernel used for geometric correction. Note the significantly larger negative lobe amplitude compared to PCC and that the MTFC kernels are not constrained to one for zero shift or zero at ±1 pixel shift (as is PCC). They, therefore, produce a sharpening effect at all resampling locations, even when aligned exactly at an input pixel. (Image data and MTFC kernel provided by Jim Storey, USGS EROS/SAIC.)*

than does PCC, implying greater high-frequency boost. In Fig. 7-22, we see the characteristic sharpening achieved by MTFC and the high-pass appearance of the difference with the unprocessed image.

7.4   Noise Reduction

The basic types of image noise were introduced in Chapter 4. If the noise is severe enough to degrade the quality of the image, i.e., to impair extraction of information, an attempt at noise suppression is warranted. Analysis of the image and noise structure is necessary before applying a noise suppression algorithm. In some cases, sensor calibration data and even test images may exist that are sufficiently comprehensive to estimate noise parameters. Unfortunately, however, such data are often incomplete or the noise is unexpected (for example, interference from other equipment or external signal sources). We must then learn as much as we can about the noise from the noisy image itself; some examples are given in the following sections. Periodic noise is relatively easier to diagnose and characterize from the noisy imagery than is random noise. We have attempted to neatly categorize image noise in the following discussion, but many variations exist and may occur simultaneously in the same image.

Since most image noise originates at the detectors or in the electronics of the sensor, its characteristics are defined at individual pixels or within scan lines. It is generally best, therefore, to try to remove the noise before any resampling of the image for geometric correction. Resampling will "smear" the noise into neighboring pixels and lines, so that system models for the noise will no longer apply, making it more difficult to remove.

7.4.1 Global Noise

Global noise is characterized by a random variation in DN at every pixel. Low-pass spatial filters can reduce the variance of such noise, particularly if it is uncorrelated from pixel-to-pixel, by averaging several neighboring pixels. Unfortunately, the variance of the noiseless part of the image, i.e., the signal, will also be reduced, albeit by a lesser amount because of the intrinsic spatial correlation present in the signal (Chapter 4). More complex algorithms that simultaneously preserve image sharpness and suppress noise are known as *edge-preserving smoothing* algorithms (Chin and Yeh, 1983; Abramson and Schowengerdt, 1993) and are discussed in the following sections.

Sigma filter

The fundamental conflict in spatial noise smoothing is the trade-off between noise averaging and signal averaging. A way is needed to separate the two components *before* averaging. In the sigma filter (Lee, J., 1983), they are separated by contrast. A moving-average window is used, as described in Chapter 4. At each output pixel, only those input pixels having *DN*s within a specified threshold of the center pixel DN_c are averaged (Fig. 7-23). A fixed threshold, Δ, can be set in terms of the *global* noise *DN* standard deviation, σ,

$$\Delta = k\sigma, \tag{7-19}$$

resulting in an acceptable range about the center pixel DN_c of,

$$DN_c \pm k\sigma. \tag{7-20}$$

Lee suggested a *k* value of two, since that corresponds to 95.5% of the data from a Gaussian distribution (the noise is assumed additive and Gaussian distributed).

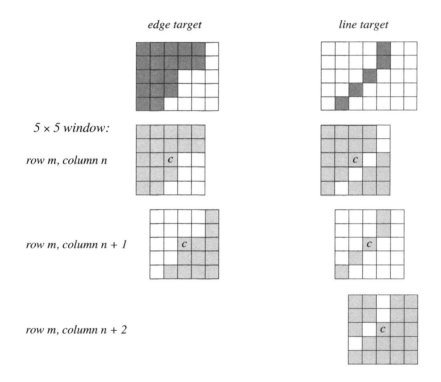

FIGURE 7-23. *The behavior of the sigma filter near edges and lines. The image data are noisy with a standard deviation of σ, but the difference between the mean DNs of the target and background is greater than $2k\sigma$. Therefore, the 5×5 sigma filter calculates the average of only the light grey areas as the window moves across the target, avoiding averaging of both the target and background. The "pool" of pixels from which the filter calculates an average can take any irregular, even disconnected, shape within the 5×5 window. If the target-to-background contrast is less than $2k\sigma$, there will be some averaging of the target and background.*

If the noise standard deviation is signal-dependent, the threshold can be made adaptive by using the *local* standard deviation,

$$\Delta_{local} = k\sigma_{local}. \tag{7-21}$$

However, the filter then becomes sensitive to higher contrast signal features, and they will be smoothed accordingly.

Nagao-Matsuyama filter

The sigma filter has no directional sensitivity. Many spatial features of interest are linear, at least over local regions. The Nagao-Matsuyama filter is designed to adapt to linear features at different orientations (Nagao and Matsuyama, 1979). A 5×5 window is used, and at each location of the window, the *DN* mean and variance in each of nine subwindows is calculated (Fig. 7-24). The mean of the subwindow with the lowest varaiance is used for the output *DN*. Thus, the filter finds the most homogeneous local region (of the nine) and uses its average for the output.

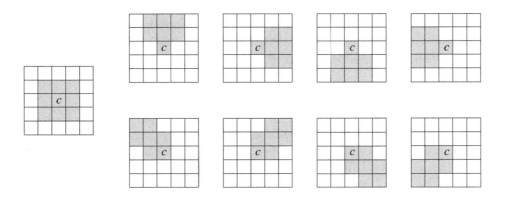

FIGURE 7-24. *The nine subwindows used to calculate local DN variance in the Nagao-Matsuyama algorithm. The pixels in the subwindow with the lowest DN variance are averaged for the output pixel at c.*

To illustrate these methods for reducing global noise, a *Side-Looking Airborne Radar (SLAR)* image will be used (Fig. 7-25). The image has the global, random speckle noise associated with synthetic aperture radar data. A 5×5 LPF smooths the noise, but also smooths the signal component of the image by an unacceptable amount. The global sigma filter achieves considerable smoothing of the noise, with much less signal smoothing. The small bright features and street patterns in the city are generally better preserved by the sigma filter. The sigma filter can average any number of pixels, and may not average at all for high-contrast point-like objects against a background. The Nagao-Matsuyama filter is better than the sigma filter at retaining linear features through the smoothing process. Small details in the city are lost in the seven-pixel averaging, but linear features such as the airport runways at the right are better preserved than with the sigma filter.

7.4.2 Local Noise

Local noise refers to individual bad pixels and lines, typically caused by data transmission loss, sudden detector saturation or other intermittent electronic problems. 1-D and 2-D array pushbroom sensors can have bad detectors that affect every image at the same pixel locations. The pixels affected by the noise usually have zero *DN*, indicating data loss, or maximum *DN*, indicating saturation.[5] An example is shown in Fig. 7-26. A 3×1 median filter removes the scanline noise, but also alters many other pixels, as shown in Fig. 7-26. A *conditional median filter* can be implemented by comparing each original image pixel with its immediate neighbors above and below. For example, if the current pixel differs from the average of its two neighbors (above and below) by more than a specified threshold, it is replaced by the median of the three pixels. Otherwise, it is unchanged. For the MSS example, this process results in a conditional mask that specifies which pixels to replace. This general approach of conditional masking, called *noise cleaning* (Schowengerdt, 1983; Pratt, 1991), results in significantly less overall change in the image, while still removing most of the noise. It requires two steps, detection of the noisy pixels and replacement of the detected noisy pixels by estimated good pixels.

Detection by spectral correlation

As mentioned in Chapter 5, the PCT can isolate uncorrelated image features in the higher-order PCs. This is not only useful for change detection in multitemporal images, but also for isolation of uncorrelated noise in multispectral images. Two examples are presented here, one with TM bands 2, 3, and 4 (green, red, and NIR) from the Marana agricultural scene (Fig. 7-27, Plate 9-1) and the other with ALI bands 1, 2, and 3 (blue, green, and red) from the Mesa scene (Fig. 7-28, Plate 8-1). A PCT of the TM imagery shows some local line noise clearly isolated into PC_3 because the noise occurs only in band 2 (Fig. 7-27). It can then be removed, if desired, by leaving PC_3 out of an inverse PCT. That would also, of course, lose any image information in PC_3. Alternatively, the PC_3 image could be processed with a median filter or some other noise-cleaning algorithm as described above, followed by an inverse PCT of PC_1, PC_2, and the cleaned PC_3. A PCT of the ALI image reveals vertical (in-track) noise in PC_3. This noise is rather periodic and could be reduced using the Fourier techniques described later in this chapter.

Some of the research papers on techniques for reducing global and local random image noise are noted in Table 7-4. Comparisons of various algorithms are provided in a number of papers, including Chin and Yeh (1983), Mastin (1985), and Abramson and Schowengerdt (1993). As in

---

5. Local pixel noise is called "salt noise" if the bad pixels are bright, "pepper noise" if the bad pixels are dark, and "salt and pepper noise" if both cases are present! It is also referred to as "impulse noise."

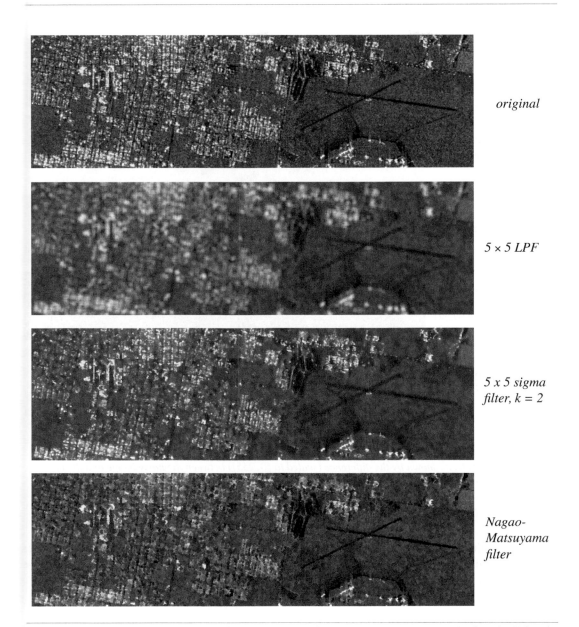

original

5 × 5 LPF

5 x 5 sigma
filter, k = 2

Nagao-
Matsuyama
filter

FIGURE 7-25. *Speckle noise filtering of a SLAR image of Deming, New Mexico, acquired in the X-band with HH polarization on July 1, 1991, from 22,000 feet altitude. The GSI is about 12 × 12 meters. Metallic objects such as cars, metal roofs, and power lines show a high signal ("return"). (The SLAR image is from the U.S. Geological Survey's Side-Looking Airborne Radar (SLAR) Acquisition Program 1980–1991 and is available on CD-ROM. The processed images are courtesy of Justin Paola, Oasis Research Center.)*

scanline noise

3 × 1 median
filtered

difference

conditional
mask
(threshold = 90DN)

conditional
median filtered

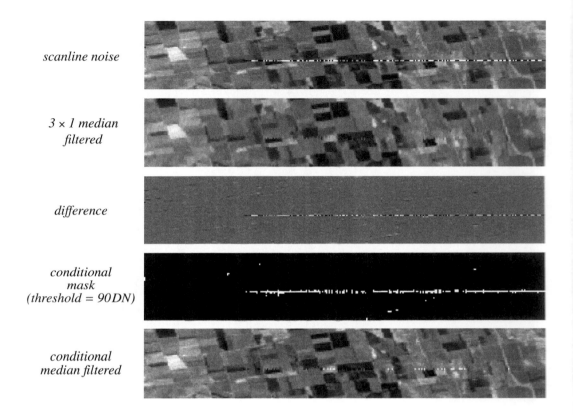

*FIGURE 7-26. Local line noise in an MSS image. A 3 × 1 median filter removes the noise, but also changes
many good pixels, as shown by the difference (scaled to [0, 255]) between the noisy image and the filtered
result. A mask can be derived as explained in the text and used to restrict the pixels replaced by the median
filter. In this example, the regular median filter replaces 43% of the image pixels, while the conditional
median filter replaces only 2%.*

most noise filtering, no single algorithm has proven to be superior in all cases. Most procedures
require specification of image-dependent parameters and, in practice, interactive "tuning" of those
parameters to achieve satisfactory results.

7.4.3 Periodic Noise

Global periodic noise (commonly called *coherent* noise) is a spurious, repetitive pattern that has
consistent characteristics throughout an image. One source is electronic interference from data
transmission or reception equipment; another is calibration differences among detectors in whiskbroom or pushbroom scanners. The consistent periodicity of the noise leads to well-defined spikes

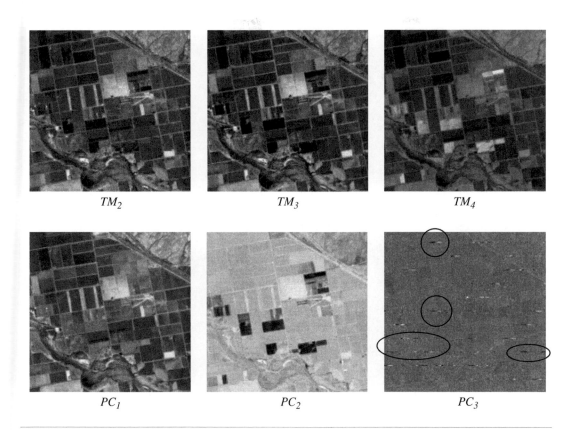

FIGURE 7-27. *Application of the PCT to noise isolation in a TM multispectral image. The scanline noise is isolated into PC3; some examples are circled. The noise could be removed by setting PC₃ equal to a constant, followed by an inverse PCT. The local noise in this example has a maximum range of about ±10DN. The PCs have been individually contrast stretched, which accentuates the noise amplitude relative to the image amplitude. On inspection of the original three bands, the noise is seen to occur only in band 2, explaining why it is well isolated by the PC transformation.*

in the Fourier transform of the noisy image (Chapter 6). If the noise spikes are at a sufficient distance from the image spectrum (i.e., the noise is relatively high frequency), they may be removed by simply setting their Fourier amplitude to zero. The filtered spectrum can then be inverse Fourier transformed to yield a noise-free image. Spatial domain convolution filters can be designed that produce the same result; large spatial windows are required to achieve localized frequency domain filtering, however (Pan and Chang, 1992).

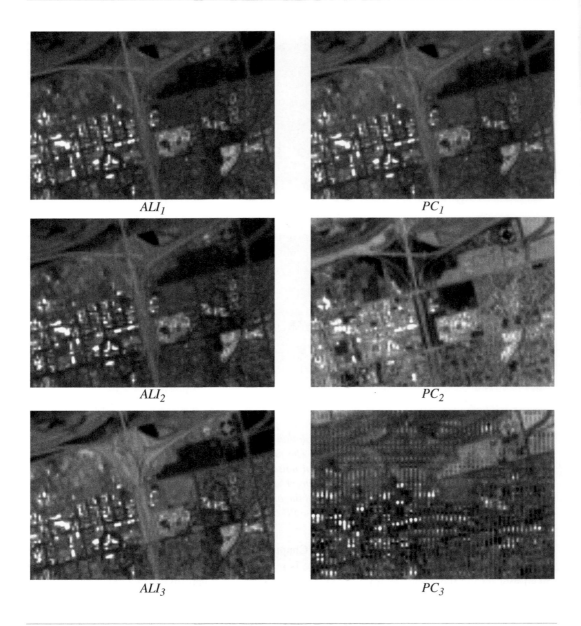

FIGURE 7-28. A second example of the ability of the PCT to isolate noise, in this case in an ALI image. The severe vertical noise in PC3 is not evident in the spectral bands. Again, the PCs have been individually contrast stretched, which accentuates the amplitude of the noise. The eigenvalue λ_3 is only 6.7 compared to 49.4 for λ_2 and 1585 for λ_1, i.e., the noise variance is less than 0.5% of the total image variance.

TABLE 7-4. Some research papers on reducing random image noise.

sensor	type of noise	reference
scanned aerial photo	global	Nagao and Matsuyama, 1979
	global	Lee, 1980
	global, local	Lee, 1983
HYDICE	bad detectors	Kieffer, 1996
Seasat SAR	global	Lee, 1981
TM with simulated noise	local	Centeno and Haertel, 1995
Viking Mars Orbiter	global, local	Eliason and McEwen, 1990

If the frequency of the noise falls within the image spectrum, then valid image structure will also be removed if the Fourier domain noise spikes are simply set to zero. Although this is usually not a serious problem, in practice it can be partially ameliorated by replacing the noise spikes in the Fourier domain with values interpolated from surrounding noise-free areas. Both the real and imaginary components of the spectrum must be interpolated.

Local periodic noise is periodic, but the noise amplitude, phase, or frequency varies across the image. One approach to this problem is to estimate the local amount of noise and remove only that amount at each pixel. For example, the global noise pattern may be obtained by isolating the Fourier domain noise spikes and inverse Fourier transforming them to the spatial domain. The local spatial correlation between this global noise pattern and the noisy image is then calculated for every image pixel (see Chapter 8 for a discussion of spatial correlation). At each pixel of the noisy image, a weighted fraction of the noise pattern is then subtracted; the weight is proportional to the magnitude of the local correlation. In areas where the noise component dominates, a correspondingly larger fraction of the noise pattern is removed and conversely, in areas where the noise is weak and the image dominates, a smaller fraction of the noise pattern is removed. A simplified version of this procedure, implemented entirely in the spatial domain, was used in Chavez (1975) to remove the single frequency, variable amplitude noise in Mariner 9 images of Mars.

7.4.4 Detector Striping

Unequal detector sensitivities and other electronic factors cause line-to-line striping in whiskbroom scanner imagery. If the striping is due to detector calibration differences, it has a periodicity which is equal to the number of detectors/scan, e.g., 16 in TM or 10, 20, or 40 in MODIS imagery. In pushbroom imagery, there is no scan periodicity, since each column of the image is acquired simultaneously by each of hundreds or thousands of detectors in the cross-track array. However, other types of periodic noise have been observed, e.g., in SPOT data, probably arising from readout of the CCD arrays. A variety of techniques have been applied to correct striping and coherent periodic noise (Table 7-5).

TABLE 7-5. Some of the published work on sensor-specific periodic noise reduction.

sensor	type of noise	reference
AVHRR	coherent	Simpson and Yhann, 1994
AVIRIS	coherent	Curran and Dungan, 1989; Rose, 1989
Fuyo-1	striping, periodic	Filho *et al.*, 1996
GOES	striping	Weinreb *et al.*, 1989; Simpson *et al.*, 1998
HYDICE	spectral jitter	Shetler and Kieffer, 1996
Hyperion	striping	Bindschadler and Choi, 2003
MSS	striping striping striping within-line, coherent striping striping striping	Chavez, 1975 Horn and Woodham, 1979 Richards, 1985 Tilton *et al.*, 1985 Srinivasan *et al.*, 1988 Wegener, 1990 Pan and Chang, 1992
SPOT	near in-track striping cross-track striping 2 × 2 "chessboard"	Quarmby, 1987 Büttner and Kapovits, 1990 Westin, 1990
TIMS	scanline, periodic	Hummer-Miller, 1990
TM	striping banding banding banding banding striping, coherent, banding	Poros and Peterson, 1985 Fusco *et al.*, 1986 Srinivasan *et al.*, 1988 Crippen, 1989 Helder *et al.*, 1992 Helder and Ruggles, 2004
VHRR	aperiodic striping	Algazi and Ford, 1981

Correction for detector striping, called *destriping*, should be done before any geometric corrections, while the data array is still aligned to the scan direction. The models discussed in Chapter 4 can then be applied. If only geometrically-processed data are available, then the data array will not be orthogonal to the scan direction, i.e., each original scan line may cross several of the data array lines if there are applied rotational geometric corrections (see Fig. 7-1), and it would be much more difficult to remove the striping noise.

Striping is not always the same across an image. Thus, the model with fixed characteristics described in Chapter 4 is not completely correct. The degree of correction achieved with that model, however, is often adequate and more complex models are unnecessary.

Global, linear detector matching

Over a large image consisting of millions of pixels, the data from each individual detector should have nearly the same *DN* mean and standard deviation. This realization provides a simple rationale for destriping. One of the detectors is chosen as a reference and its global mean *DN* and standard deviation are calculated. The global *DN* means and standard deviations in each of the other detectors are then linearly adjusted to equal the values for the reference detector. The transformation for detector *i*,

$$DN_i^{new} = \frac{\sigma_{ref}}{\sigma_i}(DN_i - \mu_i) + \mu_{ref} \qquad (7\text{-}22)$$

is applied to all pixels from that detector. After this transformation, each detector will produce the same global *DN* mean and standard deviation as that produced by the reference detector.[6] In effect, a linear gain and bias correction has been used to equalize the sensitivities of all detectors. An example is shown in Fig. 7-29.

Nonlinear detector matching

In some cases, a linear *DN* correction is not sufficient, and a nonlinear correction may be indicated. An algorithm for matching two image *DN* distributions was presented in Chapter 5 in the context of contrast matching. Extending this scheme to the destriping problem, we select one detector (or the average of all detectors) as the reference, calculate the cumulative distribution functions $CDF_{ref}(DN_{ref})$ and $CDF_i(DN_i)$, and do the following transformation,

$$DN_i^{new} = CDF_{ref}^{-1}[CDF_i(DN_i)]. \qquad (7\text{-}23)$$

The result is that the *CDF* of the data from each detector matches that from the reference detector, and nonlinearities in the detector calibrations are accounted for by Eq. (7-23) (Horn and Woodham, 1979).

Statistical modification

The use of a deterministic truncation rule in converting floating point calculated *DN*s to integer *DN*s can result in residual striping. One solution is to use the fractional part of the floating point *DN* as a probability measure controlling whether to round up or down in the conversion to integer. Thus, any systematic round-off striping is converted to a random variation, which is less visually apparent and perhaps more statistically precise (Bernstein *et al.*, 1984; Richards and Jia, 1999).

---

6. The global *DN* mean and standard deviation, averaged over all detectors, can also be used. That would perhaps be a little more robust, since the possibility of picking a reference detector with some peculiarity would be avoided.

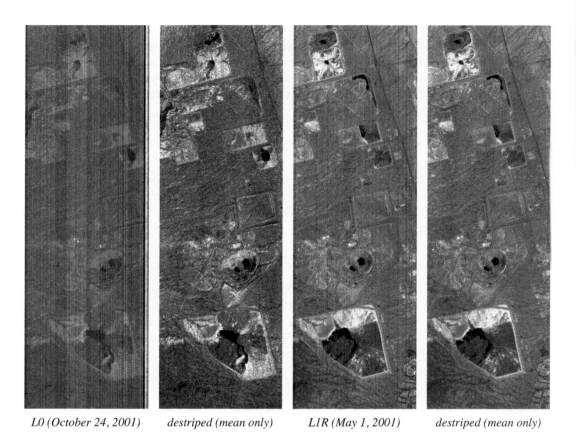

L0 (October 24, 2001)   destriped (mean only)   L1R (May 1, 2001)   destriped (mean only)

FIGURE 7-29. *Detector striping and removal in Hyperion images of large copper mines and waste ponds south of Tucson, Arizona. This is band 128, in the middle of a strong water absorption band at 1427nm, so the at-sensor radiance is low and easily degraded by any detector calibration errors. By adjusting the global means of the 256 pushbroom detectors to be equal (Eq. (7-22)), the striping is effectively removed, and the ground-reflected signal is apparent. Equalization of the standard deviations (gains) of all detectors was unnecessary in this case. A similar type of correction is applied in producing level 1R data, using calibration data rather than image data (Pearlman et al., 2003), but some residual striping remains, which can be removed by adjusting the detector data means, as shown.*

Spatial filter masking

Fourier domain filtering can be effective for destriping of relatively small images. The approach is to examine the amplitude or power spectrum of the image to locate the noise frequency components, design a blocking ("notch") filter[7] to remove them, apply the filter to the noisy image spectrum in the frequency domain, and finally calculate an inverse Fourier transform to obtain the corrected image (Fig. 7-30). This approach to removing periodic noise from images dates from the earliest spacecraft missions (Rindfleisch *et al.*, 1971) and has been applied to MSS (Pan and Chang, 1992), TM (Srinivasan *et al.*, 1988), and Fuyo-1 imagery (Filho *et al.*, 1996). A concise introduction to Fourier domain filtering of spatial data is given in (Pan, 1989).

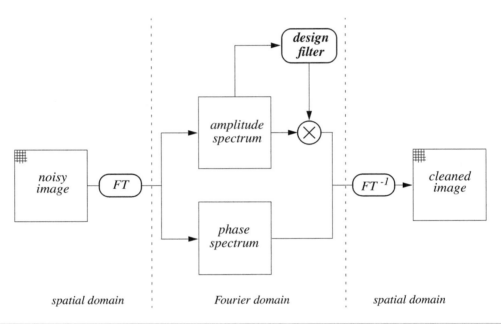

FIGURE 7-30. *Algorithm flow diagram for Fourier amplitude filtering. The "design filter" step can be manual or semi-automated, as described in the text.*

A 256-by-256 image block from an early TM band 1 image over the Pacific Ocean is shown in Fig. 7-31. The signal level is low, so the striping noise is readily apparent, both in the image and in the amplitude spectrum, where it appears as the column through zero frequency. A filter that blocks

---

7. A Gaussian amplitude *notch filter*, centered at frequencies $(\pm u_0, \pm v_0)$ is given by, $W(u, v) = 1 - e^{-[(u - u_0)^2 / \sigma_u^2 + (v - v_0)^2 / \sigma_v^2]^{1/2}}$. The width of each notch is proportional to (σ_u, σ_v). This is an inverted *Band-Pass Filter (BPF)*.

that column, except at low frequencies, appears to remove the striping noise and retain the lower, image frequency components. If we also block the pair of high-frequency columns on either side of the center noise spike, a more subtle, within-line coherent noise is also removed.

A practical question that may arise is, can the same filter remove noise from another part of the image? A second image block over downtown San Francisco (Fig. 7-32) was selected from the same scene. The amplitude spectrum of this block shows a more complex structure, as expected, but with similar noise components. The filter derived previously removes this noise without appearing to harm the image. The noise pattern looks entirely different than that for the ocean block (Fig. 7-31); the frequency components are the same, which is why the same notch filter works, but their amplitudes are different.

A more automated and conservative approach to designing the noise filter is possible for some images. When the frequency content of the scene is low (such as for the example ocean image segment) and the noise is primarily at higher frequencies, we can automatically find a noise filter with the following steps:

- Multiply the amplitude spectrum of the image by a "soft" high-pass frequency domain filter, such as an inverted Gaussian. This reduces the amplitude of the lower frequency components containing the scene content and removes the zero frequency component.

- Threshold the result at a reasonable value.

- Invert and scale the result to zero for the noise components and one for all others, thus creating a noise mask.

- Apply the noise mask to the image spectrum as previously shown and calculate the inverse Fourier transform.

There are two places for manual intervention in this procedure, one for the width of the high-pass filter in step 1, which should be as small as possible while still removing most of the scene content, and the other for the threshold of step 2, which determines the extent of noise removal. An example for our image is shown in Fig. 7-33. The high-pass filter has a 1/e width of 0.05 cycles/pixel, i.e., one-tenth of the highest frequency, and the threshold applied to the modified spectrum is simply midway between the minimum and maximum amplitude values. These parameters remove most of the noise in the image, but less than in Fig. 7-31.

Debanding

Banding refers to a striping-like noise that affects each multidetector scan of a system like TM or ETM+. The noise is aggravated by bright scene elements, such as clouds, that cause detector saturation for part of the scan, and depends on the scan direction. Noise removal is complicated in geometrically-corrected imagery by the fact that relative detector calibration algorithms like those used for destriping are not valid if the data have been resampled in-track. The Fourier filter approach can be used, although the striping and banding components may no longer align with single columns in the frequency domain if image rotation was included in the geometric corrections.

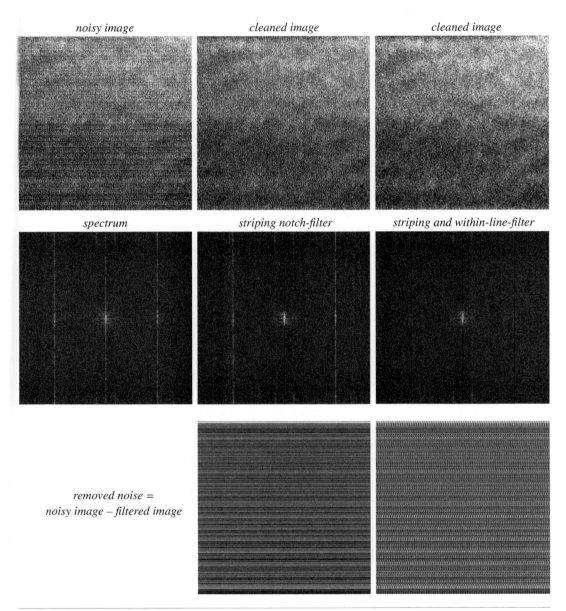

FIGURE 7-31. *Fourier amplitude filtering to remove striping from a nearly uniform ocean area in a TM image acquired on December 31, 1982. Blocking the primary striping noise component results in a cleaner image, and blocking the three periodic noise components results in further improvement. The noise DN standard deviation is only 0.45 and 0.77, respectively; the mean DN is zero in both cases. An additive noise model is assumed.*

noisy image   *filtered image*

filtered spectrum   *removed noise pattern*

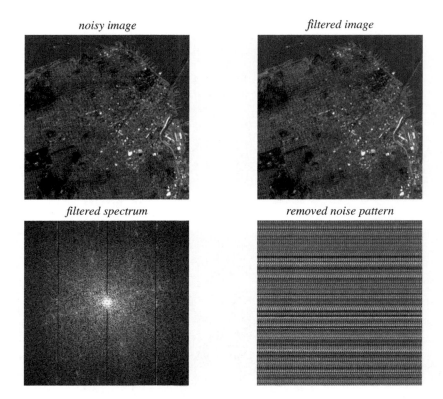

FIGURE 7-32. Application of the filter derived from the ocean area (Fig. 7-31) to another part of the same TM image. The Fourier spectrum is more complex, consistent with the detail in the image, but the noise characteristics are similar to those of the ocean area. The filtered image has no obvious artifacts due to the noise filter, even though the removed noise pattern is considerably different from that in Fig. 7-31. As explained in the text, this difference is in the amplitudes of the frequency components, not their frequencies. The noise DN standard deviation in this case is 0.69. Because this image has higher frequency content than the ocean segment, some image frequency components are also being removed here.

A simple, multistage convolution filter procedure has been described for this problem (Crippen, 1989). The process consists of three spatial filtering steps to isolate the noise from the image structure, followed by subtraction of the noise from the original image (Fig. 7-34). The 1 row × 101 column low-pass filter (LPF) isolates the low-frequency signal and scan noise from the high-frequency image components. The 33 row × 1 column high-pass filter (*HPF*) isolates the relatively high-frequency banding noise from the low-frequency signal. The use of 33 lines (rather than 32, the period of the TM forward/backward scans) arises from the adjustment of the original 30 m TM

spectrum of HP-filtered image   noise filter

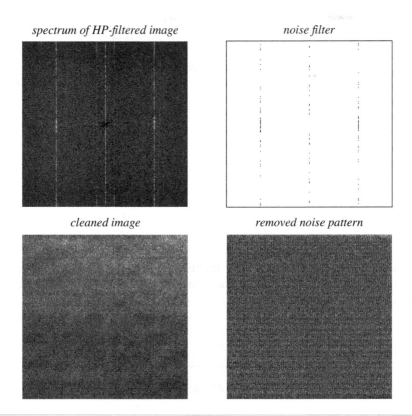

cleaned image   removed noise pattern

FIGURE 7-33. *Automatic filter design for striping. The noisy image is first high-pass filtered to remove the lower spatial frequency components. The Fourier amplitude spectrum of the filtered image is then thresholded to yield the noise filter. The filtered image is comparable to those of Fig. 7-31, but the threshold parameter used here results in somewhat less noise removed.*

---

pixel to 28.5 m in the system-corrected product. The last filter, a 1 row × 31 column LPF, suppresses diagonal artifacts that can be introduced by the previous filter. Finally, the isolated noise pattern is subtracted from the original image.

This debanding technique is illustrated with a TM band 4 image (Fig. 7-35). The three convolution filters are applied as prescribed, with the exception of masking the water (low *DN*) pixels, and the resulting isolated noise pattern is subtracted from the original image. The water masking step is essential to avoid edge artifacts at the boundary between land and water for this particular image, but appeared to be unnecessary in Crippen's examples. Most of the banding noise has been removed, and one can now see flow patterns in the Bay which were poorly depicted in the original data. To process other bands, the same mask can be used; the NIR bands work well for creating the mask because of the low reflectance of water in the NIR. This technique is largely heuristic and

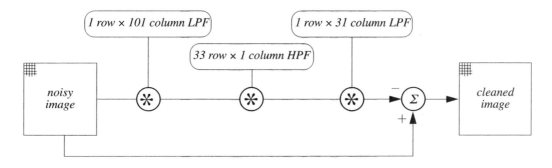

FIGURE 7-34. *The convolution filter algorithm proposed in Crippen (1989) to remove banding noise in system-corrected TM imagery. The series of filters produces an estimate of the noise pattern which is subtracted from the noisy image.*

cosmetic in nature, with no calibration basis for its derivation, unlike the *CDF*-based destriping algorithms. Another approach to debanding, based on analysis of TM engineering and image data, is presented in Helder and Ruggles (2004).

7.5   *Radiometric Calibration*

There are a number of important reasons to calibrate remote sensing data. The raw sensor DNs are simply numbers, without physical units. Each sensor has its own gains and offsets applied to the recorded signals to create the *DNs*. To do inter-sensor data comparisons, they must be converted to at-sensor radiances (Chapter 2). This step is *sensor calibration*. If we desire to compare surface features over time, or to laboratory or field reflectance data, corrections must be made for atmospheric conditions, solar angles, and topography. This is *atmospheric, solar,* and *topographic correction.* We are calling this entire calibration and correction process *radiometric calibration*.

There are several levels of radiometric calibration (Fig. 7-36). The first converts the sensor *DNs* to at-sensor radiances and requires sensor calibration information (e.g., EOS level 1B, Table 1-7). The second is transformation of the at-sensor radiances to radiances at the earth's surface. This level is much more difficult to achieve, since it requires information about the view-path atmospheric conditions at the time and locations of the image and sensor. That information can be in different forms, ranging from a simple categorization of the atmospheric condition as one of several "standard atmospheres," to estimates of certain parameters such as path radiance from the image data itself, to (ideally) coincident ground measurements. The combination of *in-image* data, i.e., the information contained within the image itself, with atmospheric modeling is increasingly used (Teillet and Fedosejevs, 1995), particularly for hyperspectral imagery (Gao *et al.*, 1993); (Goetz *et al.*, 2003). The final level of calibration to surface reflectance is achieved by correction for

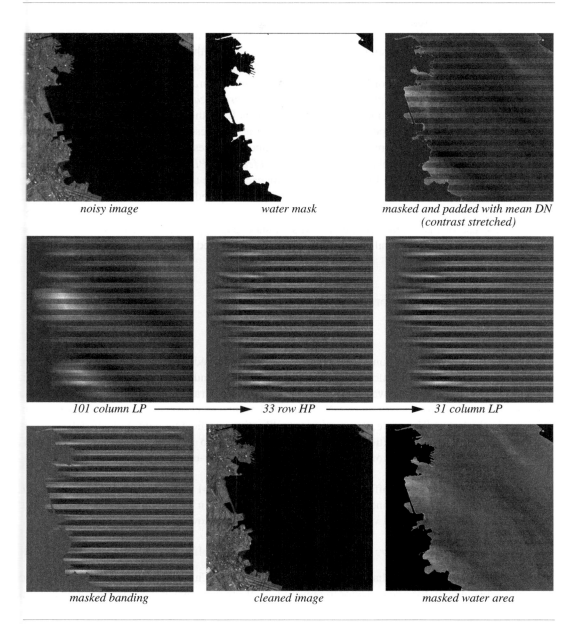

noisy image   water mask   *masked and padded with mean DN (contrast stretched)*

101 column LP ⟶ *33 row HP* ⟶ *31 column LP*

masked banding   cleaned image   masked water area

FIGURE 7-35. *Landsat TM band 4 of San Francisco Bay from August 12, 1983, shows banding, probably caused by a bright fog bank just off the left edge of the image (see Fig. 4-14 for another part of the same scene). A mask for water pixels is made by thresholding the image at a DN of 24, and the zeros in the mask are replaced by the mean DN of the water area to minimize edge artifacts from later processing. The remaining images show the cumulative result of the three filters, masked noise, and final corrected image.*

topographic slope and aspect, atmospheric path length variation due to topographic relief, particularly in mountainous terrain, solar spectral irradiance, solar path atmospheric transmittance, and down-scattered "skylight" radiance (e.g. EOS level 2, Table 1-7; see also Chapter 2).

Because of the complexity of full remote-sensing image calibration, there has been considerable interest in image-based techniques that provide *relative normalization* in certain restricted applications, such as multitemporal comparisons among images from the same sensor, or comparison of hyperspectral image data to spectral reflectance libraries, as described later. One example of an image-based technique for normalization of multitemporal images is the use of man-made features (concrete, asphalt, building roofs) as *pseudo-invariant features (PIFs)* (Schott *et al.*, 1988). A linear transform determined from the PIFs is performed on each image to normalize its *DN*s to those of a reference image. In Schott *et al.* (1988), the technique compared favorably to the *CDF* reference stretch discussed in Chapter 6. Of course, the spatial resolution of the imagery must be high enough to allow the use of man-made features as PIFs. A statistical approach has also been used to find invariant pixels in a bi-temporal image pair, which are then used to determine the *CDF* reference stretch (Canty *et al.*, 2004).

7.5.1 Multispectral Sensors and Imagery

Sensor calibration

There is some potential for confusion in the terminology of radiometric calibration. The sensor engineer uses Eq. (3-30) for calculations; here the sensor gain and offset have units of *DN*-per-unit radiance. This provides a "forward" calculation from predicted radiance to output sensor *DN*. The user of the data, however, receives *DN* values and wants to convert them to radiances in an "inverse" calculation. Therefore, the user wants calibration coefficients with units of radiance-per-*DN*. Unfortunately, these coefficients are also often referred to as "gain" and "offset" (EOSAT, 1993). To avoid any confusion here, we will call the user-oriented calibration coefficients, "cal_gain" and "cal_offset."

The calibration gains and offsets for the Landsat-4 and -5 TM systems, measured preflight, and the gains and offsets for the Landsat-7 ETM+ system are given in Table 7-6. These coefficients can be applied to Level 0 TM pixel values in each band, DN_b, as follows,

$$at\text{-}sensor: \quad L_b^s = \text{cal\_gain}_b \cdot DN_b + \text{cal\_offset}_b \tag{7-24}$$

to produce band-integrated radiance values, L_b^s. Level 1 data (Chapter 1) are already calibrated to radiance units (except for a scale factor to prevent loss of precision when the data are stored as 8-bit integers).

Although the sensor gain and offset are often assumed to be constant throughout the sensor's life, they can and usually do change over time. The Remote Sensing Group of the Optical Sciences Center, University of Arizona, measured the gain response of the Landsat-5 TM since 1984 (Thome *et al.*, 1994). The gain shows a generally slow, but steady decrease over that time in most TM bands

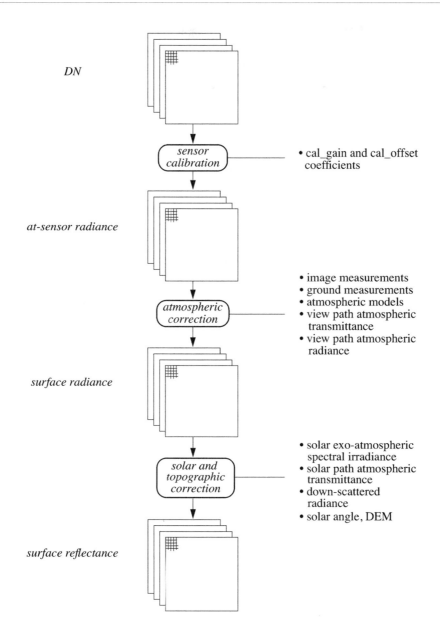

FIGURE 7-36. Data flow for calibration of remote sensing images to physical units.

(Fig. 7-37). This degradation in sensor performance with time in orbit is usually attributed to material deposition on the sensor optics caused by outgassing from the system in vacuum. Similar degradations have been reported for the AVHRR system (Rao and Chen, 1994).

TABLE 7-6. *Pre-flight measurements of the TM calibration gain and offset coefficients for Landsat-4 and -5 are calculated using the procedure provided in EOSAT (1993). Landsat-7 ETM+ values are for the low gain setting and are calculated from the spectral radiance range table in NASA (2006). Gain is given here in radiance/DN units and offset is in radiance units of* $W\text{-}m^{-2}\text{-}sr^{-1}\text{-}\mu m^{-1}$.

band	Landsat-4		Landsat-5		Landsat-7	
	cal_gain	cal_offset	cal_gain	cal_offset	cal_gain	cal_offset
1	0.672	−3.361	0.642	−2.568	1.176	−6.2
2	1.217	−6.085	1.274	−5.098	1.205	−6.4
3	0.819	−4.917	0.979	−3.914	0.939	−5.0
4	0.994	−9.936	0.925	−4.629	0.965	−5.1
5	0.120	−0.7208	0.127	−0.763	0.190	−1.0
6	0.0568	+1.252	0.0552	+1.238	0.067	0.0
7	0.0734	−0.367	0.0677	−0.0338	0.066	−0.35
8	–	–	–	–	0.972	−4.7

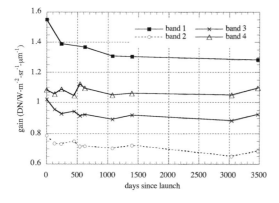
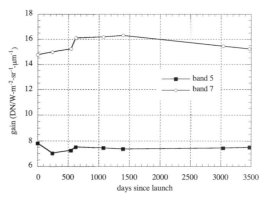

FIGURE 7-37. *Time history of sensor gain for the Landsat-5 TM sensor in the non-thermal bands. Note all bands show a decrease, except for band 5. The value at zero days since launch is from the preflight laboratory calibration. The remainder are measured using a field calibration procedure, which includes atmospheric measurements and modeling, and reflectance measurements of a bright, spectrally and spatially uniform area. These data are described in Thome et al. (1994); ETM+ calibration monitoring results for more than four years are presented in Markham et al. (2004).*

Atmospheric correction

The modeling of Chapter 2 showed that the atmosphere plays a complex role in optical remote sensing. As shown there with MODTRAN simulations, the atmosphere dramatically alters the spectral nature of the radiation reaching the sensor. It is impossible to apply the complete model to every pixel in a typical remote-sensing image. Independent atmospheric data are almost never available for any given image, and certainly not over the full *FOV*. So, how are we to correct the imagery for atmospheric effects? A robust approach to the problem is to estimate atmospheric parameters *from the image itself*. The parameters can then be refined with any available ancillary measurements and iterative runs of an atmospheric modeling program to achieve consistency between the atmospheric parameters and the image data. Any atmospheric correction that relies on physical modeling of the atmosphere first requires correction of the sensor *DN*s to radiances at the sensor, i.e., sensor calibration, as discussed in Sect. 7.5.1.

This "bootstrap" approach has been the most common type of atmospheric correction since the early days of Landsat (Table 7-7). A usual simplification is to assume that the down-scattered atmospheric radiance, E_λ^d, in Eq. (2-10) is zero, so that, in band-integrated form,

$$at\text{-}sensor:\quad L_b^s(x, y) \cong L_b^{su}(x, y) + L_b^{sp} . \tag{7-25}$$

Assuming this equation is an equality and referring to Eq. (2-8), we can rewrite Eq. (7-25) as,

$$at\text{-}sensor:\quad L_b^s(x, y) = \tau_{vb} L_b(x, y) + L_b^{sp} \tag{7-26}$$

and solve for the surface radiance, L_b, in each band,

$$earth's\ surface:\quad L_b(x, y) = \frac{L_b^s(x, y) - L_b^{sp}}{\tau_{vb}} . \tag{7-27}$$

In the simplest corrections for the visible wavelength region, the effort is focused on estimating the upwelling atmospheric path radiance, L_b^{sp}, with the view path transmittance assumed to be one, or at least a constant over wavelength. To a first approximation, this emphasis is not unreasonable, since the path radiance is the dominant atmospheric effect in the visible spectral region. The approach most commonly used requires identification of a "dark object" within the image, estimation of the signal level over that object, and subtraction of that level from every pixel in the image. Candidate dark objects are deep, clear lakes (Ahern et al., 1977), shadows (if they contain a low reflectance object), and asphalt paving (Chavez, 1989). This procedure is termed *Dark Object Subtraction (DOS)*. The rationale is that the only possible source for an observed signal from a dark object is the path radiance. The major weaknesses in this approach are in identifying a suitable object in a given scene, and once one is located, in the assumption of zero reflectance for that object (Teillet and Fedosejevs, 1995). Also, the DOS technique cannot correct for the view path transmittance term in Eq. (7-27), which is especially important in TM bands 4, 5, and 7.

TABLE 7-7. Examples of atmospheric correction techniques for multispectral remote-sensing images. Some of the techniques require the tasseled cap and NDVI spectral transforms described in Chapter 5. The items are ordered chronologically, rather than by sensor, to better convey the history of development in this field.

sensor	approach	reference
MSS	band-to-band regression	Potter and Mendolowitz, 1975
MSS	all-band spectral covariance	Switzer *et al.*, 1981
airborne MSS	band-to-band regression	Potter, 1984
AVHRR	iterative estimation	Singh and Cracknell, 1986
MSS, TM	DOS with exponential scattering model	Chavez, 1988
TM	DOS with exponential scattering model, downwelling atmospheric radiance measurements	Chavez, 1989
TM	pixel-by-pixel tasseled cap haze parameter	Lavreau, 1991
AVHRR	DOS, NDVI, AVHRR band 3	Holben *et al.*, 1992
airborne TMS, Landsat TM	ground and airborne solar measurements, atmospheric modeling code	Wrigley *et al.*, 1992
TM	comparison of ten DOS and atmospheric modeling code variations with field data	Moran *et al.*, 1992
TM	dark target, modeling code	Teillet and Fedosejevs, 1995
TM (all bands)	atmospheric modeling code, region histogram matching	Richter., 1996a; Richter, 1996b
TM	DOS with estimated atmospheric transmittance	Chavez, 1996
TM	dark target, atmospheric modeling code	Ouaidrari and Vermote, 1999
TM, ETM+	empirical line method, single target, ground measurements	Moran *et al.*, 2001
TM	water reservoirs, comparison of 7 methods for 12 dates	Hadjimitsis *et al.*, 2004
AVHRR	2-band PCT used to separate aerosol components	Salama *et al.*, 2004

More sophisticated approaches use dark target radiance estimation, combined with atmospheric modeling. For example, Chavez has used a look-up table of atmospheric scattering dependencies (Table 7-8) to estimate the path radiance in different bands (Chavez, 1989). The model used for a given image is determined by an estimated "haze" value from a blue or green waveband. Teillet and Fedosejevs have recommended the use of a dark target, with either a known or assumed reflectance, and detailed atmospheric modeling with the 5S code (Tanre *et al.*, 1990), but with no atmospheric measurements required, in a procedure that may be applied in a production environment (Teillet and Fedosejevs, 1995). They also make the point that significant errors can result from assuming the dark object has zero reflectance, even if the actual reflectance is only 0.01 or 0.02.

TABLE 7-8. *The discrete characterization of atmospheric conditions used by Chavez (1989).*

atmospheric condition	relative scattering model
very clear	$\lambda^{-4.0}$
clear	$\lambda^{-2.0}$
moderate	$\lambda^{-1.0}$
hazy	$\lambda^{-0.7}$
very hazy	$\lambda^{-0.5}$

Solar and topographic correction

To find the spectral reflectance of every pixel in an image requires more external information and further correction. Using Eq. (2-10) and solving Eq. (7-27) for reflectance in terms of the surface radiance, we obtain,

$$\rho_b(x, y) = \frac{\pi L_b(x, y)}{\tau_{sb} E_b^0 \cos[\theta(x, y)]} \tag{7-28}$$

Thus, we need the surface radiance (as obtained above), solar path atmospheric transmittance, exo-atmospheric solar spectral irradiance, and the incident angle (which depends on the solar angles and topography).[8] Again, in the visible spectral region, the atmospheric transmittance is reasonably predictable for different atmospheric conditions. Ideally, however, it should be measured at the time of image acquisition. The exo-atmospheric solar spectral irradiance is a well-known quantity and for flat terrain the incident angle is readily obtained. For non-flat terrain, calculation of the incident angle at each pixel requires a DEM at the same *GSI* as the sensor image.

We have assumed in all our derivations that the earth's surface was a Lambertian reflector, i.e., that it reflects equally in all directions. There is considerable evidence that this is not true for some types of surfaces, such as forests. A number of other analytical reflectance functions have been proposed, including the *Minnaert model*,

$$L_\lambda = L_n(\cos\theta)^{k(\lambda)}(\cos\phi)^{k(\lambda)-1} \tag{7-29}$$

---

8. Remember that the atmospheric down-scattered radiance term was assumed to be zero earlier in this derivation.

where ϕ is the radiation exitance angle to the surface normal vector, k is the empirical Minnaert constant, and L_n is the radiance if both θ and ϕ are zero. If k is one, Eq. (7-29) reduces to a Lambertian model. The constant k is typically found empirically by regression of $\log(L_\lambda \cos\phi)$ against $\log(\cos\theta\cos\phi)$. The Minnaert model has been compared to the Lambertian model for Landsat MSS and TM imagery of mountainous, forested areas (Smith *et al.*, 1980; Colby, 1991; Itten and Meyer, 1993).

Image examples

An example of calibration and correction of TM imagery for sensor response and atmospheric scattering is shown in Fig. 7-38. The pixel *DN* values are first calibrated to radiances using Table 7-6 and Eq. (7-24). Band 1 becomes darker in the figure, reflecting the correction for its relatively high gain compared to bands 2 and 3. However, its relatively higher atmospheric scattering level remains. Then, an atmospheric scattering component is estimated from the lowest values within the Briones Reservoir and subtracted from every pixel. After this correction, band 1 becomes quite dark compared to bands 2 and 3. This final set of sensor- and path radiance-calibrated images displays the correct relative scene radiances (to the extent that the print density of the figure is a linear function of the pixel values).

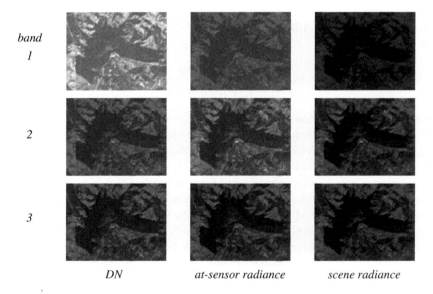

FIGURE 7-38. *Example calibration and path radiance correction of TM bands 1, 2, and 3 using sensor calibration coefficients and DOS. All images are displayed with the same LUT to maintain their relative brightness and contrast. The first column shows the original imagery, the second column shows the effect of correction for the sensor pre-flight gain and bias coefficients, and the third column shows the effect of a subtractive correction for estimated atmospheric path radiance. The latter is determined from the darkest pixels within the Briones Reservoir. Atmospheric transmittance is assumed to be the same in each band.*

Another example of the same calibration process is shown in the Cuprite, Nevada, TM image of Fig. 7-39. All six non-TIR bands are shown. Note how the calibration to scene radiance makes bands 5 and 7 appear dark. Their relatively high sensor gain has been removed, and the scene radiance still includes the solar irradiance factor, which falls off significantly in the SWIR. The latter can be removed by dividing each band by the corresponding TM band-integrated solar exo-atmospheric irradiance, which would leave the solar path atmospheric transmittance and topographic shading as the remaining external influences. Scene-based techniques can also be used, as described next for hyperspectral imagery.

The correction of SPOT imagery for topographic and reflectance variation in rugged terrain is shown in Plate 7-1 from Shepherd and Dymond (2003). Corrections were applied for the cosθ variation in incident irradiance and for reflectance variation as a function of incident and view angles. These angles were obtained from a co-registered DEM. Angular reflectance variation has not been described in detail here (see Chapter 2), but can be important for certain types of terrain cover such as forests and in mountainous areas. It essentially represents the departure of materials from the perfectly diffuse Lambertian reflectance model. Some of this effect is due to the subpixel geometrical relationships between trees, their spacing, and shadows (Gu and Gillespie, 1998). These subpixel variations are spatially integrated by the sensor and appear as angular variations in the pixel data.

7.5.2 Hyperspectral Sensors and Imagery

Atmospheric transmittance and absorption correction and sensor calibration of hyperspectral imagery are more difficult than they are for multispectral imagery because of hyperspectral imagery's substantially higher spectral resolution. The specific reasons for this extra difficulty include the following:

- Hyperspectral sensor bands coinciding with narrow atmospheric absorption features or the edges of broader spectral features will be affected by the atmosphere differently than the neighboring bands.

- The band locations in imaging spectrometer systems are prone to small wavelength shifts under different operating conditions, particularly in airborne sensors.

- Many analysis algorithms for hyperspectral data require precise absorption band-depth measurements exactly at the wavelength of maximum absorption.

- From the computational standpoint alone, the calibration problem is much greater for hyperspectral systems.

Sensor calibration

Hyperspectral sensors are particularly sensitive to spectral calibration. In systems with a 2-D pushbroom array, such as HYDICE, a narrow optical aperture (slit) is imaged onto the focal plane. Aberrations in the optical system can cause that image to curve, leading to an effect commonly

band
1

2

3

4

5

7

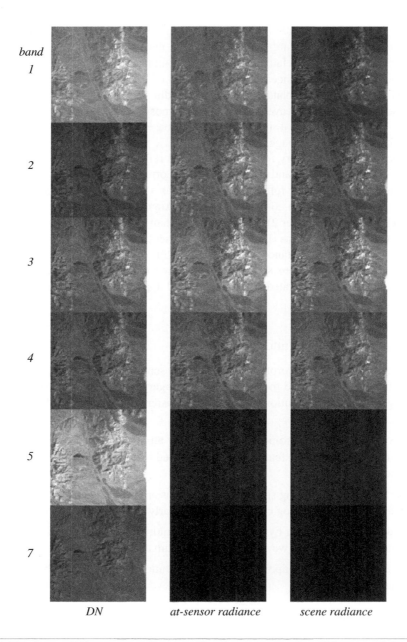

DN   at-sensor radiance   scene radiance

FIGURE 7-39.   Sensor DN-to-scene radiance calibration for a Landsat TM scene of Cuprite, Nevada, acquired on October 4, 1984. Again, atmospheric spectral transmittance is assumed constant across all bands.

called "smile." It results in an in-track shift of the spectrum as a function of cross-track detector number. Various image-based techniques that rely on known atmospheric molecular absorption bands have been used for determination of the smile error (Neville *et al.*, 2003; Perkins *et al.*, 2005). If the effect is unchanging from image to image, then a one-time careful calibration can be used. Correction can be done by in-track interpolation of the spectral data. Airborne hyperspectral sensors also can have problems arising from environmental influences, such as change in spectral calibration with altitude due to atmospheric pressure, molecular constituent, and temperature changes (Basedow *et al.*, 1996).

Atmospheric correction

High-resolution atmospheric modeling programs can be used to correct hyperspectral imagery for atmospheric effects. For example, a detailed correction of AVIRIS data using inversion of the 5S program's atmospheric calculations is described in Zagolski and Gastellu-Etchegorry (1995), and a comparison of 5S and the LOWTRAN-7 atmosphere modeling program for correction of AVIRIS data is described in Leprieur *et al.* (1995). An advantage of imaging spectrometers over broadband sensors such as TM is that, by virtue of their high spectral resolution, it is possible to estimate atmospheric water vapor content from absorption bands and use that information to facilitate calibration of the imagery to reflectance (Gao and Goetz, 1990; Carrere and Conel, 1993; Gao *et al.*, 1993). Moreover, the spectral information is available at every pixel, so it is at least theoretically possible to do *pixel-by-pixel* atmospheric corrections. A number of the programs of this type are listed in Table 7-9.

TABLE 7-9. *Atmospheric modeling and correction software programs. Some are commercial products and most rely on data or models from MODTRAN or 6S (Vermote et al., 1997).*

program	reference	remarks
ACORN (Atmospheric CORrection Now)	Miller, 2002	based on MODTRAN-4
ATCOR (ATmospheric CORrection)	Richter and Schlapfer, 2002	ERDAS Imagine
ATREM (ATmospheric REMoval)	Gao *et al.*, 1993	
FLAASH (Fast Line-of-sight Atmospheric Analysis of Spectral Hypercubes)	Adler-Golden *et al.*, 1999; Matthew *et al.*, 2000	RSI ENVI
HATCH (High-accuracy ATmospheric Correction for Hyperspectral data)	Goetz *et al.*, 2003; Qu *et al.*, 2003	improved ATREM
Tafkaa	Montes *et al.*, 2003	based on ATREM

One of the early attempts at using in-image hyperspectral information to aid in atmospheric correction was described in Gao *et al.* (1993). A key component of that approach is ratioing of spectral bands within and on either side of water vapor absorption bands in the NIR to estimate atmospheric transmittance in the water band, and thereby the total amount of atmospheric water vapor in the view path to each pixel. Water vapor is a key atmospheric component, as it can vary spatially, temporally, and with altitude more than other constituents. With this approach, it is estimated from the in-image data and input to the atmospheric model code.

We'll demonstrate part of this process for atmospheric correction with AVIRIS imagery (Plate 7-2). The two absorption bands of interest are centered at 940nm and 1140nm, and are about 50nm wide. A correction is first made for exo-atmospheric solar spectral irradiance (Chapter 2), which is the reason for the general decrease in AVIRIS radiance from 800 to 1300nm (Fig. 7-40). As was shown in Chapter 2, the solar irradiance is a multiplicative factor in at-sensor radiance, so the correction is to divide the at-sensor radiance by the known solar irradiance. That results in the *apparent reflectance*, which does not have the decreasing trend toward longer wavelengths (Fig. 7-40).

The average apparent reflectance of the water vapor absorption bands is calculated over 50nm and of the background spectral regions over 30nm. The result clearly shows the lower transmittance within the absorption band (Fig. 7-41). To estimate that relative transmittance, the ratio of the band-averaged apparent reflectance to the background-averaged apparent reflectance is calculated for both absorption bands and the result averaged (Fig. 7-42). Using that image-derived value, the water vapor amount and total transmittance of all molecular gases in the atmosphere are estimated from a pre-calculated table generated by the 5S atmospheric modeling code (Tanre *et al.*, 1990). That total transmittance is then used to estimate surface reflectance. An at-sensor radiance image and retrieved reflectance image for this type of atmospheric correction are shown in Plate 7-2. This approach has been improved with more accurate molecular absorption databases and a smoothness criterion for the retrieved surface reflectance spectrum (Qu *et al.*, 2003). A similar approach is also used to produce MODIS atmospheric products (King *et al.*, 2003).

These examples do not account for the influence of surrounding surface areas on the pixel of interest, the atmospheric *adjacency effect* (Chapter 2). A MODTRAN-based approach to correction for the atmospheric adjacency effect in HYDICE imagery is described in Sanders *et al.* (2001). Another effect not accounted for in most programs is shadows from clouds. Because of the partial transmittance of most clouds and atmospheric scattering, shadows contain some surface reflectance information. An interesting approach to removing cloud shadows ("de-shadowing") from multi- and hyperspectral imagery is presented in Richter and Muller (2005) and an example is shown in Plate 7-4.

Normalization techniques

A number of empirical techniques have been developed for calibration of hyperspectral data (Table 7-10). They produce relative calibrations in an empirical way, without the explicit use of atmospheric data and models; for this reason, they are more properly referred to as *normalization* techniques, rather than calibration techniques. For example, the empirical line method adjusts the data for an arbitrary gain and offset, and therefore matches the linear model relating at-sensor radiance and reflectance.

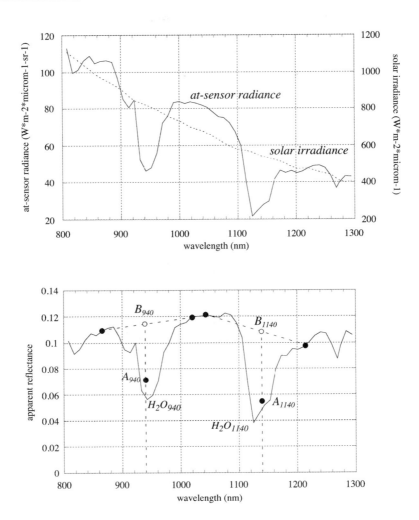

FIGURE 7-40. *The at-sensor radiance of a soil area in the AVIRIS image of Plate 7-2 is shown at the top with the solar exo-atmospheric irradiance curve. Dividing the AVIRIS radiance by solar irradiance gives what is called "apparent reflectance" (Gao et al., 1993), essentially the partially-calibrated spectrum shown below. Three spectral band regions are selected for each water vapor band, two on either side and one in the middle. The average of three AVIRIS bands are calculated for each of the out-of-band points, which are then averaged to give point B. Five AVIRIS bands are averaged for the in-band point A. The ratio of the apparent reflectance at point A to that at point B is assumed to approximate the transmittance of the atmosphere in the water band.*

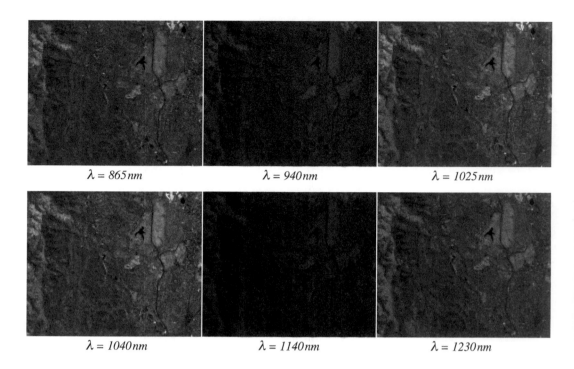

λ = 865 nm   λ = 940 nm   λ = 1025 nm

λ = 1040 nm   λ = 1140 nm   λ = 1230 nm

FIGURE 7-41. The averaged apparent reflectance images for the two water vapor absorption bands and the surrounding background spectral regions are shown. The data are clearly lower in the water vapor absorption bands than in the background bands, which is interpreted as resulting from lower atmospheric transmittance due to water vapor absorption rather than any difference in surface reflectance.

The different normalization procedures can be compared in terms of whether or not they compensate for the various external factors in remote-sensing radiometry (Table 7-11). On this basis, only the residual image approach appears to correct, at least in a heuristic fashion, for all major factors; the empirical line method is the next best technique, but requires extensive field measurements. Not all remote-sensing applications require full calibration to surface reflectance, however.

An AVIRIS image of Cuprite, Nevada, will be used to illustrate the different normalization procedures. This area has been extensively measured, studied, and mapped because of the presence of hydrothermally-altered minerals with little vegetative cover. These minerals have characteristic molecular absorption bands in the SWIR spectral region between 2 and 2.4 μm (Chapter 1), and we will restrict our analysis to that spectral region.

A color composite of AVIRIS bands (at-sensor radiances) at 2.1, 2.2, and 2.3 μm is shown in Plate 9-3. The colors are a clue that there is substantial minearological variation in this area, but the signatures are obscured by topographic shading and variation in surface brightness. Three sites are selected to represent the minerals alunite, buddingtonite, and kaolinite, and three single pixels in

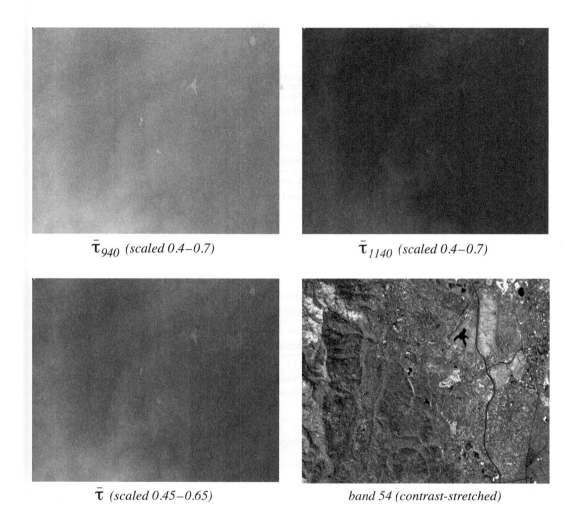

$\bar{\tau}_{940}$ *(scaled 0.4–0.7)* $\bar{\tau}_{1140}$ *(scaled 0.4–0.7)*

$\bar{\tau}$ *(scaled 0.45–0.65)*   *band 54 (contrast-stretched)*

FIGURE 7-42. *Dividing the center image by the average of the two on either side in Fig. 7-41 produces these estimated atmospheric transmittance maps (top), which were then averaged to produce the final estimated transmittance map (bottom). The band 54 image is shown for comparison—note the spatial variation in transmittance that appears partially correlated with elevation—the Russian Ridge area at the left of the image, which has elevations as high as 700 m, has higher atmospheric transmittance than the lower areas on the right, which have elevations as low as 10m. The average transmittance over the area is 0.55 ± 0.017.*

TABLE 7-10.   Empirical normalization techniques for hyperspectral imagery that has previously been calibrated to at-sensor radiances.

technique	description	reference
residual image	scale each pixel's spectrum by a constant such that the value in a selected band equals the maximum value in that band for the entire scene subtract the average normalized radiance in each band over the entire scene from the normalized radiance in each band	Marsh and McKeon, 1983
continuum removal	generate a piecewise-linear or polynomial continuum across "peaks" of image spectrum and divide each pixel's spectrum by the continuum	Clark and Roush, 1984
internal average relative reflectance (IARR)	divide each pixel's spectrum by the average spectrum of the entire scene	Kruse, 1988
empirical line	band-by-band linear regression of pixel samples to field reflectance spectra for dark and bright targets	Kruse *et al.*, 1990
flat-field	divide each pixel's spectrum by the average spectrum of a spectrally-uniform, high-reflectance area in the scene	Rast *et al.*, 1991

TABLE 7-11.   Comparison of hyperspectral image normalization techniques in terms of their ability to compensate for various physical radiometric factors.

technique	view path radiance	topography	solar irradiance	solar path atmospheric transmittance
residual images	yes	yes	yes	yes
continuum removal	no	no	yes	no
IARR	no	no	yes	yes
empirical line	yes	no	yes	yes
flat-field	no	no	yes	yes

different locations were selected as bright targets to use for the flat-field normalization. The identification of the mineral sites is aided by maps published in Kruse *et al.* (1990). Three-by-three pixel areas are averaged to reduce variability and the average spectral signature plotted in Fig. 7-43. Notice how these curves are largely dominated by the solar spectral irradiance. The atmospheric CO_2 absorption band at 2060nm is also apparent.

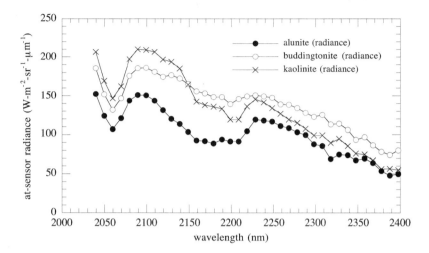

FIGURE 7-43. *AVIRIS at-sensor spectral radiances for the three mineral sites at Cuprite. The prominent absorption feature on the left is a CO_2 atmospheric band (compare to Fig. 2-4). The dips in the spectra between 2100 and 2200 nm are the characteristic molecular absorption bands of each mineral. The overall decrease in the radiance from 2000 to 2400 nm is primarily due to the decrease in solar irradiance.*

The flat-field and IARR-normalized versions are calculated by dividing each band of the image by the bright target *DN* or the average *DN* of the band, respectively (Fig. 7-44). The atmospheric absorption features are nicely removed by either of these procedures, and the mineral absorption features are unaffected. They achieve this discrimination because the spectrum of every pixel in the scene is modified equally by the atmospheric transmittance, while the various mineral absorption features either tend to cancel each other in the full-scene averaging (IARR), or are not present to begin with in the correction function (flat-field).

These normalization techniques accomplish a significant correction of the hyperspectral radiance spectra, to the point where attempts at matching them to laboratory or field spectra may be successful. The data in Fig. 1-8 are replotted in Fig. 7-45 for comparison to the at-sensor radiances and IARR-corrected values. These two techniques produce similar results for this image, indicating that the average scene spectrum is nearly "white" (Fig. 7-46). Such a fortuitous situation is unlikely for more heterogeneous scenes, such as the one of Palo Alto (Plate 1-3).

The continuum removal procedure is somewhat different from the scene normalization techniques. In this case, the spectra of individual pixels are each adjusted by dividing them by the continuum. The modified spectra exhibit a flat background because of this operation, but also retain the atmospheric absorption features. In our case, the average spectra from each mineral site were normalized in this way (Fig. 7-47).

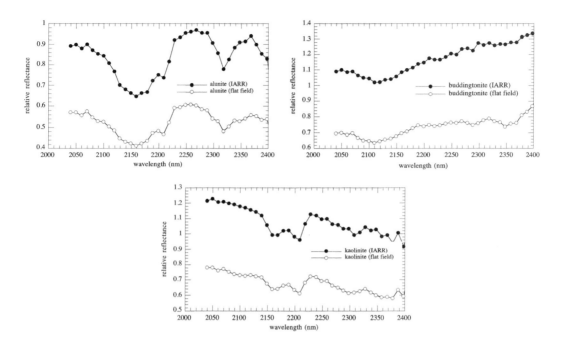

FIGURE 7-44. *Mineral spectra after normalization by the flat-field and IARR techniques. Note the IARR result is greater than the flat-field result, as expected.*

Image examples

The AVIRIS images in Plate 7-2 were corrected for the atmosphere using an ATREM-like procedure involving water vapor estimation by band ratioing, as discussed earlier. A single measured target reflectance in the scene (a bare soil field at Stanford University) was also used to supplement the atmospheric modeling process. The natural color composites shown clearly demonstrate the removal of blue path radiance scattered light from the data.

Two natural color composites of AVIRIS data are shown in Plate 7-3. These data are normalized first for path radiance by the DOS method and then for incident solar irradiance and solar path atmospheric transmission by division by the DOS-corrected spectral values of a bright target in each image. The latter step is an approximate correction only to the extent that the bright target has an intrinsically "white," or spectrally neutral, spectrum. This example demonstrates the robustness of normalization techniques based only on in-image data.

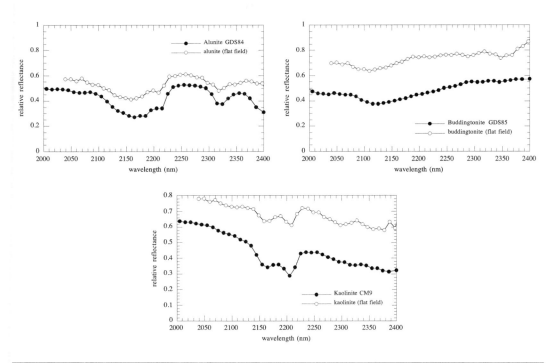

FIGURE 7-45. *Comparison of mineral reflectances and flat-field-normalized relative reflectances from the 1990 AVIRIS image.*

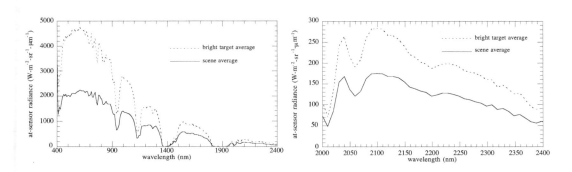

FIGURE 7-46. *Comparison of the bright target and average scene spectra for the Cuprite scene. The full VSWIR spectrum (left) and the expanded SWIR region (right) are shown.*

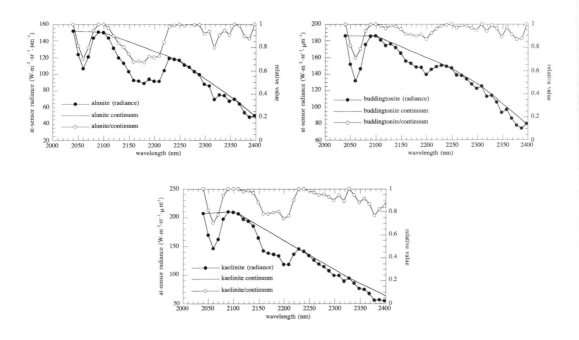

FIGURE 7-47. *Mineral spectra adjusted by continuum removal. The continuum was defined manually as a piecewise-linear envelope enclosing the radiance spectra.*

This completes our discussion of radiometric correction and calibration of remote-sensing imagery. It is probably one of the most logistically difficult tasks in remote sensing, and for that reason is too often ignored or only partially achieved with image-based techniques. As was pointed out in Chapter 2, however, many types of remote-sensing analyses rely only on *relative* signature discrimination, and therefore do not require *absolute* radiometric calibration.

7.6   Summary

Remote-sensing images require various types of systematic corrections for noise and geometric distortion. They also require radiometric calibration for accurate comparisons between dates and sensors. The important aspects of these operations are:

- Rectification of system-corrected products to maps can be implemented over limited areas with moderate topographic relief by global polynomial functions.

· Resampling for geometric correction alters the local spatial and global radiometric properties of the image. Nearest-neighbor resampling does not introduce any new pixel spectral vectors into the data, bilinear resampling introduces new vectors within the original *DN* range, and cubic resampling introduces new vectors within and outside the original *DN* range.

· Noise correction is necessary only if the noise will affect information extraction. It is best done before any resampling of the imagery.

· Local noise is removed with a variety of spatial filters, while global periodic noise is removed with Fourier filters. Unequal detector calibration noise, or striping, can be corrected by a global image-based adjustment of each detector's response. Uncorrelated noise in multispectral images may be removed by spectral decorrelation techniques, such as the PCT.

· Full calibration of remote sensing imagery requires sensor calibration and atmospheric and topographic effects correction. There is a wide assortment of techniques available, ranging from simple (DOS for path radiance correction) to complex (water vapor estimation from hyperspectral image data for radiative transfer modeling).

· Hyperspectral imagery presents special calibration challenges, but also provides some atmospheric data that is useful for calibration.

In the next chapter, we look at the spatial registration of images and some techniques for image fusion. Resampling and relative radiometric calibration play important roles in both topics.

7.7 Exercises

Ex 7-1. Specify a mathematical form for the frequency domain Gaussian notch-filter for:
  · in-track striping in pushbroom imagery
  · electronic interference noise at 45° to the cross-track direction
  · cross-track periodic noise with a period of four pixels

Ex 7-2. Show in detail how Eq. (7-29) and linear regression are used to find the Minnaert constant k.

Ex 7-3. Explain how each hyperspectral normalization algorithm in Table 7-11 achieves the indicated corrective capabilities.

Ex 7-4. Show that convolution of a row of pixels with the triangle weighting function of Fig. 7-14 and with calculation of output values between the original pixels, i.e., resampling, is equivalent to linear interpolation.

Ex 7-5. Explain how up to 94% of the resampled image pixel vectors are created by resampling in Fig. 7-18. Why could it be less than 94%?

Ex 7-6. Given the *DN* values of four neighboring pixels, find the *DN* of the resampled pixel at $\times$ using bilinear resampling:

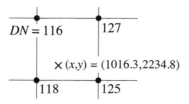

Ex 7-7. The resampling scheme described in Fig. 7-13 is "row-first," i.e. the intermediate pixels at E and F are calculated in neighboring rows. Show that a "column-first" procedure produces the same end result for $DN(x,y)$ using the example in Ex. 7-6.

CHAPTER 8

Image Registration and Fusion

8.1   Introduction

Spatial registration of multidate or multisensor images is required for many applications in remote sensing, such as change detection, the construction of image mosaics, DEM generation from stereo pairs, and orthorectification. Registration is the process which makes the pixels in two images precisely coincide to the same points on the ground. Although the images are then in *relative* registration, the *absolute* ground coordinates of the pixels may be unknown. Registration is easier to achieve with imagery that has been previously corrected for sensor and orbit distortions. Once registered, the images can be combined, or *fused*, in a way that improves information extraction. Examples of fusion include extraction of DEMs from stereo image pairs and compositing of images with different spatial and spectral resolutions.

8.2   What Is Registration?

Overlapping coverage from multiple images is obtained in several ways:

- the same satellite sensor on revisits, either in the same orbit (Table 1-5) or by pointing off-nadir in different orbits
- neighboring orbits of the same satellite sensor, which are typically several days apart
- different satellite sensors
- satellite and airborne sensors

These possibilities are illustrated in Fig. 8-1. It is difficult, but not impossible, to obtain coverage on the same date from multiple sensors. The concept of a satellite "train" described in Chapter 1, where several satellite sensors are in the same orbit and separated by a few minutes, allows multiple image (or other data) collections within a short time period. To take advantage of multitemporal images, they must first be registered, pixel-by-pixel.

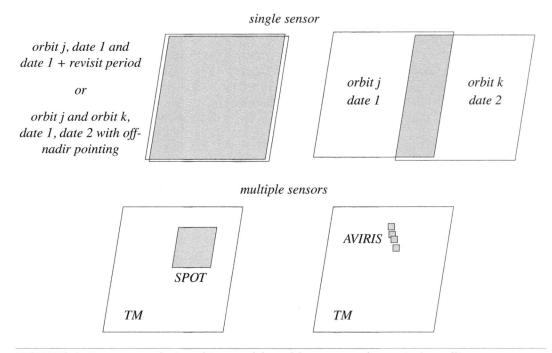

FIGURE 8-1. *Four ways to obtain multitemporal data of the same area from a single satellite sensor or two different sensors. The common ground coverage of the images is shown after they have been co-registered.*

To register such images, we need to establish a coordinate transformation that relates the pixel coordinates (row and column) of one image to those of the other, or relates both to a common reference. This can be done with a few GCPs and a suitable distortion model, as described in Chapter 7. Such an approach is often sufficient for satellite imagery of areas with little to moderate topographic relief, but the manual identification and measurement of GCPs is tedious, particularly in a production environment. Moreover, topographic variations in aerial or high-resolution satellite imagery are usually at spatial frequencies that are too high to be accurately modeled by global polynomials. We, therefore, need ways to increase the GCP density so that lower-order models, such as piecewise polynomials, may be used. Solutions to these problems are the subject of the first part of this chapter.

8.3   Automated GCP Location

Automated GCP location in two images consists of two steps. The first *extracts spatial features* from each image. Then, the features are paired by *correspondance matching*. Different types of spatial features have been used, including points, lines, and regions (Table 8-1). The success of the process depends, in part, on the similarity of the features in the two images. Temporal, view angle, and sensor differences all adversely affect correspondence matching of GCP features.

8.3.1 Area Correlation

Manual determination of single-pixel GCPs was described in Chapter 7. The local spatial context of the GCP is visually and mentally incorporated in that process. Similarly, small areas in each image ("chips" or "patches") can serve as the spatial features for automated registration. It is not necessary to precisely specify the location of corresponding chips in both images, because spatial cross-correlation will be used to determine the required shift for registration. Thus, the most tedious part of manual GCP selection is automated. The chips should be small enough that a simple shift is sufficient for registration of each pair of chips, i.e., we do not want internal distortions within a chip.[1] The differences between the shifts determined from the different chips define any global rotation, skew, or other misalignments between the full images (Chapter 7). Some suggestions for a semi-automated approach that combines the benefits of fully manual and fully automated GCP selection have been made in Kennedy and Cohen (2003).

---

1. However, this is unavoidable in the presence of high terrain relief or large relative rotation between the two images. Any internal distortions in one image chip relative to its mate in the other image can be thought of as adding to the dissimilarity between the two features.

TABLE 8-1. Some examples of image registration work. Surveys of image registration techniques are provided in Brown (1992), Fonseca and Manjunat (1996) and Zitova and Flusser (2003).

image types	features	reference
airborne MSS band/band, Apollo 9 SO65 photo band/band	area correlation using FFT	Anuta, 1970
simulated	area correlation	Pratt, 1974
MSS/airborne	edges, shapes	Henderson *et al.*, 1985
HCMM day/night, MSS/TMS	regions	Goshtasby *et al.*, 1986
airborne scanner	points	Craig and Green, 1987
TM/MSS, MSS/TMS	points	Goshtasby, 1988; 1993
multitemporal TM	points	Ton and Jain, 1989
TM/SPOT	regions	Ventura *et al.*, 1990
multitemporal TM	area correlation	Scambos *et al.*, 1992
SPOT/SPOT,MSS/MSS, MSS/SPOT,TM/SPOT	wavelet transform	Djamdji *et al.*, 1993b
balloon-borne photos	wavelet transform	Zheng and Chellappa, 1993
TM/SPOT	regions	Flusser and Suk, 1994
airborne/airborne, TM/SPOT,TM/Seasat	contours	Li *et al.*, 1995
airborne/airborne	points	Liang and Heipke, 1996
airborne/airborne	FFT phase correlation	Reddy and Chatterji, 1996
multitemporal TM	edge contours	Dai and Khorram, 1999
multitemporal AVHRR	wavelet transform (parallel implementation)	Le Moigne *et al.*, 2002
TM/IRS pan/SAR	mutual information	Chen *et al.*, 2003
multitemporal TM, ETM+/IKONOS	area correlation	Kennedy and Cohen, 2003
TM/TM (uncorrelated bands)	Fisher information	Wisniewski and Schowengerdt, 2005
TM/ERS SAR	edge contours	Hong and Schowengerdt, 2005

To find GCPs in the two images, an N-by-N "target" chip T ("template") is selected in the reference image and an M-by-M "search" chip S, with M greater than N, is selected in the other image. A "template match" spatial similarity metric is calculated by sliding the target chip over the central L-by-L region of the search area (Fig. 8-2), multiplying the two arrays pixel-by-pixel, and summing the result for each shift location (i,j),

$$\sum_{m=1}^{N}\sum_{n=1}^{N} T_{mn}S_{i+m,j+n} \tag{8-1}$$

The target and search areas do not have to be square; the only requirement is that the search area be larger than the target area. To prevent false correlation peaks arising from changes in the image DN over the search area, this metric is usually normalized in the following way to give the *normalized cross-correlation* (Hall, 1979; Rosenfeld and Kak, 1982; Pratt, 1991; Gonzalez and Woods, 1992),

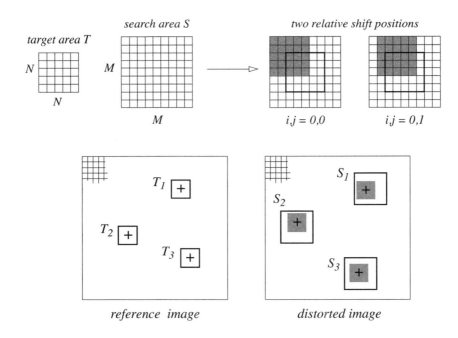

FIGURE 8-2. *Area correlation for image registration. A 5 × 5 pixel target area, T, and a 9 × 9 search area, S, are shown at the top. The DN arrays are correlated by "sliding" the target patch over the search area and calculating Eq. (8-2) at each possible shift position (only the first two are shown; the bold square shows the L × L possible shift locations). In this example, N and L are each five. At the bottom, we have a distorted image which is to be registered to the reference image. Three patches are used to find the distorted image coordinates at which maximum cross-correlation is obtained, possibly at fractional row and column values. The crosses in the distorted and reference images then represent the same ground point which is, in effect, a GCP. With these three GCPs, a global affine transformation can be performed.*

$$R_{ij} = \left[\sum_{m=1}^{N} \sum_{n=1}^{N} T_{mn} S_{i+m,j+n} \right] / K_1 K_2 ,$$ (8-2)

where

$$K_1 = \left[\sum_{m=1}^{N} \sum_{n=1}^{N} T_{mn}^2 \right]^{1/2} \quad \text{and} \quad K_2 = \left[\sum_{m=1}^{N} \sum_{n=1}^{N} S_{i+m,j+n}^2 \right]^{1/2} .$$ (8-3)

This normalization will remove any local gain differences in the two images, thus increasing their similarity. Since K_1 is the same for all shifts, it can be ignored in locating the relative maximum correlation. Subpixel precision can be achieved by interpolating the $L \times L$ correlation surface to estimate the point of maximum correlation, which presumably indicates the shift needed to register the two chips. In the calculation of Eq. (8-2), the coordinate system is "local," i.e., (m,n) is defined relative to the same pixel in each of the patches. One must keep track of the "global" coordinates, of course, in the original image frames.

Sometimes the mean DN within each chip is subtracted from the data before multiplication, yielding the *cross-correlation coefficient*,[2]

$$r_{ij} = \frac{\sum_{m=1}^{N} \sum_{n=1}^{N} (T_{mn} - \mu_T)(S_{i+m,j+n} - \mu_S)}{\left[\sum_{m=1}^{N} \sum_{n=1}^{N} (T_{mn} - \mu_T)^2 \right]^{1/2} \left[\sum_{m=1}^{N} \sum_{n=1}^{N} (S_{i+m,j+n} - \mu_S)^2 \right]^{1/2}} .$$ (8-4)

The benefit gained is that any local additive bias differences between the two images are removed, further increasing their similarity. An example application, in which the correlation is successful even though the two images, eight months apart, are not radiometrically matched prior to the operation, is shown in Fig. 8-3.

The spatial cross-correlation requires on the order of $N^2 L^2$ operations. The numerator in Eq. (8-2) can be computed via the *Fast Fourier Transform (FFT)*, which may be more efficient than direct spatial correlation (Anuta, 1970). Techniques that significantly increase the speed of spatial domain correlations have been described in Barnea and Silverman (1972) and applied to Landsat images in Bernstein (1976). These *Sequential Similarity Detection Algorithms (SSDAs)* use a small number of randomly-located pixels within the target and search areas to quickly find the approximate point of registration, followed by full calculations at shifts in the vicinity of the estimate for

2. Note this is defined the same for the two corresponding $N \times N$ pixel areas in the target and search windows as the spectral correlation coefficient in Eq. (4-17).

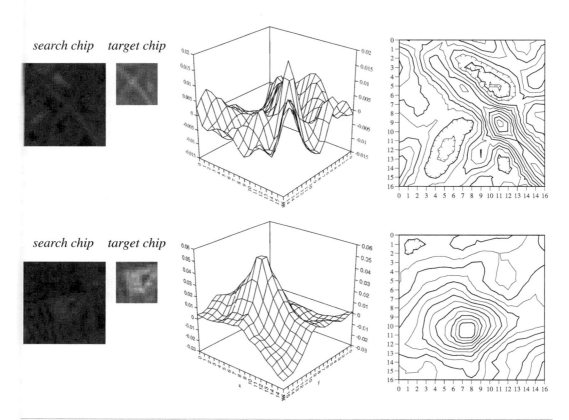

FIGURE 8-3. *Example cross-correlation coefficient surfaces obtained from multitemporal TM image chips of San Jose, California, acquired on December 31, 1982 (search chip), and August 12, 1983 (target chip). Note how the correlation surface shape mimics that of the target, particularly for the highway intersection. The cross-correlation peaks can be estimated to a fractional pixel shift by fitting a 2-D function such as a bivariate polynomial.*

precise registration. The spatial similarity metric used in this case is the sum of the absolute difference, pixel-by-pixel, for different shifts,

$$D_{ij} = \sum_{m=1}^{N} \sum_{n=1}^{N} \left| T_{mn} - S_{i+m,\,j+n} \right|. \tag{8-5}$$

A running subtotal of D_{ij} at each shift is compared to a threshold. If the threshold is exceeded, even before all N pixels have been included in the sum, it is assumed that the current shift is not a candidate for the point of registration, and the algorithm moves to the next candidate location.

Finally, we note that the correspondence matching aspect of registration is intrinsically solved by the manual specification of areas T and S, which implies that they contain the same ground area.

To automate the process further, correspondance matching must be explicitly addressed (Ton and Jain, 1989; Ventura *et al.*, 1990; Liang and Heipke, 1996).

Relation to spatial statistics

In Chapter 4, we described several measures of spatial similarity, such as the covariance and semi-variogram functions. While not defined in exactly the same way, the correlation coefficient of Eq. (8-4) is clearly another measure of similarity, in this case between two different but similar images. A useful way to think of area correlation is that if the two images are identical, the correlation becomes a spatial statistic on a single image and behaves similarly, with a peak for zero lag (shift) and a decrease as the lag increases. This is the limiting and optimal case for correlation between two images; in reality, they always differ to some degree. The more dissimilar the two images, the lower the correlation, and for quite different images, *there is no guarantee of a peak in the correlation at the point of registration*. Particularly difficult examples are registration of optical and radar images (Hall, 1979) or multitemporal images from different seasons. It is for these reasons that derived spatial features based on scene regions, or the boundaries between regions, are of interest for registration. Another promising approach to registration of dissimilar images uses information theory measures, such as entropy or Fisher information, calculated from the joint probability distribution of the two images (Wisniewski and Schowengerdt, 2005).

8.3.2 Other Spatial Features for Registration

Registration of two images can be viewed as a scale-space problem ranging from global registration (where we make a coarse estimate of the transformation, encompassing global rotation, scale difference and offset) to a local registration (where we refine the global model to accomodate high spatial frequency distortions, such as topographic parallax). The scale-space transform techniques discussed in Chapter 6 are therefore applicable to this problem. Zero-crossings of operators such as the Laplacian-of-Gaussian (LoG) filter have been used as spatial features in autonomous registration (Greenfield, 1991; Schenk *et al.*, 1991).

A wavelet transform technique for automatically locating numerous GCPs in two images has been developed and applied to SPOT registration with MSS and TM (Djamdji *et al.*, 1993b). The features used are local DN maxima in the thresholded high-frequency components (see Chapter 6), which occur at major image features such as rivers, topographic ridges or valleys, or roads. Typically, over 100 GCPs are automatically generated by the algorithm, depending on the scene content. The correspondence matching of GCPs in the two images is then achieved (in an unspecified way in the previous reference) and a global polynomial defined using the techniques of Chapter 7. A registration accuracy of about 0.6 pixel (of the image with the larger *GSI*) is claimed.

Another scale-space approach to registration of images with similar *GSI*s operates directly in the spatial domain. A resolution pyramid is created for each image, and coarse disparity estimates are formed at the highest levels using area correlation. These estimates are propagated by scaling to the next lower level to constrain the correlation search area. The estimated disparity is refined as the analysis propagates down the pyramid. This approach is described in detail next.

8.4   Orthorectification

As discussed in Chapter 3, ground objects at different elevations exhibit displacements (disparities) in their image positions. Using the principles of photogrammetry, these disparities can be used to estimate the elevation differences between points. This information can then be used to make the image *orthographic*, with every point in its correct location relative to other points, regardless of elevation. It is as if all perspective distortion is removed, and every point on the ground is viewed from directly above.

Correction of remote-sensing imagery for terrain displacement requires a *Digital Elevation Model (DEM)*, consisting of a spatial grid of elevation values. Traditionally, DEMs have been created with analytical stereo plotters that can digitize elevation profiles extracted manually from stereo aerial photograph pairs. These DEMs tend to be at a lower resolution than the aerial photography. In the last few years, practical techniques have been developed for precise local registration of digital image stereo pairs. Such techniques can construct a high-resolution DEM at approximately the same *GSI* as the imagery.

8.4.1 Low-Resolution DEM

The U.S. Geological Survey has assembled a large database of DEMs within the United States at *GSI*s of 240 m, 30 m, and 10 m. The 30 m DEMs are used to rectify aerial photographs and produce *digital orthophotoquads* corresponding to existing 7.5 minute topographic map quadrangles as part of the National Spatial Data Infrastructure program. We will follow the description of the image processing aspects of this process presented in Hood *et al.* (1989).

The overall process is depicted in Fig. 8-4. For aerial photography, the camera model consists of an *internal orientation model* that relates the scanned image coordinates to the camera reference frame given by the fiducial marks, and an *external orientation model* that describes the 3-D perspective geometry of a frame camera. If the camera and scanner geometry are good, the internal model can be simply an affine transformation (Chapter 7). The external model is governed by the *collinearity equations* (Wolf and Dewitt, 2000; Mikhail *et al.*, 2001),

$$x = -f \cdot \frac{m_{11}(X_p - X_0) + m_{12}(Y_p - Y_0) + m_{13}(Z_p - Z_0)}{m_{31}(X_p - X_0) + m_{32}(Y_p - Y_0) + m_{33}(Z_p - Z_0)}$$

$$y = -f \cdot \frac{m_{21}(X_p - X_0) + m_{22}(Y_p - Y_0) + m_{23}(Z_p - Z_0)}{m_{31}(X_p - X_0) + m_{32}(Y_p - Y_0) + m_{33}(Z_p - Z_0)}$$

$$(8\text{-}6)$$

where (x,y) are the photograph coordinates in the fiducial system, (X_p, Y_p, Z_p) and (X_0, Y_0, Z_0) are the coordinates of a ground point p and the camera station, respectively, m_{ij} are the elements of the transformation matrix (which contain information about the tilt of the camera relative to the terrain), and f is the camera focal length. The same equations apply to pushbroom imaging, with the x (in-track) coordinate set to zero and a time dependence introduced in the parameters on the right side of Eq. (8-6) (Westin, 1990).

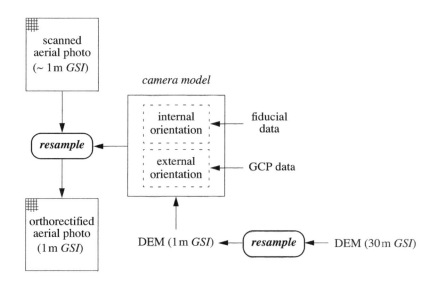

FIGURE 8-4.   *The process used to create a digital orthorectified photograph (commonly called an "orthophoto") from aerial photography and a pre-existing low-resolution DEM. The target GSI of the final product is 1 m, so the DEM must be resampled to estimate elevations on that grid. All coordinate transformations are calculated in the camera model and only one resampling of the image is necessary.*

In practice, the coefficients m_{ij} are found by a small set (at least three) of GCPs with known (X_p, Y_p, Z_p) and (x,y). The original low-resolution DEM is resampled, typically with bilinear interpolation (Hood *et al.*, 1989), to the desired high-resolution grid, and each DEM point is transformed by the collinearity and internal orientation equations to obtain the corresponding scanned photograph coordinate. Since that coordinate is not likely to be at an existing scanned image pixel, a new pixel has to be resampled between the scanned image pixels.

To illustrate, a scanned aerial frame is shown in Fig. 8-5. The *GSI* of this image is about 1.2 m. The 30 m DEM, previously calculated from other aerial stereo photography, is shown in Fig. 8-6, along with a shaded relief image with approximately the same solar angles as those for Fig. 8-5. The orthorectified aerial photograph is compared to a topographic contour map in Fig. 8-7. The rectification of the photograph can be verified by examining the two crossed roads. In the original photograph they are curved because of perspective distortion caused by terrain relief; after rectification they are straight as in the map.

8.4.2 High-Resolution DEM

If a high-resolution DEM with the desired final *GSI* is available, the previous process can be used, with the intermediate DEM resampling step eliminated. However, there are few DEMs at the spatial resolutions of aerial or high-resolution (5 m or less *GSI*) satellite imagery, and for many areas of the

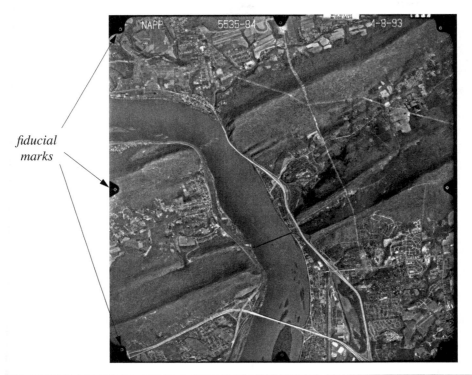

fiducial
marks

FIGURE 8-5. *The full frame as scanned from a National Aerial Photography Program (NAPP) aerial photograph of Harrisburg, Pennsylvania, acquired on April 8, 1993. The eight marks (white circles in black background) around the border are the camera fiducial marks, exposed onto every frame. They can be used to orient the internal camera model that relates the scan coordinates to the photographic image coordinates. (The image and DEM data in this example are courtesy of George Lee and Brian Bennett of the U.S. Geological Survey, Western Mapping Center.)*

world, no DEMs exist. Using area correlation as described at the beginning of this chapter, it is possible to calculate a DEM from a stereo image pair with a *GSI* that is about the same as that of the imagery. The basic equation describing the dependence of image disparity on elevation differences was derived in Chapter 3 and is repeated here,

$$\Delta Z \cong \Delta p \frac{H^2}{fB} = \Delta p \times \frac{H}{f} \times \frac{H}{B}. \tag{8-7}$$

The relative elevation difference of two image points is proportional to their relative disparity, Δp. If we perform area correlations over many candidate GCPs, we can then calculate a spatially-dense DEM. To increase the density of GCPs and reduce the computation necessary at full resolution, an image pyramid is used, as described next. A brief survey of research in DEM extraction from

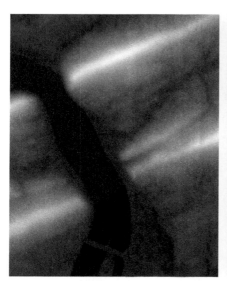 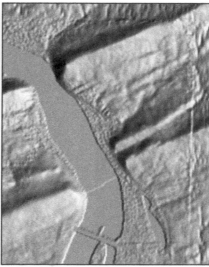

FIGURE 8-6. The Harrisburg NE quarter quad DEM (left) and a shaded relief representation (right). As we have seen before, the shaded relief image is similar to the remote-sensing image (compare to Fig. 8-5).

remote-sensing imagery is given in Table 8-2. Stereo images can be acquired with satellite sensors in two ways: overlap in images from adjacent orbits[3] or images taken at different view angles by a pointable sensor such as those on commercial high-resolution imaging satellites (Chapter 1).

Hierarchical warp stereo

The local neighborhood correlation of two images at one location tells us nothing about the global distortion between them. To measure that, we would have to do neighborhood correlations at several locations across the image. If the distortion is complex with abrupt changes, as might be caused by topographic relief, then the local correlation would have to be done at every pixel of the two images. Although that is possible, we like to avoid such intensive computations. Furthermore, the SNR of the correlation calculation can be improved by using larger window areas, but that would only further increase the computational burden.

We will describe the *Hierarchical Warp Stereo (HWS)* approach as a detailed example of DEM calculation from digital stereo image pairs (Quam, 1987; Filiberti *et al.*, 1994). The process is depicted in Fig. 8-8. The "warp image" is to be registered to the "reference image." A Gaussian pyramid, as described in Chapter 6, is first created for each image (*REDUCE*). The low-resolution layers are used to find coarse disparities which are refined at the next level (*REFINE*). Correlations

---

3. This *sidelap* amounts to about 30% of the cross-track *GFOV* for TM images at 35° latitude (Ehlers and Welch, 1987).

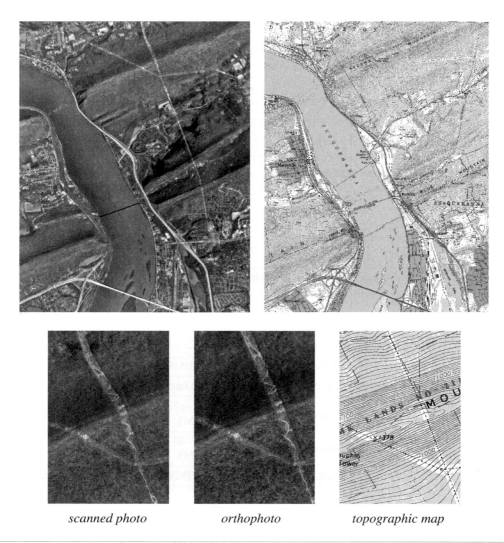

scanned photo   orthophoto   topographic map

FIGURE 8-7. *The digital orthophoto and the corresponding portion of the quadrangle map. Note the precise alignment, independent of elevation. The topographic map image was scanned and resampled to a 2.4m GSI from the original 1:24,000 scale paper map (USGS, 1995). Evidence of the orthographic projection achieved by the digital orthophoto process is shown at the bottom.*

are localized to the most likely region by transforming the coordinates of the warp image using the current disparity estimates (*warp*). The local area correlations are then performed for every pixel's neighborhood as described earlier (*match*), and the calculated disparities are added to the input disparities, which are propagated by scaling to the next, higher-resolution level (*EXPAND*). A scheme

TABLE 8-2.   Example approaches to determining elevation from stereo remote-sensing imagery. The elevation accuracy achieved is generally on the order of the sensor GSI, but depends on the topography. For example, the ETM+ accuracy quoted is for an application in rugged mountainous terrain.

sensor	technique	elevation accuracy (m)	reference
aerial photos	patch correlation	–	Panton, 1978
ASTER	hierarchical area match	10–30 RMS	Eckert *et al.*, 2005
LISS-II	area correlation	34–38 RMS	Rao *et al.*, 1996
ETM+ pan	sensor model, photogrammetry	92	Toutin, 2002
MISR	local maxima, area correlation	500–1000	Muller *et al.*, 2002
MTI	multiple images, HWS, sensor model, photogrammetry	6 RMS	Mercier *et al.*, 2003; Mercier *et al.*, 2005
QuickBird	sensor model, photogrammetry, area correlation	0.5–6.5	Toutin, 2004
SIR-B	HWS on a massively parallel processor	–	Ramapriyan *et al.*, 1986
SPOT	film images on analytical stereoplotter	7.5 RMS	Konecny *et al.*, 1987
	film images on analytical stereoplotter	7 maximum	Rodriguez *et al.*, 1988
	edge matching, area correlation	12–17 RMS	Brockelbank and Tam, 1991
	edge point matching and correlation	26–30 RMS	Tateishi and Akutsu, 1992
TM	area correlation	42 RMS	Ehlers and Welch, 1987

needs to be included to fill areas in the disparity map ("holes") where the spatial correlation is too low to be useful. That occurs in image areas that have little spatial detail or contrast. Holes can be filled by an interpolation routine that uses the nearest valid disparity values.

To demonstrate the HWS algorithm, the stereo pair of aerial photographs in Fig. 8-9 is used. Ground control is not available for the raw disparity map extraction, and the focal length and altitude of the camera are not known. A target area of 13 × 13 pixels and a search area of 29 × 29 pixels were used, allowing for a maximum disparity of ±8 pixels (Filiberti *et al.*, 1994). The calculated disparity map appears similar to a DEM derived by interpolation of a contour map, but there is a low-to-high trend, or tilt, in the disparity map (Fig. 8-10).[4] The trend can be removed and the

---

4. A camera bubble level imaged in the border of each aerial frame also indicated that there was relative airplane pitch in the flight direction between the two frames.

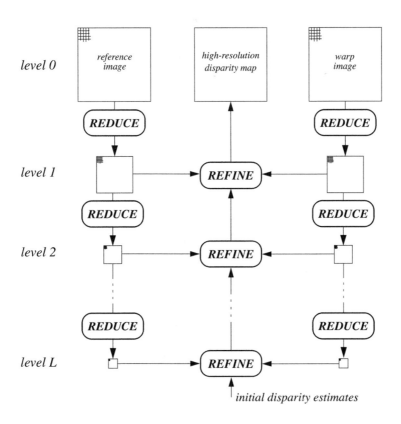

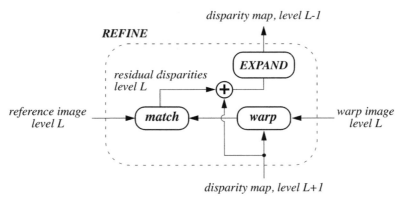

FIGURE 8-8. The HWS algorithm. At each level of the pyramid, the REFINE operation improves the resolution and accuracy of the disparity map. The match operation is local area correlation, centered on every pixel (after Quam, 1987).

disparities calibrated to elevation values by fitting a planar function through three corners of the two maps and adjusting the disparity data to match the topographic data at those three points. Such a correction procedure is not generally recommended; the internal and external camera orientation process described for orthorectification should be followed when fiducial and GCP data are available. We have, in effect, achieved approximate orientation *after* DEM generation by referencing disparities to the existing topographic map.

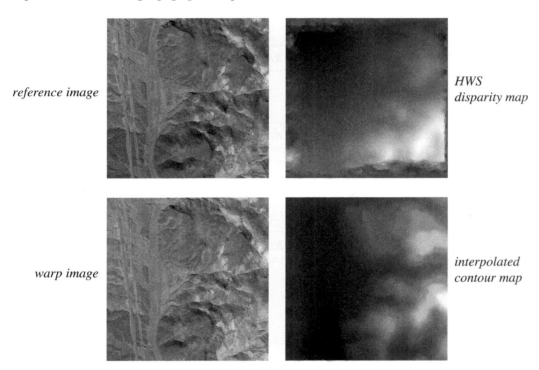

reference image

HWS disparity map

warp image

interpolated contour map

FIGURE 8-9. Stereo pair of aerial photos acquired at 12:20 P.M. on November 13, 1975, over Cuprite, Nevada. The flight direction is vertical, so the primary parallax is also vertical. However, the photographs are uncontrolled with respect to ground coordinates and in fact are both tilted in the flight direction relative to the terrain horizontal plane. The "noise" around the border of the disparity map is outside the region of overlap between the two photographs. A gridded DEM obtained by interpolation of 40-foot elevation contours on a 1:24,000 topographic map is also shown. (Aerial photography courtesy of Mike Abrams, Jet Propulsion Laboratory, and photo scanning courtesy U.S. Geological Survey, Western Mapping Center; HWS results and interpolated contour map produced by Daniel Filiberti, University of Arizona.)

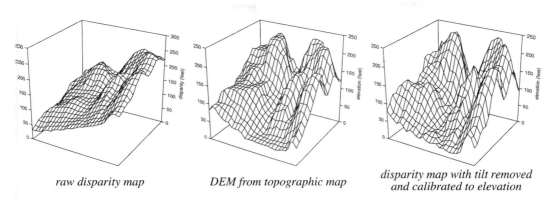

raw disparity map   DEM from topographic map   disparity map with tilt removed
                                                                            and calibrated to elevation

FIGURE 8-10.   The raw disparity map shows the effect of uncorrected camera tilt relative to the
topographic map DEM. The z-coordinates at three corners of both maps were used to make a planar
adjustment to the HWS-derived DEM, producing the corrected map. The structure in the HWS-derived
DEM now clearly matches that in the topographic map. The plots are for a 300 × 300 pixel area in the
lower right of the full DEM (Fig. 8-9) and are subsampled by a factor of about 15 to produce clear surface
plots.

Since we already have a topographic contour map in this case, why go to the effort of processing the scanned aerial photographs with the HWS algorithm? The answer is that with a correlation
process, such as HWS, the DEM is derived directly from the digital imagery, with little manual
intervention compared to the analytical stereo plotter approach. The derived DEM is also at a
higher ground resolution, as evidenced in Fig. 8-11. Finally, generation of the dense, gridded DEM
from the topographic contour map requires digitizing and interpolation between the contours,
which further degrades the already low-resolution topographic data.

8.5   *Multi-Image Fusion*

Remote-sensor systems are designed within often competing constraints, among the most important being the trade-off between *GIFOV* and signal-to-noise ratio (*SNR*). Since multispectral, and to
a greater extent hyperspectral, sensors have reduced spectral bandwidths compared to panchromatic
sensors, they typically have a larger *IFOV* in order to collect more photons and maintain image
SNR. Many sensors such as SPOT, ETM+, IKONOS, OrbView, and QuickBird have a set of multispectral bands and a co-registered higher spatial resolution panchromatic band. With appropriate
algorithms it is possible to combine these data and produce multispectral imagery with higher
spatial resolution. This concept is known as multispectral or multisensor *merging*, *fusion*, or *sharpening* (of the lower-resolution image).

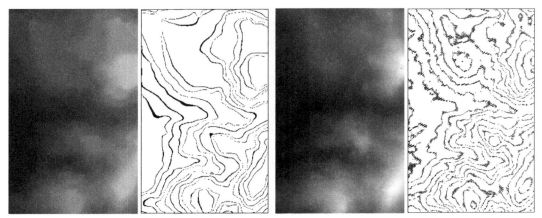

interpolated topographic map   *HWS-derived DEM*

FIGURE 8-11. Subsets of the interpolated topographic map and HWS-derived DEM, corrected for tilt and z-scale as described in the text (compare to Fig. 8-9). A contouring algorithm was run on both DEMs. The greater detail in the HWS-derived DEM is obvious. That does not mean, of course, that it is necessarily more accurate; it should be evaluated against a dataset with higher resolution than the interpolated contour map to evaluate absolute accuracy.

The low- and high-resolution images must be geometrically registered prior to fusion. The higher-resolution image is used as the reference to which the lower-resolution image is registered. Therefore, the lower-resolution image must be resampled to match the *GSI* of the higher-resolution image. The GCP approach (Chapter 7) or automated registration approaches described in this chapter can be used for this step. The dominant relative distortion is the *GSI* difference, but shift and rotation, and possibly higher-order distortions, may also be present. If the low- and high-resolution images come from the same sensor, e.g,. ALI multispectral and pan bands, the data are already (or at least should be) registered. The only step needed is to resample the multispectral bands to the *GSI* of the pan band.

Effective multisensor image fusion also requires radiometric correlation between the two fused images, i.e., they must have some reasonable degree of similarity. Artifacts in fused images arise from poor spectral correlation. A problem for some sensors is lack of coverage of the NIR by the pan band (Table 8-3). If the panchromatic band's sensitivity does not extend into the near IR, images with vegetation will show good correlation between the visible bands and panchromatic band, but poor correlation between the NIR band and panchromatic band. CIR fusion composites from such imagery can have artifacts. Fusion techniques can also be applied to diverse sensor imagery, for example, satellite multispectral imagery and aerial photography or SAR, but the same concerns on radiometric correlation apply and are even more problematic. One approach in this application is to use an intermediate thematic classification of one of the images to decouple the radiometric values in the two images from the fusion process (Filiberti and Schowengerdt, 2004).

TABLE 8-3. *The panchromatic band response range for several sensors.*

sensor	minimum wavelength (nm)	maximum wavelength (nm)
ALI	480	690
ETM+	500	900
IKONOS	526	929
OrbView	450	900
QuickBird	450	900
SPOT-5	480	710

Many different sharpening algorithms have been demonstrated (Table 8-4). They generally fall into two categories, feature space and spatial domain techniques. We describe and compare them in the following sections.

TABLE 8-4. *Multisensor and multispectral image fusion experiments .*

low resolution image	high resolution image	*GSI* ratio	technique	reference
MSS	airborne SAR	–	FCC	Daily *et al.*, 1979
MSS	Seasat SAR	3:1	HFM	Wong and Orth, 1980
simulated from MSS-1 and -4	MSS-2	5:1, 3:1	HBF	Schowengerdt, 1980
MSS HCMM	RBV MSS	2.7:1 7.5:1	HSI	Haydn *et al.*, 1982
simulated SPOT ms	simulated SPOT pan	2:1	HFM HFM, LUT	Cliche *et al.*, 1985 Price, 1987
SPOT ms	SPOT pan	2:1	HFM IHS radiometric	Pradines, 1986 Carper *et al.*, 1990 Pellemans *et al.*, 1993
TM	aerial photograph	7:1	image addition	Chavez, 1986
TM	airborne SAR	2.5:1	HSI	Harris *et al.*, 1990
TM-6	TM-1–5, 7	4:1	HBF LUT	Tom *et al.*, 1985 Moran, 1990
TM, SPOT ms	SPOT pan	3:1, 2:1	HSI	Ehlers, 1991

TABLE 8-4. Multisensor and multispectral image fusion experiments (continued).

low resolution image	high resolution image	GSI ratio	technique	reference
TM	SIR-B SAR	2.3:1	HSI	Welch and Ehlers, 1988
TM	SPOT pan	3:1	HSI HBF, HSI, PCT HSI, PCT HFM	Welch and Ehlers, 1987 Chavez *et al.*, 1991 Shettigara, 1992 Munechika *et al.*, 1993
AVIRIS	aerial photograph	6:1	HFM	Filiberti *et al.*, 1994
TM	SPOT pan	3:1	wavelet transform	Yocky, 1996
SPOT ms	SPOT pan	2:1	wavelet transform	Garguet-Duport *et al.*, 1996
TM	SPOT pan	3:1	wavelet transform, HSI, PCT	Zhou *et al.*, 1998
TM thermal band	TM ms	4:1	unconstrained and weighted least squares unmixing	Zhukov *et al.*, 1999
TM	SPOT pan	3:1	HFM	Liu, 2000
IKONOS ms	IKONOS pan	4:1	wavelet transforms	Ranchin *et al.*, 2003
ETM+ ms	ETM+ pan	2:1	adaptive HFM	Park and Kang, 2004
ETM+ ms AVIRIS	ETM pan SAR	2:1 2:1	unconstrained and weighted least squares unmixing	Filiberti and Schowengerdt, 2004
IKONOS ms	IKONOS pan	4:1	wavelet transforms	Gonzalez-Audicana *et al.*, 2005
ETM+ ms	ETM+ pan	2:1	Fourier transforms	Lillo-Saavedra *et al.*, 2005

8.5.1 Feature Space Fusion

As discussed in Chapter 5, we can transform a multispectral image into a new space in which one image represents the correlated component, for example, PC_1 in the space created by a Principal Components Transform (PCT), or intensity in a space created with a Color-Space Transform (CST). Feature space fusion techniques replace this component with the higher resolution image and transform the result back to the image space (Fig. 8-12).

It is important in both approaches that the original and replacement components are radiometrically correlated. For that reason, a "match" operation is included in Fig. 8-12. It often is not sufficient to linearly scale the minimum and maximum DNs if the two image histograms have different

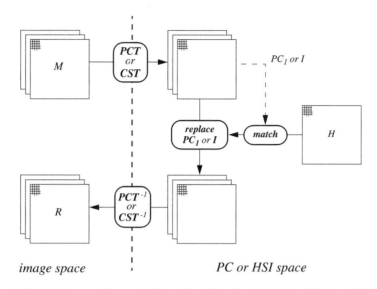

image space   *PC or HSI space*

FIGURE 8-12. *Image fusion using feature space component replacement, in this case either the first principal component or the intensity component. The multispectral image M has been previously registered to the high-resolution image H. A CST is limited to three bands by definition, but the PCT is not.*

---

shapes; a *CDF* reference stretch (Chapter 5) is more useful in that case and will partially compensate for weak correlation between the original component and the high-resolution replacement.

8.5.2 Spatial Domain Fusion

The idea in this case is to somehow transfer the high-frequency content of the higher-resolution image to the lower resolution image. One example of this approach was presented in Schowengerdt (1980). To simulate lower resolution bands, the 80 m bands 1 and 4 of the Landsat MSS were degraded by spatial averaging and down-sampled to either 240 m or 400 m. They were then resampled to the original 80 m *GSI* and combined with the 80 m band 2 by the following pixel-by-pixel operation,

$$R_{ijk} = M_{ijk} + HPF[H_{ij}] \tag{8-8}$$

where R_{ijk} is the reconstructed (fused) image in band k, M_{ijk} is the lower-resolution multispectral image in band k resampled to the *GSI* of the higher-resolution image, and $HPF[H_{ij}]$ is a high-pass version of H_{ij}, the higher-resolution image (in this case band 2). Note Eq. (8-8) is like the high-boost filter (*HBF*) of Eq. (6-6), except that the image component M_{ijk} and the high-pass component $HPF[H_{ij}]$ come from two different images. The success of the reconstruction in Eq. (8-8) is thus

expected to depend on the degree of radiometric correlation between the high-frequency components of the two images. Although the color composite of R_{ij1}, R_{ij4}, and M_{ij2} was visually sharper
than the composite of the degraded bands 1 and 4 with the original band 2, there were artifacts
wherever band 2 was negatively correlated with these bands, in particular near vegetation/soil
boundaries. A heuristic, spatially-variable, multiplicative weighting function, K_{ijk}, was included to
improve the fusion in these regions,

$$R_{ijk} = M_{ijk} + K_{ijk}HPF[H_{ij}]. \qquad (8\text{-}9)$$

High frequency modulation

In the *High Frequency Modulation (HFM)* algorithm, the higher-resolution panchromatic image, H,
is multiplied by each band of the multispectral, lower-resolution image, M, and normalized by a
low-pass filtered version of H, $LPF[H]$, to estimate an enhanced multispectral image in band k,

$$R_{ijk} = M_{ijk}H_{ij}/LPF[H_{ij}] \qquad (8\text{-}10)$$

Thus, the algorithm assumes that the enhanced (sharpened) multispectral image in band k is simply
proportional to the corresponding higher-resolution image at each pixel. The constant of proportionality is a *spatially-variable gain factor*, K_{ijk},

$$K_{ijk} = M_{ijk}/LPF[H_{ij}], \qquad (8\text{-}11)$$

making Eq. (8-10),

$$R_{ijk} = K_{ijk}H_{ij}. \qquad (8\text{-}12)$$

Using the two-component spatial frequency image model introduced at the beginning of Chapter 6,

$$H_{ij} = LPF[H_{ij}] + HPF[H_{ij}], \qquad (8\text{-}13)$$

we can rewrite Eq. (8-10) as,

$$\begin{aligned} R_{ijk} &= M_{ijk}(LPF[H_{ij}] + HPF[H_{ij}])/LPF[H_{ij}] \\ &= M_{ijk}(1 + HPF[H_{ij}]/LPF[H_{ij}]) \qquad (8\text{-}14) \\ &= M_{ijk} + K_{ijk}HPF[H_{ij}] \end{aligned}$$

which is the same form as Eq. (8-9), with a specific spatial weighting function, K_{ijk}, given by
Eq. (8-11). Thus, the HFM technique is equivalent to calculating a high-boost filtered version of M,
with the high-frequency components coming from the higher-resolution image H and weighted by
K. The HFM process is shown in Fig. 8-13, and a natural color composite of the same image is
shown in Plate 8-2.

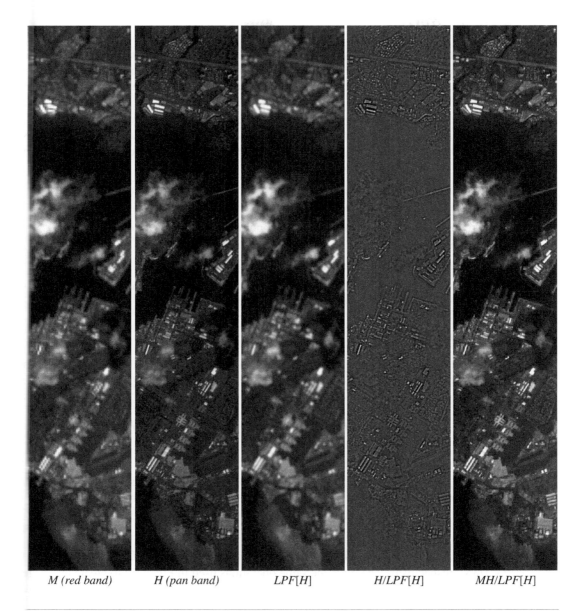

| M (red band) | H (pan band) | LPF[H] | H/LPF[H] | MH/LPF[H] |

FIGURE 8-13. *The sequence of processing to fuse two images using HFM (Eq. (8-10)). The image of Pearl Harbor, Hawaii, is from the ALI; the red band has a 30m GSI and the pan band has a 10m GSI. Note that the DNs in the ratio, H/LPF[H], vary about a mean value of one, which creates the sharpening effect at contrast edges in the fused image. A natural color fusion with this image is shown in Plate 8-2. (ALI image courtesy George Lemeshewsky, U. S. Geological Survey.)*

Filter design for HFM

Low- and high-pass filters must be applied to the higher-resolution image to implement spatial domain fusion. These filters cannot be arbitrary because they define the radiometric normalization between the two images. For example, in Eq. (8-10), it would seem reasonable to match the *LPF* to the effective *PSF* of the multispectral image relative to that of the higher-resolution panchromatic image. This intuitive relationship can be derived mathematically (Schowengerdt and Filiberti, 1994). Essentially, using the net relative *PSF* between the lower- and higher-resolution images maintains the local radiance from the original to the reconstructed images. The net *PSF* consists of three components, *all sampled at the smaller GSI,*

- the *GIFOV* of the lower-resolution sensor, e.g., a 3×3 square *PSF* when merging 10 m panchromatic and 30 m multispectral ALI images
- the remaining lower-resolution sensor *PSF* components, including image motion, optics, etc. (see Chapter 3)
- the resampling function used to register the two images (see Chapter 7)

These three components are convolved with each other to obtain the net *PSF*, as explained in Chapter 3, which should be used to calculate the *LPF*[*H*] component in Eq. (8-10).

Sharpening with a sensor model

An interesting approach that combines HFM with simultaneous restoration of the multispectral data (Chapter 7) was developed and applied to ALI imagery by Lemeshewsky (2005). The iterative equation for this technique is,

$$R_{ijk}^{m+1} = R_{ijk}^m + \alpha_1 LPF[\uparrow M_{ijk} - \downarrow (PSF_k * R_{ijk}^m)] + \alpha_2 HPF[H_{ij} - R_{ijk}^m]$$
$$= R_{ijk}^m + \Delta_M^m + \Delta_H^m$$

$$(8\text{-}15)$$

which defines the estimate of a higher-resolution multispectral band R_{ijk}^{m+1} at iteration $m + 1$ in terms of the estimate R_{ijk}^m at iteration m; the initial estimate is the resampled multispectral image,

$$R_{ijk}^0 = LPF[\uparrow M_{ijk}].$$

$$(8\text{-}16)$$

The symbols $\uparrow$ and $\downarrow$ represent *up-sampling* (insertion of zeros) by $3 \times$ and *down-sampling* (decimation) by $1/3 \times$, respectively. These are the same operations used in Fig. 6-32 to create an image pyramid. The spatial convolution operator *LP* used to interpolate at the zeros for resampling of the ALI multispectral bands can be either the nearest-neighbor or bilinear resampling function,

$$\text{nearest-neighbor: } 1/9 \cdot \begin{bmatrix} +1 & +1 & +1 \\ +1 & +1 & +1 \\ +1 & +1 & +1 \end{bmatrix}, \text{bilinear: } 1/8.92 \cdot \begin{bmatrix} +0.11 & +0.22 & +0.33 & +0.22 & +0.11 \\ +0.22 & +0.44 & +0.66 & +0.44 & +0.22 \\ +0.33 & +0.66 & +1 & +0.66 & +0.33 \\ +0.22 & +0.44 & +0.66 & +0.44 & +0.22 \\ +0.11 & +0.22 & +0.33 & +0.22 & +0.11 \end{bmatrix} \quad (8\text{-}17)$$

which are determined by the $3 \times$ difference in *GSI* between the pan and multispectral bands. The HP convolution filter was a discrete Laplacian filter (Chapter 6),

$$1/256 \cdot \begin{bmatrix} -1 & -4 & -6 & -4 & -1 \\ -4 & -16 & -24 & -16 & -4 \\ -6 & -24 & +220 & -24 & -6 \\ -4 & -16 & -24 & -16 & -4 \\ -1 & -4 & -6 & -4 & -1 \end{bmatrix} \cdot \qquad (8\text{-}18)$$

PSF_k is the sensor *PSF* for band k, sampled at 10m (see Chapter 7 for a description of the components of the ALI *PSF*). The two parameters α_1 and α_2 are adjusted to achieve convergence of the algorithm in a reasonable number of iterations; values of 1 and 0.5, respectively, were used in Lemeshewsky (2005).

It can be seen that Eq. (8-15) is an iterative change of the sharpened multispectral band by the addition of two terms,

- Δ_M – the difference between the original lower-resolution multispectral band and the convolution of the sensor *PSF* with the current, sharpened multispectral band (similar to *iterative deblurring* used for image restoration (Biemond *et al.*, 1990)).

- Δ_H – the difference between the higher-resolution pan band and the current, sharpened multispectral band (similar to a *regularization* term used in image restoration (Biemond *et al.*, 1990)).

As the algorithm proceeds from iteration to iteration, the incremental change terms, Δ_M and Δ_H, decrease in magnitude, indicating convergence. That drives the solution to preserve the lower spatial frequency spectral content of the multispectral band and the higher spatial frequency spatial content of the pan band, respectively. No explicit histogram matching, as needed in feature space fusion, is required by this algorithm.

The algorithm tends to converge within 5 to 7 iterations and can produce images that are visually sharper than those produced by HFM fusion (Plate 8-3). The technique with $(\alpha_1 = 1, \alpha_2 = 0.5)$ was compared to restoration-only $(\alpha_1 = 1, \alpha_2 = 0)$, HP-only $(\alpha_1 = 0, \alpha_2 = 0.5)$, and HFM fusion of simulated ALI images in Lemeshewsky (2005); the iterative approach generally produced the lowest RMS error. A detailed quantitative comparison of this approach and the others has not been made on a wide range of real imagery, however.

8.5.3 Scale-Space Fusion

Fusion of multiresolution imagery seems to be a natural application of the scale-space concepts discussed in Chapter 6. In particular, the wavelet transform has been used to fuse SPOT multispectral and panchromatic imagery (Garguet-Duport *et al.*, 1996) and Landsat TM and SPOT panchromatic imagery (Yocky, 1996). In combining the SPOT multispectral and panchromatic images, the pan image is first reference stretched three times, each time to match one of the multispectral band histograms. The first level of the wavelet transform (since the *GSI*s differ by a factor of two) is computed for each of these modified pan images. A synthetic level 1 multispectral wavelet decomposition is constructed using each original multispectral image (20 m *GSI*) and the three high-frequency components (L_xH_y, H_xL_y, H_xH_y; see Chapter 6) from the corresponding modified pan image. These components are also at 20 m *GSI* because of the down-sampling in the wavelet transform. The inverse wavelet transform is then calculated, resulting in three reconstructed multispectral images at 10 m *GSI*. A similar approach was used to fuse SPOT panchromatic and TM multispectral images, except that a five level wavelet pyramid was constructed for both images (after they were registered and resampled to the same *GSI*) and all the SPOT data at the highest level was replaced by all the TM data at the same level (Yocky, 1996). Both of these wavelet-based techniques are similar to HFM, with the wavelet algorithm used to perform the necessary scale resampling.

8.5.4 Image Fusion Examples

An ALI image of Mesa, Arizona, acquired on July 27, 2001, is used to demonstrate spectral space fusion (Plate 8-1). Since the ALI pan band spectral response includes the visible bands and not the NIR band (Table 8-3), a natural color composite is expected to work better than a CIR composite for the fusion. A hexcone CST was used to generate the hue, saturation, and value components from ALI bands 3, 2, and 1 (corresponding to TM bands 3, 2, and 1). The pan band was then matched using the *CDF* reference stretch (Chapter 5) to the value component (Fig. 8-14). The DN transformation was not major as the two images were already well correlated. Finally, the inverse CST was performed on the hue, saturation, and pan components. The result is a merge of the pan and visible bands, with little evidence of artifacts (Plate 8-1).

   To demonstrate the importance of radiometric correlation between the pan band and the replaced component value, a fusion with a CIR composite made from ALI bands 4, 3, and 2 (corresponding to TM bands 4, 3 and 2) was also done. The same process was used as before; the original and resulting fused products are shown in Plate 8-1. Note how the intense red of the vegetated areas is significantly muted in the fused result. This is a result of trying to match the pan band to the value component. From Fig. 8-15, it is clear that due to the vegetation signal in the NIR there is much less correlation than there was for the natural color composite and that the pan band cannot be matched well by the reference stretch algorithm to the replaced value component.

   Finally, a PCT fusion comparison is made. The original and matched pan band versus PC_1 scattergrams and pan band transformation are shown in Fig. 8-16. For this particular image, the vegetation signature was effectively captured completely in PC_2, allowing the relatively high correlation

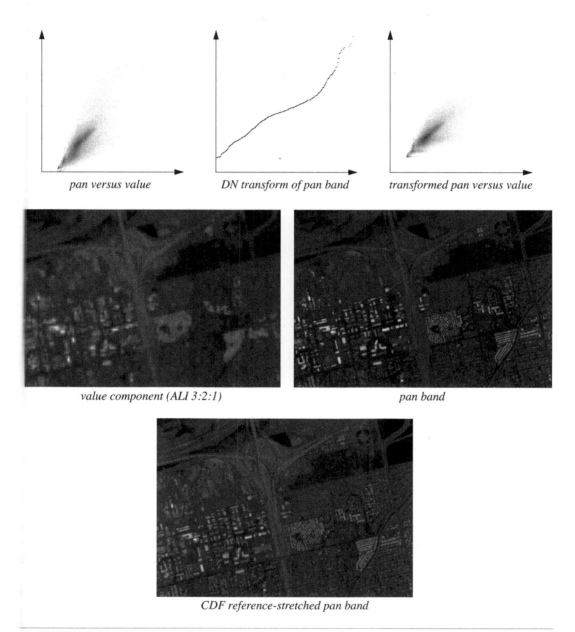

pan versus value   *DN transform of pan band*   *transformed pan versus value*

value component (ALI 3:2:1)   *pan band*

CDF reference-stretched pan band

FIGURE 8-14. *The scattergram of the original pan band versus the value component of the visible bands 3, 2, and 1 shows high correlation ($\rho = 0.817$). The CDF reference stretch improves that correlation slightly ($\rho = 0.828$). The only visible change to the pan band after matching is a reduction of contrast in the low-to-mid DN range.*

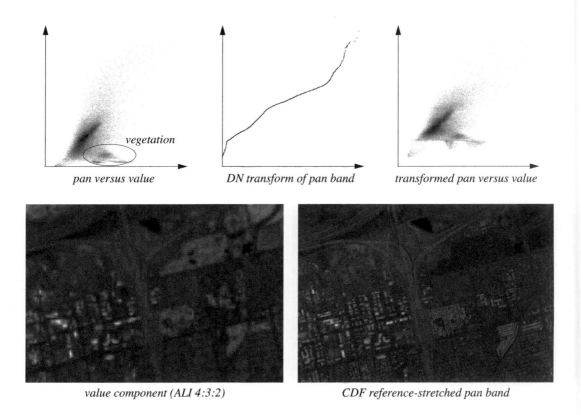

pan versus value   *DN transform of pan band*   *transformed pan versus value*

value component (ALI 4:3:2)   *CDF reference-stretched pan band*

FIGURE 8-15. *The scattergram of the original pan band versus the value component of the NIR band 4 and visible bands 3 and 2 shows much less correlation ($\rho = 0.454$) than for the visible bands alone (Fig. 8-14). The CDF reference stretch improves that correlation only slightly ($\rho = 0.46$) but does not remove the vegetation-caused differences. The large features in the upper right of the image are vegetated areas (see Plate 8-1) and are not matched well by the reference stretch.*

between PC_1 and the pan band. The PCT of the original color composite bands (both natural color and CIR) was performed, PC_1 was replaced by a matched pan band, and the inverse transform of the modified PC images calculated. The PCT clearly works better than a CST for fusion of the CIR bands for this image (Plate 8-1).

The spatial domain fusion algorithms, HFM and iterative sharpening, are demonstrated on another ALI image in Plates 8-2 and 8-3. They produce comparable results, but the iterative technique appears to produce some extra sharpening. A thorough comparison of these two techniques has not been published in the literature to date.

In summary, the quality of fused multisensor images will be improved by attention to the following points:

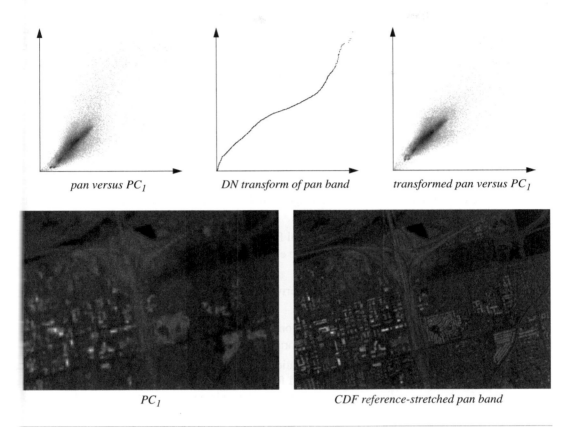

pan versus PC₁   *DN transform of pan band*   *transformed pan versus PC₁*

PC₁   *CDF reference-stretched pan band*

FIGURE 8-16. The original pan band and the PC₁ of bands 4, 3, and 2 are highly correlated (ρ = 0.823).

- To reduce scene-related factors, the images should be as close to the same date and time as possible, and the area should have little topographic relief.

- To reduce sensor-related factors, the spectral band of the higher-resolution image should be as similar as possible to that of the replaced lower-resolution component

- To reduce residual radiometric artifacts, the higher-resolution image should be globally contrast matched to the replaced component. This can be accomplished by relative radiometric calibration of the two sensors or by a reference contrast stretch.

These factors are less important when the fused images are from regions of the spectrum with different remote-sensing phenomenologies (Chapter 2). Then there is no *a priori* reason to assume radiometric correlation between the images. For example, fusion of low-resolution thermal imagery and higher-resolution, multispectral visible imagery can provide useful interpretation clues (Haydn *et al.*, 1982). A CST fusion approach, with the thermal image replacing the hue component of a three-band multispectral image, is particularly effective. Another common example is the use of relatively high-resolution SAR imagery as a replacement for the intensity component of three bands of optical imagery. The goal in both of these examples is combination and qualitative visualization of multiple datasets. Fusion of disparate data for quantitative information extraction is more demanding and difficult.

8.6   Summary

Image-to-image registration is required for many types of remote-sensing analyses. Tools for automating and improving the accuracy of the registration process beyond manually-selected GCPs were described in this chapter. Registered images can then be fused for information extraction. The major contributions of this chapter are:

- Pixel-by-pixel registration requires spatial feature extraction and matching in both images. Scale-space data structures are an efficient way to include large and small scale distortions.

- High-resolution elevation maps can be derived by pixel-by-pixel registration of stereo images.

- The quality of multiresolution image fusion depends strongly on the radiometric correlation between the fused images. Weak correlation results in poor fusion quality.

In the next chapter, the production of thematic maps from remote-sensing images is discussed. The mathematical and algorithmic tools needed for this process are altogether different than those described so far, but the quality of the resulting maps depends on many of the factors described in preceding chapters.

8.7   Exercises

Ex 8-1. Why is it impossible to obtain stereo coverage from the bottom portion of one Landsat TM scene and the upper portion of the next scene, even though they overlap?

Ex 8-2. The cross-correlation coefficient of Eq. (8-4) is commonly used to register image patches. Which environmental and calibration factors in remote-sensing imagery are removed by this normalization?

Ex 8-3. Derive a mathematical expression in image space for PCT fusion of a lower-resolution multispectral image and a higher-resolution panchromatic image. For simplicity, assume they are registered and have the same *GSI*, even though the *GIFOV*s are different.

Ex 8-4. Generate data flow diagrams, similar to Fig. 8-12, for the two wavelet-based fusion techniques described in Sect. 8.5.3.

Ex 8-5. Why do you think the vegetation signature was so effectively "captured" in PC_2 for the ALI image in Plate 8-1, as stated in reference to Fig. 8-16?

Ex 8-6. In the caption of Fig. 8-13, it's stated that " . . . the DNs in the ratio, H/*LPF*[H], vary about a mean value of one, . . . " Explain.

CHAPTER 9

Thematic Classification

9.1   Introduction

A *thematic map* shows the spatial distribution of identifiable earth surface features; it provides an *informational description* over a given area, rather than a data description. Image classification is the process used to produce thematic maps from imagery. The themes can range, for example, from categories such as soil, vegetation, and surface water in a general description of a rural area, to different types of soil, vegetation, and water depth or clarity for a more detailed description. It is implied in the construction of a thematic map from remote-sensing imagery that the categories selected for the map are distinguishable in the image data. As described in the previous chapters, a number of factors can cause confusion among spectral signatures, including topography, shadowing, atmospheric variability, sensor calibration changes, and class mixing within the *GIFOV*. Although some of these effects can be modeled, some cannot (with any reasonable amount of effort), and so they must be treated simply as statistical variability. In this chapter, we will look at classification algorithms, and in particular, how their performance depends on the physical and data models discussed in Chapters 2, 3, and 4.

9.2   The Classification Process

Traditionally, thematic classification of an image involves several steps, as shown in Fig. 9-1:

  • *Feature extraction* — Transformation of the multispectral image by a spatial or spectral transform to a feature image. Examples are selection of a subset of bands, a PCT to reduce the data dimensionality, or a spatial smoothing filter. This step is optional, i.e., the multispectral image can be used directly, if desired.

  • *Training* — Selection of the pixels to train the classifier to recognize the desired *themes*, or *classes,* and determination of decision boundaries which partition the feature space according to the training pixel properties. This step is either *supervised* by the analyst or *unsupervised* with the aid of a computer algorithm.

  • *Labeling* — Application of the feature space decision boundaries to the entire image to label all pixels. If the training was supervised, the labels are already associated with the feature space regions; if it was unsupervised, the analyst must now assign labels to the regions. The output map consists of one label for each pixel.

The end result is a transformation of the numerical image data into descriptive labels that categorize different surface materials or conditions. By virtue of the labeling process, we have presumably converted the data into a form that has informational value.

A large reduction in data quantity also takes place during classification; the multispectral image, consisting of several to hundreds of bands with at least 8 bits/pixel/band, is reduced to a map consisting of as few as a dozen or so category labels.[1] The map can, therefore, be stored with binary encoding as a single band file with less than 8 bits/pixel.Consequently, classification is sometimes used as a compression tool for efficient data transmission (Fig. 9-2). The look-up table from pixel vectors to labels is called a *codebook*, which is used to encode the data at the transmitter and decode the data at the receiver. If the codebook consists of a table linking each unique pixel feature vector to a unique label, then the process is *lossless*; the labels received at the receiver can be decoded exactly into the corresponding pixel vectors. If different pixel vectors are mapped to the same label, the process is *lossy.* The performance of the compression system is then judged by how well it preserves the original numerical content of the image.

Thus, from this perspective, we can say the classifier produces an *approximation* to the original image, which naturally leads to evaluating the classifier on the basis of the accuracy of that approximation. We will use this approach to compare classification results in many of the examples in this chapter. It provides an objective, numerical evaluation criterion which avoids difficulties with specification of appropriate test sites for estimating the *label accuracy* of the thematic map. The latter approach is widely used in classification applications, however, because analysts are interested in the label accuracy of the map *per se*. This topic is not covered here; the reader is referred to Landgrebe (2003) or Jensen (2004) for discussions of classification accuracy evaluation.

---

1. Some complex and detailed mapping projects may approach as many as 30 or 40 categories for a single dataset, but this is relatively uncommon.

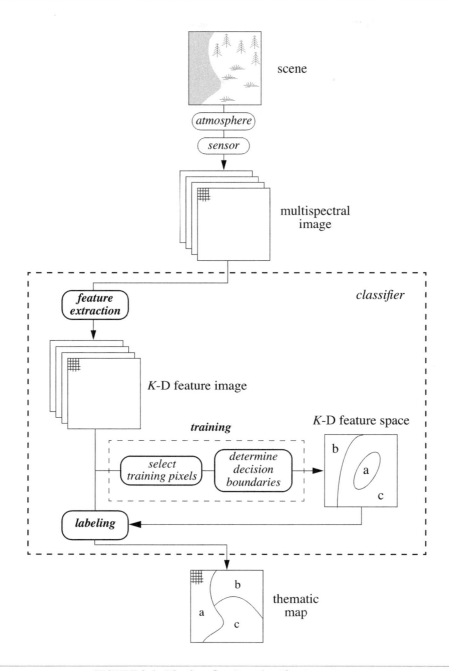

FIGURE 9-1. *The data flow in a classification process.*

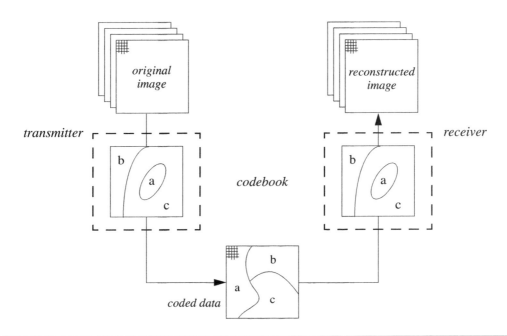

FIGURE 9-2.   *Classification as a data compression technique. The training step of Fig. 9-1 has been previously performed to create the codebook. The left column is the encoding stage that takes place at the source transmitter, and the right column is the decoding stage that takes place at the receiver. The decoded image may or may not be a perfect reconstruction, as discussed in the text.*

9.2.1 The Importance of Image Scale and Resolution

In traditional photointerpretive mapping from aerial photographs, the *geometric scale* of the photographs, defined as the ratio of flight altitude (above ground level) to camera focal length, is of primary importance. A 1:24,000 scale aerial photograph reveals more surface detail than does a 1:48,000 scale photograph taken with the same camera system at twice the altitude. Mapmakers have come to associate the type of information that can be reliably extracted from aerial photographs with the photographic scale (Avery and Berlin, 1992).

With digital satellite imagery, the situation is essentially the same, except that scale is not an explicit characteristic of the data. Once the system altitude and sensor parameters (focal length, detector size, etc.) are fixed in the design process, the relevant measures for image analysis are the *GSI* and *GIFOV*. Clearly, the degree and type of information that can be extracted from digital imagery depends on the *GSI* and *GIFOV* (see Chapter 4). For example, a *GSI* and *GIFOV* of one meter per pixel, typical of commercial satellite systems, may allow identification of larger road

~ehicles *by type*, a task which is completely out of the question for TM imagery.² Similarly, it is
~ossible to map some features of urban areas with TM data, but not with AVHRR data.

A widely-used, hierarchical land-use/land-cover classification system was published in
Anderson *et al.* (1976). It consists of multiple levels as shown for an "urban" class in Table 9-1; the
~ull set of level I and II categories are in Table 9-2. Level III categories, or even level IV categories,
~f sufficient information is available, can be adapted to particular applications or regions, within the
~ramework of the fixed-category structure in levels I and II as defined by the U.S. Geological Sur-
~ey (Avery and Berlin, 1992). The Anderson classification scheme is not the only one by any
~eans, but is representative of this common approach to thematic labeling.

*TABLE 9-1. An example category in a 3-level Anderson land-cover/land-use classification scheme.
Levels I and II are fixed, but level III can be adapted to a particular application.*

Level I	Level II	Level III
1 urban/built-up land	11 residential	111 single family units 112 multifamily units 113 group quarters 114 residential hotels 115 mobile home parks 116 transient lodgings 117 other

The *GIFOV* required to map particular types of objects, for example, roads (Benjamin and Gay-
dos, 1990) and general temperate land cover (Markham and Townshend, 1981) has been evaluated.
It is generally accepted that the Anderson Level I categories can be reliably mapped using 80 m
Landsat MSS imagery and Level II categories with 30 m TM and 20 m SPOT multispectral imagery
(Lillesand *et al.*, 2004). Level III requires 5 m SPOT panchromatic or higher-resolution imagery.

9.2.2 The Notion of Similarity

Similarity between pixels, or groups of pixels, is a fundamental concept behind many image-pro-
cessing algorithms. In classification, for example, we want to label areas on the ground that have
similar physical characteristics. We do this by grouping data with similar characteristics, namely
class signatures. A major question is, therefore: how well do the class data signatures in the image
correspond to the class physical characteristics that actually distinguish one category from another?
An example of the type of semantic problem that can arise is in the mapping of urban land-*use*
classes, such as "urban residential" or "light industrial." They typically are composed of several
land-*cover* types (such as vegetation, pavement, and different types of roof coverings), each with a

2. The military community has long used a guideline of three to five "resolution elements" across the linear
   dimension of an object as a threshold for visual target identification; for example, distinguishing a tank
   from a truck.

TABLE 9-2. *A level I and II land-use and land-cover classification hierarchy (Anderson et al., 1976).*

Level I	Level II
1 urban/built-up land	11 residential 12 commercial and services 13 industrial 14 transportation, communication, and utilities 15 industrial and commercial complexes 16 mixed urban/built-up land 17 other urban/built-up land
2 agricultural land	21 cropland and pasture 22 orchards, groves, vineyards, nurseries, and ornamental horticultural areas 23 confined feeding operations 24 other agricultural land
3 rangeland	31 herbaceous rangeland 32 shrub and brush rangeland 23 mixed rangeland
4 forest land	41 deciduous forest land 42 evergreen forest land 43 mixed forest land
5 water	51 streams and canals 52 lakes 53 reservoirs 54 bays and estuaries
6 wetland	61 forested wetland 62 non-forested wetland
7 barren land	71 dry salt flats 72 beaches 73 sandy areas other than beaches 74 bare exposed rock 75 strip mines, quarries, and gravel pits 76 transitional areas 77 mixed barren land
8 tundra	81 shrub and brush tundra 82 herbaceous tundra 83 bare ground tundra 84 wet tundra 85 mixed tundra
9 perennial snow or ice	91 perennial snowfields 92 glaciers

different spectral signature. Within an area of particular land-use, therefore, several spectral classes occur, resulting in a heterogeneous spectral signature whose characteristics depend on the proportions of each of the component land-cover types, which in turn changes from pixel-to-pixel. Thus, for land-use mapping, we must look for more complex relationships between the physical measurements, the pixel-by-pixel multispectral image, and the map classes of interest. Sometimes, remote-sensing data are insufficient for the task, and additional, so-called *ancillary* information is required.

From the discussions and examples of previous chapters, it is clear that the spectral signature of a given surface material is not characterized by a single, deterministic spectral vector, but rather a *distribution* of vectors. In Fig. 9-3, we show two possible cases for the relative distributions of training data from three classes. *To a large extent, our ability to perform an accurate classification of a given multispectral image is determined by the extent of overlap between class signatures.* An optimum compromise can be achieved with the maximum-likelihood, or Bayes, classifier which minimizes the total error in the classification if our estimate of the underlying probability distributions is correct. Another approach accepts the fact that class signatures overlap and expresses that as likelihoods of membership in each class, i.e., a *fuzzy* classification, as explained next.

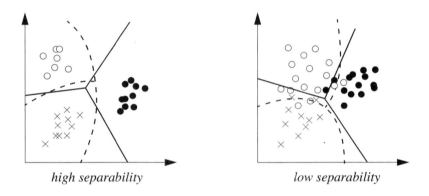

high separability   low separability

FIGURE 9-3. *Two possible sets of training data for three classes in feature space and candidate decision boundaries for class separation. If the training classes are highly separable, there are many potential decision boundaries that can separate the classes without error, e.g. either the solid or dashed lines. If the training data from different classes overlap, then the exact form of the decision boundary is critical to the resulting classification error.*

9.2.3 Hard Versus Soft Classification

The notion of a thematic map, as traditionally used in geography, geology, and other earth science disciplines, presumes that every spot on the ground can be labeled as belonging to one, and only one, category. Such discrete categorization is convenient and appealing in its simplicity, but is not a particularly accurate protrayal of real landscapes and, in fact, is inconsistent with the high-resolution, numerical nature of remote-sensing data.

Remote sensing provides, in effect, a continuous labeling function with as many as 2^{QK} possible categories for data with Q bits per pixel per band and K bands. When this nearly continuous measurement space is compressed by classification into relatively few, discrete labels, we ignore a large amount of the information contained in the data in order to obtain the relative simplification of the thematic map.[3]

Most classification algorithms produce a "likelihood" function for the assignment of a class label to each pixel. A hard classification is produced by selecting that class label with the greatest likelihood of being correct (the *Winner-Take-All (WTA)* algorithm, as it is termed in the artificial neural network literature). The feature space decision boundaries for a *hard* classification are well-defined (Fig. 9-4). If the likelihood values are retained, however, allowing for multiple labels at each pixel, a *soft* classification is obtained. The feature space decision boundaries for a soft classification may be thought of as ill-defined, or *fuzzy*. What do the likelihood values represent? The most obvious interpretation is that they represent the relative proportions of each category within the spatially- and spectrally-integrated multispectral vector of the pixel. We will introduce classification as a "hard" algorithm initially because that is the traditional view. Later in this chapter "soft" decision algorithms are addressed.

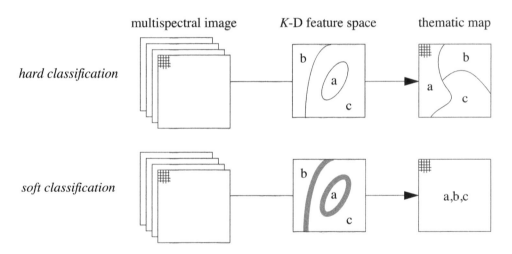

FIGURE 9-4. *One way to view the difference between hard and soft classification. In both cases pixel spectral vectors from the multispectral image are passed through the decision space for labeling. In a hard classification the decision is "winner-take-all," with only one label being permitted at each pixel. In a soft classification, the decision is multivalued, with the possibility of more than one label per pixel. Each label has an associated likelihood of being correct. These likelihoods can be interpreted in a number of ways, one of which is that they indicate the proportion of each category within the pixel, à la the mixing model discussed in Sect. 9.8.1.*

---

3. The actual number of *distinguishable* data categories is somewhat reduced by sensor noise.

9.3 Feature Extraction

The multispectral image data can be used directly in a classification, but it contains all the various external influences, such as atmospheric scattering and topographic relief, described in earlier chapters. Also, the data are often highly correlated between spectral bands, resulting in inefficient analysis. Furthermore, image-derived features, such as measures of spatial structure, may provide more useful information for classification. Thus, it is prudent to consider various pre-classification transformations to extract the greatest amount of information from the original image. Most of the techniques described in Chapters 5, 6, and 7 are useful in that regard. As pointed out in a number of cases, some features tend to suppress undesired variability in remote-sensing signatures, e.g., spectral ratios can suppress topographic variation. In general, it is wise to use such features because they allow the classifier to better distinguish spectral classes. Separability analysis, discussed later, provides a way to determine the most effective subset of a given set of features.

9.4 Training the Classifier

In order to classify an image into categories of interest, the classification algorithm needs to be trained to distinguish those categories from each other. Representative category samples, known as *prototypes*, *exemplars*, or simply *training samples*, are used for this purpose. After the classifier is trained to "recognize" the different categories represented by the training samples, the "rules" that were developed during training are used to label all pixels in the image as one (hard classification) or more (soft classification) of the training categories.

If we are only interested in classifying a single scene relative to training signatures derived from pixels within that scene, atmospheric factors are not particularly critical, as long as they are consistent over the whole scene. They then influence the training and unknown pixel spectral vectors equally and do not affect their relative positions in spectral space. If the atmosphere varies significantly across the area of interest (for example, due to haze or smoke), then spatially-dependent correction is needed, as demonstrated in Richter (1996a). If training data from one image are to be used to classify a second image, then either both images should be corrected for atmospheric effects (Chapter 7), or the second should be normalized to the first (Chapter 5).

The training of a classification algorithm can be either *supervised*, in which case the prototype pixel samples are already labeled by virtue of ground truth, existing maps, or photointerpretation, or *unsupervised*, in which the prototype pixels are not labeled, but have been determined to have distinguishing intrinsic data characteristics.[4]

---

4. Ground "truth" refers to any knowledge of an area which is, for all practical purposes, certain. It implies that field visits have been made to specific sites and that the surface materials at those sites have been accurately mapped and measured. It is sometimes used, less appropriately, to describe knowledge from other sources, such as aerial photographs.

9.4.1 Supervised Training

The themes of interest are determined by the application. In geology, one may wish to map different mineral types. In forestry, the themes could be different tree species, or perhaps healthy and diseased trees. Agricultural classes would normally include different crop types, fallow fields, or amount of soil moisture. For supervised training, the analyst must select representative pixels for each of the categories. If the multispectral image contains sufficiently distinct visual cues, it may be possible to find suitable training areas by visual examination. Frequently, however, one must resort to additional sources of information, such as field data or existing maps, to find representative areas for each class. The process of finding and verifying training areas can, therefore, be rather labor intensive.

With supervised training, it is important that the training area be a homogeneous sample of the respective class, but at the same time include the range of variability for the class. Thus, more than one training area per class is often used. If there is considerable within-class variability, the selection of training sites can be laborious, and it is impossible to be entirely certain that a comprehensive set of training samples for each class has been specified. Example training sites for three visually distinguishable classes in a TM image are shown in Plate 9-1. These classes and their training pixels are used in many examples throughout this chapter.

In many cases it is impossible to obtain homogeneous sites. A common problem is sparse vegetation, which complicates attempts to map both vegetation and soils. One technique for improving training data under these conditions is to "clean" the sites of outlying pixels (in feature space) before developing the final class signatures (Maxwell, 1976). This can be done by classifying the training pixels according to their given signatures. Some training pixels will likely be misclassified, or at least have a low likelihood of being in the correct class. Those pixels are then excluded from the training set and the class signatures are recalculated from the remaining pixels. Another approach excludes those pixels within the training sites that do not meet certain spatial and spectral homogeneity criteria (Arai, 1992).

In certain applications, the nature of the classes makes manual training site delineation difficult, for example, an "asphalt" class consisting of narrow roads, or a "water" class for a river. In these cases, a semi-automated delineation is useful. Efficient tools have been developed that use *region growing* algorithms to define an irregular site, starting from a seed pixel specified by the analyst (Buchheim and Lillesand, 1989). The concepts behind region-growing algorithms are discussed in Sect. 9.7.

Separability analysis

Once the training pixels are selected, the *features* to be used by the classifier should be specified. These may be all or a subset of the original multispectral bands, or derived features, such as the principal components (Chapter 5). Since supervised training is driven by the desired map themes, and not by characteristics of the data itself, there is no assurance that the classes will actually be distinguishable from one another. A *separability* analysis can be performed on the training data to estimate the expected error in the classification for various feature combinations (Swain and Davis, 1978; Landgrebe, 2003). The results may suggest that some of the initial features be dropped before classification of the full image.

Finding an appropriate definition for interclass separability is not trivial. Simple measures of the separation of the means, such as the *city block*, *Euclidean*, and *angular* distances would at first seem to be good candidates (Fig. 9-5, Table 9-3). They do not account, however, for overlap of class distributions due to their variances and are, therefore, not particularly good as measures of separability. The *normalized city block* measure is better, in that it is proportional to the separation of the class means and inversely proportional to their standard deviations. If the means are equal, however, it will be zero regardless of the class variances, which does not make sense for a statistical classifier based on probabilities. For this reason, probability-based measures have also been defined.

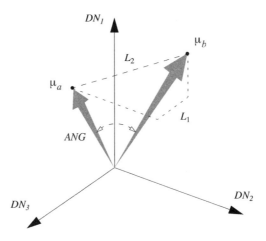

FIGURE 9-5. *The L_1, L_2, and ANG distance measures depicted for two vectors in 3-D. Note that ANG, the arc-cosine of the normalized inner (dot) product of the two vectors, is independent of the length of either vector. This property makes it useful for classification of data that is not corrected for topographic shading (refer to Fig. 4-37). Although this diagram shows two class mean vectors, these three distance measures can be applied to any two vectors.*

The *Mahalanobis* separability measure is a multivariate generalization of the Euclidean measure for normal distributions. It is always zero if the class means are equal. The *divergence* and *Bhattacharyya distance* measures avoid this problem. Divergence equals zero only if the class means *and* covariance matrices are equal. A problem remains with both distance measures, however, in that they increase without bound for large class separations, and do not asymptotically converge to one, as does the probability of correct classification. A *transformed divergence*, based on the ratio of probabilities for classes *a* and *b*, does exhibit this latter behavior. The *Jeffries-Matusita distance* depends on the difference between the probability functions for *a* and *b* and is similar to the probability of correct classification for large class separations, but requires more computation than does transformed divergence (Swain and Davis, 1978).

TABLE 9-3. *Distance measures between the means of two distributions in feature space. The city block, Euclidean, and angular measures ignore the variances of the distributions, the normalized city block includes 1-D variances, and the last five measures assume normal class distributions with K-D covariances. All of these distance measures are scalars. Derivations of the normal distribution-based distance measures can be found in many books on statistical pattern recognition, including Duda et al., (2001), Swain and Davis (1978), Richards and Jia (1999), and Landgrebe (2003).*

name	formula
city block	$L_1 = \|\boldsymbol{\mu}_a - \boldsymbol{\mu}_b\| = \sum_{k=1}^{K} \|m_{ak} - m_{bk}\|$
Euclidean	$L_2 = \|\boldsymbol{\mu}_a - \boldsymbol{\mu}_b\| = [(\boldsymbol{\mu}_a - \boldsymbol{\mu}_b)^T (\boldsymbol{\mu}_a - \boldsymbol{\mu}_b)]^{1/2} = \left[\sum_{k=1}^{K} (m_{ak} - m_{bk})^2\right]^{1/2}$
angular	$ANG = \mathrm{acos}\left(\dfrac{\boldsymbol{\mu}_a^T \boldsymbol{\mu}_b}{\|\boldsymbol{\mu}_a\| \|\boldsymbol{\mu}_b\|}\right)$
normalized city block	$NL_1 = \sum_{k=1}^{K} \dfrac{\|m_{ak} - m_{bk}\|}{(\sqrt{c_{ak}} + \sqrt{c_{bk}})/2}$
Mahalanobis	$MH = \left[(\boldsymbol{\mu}_a - \boldsymbol{\mu}_b)^T \left(\dfrac{C_a + C_b}{2}\right)^{-1} (\boldsymbol{\mu}_a - \boldsymbol{\mu}_b)\right]^{1/2}$
divergence	$D = \dfrac{1}{2} tr\left[(C_a - C_b)(C_b^{-1} - C_a^{-1})\right] + \dfrac{1}{2} tr\left[(C_a^{-1} + C_b^{-1})(\boldsymbol{\mu}_a - \boldsymbol{\mu}_b)(\boldsymbol{\mu}_a - \boldsymbol{\mu}_b)^T\right]$
transformed divergence	$D^t = 2[1 - e^{-D/8}]$
Bhattacharyya	$B = \dfrac{1}{8} MH + \dfrac{1}{2} \ln\left[\dfrac{C_a + C_b}{2\|C_a\|^{1/2} \|C_b\|^{1/2}}\right]$
Jeffries-Matusita	$JM = \left[2(1 - e^{-B})\right]^{1/2}$

Separability can be used to determine the combination of features that is best, on average, at distinguishing among the given classes. A measure of separability is typically computed for all possible pairs of classes[5] and for all combinations of q features, out of K total features (Landgrebe, 2003; Jensen, 2004). The average separability over all class pairs is computed, and the subset of features that produces the highest average separability is found. One may then use that subset for classification and save computation time in the classification stage. It is natural, but not necessary, to match the separability measure for the feature subset analysis to that used by the classifier. For example, one would use Euclidean distance if the nearest-mean algorithm is to be used for classification, or Mahalanobis distance if the Gaussian maximum-likelihood algorithm is to be used.

9.4.2 Unsupervised Training

For unsupervised training, the analyst employs a computer algorithm that locates concentrations of feature vectors within a heterogeneous sample of pixels. These so-called *clusters* are then assumed to represent classes in the image and are used to calculate class signatures. They remain to be identified (labeled), however, and may or may not correspond to classes of interest to the analyst. Supervised and unsupervised training thus complement each other; in the former, the analyst imposes knowledge on the analysis to constrain the classes and their characteristics, and in the latter, an algorithm determines the inherent structure of the data, unconstrained by external knowledge.

The term "cluster" is misleading in most cases. It implies that there is a distinct grouping of pixel vectors in a localized region of the multidimensional data space. This is seldom the case. In order to create a comprehensive set of classes that span the full data space, a large training sample must be used. Due to subpixel class mixing (Sect. 9.8), sensor noise, topographic shading, and other factors, the data distribution is typically diffuse (see simulated scattergrams in Chapter 4 and real scattergrams in Chapter 5). Competitive algorithms, such as K-means (described next), find an optimal partitioning of the data distribution into the requested number of subdivisions. The final mean vectors resulting from the clustering will be at the centroids of each subdivision.

In defining image areas for unsupervised training, the analyst does not need to be concerned with the homogeneity of the sites. Heterogeneous sites can be purposely chosen to insure that all classes of interest and their respective within-class variabilities are included. Even the full image (perhaps subsampled to reduce computation time) may be used in the clustering algorithm for a "wall-to-wall" description. The assignment of identifying labels to each cluster may be done by the analyst after training or after classification of the full image. Because unsupervised training does not require any information about the area, beyond that in the image itself, it can be useful to delineate relatively homogeneous areas for potential supervised training sites.

---

5. There are $M(M-1)/2$ possible pairs out of M classes. In general, for R classes out of a total of M classes, there are (ignoring order) $M!/[R!(M-R)!]$ possible combinations.

K-means clustering algorithm

One of the more common clustering methods is the *K*-means algorithm (Duda and Hart, 1973). In Fig. 9-6 we show the application of the *K*-means algorithm to a two-dimensional set of idealized data. In the first step of the algorithm, an initial mean vector ("seed" or "attractor") is arbitrarily specified for each of *K* clusters.[6] Each pixel of the training set is then assigned to the class whose mean vector is closest to the pixel vector, thus forming the first set of decision boundaries.[7] A new set of cluster mean vectors is then calculated from this classification, and the pixels are reassigned accordingly. In each iteration, the *K* means will tend to gravitate toward concentrations of data within their currently-assigned region of feature space. The iterations are continued until there is no significant change in pixel assignments from one iteration to the next. The criterion for ending the iterative process can be defined in terms of the *net mean migration* from one iteration to the next. Specifically, this is the magnitude change of the mean vectors from iteration $i - 1$ to iteration i, summed over all *K* clusters,

$$\Delta\mu(i) = \sum_{k=1}^{K} \left| \mu_k^i - \mu_k^{i-1} \right|. \tag{9-1}$$

The final, stable result is not sensitive to the initial specification of seed vectors, but more iterations may be required for convergence if the final vectors are not close to the seed vectors. A typical iteration sequence for the net migration of the cluster means is shown in Fig. 9-7. The final cluster mean vectors may be used to classify the entire image with a minimum-distance classifier in one additional pass, or the covariance matrices of the clusters may be calculated and used with the mean vectors in a maximum-likelihood classification.

The number of ways to determine clusters in data has been limited only by the ingenuity of researchers in defining clustering criteria. Clustering of image data is discussed in a number of books (Duda *et al.*, 2001; Jain and Dubes, 1988; Fukunaga, 1990), and Fortran computer programs can be found in Anderberg (1973) and Hartigan (1975). All commonly-used algorithms employ iterative calculations to find decision boundaries for the given data and optimization criteria. The *ISODATA* algorithm (Ball and Hall, 1967) is a common modification of the *K*-means algorithm and includes merging of clusters if their separation is below a threshold, and splitting of a single cluster into two clusters if it becomes too large. Unfortunately, as the number of capabilities grows in a cluster algorithm, so does the list of parameters that must be set by the user!

Clustering examples

The idealized data distribution of Fig. 9-6 is not particularly realistic. Most of the time, image data does not exist in such distinct clusters, but rather in a large "blob," which may contain only hints of internal concentrations. We will use bands 3 and 4 of a Landsat TM image to illustrate the process

6. Not to be confused with *K* data dimensions. The use of *K* for the number of clusters is common convention.
7. This is the *nearest-mean algorithm* discussed in Sect. 9.6.4.

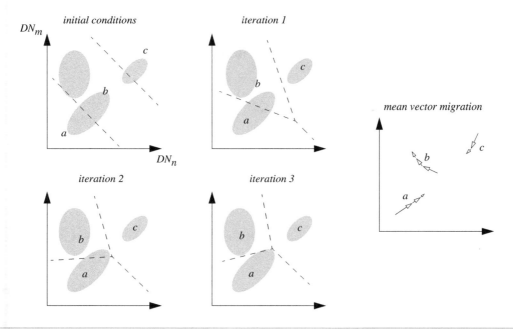

FIGURE 9-6. *An idealized data distribution during three iterations of the K-means clustering algorithm with the nearest-mean decision criterion. The data are depicted as three distinct clusters, with the current, estimated mean vector for each class located at a, b, and c. The initial "seed" locations are equidistant along the feature space diagonal. The mean vector migration depicts the movement of the cluster means from iteration to iteration.*

of clustering on actual image data. The area is near Marana, Arizona, and contains irrigated agricultural fields, most of which are fallow. Surrounding the agricultural area is desert, with a mixture of soils and thin vegetation cover (Plate 9-1). The image scattergram reveals that the two bands are strongly correlated because of the large areas of bare soil, ranging from dark to light. The few fields that have a growing crop appear in the upper left of the band 4 versus band 3 scattergram (see Chapter 4 for a discussion). Because the area of growing crop is relatively small, a clustering algorithm will be dominated by the soil pixels. In this example, the entire image is used by the clustering algorithm.

In Fig. 9-8, we see the results of a *K*-means clustering algorithm (using the Euclidean distance measure) for different values of *K*. The initial cluster seeds were distributed along the first principal component axis of the data. As predicted, the cluster mean locations are dominated by the large population of soil pixels, and not until six clusters have been requested do we see the appearance of a unique cluster for the crop pixels. This cluster remains and shows some movement toward the vegetation part of the spectral space when seven clusters are used.

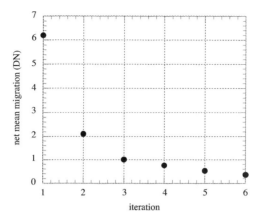

FIGURE 9-7. *Typical behavior of the net mean migration from one iteration to the next in the K-means algorithm. The plot is for real image data.*

Imagine every cluster label in the clustered image is replaced by the average DN of the cluster; the result is an approximation to the original DN image (as in Fig. 9-8). Thus, one may view image clustering as an attempt to "fit" the data with relatively few, distinct signatures (the cluster means).[8] With only two clusters, the fit is quite poor, but improves with each additional cluster. In Fig. 9-9, the residual magnitude error at every pixel (the magnitude difference between the pixel DN and the average DN of the cluster it's assigned to) is shown for each case in Fig. 9-8. We see that the overall fit steadily improves as the number of clusters increases, but that the largest error (in the active crop fields) remains high until six clusters are used, when the error drops dramatically. This behavior is clearly apparent if the residual error is plotted against the number of clusters (Fig. 9-9).

9.4.3 Hybrid Supervised/Unsupervised Training

Because supervised training does not necessarily result in class signatures that are numerically separable in feature space, and unsupervised training does not necessarily result in classes that are meaningful to the analyst, a hybrid approach might achieve both requirements. Unsupervised training is first performed on the data and an unlabeled cluster map of the training area is produced using the obtained clusters. Typically, a large number of clusters, say 50 or more, is used to insure adequate data representation. The analyst then evaluates the map with field survey data, aerial photographs, and other reference data and assigns class labels to the clusters. Normally, some clusters must be subdivided or combined to make this correspondence; the analyst might also choose to modify the class labels based on the clustering results. The labeled cluster data can then be used in a final supervised classification, or the labeled cluster map can be simply accepted as the final map.

---

8. This concept is the basis of a data compression technique, *vector quantization* (Pratt, 1991; Sayood, 2005), and was discussed in general terms at the beginning of this chapter.

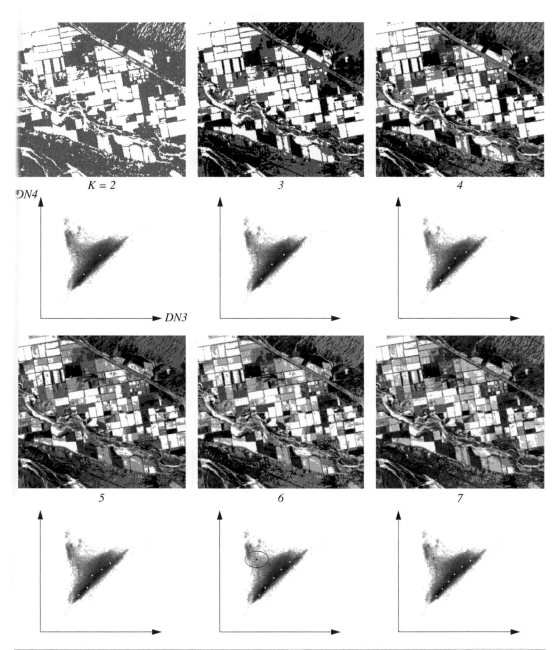

FIGURE 9-8. *Final cluster maps and band 4 versus band 3 scattergrams for different numbers of clusters. The new cluster center (circled) that first appears for K equal to 6 represents vegetation.*

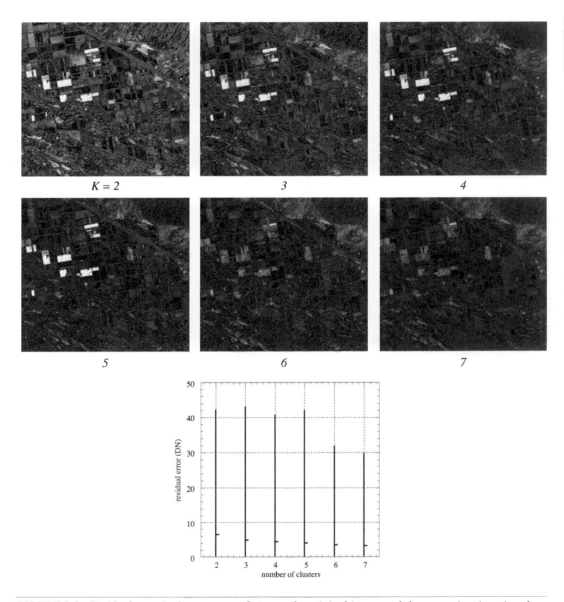

FIGURE 9-9. *Residual magnitude error maps between the original image and the approximation given by the cluster mean DNs are shown above. Note the greatly reduced error in the crop fields when six or more clusters are used. The residual error range and average (crossbar) versus the number of clusters are plotted below. Note the average error steadily decreases as the number of clusters increases, but the maximum error does not decrease until six clusters are used, where it drops 25%.*

9.5   *Nonparametric Classification*

Classification algorithms may be grouped into one of two types: *parametric* or *nonparametric*. Parametric algorithms assume a particular class statistical distribution, commonly the normal distribution, and require estimates of the distribution parameters, such as the mean vector and covariance matrix, for classification. Nonparametric algorithms make no assumptions about the probability distribution and are often considered *robust* because they may work well for a wide variety of class distributions, *as long as the class signatures are reasonably distinct*. Of course, parametric algorithms should yield good results under the same conditions, even if the assumed class distribution is invalid. See the earlier discussion about Fig. 9-3.

9.5.1 Level-Slice Classifier

This classifier, also known as a *box* or *parallelepiped classifier*, is perhaps the simplest of all classification methods. A set of K-dimensional boxes, centered at the estimated class mean vectors, are placed in K-dimensional feature space. If an unlabeled pixel vector lies within one of the boxes, it is assigned that class label (Fig. 9-10). Specification of the box limits is typically in terms of the data extent in each dimension, for example ± 1 standard deviation about the mean in each band. The delineation of the boxes can also be done directly by the analyst in feature space in an interactive manner. Since the boxes are aligned with the data axes, classification labeling of the full image can be achieved quickly with hardware or software LUTs (Chapter 1) and the resulting map viewed simultaneously while manipulating the feature space boxes. A complication occurs if a pixel vector falls within two or more boxes (the boxes can overlap unless that is explicitly prohibited). A decision on a pixel's label must then be made with another algorithm, such as the nearest-mean.[9]

By its nature, the level-slice algorithm also creates an "unlabeled" class, consisting of all pixel vectors that do not fall within any of the designated boxes. This characteristic can be used to prevent overestimation of class populations caused by outliers; the optimum box size therefore is a compromise between including as many valid pixels as possible within each box and excluding outliers and pixels from other classes. Small boxes will usually produce classifications with high confidence in the labels of the classified pixels (assuming confidence in a classification result depends inversely on the pixel vector's distance to the class mean), but also a potentially large number of unlabeled pixels.

Given that class DN distributions are not typically aligned with the K-D data axes because of correlation between spectral bands, the level-slice classifier does not fit remote-sensing data particularly well. One modification to address this problem allows hyperdimensional boxes that align with the data clusters and whose faces are parallelograms rather than rectangles, i.e. true parallelepipeds (Landgrebe, 2003).

---

9. We have a bit of semantic difficulty here—if the locations of the boxes and their sizes are dependent on the training class means and standard deviations, then we in fact have a *parametric* classifier. If the boxes are not so constrained, but are designed interactively by examination of feature space scattergrams, then we have a *nonparametric* classifier. In either case, the addition of the nearest-mean classifier to resolve conflicts is a parametric modification.

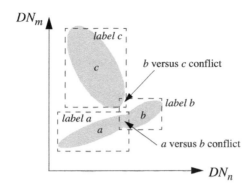

FIGURE 9-10. Level-slice decision boundaries for three classes in two dimensions. Pixel labeling within the conflict regions must be handled with an alternate decision rule, for example by the nearest-mean classifier.

9.5.2 Histogram Estimation Classifier

A classifier based on *K*-D histogram estimation was proposed in Skidmore and Turner (1988). The basic classifier consists of three steps:

- Supervised training pixels are used to construct the feature-space histogram *of each defined class* in *K*-D. The histograms are each normalized properly by the respective total number of pixels in each training class, as in Eq. (4-1), accounting for different training set sizes in each class.

- Every spectral vector "cell" in *K*-D is checked to find the class with the most histogram counts, and that class label is assigned to the cell. A LUT that maps spectral vector to class label is thus created.

- Unlabeled pixels are classified by the LUT.

The primary benefit of this classifier is that it is fast in the classification stage because of the LUT. However, a serious drawback is caused by the sparseness of the estimated class histograms. Training pixel class populations are, by definition, small samples of the overall class population. Therefore, it is unlikely that every class spectral vector will be represented in the training data, leading to empty cells in the *K*-D space. Ways to reduce this problem include rebinning the histogram cells to a coarser *DN* resolution (Skidmore and Turner, 1988) and filling of empty cells by convolution filtering in the *K*-D space (Dymond, 1993). The underlying histogram distribution can also be approximated by *Parzen estimation*, consisting of superposition of component "kernel" distributions (James, 1985; Fukunaga, 1990).

9.5.3 Nearest-Neighbors Classifier

A number of nonparametric classification schemes assign labels to unknown pixels according to the labels of neighboring training vectors in feature space. They include:

- *Nearest-neighbor*—assign the same label as that of the nearest training pixel
- *K nearest-neighbors*—assign label according to the majority label of k nearest-neighbor training pixels
- *Distance-weighted k nearest-neighbors*—assign weights to the labels of k nearest-neighbor training pixels, inversely proportional to their Euclidean distance from the unknown pixel, and assign the label with the highest aggregate weight

These and similar algorithms are described and evaluated in Hardin (1994). They tend to be slow because distances from each unknown pixel to many of the training pixels must be calculated. Modifications to speed the classification are presented in Hardin and Thomson (1992). If the training class signatures are compact and well separated, the nearest-neighbor algorithms will produce results similar to those of the parametric nearest-mean algorithm. For discussions of the relationship between the nearest-neighbor classifier and the maximum-likelihood classifier, see James (1985) and Fukunaga (1990).

9.5.4 Artificial Neural Network (ANN) Classifier

Artificial Neural Networks (ANNs) can be used as a nonparametric technique for classification. They differ significantly from the level-slice and histogram estimation algorithms in that the decision boundaries are not fixed by a deterministic rule applied to the prototype training signatures but are determined in an iterative fashion by minimizing an error criterion on the labeling of the training data. In that sense, ANNs are similar to clustering algorithms. Examples of the use of ANNs for remote-sensing image classification are listed in Table 9-4. An excellent presentation of neural network classification and its relationship to traditional statistical classification is given in Schürmann, (1996).

A basic network is shown in Fig. 9-11.[10] This network has three layers; the middle (or "hidden") layer and the output layer contain *processing elements* at each node. The input layer nodes, on the other hand, are simply an interface to the input data and do not do any processing. The input patterns are the features used for classification. In the simplest case, they are the multi-spectral vectors of the training pixels, one band per node. Other features, such as a spatial neighborhood of pixels or multitemporal spectral vectors, can also be used (Paola and Schowengerdt, 1995a).

Within each processing node, we have a summation and transformation (Fig. 9-11). At each hidden layer node, j, the following operation is performed on the input pattern, p_i, producing the output, h_j,

---

10. There are actually *many* ANN variants. We will describe only a basic network architecture.

TABLE 9-4. *Example applications of ANNs in remote-sensing image classification. Note the wide range of sensor and feature data that have been used. A survey and analysis of papers published prior to 1994 can be found in Paola and Schowengerdt (1995a).*

features	reference
aerial photograph	Kepuska and Mason, 1995; Qiu and Jensen, 2004
ASAS, multiangle	Abuelgasim *et al.*, 1996
AVHRR	Yhann and Simpson, 1995; Visa and Iivarinen, 1997; Li *et al.*, 2001; Arriaza *et al.*, 2003
multitemporal AVHRR NDVI	Muchoney and Williamson, 2001
AVHRR, SMMR	Key *et al.*, 1989
AVIRIS	Benediktsson *et al.*, 1995
ETM+	Fang and Liang, 2003
Fengyun-1C 0.6 μm, 1.6 μm and 11 μm bands	McIntire and Simpson, 2002
HyMAP	Camps-Valls *et al.*, 2004
MSS, DEM	Benediktsson *et al.*, 1990a
SPOT	Kanellopoulos *et al.*, 1992; Chen *et al.*, 1995
multitemporal SPOT	Kanellopoulos *et al.*, 1991
SPOT, texture	Civco, 1993); Dreyer, 1993
TM	Ritter and Hepner, 1990; Liu and Xiao, 1991; Bischof *et al.*, 1992; Heermann and Khazenie, 1992; Salu and Tilton, 1993; Yoshida and Omatu, 1994; Paola and Schowengerdt, 1995b; Carpenter *et al.*, 1997); Valdes and Inamura, 2000
TM, texture	Augusteijn *et al.*, 1995
TM ratios	Baraldi and Parmiggiani, 1995
airborne TM	Foody, G. M., 2004
airborne TM, airborne SAR	Serpico and Roli, 1995
multitemporal TM	Sunar Erbek *et al.*, 2004
multitemporal TM and ERS-1 SAR, SAR texture	Bruzzone *et al.*, 1999

$$\textit{hidden layer:} \quad S_j = \sum_i w_{ji}p_i \text{ and } h_j = f(S_j) \tag{9-2}$$

which is directed to each output layer node, k, where the output, o_k, is calculated,

$$\textit{output layer:} \quad S_k = \sum_j w_{kj}h_j \text{ and } o_k = f(S_k). \tag{9-3}$$

The most widely-used transformation function $f(S)$ is the *sigmoid function*, shown in Fig. 9-11. Other functions, for example a hard threshold with no gradient, may be used, but it seems that the end result of the network is not particularly sensitive to the specification of f. The shape of f can affect the convergence rate during training, however.

Back-propagation algorithm

The discrimination capability of an ANN is contained in its weights. During training, they are iteratively adjusted towards a configuration that allows the network to distinguish the prototype patterns of interest. The back-propagation algorithm minimizes the squared error over all patterns at the output of the network and was the first successful approach to training the network of Fig. 9-11 (Rumelhart *et al.*, 1986; Lippmann, 1987). It belongs to the family of iterative, gradient descent algorithms and consists of the following steps:

1. Select training pixels for each class and specify the desired output vector d_k for class k, typically $d_m = 0.9$ ($m = k$) and $d_m = 0.1$ ($m \neq k$). These are the target values that the ANN attempts to produce.

2. Initialize weights as random numbers between 0 and 1 (typically small values near zero are used).

3. Set the frequency for weight updating to one of the following:
   - after every training pixel (sequential)
   - after all training pixels in each class
   - after all training pixels in all classes (batch)

   Batch training is commonly used as it minimizes the frequency of weight updates; the remainder of the algorithm below assumes batch training.

4. Propagate the training data forward through the network, one pixel at a time.

5. After each training pixel is propagated forward through the network, calculate the output o and accumulate the total error relative to the desired output d (the 1/2 factor is a mathematical convenience),

Network architecture

input nodes (*i*)   hidden layer nodes (*j*)   output nodes (*k*)

input pattern p_i

output pattern o_k

weights w_{ji} weights w_{kj}

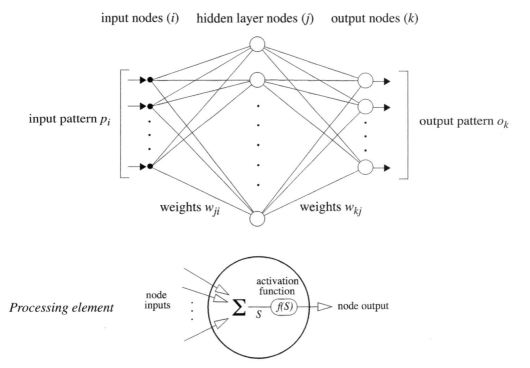

Processing element

node inputs

activation function

Σ — S — $f(S)$ → node output

Activation function

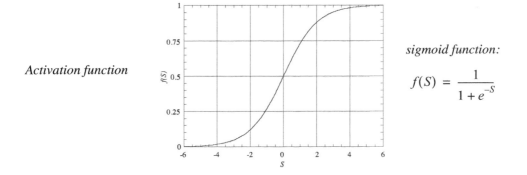

sigmoid function:

$$f(S) = \frac{1}{1 + e^{-S}}$$

FIGURE 9-11. The traditional structure of a three-layer ANN, the components of a processing element, and the sigmoid activation function.

$$\frac{\|\mathbf{\varepsilon}\|^2}{2} = \frac{1}{2}\sum_{p=1}^{P}\sum_{k}(d_k - o_k)^2 \tag{9-4}$$

Repeat over all P training patterns (pixels) for batch training.

6. After all training pixels have been used, adjust the weight w_{kj} by,

$$\Delta w_{kj} = LR\frac{\partial \varepsilon}{\partial w_{kj}} = LR\sum_{p=1}^{P}(d_k - o_k)\frac{d}{dS}f(S)\Big|_{S_k}h_j \tag{9-5}$$

where LR is a *Learning Rate* parameter used to control the speed of convergence.

7. Adjust the weight w_{ji} by,

$$\Delta w_{ji} = LR\sum_{p=1}^{P}\left\{\frac{d}{dS}f(S)\Big|_{S_j}\sum_{k}\left[(d_k - o_k)\frac{d}{dS}f(S)\Big|_{S_k}w_{kj}\right]p_i\right\} \tag{9-6}$$

8. Repeat steps 4 through 7 until ε < threshold.

From Eq. (9-5) and Eq. (9-6), we see the origin of the term "back-propagation." The errors in the output vector, o, are propagated backwards through the network links to adjust their weights (Richards and Jia, 1999).

The convergence properties of this algorithm can be problematic. First of all, the back-propagation algorithm is notoriously slow. Faster ways to obtain convergence of the weights (Go *et al.*, 2001) and architectures, such as Radial Basis Functions (RBF) networks, that can be trained faster (Bruzzone *et al.*, 1999; Foody, 2004) have been developed, and efficient preprocessors have been proposed (Gyer, 1992). Secondly, the network may not reach a "global" optimal solution that minimizes the output error, but become stuck in a "local" minimum in the error surface. A fraction of the previous iteration's Δw can be added to the current iteration's Δw to accelerate the convergence out of local minima and toward a global minimum; the added fraction is called a *momentum parameter* (Pao, 1989). Finally, because the weights are initially random, the network *does not converge to the same result in repeated runs*. It is estimated that this stochastic property may contribute 5% or more to the residual uncertainty at the end of convergence (Paola and Schowengerdt, 1997).

As noted, there are a wide assortment of ANN architectures. A survey of their use in remote-sensing classification reveals few guidelines for particular architectures or for specifying the number of hidden layers or nodes (Paola and Schowengerdt, 1995a). For a fully-connected three-layer network as in Fig. 9-11, with H hidden layer nodes, B bands (or features), and C classes, the number of free parameters is (Paola and Schowengerdt, 1995b),

$$N_{ANN} = \text{number of weights} = H(B + C). \tag{9-7}$$

This represents the *degrees-of-freedom* in the ANN. Similarly, a maximum-likelihood parametric classifier requires mean vector and covariance matrix values,

$$N_{ML} = \text{number of mean values} + \text{number of unique covariance values}$$

$$= CB + C\left(\frac{B^2 + B}{2}\right) = CB\frac{(B + 3)}{2} \ . \tag{9-8}$$

These are the degrees-of-freedom in a maximum-likelihood classifier. If we want to compare the two classifiers, a reasonable way to constrain the ANN is to make the number of weights equal to the number of parameters required by a maximum-likelihood classifier, and solve for the number of hidden layer nodes,

$$H = \frac{CB(B + 3)}{2(B + C)} \ . \tag{9-9}$$

This depends weakly on the number of classes C, but strongly on the number of bands B (Fig. 9-12). For a hyperspectral dataset of 200 bands with 10 classes, 1000 hidden layer nodes are needed to match the number of parameters in a Gaussian maximum-likelihood classifier. For TM non-TIR data, 20 nodes are sufficient, even for as many as 20 classes. The back-propagation training time is dependent on the number of hidden layer nodes, so it is desirable to use only as many nodes as needed for satisfactory classification performance. The previous analysis provides only a guideline; it does indicate that the ANNs in some published examples may be larger than needed for the problem at hand.

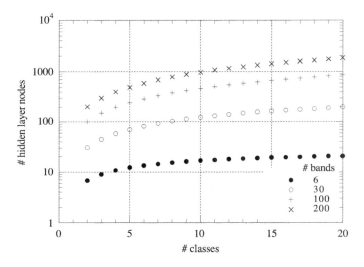

FIGURE 9-12. The number of hidden layer nodes required in a three-layer ANN to match the degrees-of-freedom in a maximum-likelihood classifier.

9.5.5 Nonparametric Classification Examples

Supervised classifications of the Marana image will be used to illustrate the differences among the nonparametric classifiers. Again, for clarity, we use only bands 3 and 4, which serve to distinguish soil and vegetation. Three well-discriminated classes and training sites are defined by visual examination of the CIR composite (Plate 9-1). In Fig. 9-13, we see the result of the level-slice classifier

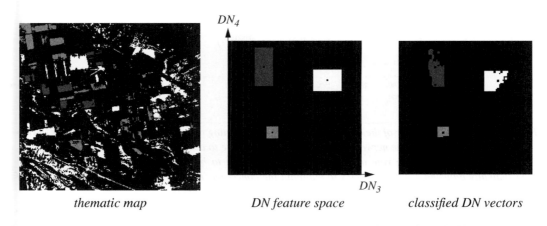

thematic map   *DN feature space*   *classified DN vectors*

FIGURE 9-13. *Level-slice classification results in the image space and the decision regions in the feature space. The class means are shown as dots in the feature space, which ranges [0,80] in both axes. The classified DN vector plot is the intersection of the class decision regions and the image scatterplot.*

with a threshold of ±4σ in each band. A large portion of the image is not classified with this threshold. A larger threshold could be used to include more image pixels, but they would be less similar to the training set and the map error would be greater.

The ANN classifier's training stage is iterative — it converges to a stable solution by small incremental changes in the system weights, which in turn determine the decision boundaries. The same image and training data were used for ANN classification with a 2-3-3 network, i.e., two input nodes (two spectral bands), three hidden layer nodes, and three output nodes (three classes). The network was trained in a "batch" mode. The mean-squared and maximum errors at the output nodes as a function of the training cycle number are shown in Fig. 9-14. There is little improvement in the training pixel classification after about 2000 cycles. The learning rate and momentum parameters change rapidly early in the process and become relatively stable once the network starts to converge at about 500 cycles.

The decision boundaries and hard classification maps are shown in Fig. 9-15 for three points during the iterative training process. At cycle 250, the crop class does not even appear in the decision space or classification map. This is evidently because of its small population relative to the dark and light soil classes. The ANN output node for "crop" has a nonzero value, but is less than that for the other two nodes at this point in the training. If the network is trained to 5000 cycles, we

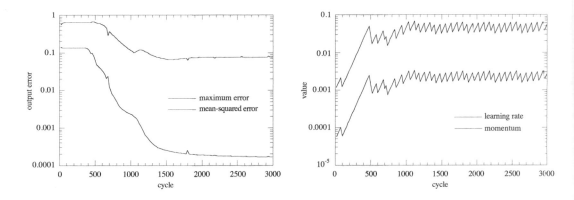

FIGURE 9-14. *The behavior of the output-node errors and learning rate and momentum parameters during training of the network. The network does not start to converge to a stable solution until about cycle 500. At this point, the learning rate and momentum do not need to be increased further; instead they are periodically adjusted every five cycles to maintain the convergence of the network. The overall shape of these curves is typical, but will vary in detail from one dataset to another.*

see that the decision boundaries change, but there is little change in the final hard map, because most of the feature space change takes place where there are few image spectral vectors (Fig. 9-15). Interestingly, though, the narrow, unpaved roads ("light soil") between the crop fields appear to be better mapped after this level of training. The mean-squared error at the output nodes drops from 0.115 at 1000 cycles to 0.074 at 5000 cycles.

The output node values for each class are shown in Fig. 9-16 and Plate 9-1 as soft classification maps. Higher *GL*s in the maps indicate larger output node values in the ANN. We can see additional evidence here that the crop class does not become competitive with the other two classes until somewhere between cycles 500 and 750. The ANN output values can be related to the spatial-spectral mixing of surface components (Foody, 1996; Moody *et al.*, 1996; Schowengerdt, 1996).

Finally, we note that the feature-space decision boundaries for ANNs are typically more adapted to the data distribution than the boundaries produced by parametric classifiers, such as Gaussian maximum-likelihood. An example for a 12 class urban land use/land cover classification of TM imagery is shown in Plate 9-2 (Paola and Schowengerdt, 1995b). We will comment more on this point later, after a review of parametric classification.

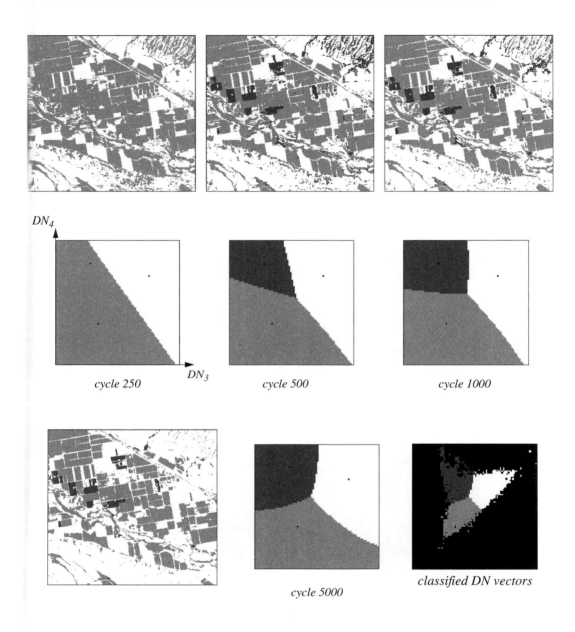

FIGURE 9-15. *Hard maps and decision boundaries at three intermediate stages and at the final stage (5000 cycles) of the back-propagation algorithm. (The processing and results in this section were provided by Justin Paola, Oasis Research Center.)*

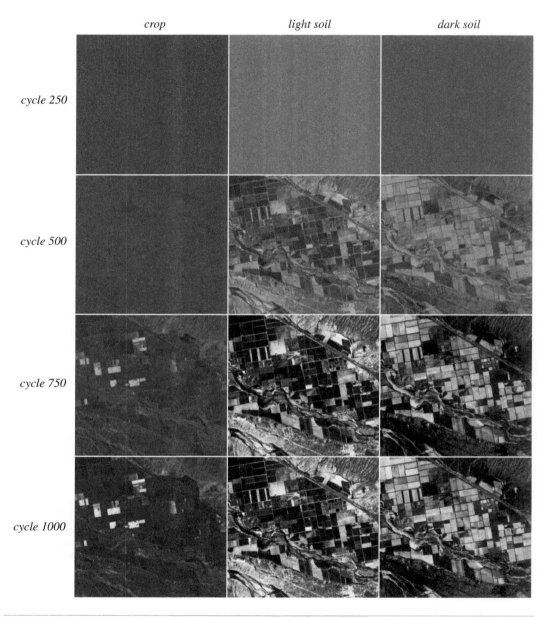

FIGURE 9-16.   *Soft classification maps at four stages of the back-propagation training process. The greylevel in each map is proportional to the class output-node value and all maps are scaled by the same LUT, so direct brightness comparisions are possible. Notice how the output-node values for all three classes increase as more training cycles are used.*

9.6   *Parametric Classification*

Parametric algorithms rely on assumptions about the form of the probability distribution for each class. The most notable example is *maximum-likelihood*, which explictly uses a probability model to determine the decision boundaries. The necessary parameters for the model are estimated from training data. A useful property of parametric classifiers, that is not possible with nonparametric classifiers, is theoretical estimation of classifier error from the assumed distributions.

9.6.1 Estimation of Model Parameters

If a sufficiently large number of representative training pixels are available in each class, we can calculate the class histograms and use them as approximations to the continuous probability density functions of an infinite sample of data (Chapter 4). These *class-conditional* probability density functions, $p(f|i)$, have unit area and describe the probability of a pixel having a feature vector f given that the pixel is in class i.

Each probability density function (histogram) may be weighted by the *a priori* probability, $p(i)$, that class i occurs in the image area of interest. These scaled probability functions, $p(f|i)p(i)$, do not have unit area. In remote sensing, the *a priori* probabilities may be estimated from external information sources such as ground surveys, existing maps, or historical data. For example, suppose the goal is to determine the proportion of crop types planted during a particular season from classification of a Landsat image. We might reasonably set the *a priori* probabilities to historical estimates of the percentage of each crop type in the area. In most cases, however, reliable *a priori* probabilities are difficult to obtain and are assumed equal for all classes. A discussion of the use of *a priori* probabilities in remote sensing is presented in Strahler (1980) and a simple example is given in Sect. 9.6.5.

To make a classification decision for a pixel, we need to know the *a posteriori* probabilities that the pixel belongs to each of the training classes i, given that the pixel has the feature vector f. This probability, $p(i|f)$, may be calculated with *Bayes' Rule*,

$$p(i|f) = \frac{p(f|i)p(i)}{p(f)} \qquad (9\text{-}10)$$

where

$$p(f) = \sum_i p(f|i)p(i). \qquad (9\text{-}11)$$

A *decision rule* may now be formed with the *a posteriori* probabilities of Eq. (9-10). It is intuitively satisfying to assign the pixel to a particular class if its *a posteriori* probability is greater than that for all other classes; this is the rule for the *maximum-likelihood* classifier. Since $p(f)$ is the same for all classes per Eq. (9-11), it can be ignored in a comparison of *a posteriori* probabilities, and we can write the *Bayes' decision rule* as,

$$\text{If } p(i|f) > p(j|f) \text{, for all } j \neq i \text{, assign pixel to class } i$$

where *p(i|f)* is found from Eq. (9-10), the training data distributions *p(f|i)*, and the *a priori* probabilities *p(i)*. In the very unlikely situation that the two *a posteriori* probabilities are exactly equal, a decision cannot be made from the class probabilities. A tie-breaking process then must be employed, such as using the classification of a neighboring, previously classified pixel or randomly choosing either class. It can be shown that the Bayes' decision rule minimizes the average probability of error over the entire classified dataset, if all the classes have normal (Gaussian) probability density functions (Duda *et al.*, 2001).

9.6.2 Discriminant Functions

The Bayes' decision rule may be restated as,

$$\text{If } D_i(f) \geq D_j(f) \text{, for all } j \neq i \text{, assign pixel to class } i$$

where the *discriminant function* for class *i*, $D_i(f)$, is given by,

$$D_i(f) = p(i|f)p(f) = p(f|i)p(i). \tag{9-12}$$

Setting *D* equal to the *a posteriori* probabilities results in a Bayes' optimal classification, but is not the only choice that produces the same result. The decision boundary does not change under any monotonic transformation, for example,

$$D_i(f) = Ap(i|f)p(f) + B = Ap(f|i)p(i) + B \tag{9-13}$$

or,

$$D_i(f) = \ln[p(i|f)p(f)] = \ln[p(f|i)p(i)] \tag{9-14}$$

are both valid discriminant functions. The latter transform is advantageous under a normal distribution assumption, as we will see next.

9.6.3 The Normal Distribution Model

To apply the parametric classification approach described previously, we must decide on an appropriate probability model. Now, for optical remote sensing,[11] it is almost universal that the *normal* (or Gaussian) distribution is chosen for the following reasons:

  · It provides a mathematically tractable, analytical solution to the decision boundaries.

  · Image samples selected for supervised training often exhibit Gaussian-like distributions.

11.The normal distribution is not appropriate for SAR image statistics.

Although normal distributions are widely assumed,[12] it is occasionally useful to remember that all of the theory and equations to this point *do not require* that assumption, and may be applied to data with *any* probability distribution. The normal distribution does lead, however, to some nice mathematical results, as described in the following.

The transformation of Eq. (9-14) is particularly useful if the class probability distributions are normal. A *K*-D, maximum-likelihood discriminant function for class *i* is then, from Eq. (4-18) and Eq. (9-14),

$$D_i(\boldsymbol{f}) = \ln[p(i)] - \frac{1}{2}\left[K\ln[2\pi] + \ln|C_i| + (\boldsymbol{f} - \boldsymbol{\mu}_i)^T C_i^{-1} (\boldsymbol{f} - \boldsymbol{\mu}_i) \right] . \qquad (9\text{-}15)$$

Only the last term must be calculated at each pixel.[13] For any two classes *a* and *b*, the decision boundary is found by setting $D_a(f)$ equal to $D_b(f)$ and solving for *f*. This is equivalent to setting,

$$\ln[p(a|\boldsymbol{f})p(\boldsymbol{f})] = \ln[p(b|\boldsymbol{f})p(\boldsymbol{f})] \qquad (9\text{-}16)$$

or,

$$p(a|\boldsymbol{f}) = p(b|\boldsymbol{f}) . \qquad (9\text{-}17)$$

The value of *f* on the decision boundary is that for which the *a posteriori* distributions are equal. To one side of the boundary the decision favors class *a* and to the other side the decision favors class *b* (Fig. 9-17). The logarithmic discriminant functions of Eq. (9-15) are quadratic functions of *f*; in two dimensions, they intersect in quadratic decision boundaries, as depicted in a general way in Fig. 9-18. A probability threshold can be used to exclude outliers, i.e., pixels with a low probability of belonging to any class, as shown later with image examples.

The total probability of classification error is given by the area under the overlapping portions of the *a posteriori* probability functions. The total probability of error is the sum of the probabilities that an incorrect decision was made on either side of the class partition. It is easy to see that the Bayes' optimal partition minimizes this error because a shift of the partition to the right or left will include a larger area from one class or the other, thereby increasing the total error.

It is instructive at this point to note again the role of the *a priori* probabilities. From Eq. (9-17) we see that the decision boundary will move to the left if $p(b)$ is greater than $p(a)$ and to the right if $p(a)$ is greater than $p(b)$. Even if reasonable estimates of the *a priori* probabilities are available, we may choose to bias them heavily if the significance of an error in one class is much greater than for others. For example, consider a hypothetical project to locate all occurrences of a rare class. The

---

12. The reader should be aware that there is a paucity of quantitative data in the literature to support this assumption. Although an arbitrary class distribution can be modeled as a *sum* of Gaussian kernel distributions by Parzen histogram estimation, such an approach is rare in remote sensing analyses. In any case, the *net* distribution for the class would not be Gaussian.

13. The quadratic term is, in fact, the Mahalanobis distance between the feature vector *f* and the distribution of class *i*; see Table 9-3.

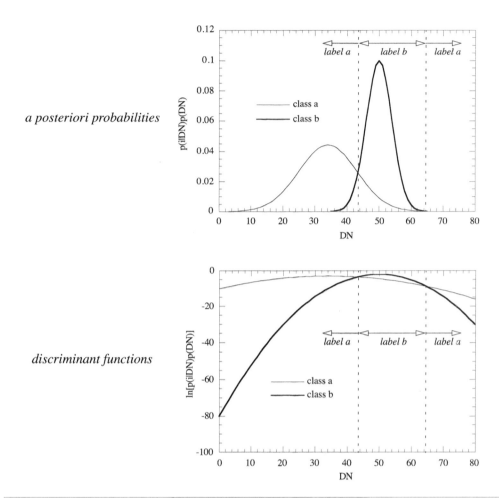

a posteriori probabilities

discriminant functions

FIGURE 9-17. *Maximum-likelihood decision boundaries for two continuous Gaussian DN distributions in one dimension. The class parameters are* $\mu_a = 34$, $\sigma_a = 9$, $\mu_b = 50$, *and* $\sigma_b = 4$. *Note the intersection of the two distributions on the right is not visible in the upper graph because of the ordinate scale, but becomes clear in the discriminant function graph below.*

actual *a priori* probability of that class would be very low, but we could assign an artificially high *a priori* probability to insure that no occurrences are missed. The penalty would be many pixels incorrectly classified as the rare class. These "false alarms" would then have to be removed by site visits or by referencing other data, such as aerial photography.

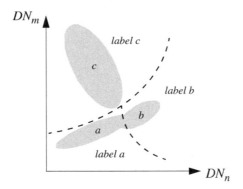

FIGURE 9-18. *Maximum-likelihood decision boundaries for three classes in two dimensions, with Gaussian distributions for each class. The boundaries are quadratics in 2-D. Their dependence on the covariance matrices of the individual classes is shown in Duda et al. (2001).*

9.6.4 The Nearest-Mean Classifier

If we assume that the covariance matrices of L classes are equal, i.e.,

$$C_i = C_0 \tag{9-18}$$

and that the *a priori* probabilities are equal,

$$p(i) = 1/L, \tag{9-19}$$

then the discriminant function of Eq. (9-15) becomes,

$$D_i(f) = A - \frac{1}{2}(f - \mu_i)^T C_0^{-1}(f - \mu_i) \tag{9-20}$$

where the constant A is given by,

$$A = \ln[1/L] - \frac{1}{2}\left[\ln[2\pi] + \ln|C_0|\right] \tag{9-21}$$

and may be ignored in a comparison of D among different classes. The right-hand term in Eq. (9-20) depends only on the feature vector f and the class mean μ_i, since C_0 is assumed to be common to all classes. The expansion of this term results in a quadratic equation in f, whose quadratic term is independent of the class i (Duda *et al.*, 2001), and may therefore be combined with the constant A for comparison among different classes. Thus Eq. (9-20) reduces to a form that is linear in f, meaning that the decision boundaries are hyperplanes (linear functions in two dimensions), in contrast to the hyperquadric functions of Eq. (9-15) in the general covariance case.

If the covariance matrices are further constrained to be diagonal, i.e., the features are uncorrelated, and to have equal variance along each feature axis, i.e.,

$$C_0 = \begin{bmatrix} c_0 & \cdots & 0 \\ \vdots & & \vdots \\ 0 & \cdots & c_0 \end{bmatrix} \qquad (9\text{-}22)$$

then

$$D_i(f) = A - \frac{(f - \mu_i)^T (f - \mu_i)}{2c_0} . \qquad (9\text{-}23)$$

The quantity $(f - \mu_i)^T (f - \mu_i)$ is the square of the L_2 distance between the vectors f and μ_i (Table 9-3). Since A and c_0 are the same for all classes, Eq. (9-23) represents the discriminant functions for the *nearest-mean*, or *minimum-distance classifier*. $D_i(f)$ will be largest (because of the minus sign) for the class i with minimum L_2, i.e., the class with the nearest mean.

The following points can be made about the nearest-mean algorithm:

- It ignores the covariance matrices of the prototype classes.
- The L_1-distance function can be used for computational efficiency, but the mathematical relationship to the maximum-likelihood algorithm would no longer apply. The nearest-mean decision boundaries for L_1 are piecewise linear and approximate the linear boundaries obtained with L_2 (Schowengerdt, 1983).

An idealized depiction of the nearest-mean decision boundaries is shown in Fig. 9-19. The nearest-mean algorithm does not create an "unlabeled" class automatically, unless an upper limit is applied to the distance from the mean. Then, pixels whose distance from the nearest class mean is greater than that threshold are assigned to the "unlabeled" class.

9.6.5 Parametric Classification Examples

The hard classification map and decision boundaries for the L_2 nearest-mean classifier for the Marana TM image and training data used earlier are shown in Fig. 9-20. No outlier threshold is implemented, so every pixel is assigned a label, unlike the level-slice classifier.

A maximum-likelihood classification of the same data is also shown in Fig. 9-20. Note the quadratic decision boundaries, characteristic of the Gaussian distribution assumption. A small threshold results in a large number of excluded pixels and shows the elliptical nature of the distribution model for each class. This high sensitivity to the threshold indicates that, for these data, the training class distributions are compact in the feature space.

To illustrate the modeling of class distributions with Gaussian probability functions, we'll use a relatively simple, single-band example. In Fig. 9-21, an MSS band 4 image of Lake Anna, Virginia, is shown with three training sites, two for water and one for vegetation. Initially, we will use only one site for water, the larger one in the lower right, and model each class with a Gaussian

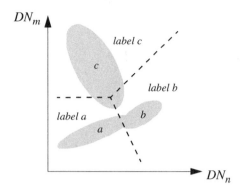

FIGURE 9-19.   Nearest-mean decision boundaries for three classes in two dimensions, using the L_2-
distance measure. Distance thresholds could be implemented either as in the level-slice algorithm, or as
circular boundaries, centered at each class mean, at a specified radius.

distribution having the respective training site mean and standard deviation. Two classifications are
shown in Fig. 9-21, nearest-mean and maximum-liklihood with equal *a priori* probabilities. The
nearest-mean classification map appears more accurate in that the lake is mapped in its entirety,
while the maximum-likelihood classification map does not correctly classify the upper arms of the
lake.

Some insight is gained into why the classifiers are producing significantly different results by
looking at the total image histogram and Gaussian models for each class (Fig. 9-22). The data are
bimodal, with high separability between the two classes. The nearest-mean decision boundary is
halfway between the two class means, while the maximum-likelihood boundary is much further to
the left because of the small standard deviation of the water class. Also, we can now see the role of
the *a priori* probabilities, which are assumed equal for the two classes. Because the water occupies
a much smaller portion of the image than vegetation, its histogram mode is smaller than that for
vegetation. However, that is not reflected by the equal *a priori* probabilities, leading to a poor
model for the data. Estimating the proportion of water to be about 5% and vegetation about 95%
and using those values as *a priori* probabilities, the *a posteriori* probabilities match the histogram
much better (Fig. 9-22). Unfortunately, this does not significantly increase the area classified as
water, because the decision boundary moves to the left by only a small amount.

The problem with the water class is that its variability within the scene is not sufficiently represented by the single training class. The width of the Gaussian distribution model for that one site is
less than the total image histogram indicates. Therefore, we include the second water site in one of
the arms of the lake, and "pool" the statistics for the two sites by averaging their DN means and
adding their DN variances. The resulting Gaussian model now has a larger variance and fits the data
histogram much better (Fig. 9-22). A second maximum-likelihood classification produces an
improved mapping of water. The nearest-mean classifier performs well in either case, because it
uses only the information about the class means and not their variances.

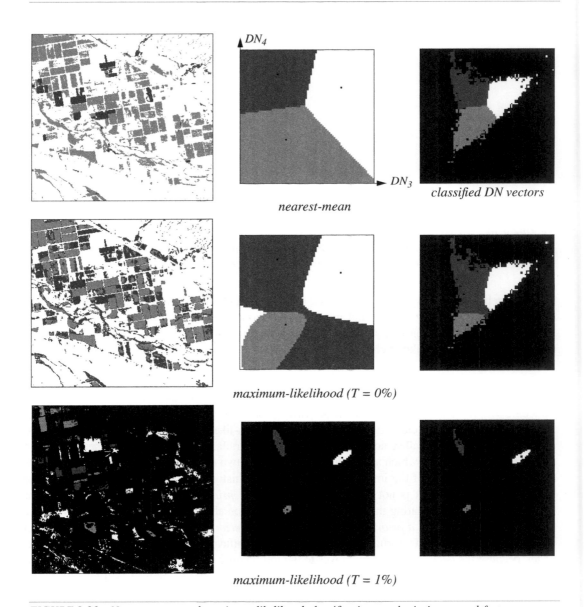

FIGURE 9-20. Nearest-mean and maximum-likelihood classification results in image and feature space. Note the disjoint decision regions for the class "light soil" if no probability threshold is used in the maximum-likelihood case. A threshold limits the distance from the mean. The threshold is specified as the percentage excluded volume under the Gaussian model distributions of each class. The fact that a low threshold of 1% results in such a large population of unlabeled pixels implies that the distributions for each class are highly localized near their means.

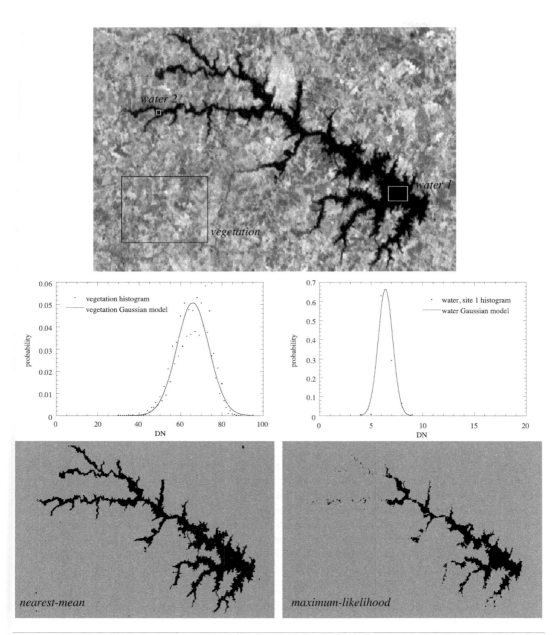

FIGURE 9-21. *Lake Anna Landsat MSS band 4 image with three training sites and initial classification maps. These sites were selected by visual interpretation of a color infrared composite of bands 4, 3, and 2. The histograms and Gaussian probability models for the vegetation and initial water sites (site 1) are each normalized to unit area. The initial classification maps are shown at the bottom.*

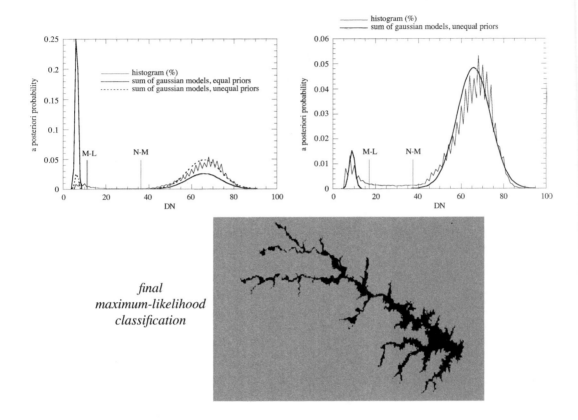

final maximum-likelihood classification

FIGURE 9-22. *The effect of a priori probabilities on the goodness-of-fit between the class Gaussian models and the data. Only one training site is used for water. The two close maximum-likelihood decision boundaries are for equal (right) and unequal (left) a priori probabilities. The final classification training models and decision boundaries are shown on the right after pooling the training statistics from the two water sites and using unequal a priori probabilities. The maximum-likelihood decision boundary has moved to the right by 6 DN from its location when only one water training site was used, while the nearest-mean decision boundary has shifted by 1.5 DN to the right. These changes result in the improved maximum-likelihood classification of water shown.*

A comparison of parametric maximum-likelihood decision boundaries to nonparametric ANN decision boundaries for a 12-class urban land use/land cover TM classification is shown in Plate 9-2. The ANN boundaries adapt to the data during the training stage and are not constrained by any parametric assumption. This can also be seen in Fig. 9-23, where the discriminant functions of the two classifiers are compared for four of the classes. These functions come from the training data, the normal model in the case of maximum-likelihood and the output node surface, adapted to the training data during back-propagation training, in the case of the ANN. In both cases, the class with the maximum discriminant value is chosen for a hard classification at each pixel. In this particular experiment, the ANN and maximum-likelihood classification maps differed at 35% of the pixels, with the ANN map being superior for certain classes (Paola and Schowengerdt, 1995b).

The search for improved classifiers continues with, for example, interest in *Support Vector Machines (SVM)* (Foody and Mathur, 2004; Camps-Valls *et al.*, 2004). Unfortunately, we cannot say that one classifier is *always* better than another. As these examples demonstrate, a classifier's performance depends on the data! If the class distributions are widely separated, most classifiers perform about the same. The maximum-likelihood classifier, however, must also have reasonably accurate covariance estimates. Given that the Gaussian assumption is appropriate and given accurate parameter estimates, maximum-likelihood theoretically produces minimum class labeling error. Nonparametric algorithms that adapt to any class distributions, like the ANN, do not have a simple theoretical basis, but are generally easier to use since less care is required by the analyst in defining and validating the training data.

9.7 Spatial-Spectral Segmentation

So far, we have not used our knowledge that there is a significant statistical correlation among neighboring pixels (Chapter 4). The classification of pixels in a spectral feature space is independent of any spatial relationships. Another approach to classification, *spatial-spectral segmentation*, incorporates spatial and spectral information into unsupervised labeling. Spatial-spectral segmentation schemes tend to work best where the scene consists of relatively homogeneous objects of at least several pixels. Example algorithms and applications are listed in Table 9-5. Most algorithms employ either a local gradient operation (Chapter 6) to detect region boundaries (edge-based) or aggregation of neighboring, similar pixels into larger regions (region-based). In the ECHO algorithm, for example, adjoining pixels are initially aggregated into small cells (e.g., 2 × 2 pixels) based on the similarity of their spectral vectors. Cells that cross spatial boundaries (edges) are detected by a threshold on the cell variance in each spectral band, and the pixels in the cell are not aggregated if an edge is detected within the cell. The homogeneous cells found at this stage are then aggregated further if they are spectrally similar to neighboring cells. The resulting spatially homogeneous areas are then classified in their entirety as *single* spectral samples, rather than pixel-by-pixel.

9.7.1 Region Growing

Spatial connectivity can be invoked to group similar pixels into an *image* region, presumed to represent a *physical* region of similar properties on the earth (Woodcock and Harward, 1992; Lobo, 1997). A threshold must be specified for the similarity of a pixel and its neighbors; if their difference is within the threshold, they are "merged" and treated as a common entity in subsequent

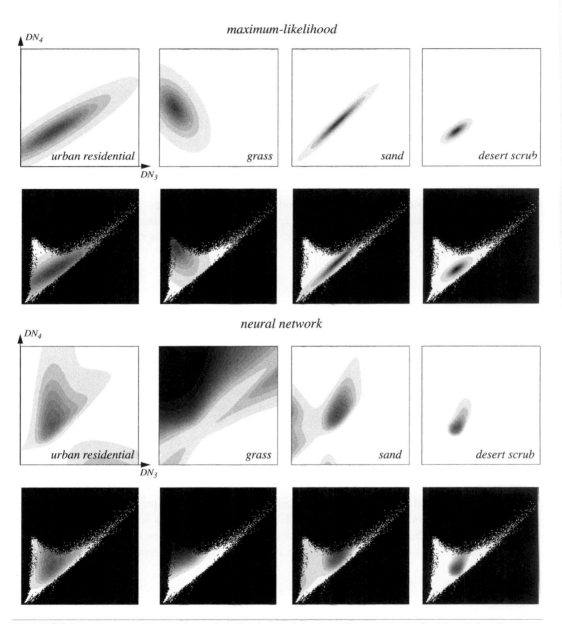

FIGURE 9-23. The natural log of the probability density functions of the maximum-likelihood classifer (top two rows) and the output surfaces of the neural network (bottom two rows) for four of the classes in Plate 9-2. The values are inverted for display; darker grey means higher value. Rows two and four show the distributions masked by the scatterplot of the data. These plots represent the discriminant functions for the classifiers (Paola and Schowengerdt, 1995b). (Rows one and three © 1995 IEEE.)

TABLE 9-5. *Some spatial-spectral segmentation algorithms, comparisons and applications.*

name	data	type	reference
–	Landsat MSS	clustering	Haralick and Dinstein, 1975
ECHO	airborne MSS, Landsat MSS	edge detection and region growing	Kettig and Landgrebe, 1976; Landgrebe, 1980
BLOB	MSS	clustering	Kauth *et al.*, 1977
–	various	clustering	Coleman and Andrews, 1979
AMOEBA	MSS, multispectral aerial	clustering	Bryant, 1979; Jenson *et al.*, 1982; Bryant, 1989; Bryant, 1990
–	simulated	pyramid structure	Kim and Crawford, 1991
–	airborne MSS, SAR	hierarchical partitioning	Benie and Thomson, 1992
–	TM	iterative region growing	Woodcock and Harward, 1992
SMAP	simulated, SPOT	pyramid structure	Bouman and Shapiro, 1994
IPRG	TM	iterative parallel region growing	Le Moigne and Tilton, 1995
IMORM	TM	iterative region merging	Lobo, 1997
–	AVHRR	ridge detector, Hough transform	Weiss *et al.*, 1998
–	IRS-1C LISS-III pan band	multiscale morphology	Pesaresi and Benediktsson, 2001
–	IRS-1B LISS-I, IRS-1D LISS-III	Markov random field	Sarkar *et al.*, 2002
CGRG	TM	guided region growing	Evans *et al.*, 2002
–	IKONOS	support vector machine	Song and Civco, 2004
–	IKONOS	various	Carleer *et al.*, 2005
–	QuickBird	region splitting/merging	Hu *et al.*, 2005

processing. Otherwise, they are retained as individual pixels. If a pixel is allowed to connect only to its neighbors above, below, right, or left, the algorithm is *four-connected*. If a pixel is allowed to connect to any of its eight neighbors, an *eight-connected* algorithm results, which is generally preferred because it includes four-connected regions as a special case. Algorithms based on four-connectedness are faster, however, since only four neighbors must be addressed.

There are many techniques for spatial segmentation (Pal and Pal, 1993). A representative and robust *region-growing* algorithm will be described (Ballard and Brown, 1982; Ryan, 1985). Pixels in the image are processed in the normal order, left-to-right (along rows) and top-to-bottom (row-by-row), which is important for this algorithm. If the data were processed right-to-left, for example, a different (but equally valid) segmentation would result. Two parameters are specified by the user, a DN difference threshold, t, and a DN standard deviation threshold, σ.

A three pixel, four-connected neighborhood is used to make decisions on whether to add a current pixel to an adjoining region, or to initialize a new region (Fig. 9-24). The algorithm consists of several case rules applied to the DNs and labels of these three pixels to handle the various relations that can arise. Case 4 provides a "filler" pixel between two diagonally-connected pixels with the same label and case 5 provides for the connection of two previously separate regions, if the pixel values warrant the change. The label of the left region is changed to that of the upper region for all pixels in the left region. The algorithm requires dynamic lists of pixels and labels that must be continuously updated; a second pass through the labeled pixels is also needed to make the final labels sequential. When finished, the output labeled map consists of a region labeled "one" in the upper left (the first pixel), progressing to the "largest" label, representing the total number of regions in the image. As the DN difference threshold, t, increases, pixels are more easily aggregated to existing regions, so there are fewer regions covering the full image. A plot of the number of regions and the average DN error in approximating the original image by the region mean DNs, for different DN difference thresholds, is also shown in Fig. 9-24.

For small amounts of pixel aggregation, the dynamic range of the label image is large, so it cannot be displayed as a grey level image with full label resolution. A useful way to present the results is to replace the label of each region by the region's average DN in the original image (an example of the compression decoding discussed in relation to Fig. 9-2). This representation is used to display the segmentation results for the Marana image in Fig. 9-25.

9.8   Subpixel Classification

We have pretended to this point that every pixel has only one labeled category, leading to a so-called *hard* classifiction. Since pixels represent a spatial average over the ground-projected spread function, it is inevitable that multiple spectral categories will be included in most of them (Chapter 4). Furthermore, remote-sensing imagery is usually resampled at some point for geometric correction and registration (Chapter 7), and this resampling, if not nearest-neighbor, induces additional spatial mixing. If the bands of a multispectral image are misregistered, mixing occurs even without resampling for registration (Billingsley, 1982; Townshend *et al.*, 1992). As the mixing proportions (class *fractions* or *abundances*, as they are sometimes called) change from pixel-to-pixel, the net spectral vector changes. This fact was recognized early on in the analysis of Landsat MSS data (Horwitz *et al.*, 1971; Nalepka and Hyde, 1972; Salvato, 1973). The advent of hyperspectral sensors has prompted renewed interest in techniques for estimating mixture components, drawing on the heritage of traditional spectroscopy (Adams *et al.*, 1993).

All natural, and most man-made surfaces, are nonuniform at some level of spatial resolution. Thus, signature mixing does not go away if one goes to higher-resoluton imagery; it simply changes. However, there is a "natural" scale to many earth surface features (Markham and

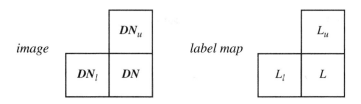

rules for merging pixels:

case	if	then
1. start new region	$\left\vert DN - DN_u \right\vert > t, \quad \left\vert DN - DN_l \right\vert > t$	$L =$ new label
2. merge with upper region	$\left\vert DN - DN_u \right\vert \leq t, \quad \left\vert DN - DN_l \right\vert > t$	$L = L_u$
3. merge with left region	$\left\vert DN - DN_u \right\vert > t, \quad \left\vert DN - DN_l \right\vert \leq t$	$L = L_l$
4. merge with upper and left region	$\left\vert DN - DN_u \right\vert \leq t, \quad \left\vert DN - DN_l \right\vert \leq t$ $L_l = L_u$	$L = L_u$
5. relabel and merge	$\left\vert DN - DN_u \right\vert \leq t, \quad \left\vert DN - DN_l \right\vert \leq t$ $L_l \neq L_u$	$L_l = L_u$ and $L = L_u$

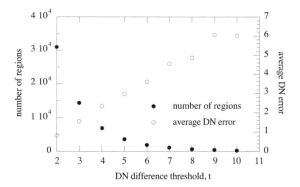

FIGURE 9-24. *The spatial neighborhood and rules used for the region-growing algorithm. The pixel being processed is the one in the lower right; it has an "upper" neighbor above and a "left" neighbor to the left. The convergence of the region growing segmentation algorithm and the associated average DN error as a function of the DN difference threshold are plotted below. The general shape of these curves is typical for any image, but the convergence rate will depend on the spatial structure and contrast of the image.*

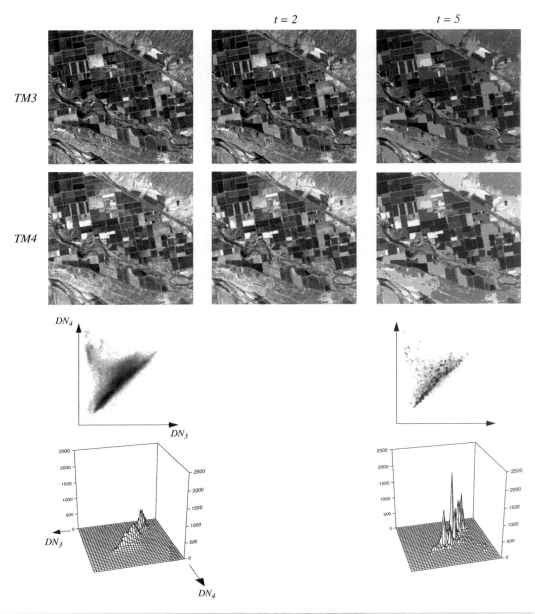

FIGURE 9-25. *Segmentation results for two DN-difference thresholds, 2 and 5. As the threshold increases, fewer but larger regions are created. The scattergrams below (original data and threshold of 5) show the tendency for pixels to group not only into contiguous spatial regions, but also into spectral clusters with distinct **DN** vectors.*

Townshend, 1981; Irons *et al.*, 1985). For example, paved roads range in width from about 3 m to 15 m or more, buildings have characteristic sizes according to their function, and shrub and tree canopy size varies characteristically by species. For specific classes, therefore, an increase in spatial resolution may reduce the percentage of mixed pixels, but mixed pixels will still occur at the boundary between objects regardless of their size or the sensor resolution. To see this, we create a simple synthetic scene consisting of buildings, streets, and grass (Fig. 9-26). The three categories are assigned different DNs, and the "scene" is convolved with various *GIFOV*s of different sizes, simulating different sensor resolutions. The result shows the expected mixing at the boundaries between objects, and that it persists, even at relatively high resolution. This will *always* be true, because all real earth scenes have spatial detail with smaller dimensions than *any* given *GIFOV*.

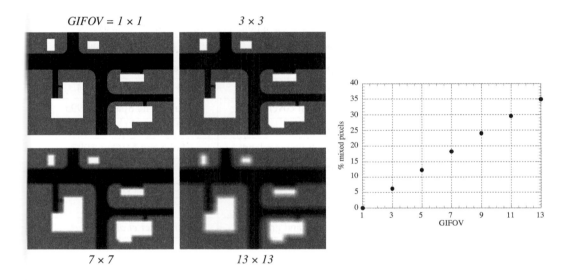

FIGURE 9-26. Simple example to illustrate spatial mixing. A synthetic scene consisting of three types of objects was created at a pixel size of one (upper left). Simulated images were then generated by spatial averaging over a range of GIFOVs. The percentage of mixed pixels for various GIFOVs is shown in the graph.

Subpixel applications in remote sensing date to the earliest days of Landsat (Table 9-6). This intriguing topic, in which we attempt to retrieve information beyond the "resolution limit" of the data, continues to be important. We look at several approaches to this problem in the following sections.

TABLE 9-6. Example applications of subpixel analyses.

data	application	reference
simulated Landsat MSS	crops and soil	Nalepka and Hyde, 1972
MSS	soil, rock, and vegetation	Marsh *et al.*, 1980
AVHRR	temperature	Dozier, 1981
TM	desert vegetation	Smith *et al.*, 1990
AVHRR, TM	forest cover	Cross *et al.*, 1991
field spectroradiometer	soil properties	Huete and Escadafal, 1991
MSS	forest, soil, shade	Shimabukuro and Smith, 1991
AVHRR, TM	vegetation/soil mapping	Holben and Shimabukuro, 1993
AVIRIS	vegetation, soil	Roberts *et al.*, 1993
simulated TM	crop types at field boundaries	Wu and Schowengerdt, 1993
simulated AVHRR	forest versus nonforest	Foody and Cox, 1994
AVHRR, TM	forest cover	Hlavka and Spanner, 1995
TM	tree, shrub cover	Jasinski, 1996
TM	desert shrub rangeland	Sohn and McCoy, 1997
simulated TM and AVHRR	directional reflectances	Asner *et al.*, 1997
TM	vegetation	Carpenter *et al.*, 1999
HYDICE, simulated targets	subpixel targets	Bruce *et al.*, 2001
AVIRIS	snow cover and grain size	Painter *et al.*, 2003
LISS-II	vegetation	Ghosh, 2004
multitemporal MODIS	crops	Lobell and Asner, 2004

9.8.1 The Linear Mixing Model

The idealized, pure signature for a spectral class is called an *endmember*. Because of sensor noise and within-class signature variability, endmembers only exist as a conceptual convenience and as idealizations in real images. The spatial mixing of objects within a *GIFOV* and the resulting linear mixing of their spectral signatures are illustrated in Fig. 9-27. Linear mixture modeling assumes a single reflectance within the *GIFOV* and is an approximation to reality in many cases; *nonlinear* mixing occurs whenever there is radiation transmission through one of the materials (such as a vegetation canopy), followed by reflectance at a second material (such as soil); or if there are multiple reflections within a material or between objects within a *GIFOV* (Borel and Gerstl, 1994; Ray and Murray, 1996; Moreno and Green, 1996). We consider only linear mixing here.

one GIFOV:

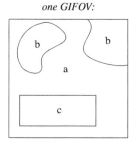

class a: 65% area, spectrum E_a
class b: 20% area, spectrum E_b
class c: 15% area, spectrum E_c

total spectrum at pixel:

$$DN = 0.65E_a + 0.20E_b + 0.15E_c$$

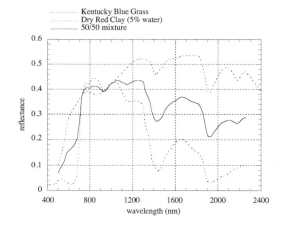

FIGURE 9-27. *The linear mixing model for a single GIFOV. The boundaries between objects can be any shape and complexity; only the fractional coverages and individual spectral reflectances are important. An example using the spectral reflectance data of Chapter 1 is also shown. The mixed soil/vegetation signature has lost some of the sharp characteristic vegetation "edge" at 700 nm.*

The spatial area of integration for mixing analysis is usually assumed to be that of the *GIFOV* of the sensor, with a uniform weighting of radiances over that area. A better model is the total system spatial response (Chapter 3), which weights the individual object radiances according to their location (Fig. 9-28). It has been shown that spatial *restoration* of multispectral images, i.e., partial correction for the sensor spatial spread function, can improve subsequent unmixing analysis (Wu and Schowengerdt, 1993) and that restoration can be included directly in the unmixing process (Frans and Schowengerdt, 1997; Gross and Schott, 1998; Frans and Schowengerdt, 1999).

Another source of mixing is strictly spectral combination of components, which can occur, for example, with minerals or water solutions (Felzer *et al.*, 1994; Novo and Shimabukuro, 1994). These *intimate mixtures* are not distinguishable in the data from the spatial-spectral mixing induced by the sensor spatial response. The end result is the same in both cases, namely a mixed spectral signal.

In two-dimensional feature space, we might have three endmembers as shown in Fig. 9-29. Now, if these endmembers are indeed "pure," and if they are an exhaustive basis for all spectral vectors in the image, the spectral vector for any pixel must lie within the *convex hull* defined by the envelope around the endmembers, as shown. The class fractions determine the location of the mixed-pixel vector in feature space. The inversion problem, termed *unmixing*, is to estimate the fractions of each class from a given pixel vector.

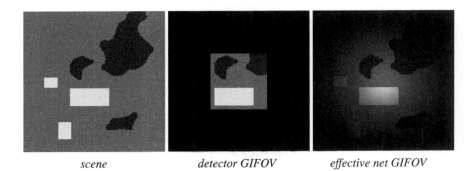

scene   *detector GIFOV*   *effective net GIFOV*

FIGURE 9-28.   *The spatial integration involved in mixing of spectral signatures. An idealized scene radiance spatial distribution is shown on the left, the region of integration corresponding to the sensor GIFOV in the center, and the region of integration corresponding to the total sensor spatial response on the right (a response similar to that of TM is assumed). Since that response is not uniform, a weighting occurs as a function of distance from the center. Although the difference between the measured signals for the two cases on the right is only about 4% in this example, it is easy to conjure a spatial distribution of objects with different reflectances that would result in a large difference.*

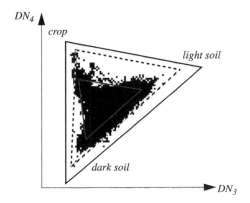

FIGURE 9-29.   *Three possible choices for endmembers for the classes "dark soil," "light soil," and "crop" in a TM NIR-red spectral space. The inner triangle is defined by the supervised class means used earlier. The middle triangle is defined by extreme pixels along each DN axis. The outer triangle contains all the pixels, but the endmembers are not actually present in the image. Only the outer triangle is consistent with Eq. (9-25).*

Linear mixing is described mathematically as a linear vector-matrix equation,

$$\boldsymbol{DN}_{ij} = \boldsymbol{E}\boldsymbol{f}_{ij} + \boldsymbol{\varepsilon}_{ij} \tag{9-24}$$

where $\boldsymbol{f}_{ij}$ is the $L \times 1$ vector of L endmember fractions for the pixel at ij, and $\boldsymbol{E}$ is the $K \times L$ endmember signature matrix, with each column containing one of the endmember spectral vectors. The lefthand side $\boldsymbol{DN}_{ij}$ is the K-dimensional spectral vector at pixel ij. The added term $\boldsymbol{\varepsilon}_{ij}$ represents the residual error in the fitting of a given pixel's spectral vector by the sum of L endmember spectra and unknown noise. The relationship in Eq. (9-24) is constrained by the assumption that an exhaustive set of endmembers (classes) has been defined, so that,

$$\sum_{l=1}^{L} f_l = 1 \tag{9-25}$$

at each pixel. This last assumption is problematic in practice, since one is never sure that a sufficient number of endmembers has been defined for a given set of data. An additional constraint can also be included at each pixel, namely that each fraction must be positive,

$$f_l \geq 0 . \tag{9-26}$$

Unmixing examples

A low-dimensional example is a good way to introduce the calculation of class fraction maps. The scatterplot in Fig. 9-29 is for TM bands 3 and 4 of the Marana image. Three endmembers are defined, "crop," "light soil," and "dark soil," which are associated with the tips of the scatterplot. We will use the model of Eq. (9-24) without the noise term, and write, for the 2-D multispectral vector at each pixel,

$$\boldsymbol{DN}_{ij} = \boldsymbol{E}\boldsymbol{f}_{ij}, \tag{9-27}$$

or, writing out each element,

$$\begin{bmatrix} DN_3 \\ DN_4 \end{bmatrix} = \begin{bmatrix} \begin{bmatrix} E_{crop3} & E_{ltsoil3} & E_{dksoil3} \\ E_{crop4} & E_{ltsoil4} & E_{dksoil4} \end{bmatrix} \end{bmatrix} \begin{bmatrix} f_{crop} \\ f_{ltsoil} \\ f_{dksoil} \end{bmatrix}. \tag{9-28}$$

This problem is underdetermined; we have three unknowns (the fraction components) but only two equations, one each for bands 3 and 4. The requirement that linear combinations of the three endmembers represent all spectral vectors in the image is expressed as

$$1 = f_{crop} + f_{ltsoil} + f_{dksoil} \tag{9-29}$$

which applies to every pixel. This constraint can be combined with Eq. (9-28),

$$\begin{bmatrix} DN_3 \\ DN_4 \\ 1 \end{bmatrix} = \begin{bmatrix} E_{crop3} & E_{ltsoil3} & E_{dksoil3} \\ E_{crop4} & E_{ltsoil4} & E_{dksoil4} \\ 1 & 1 & 1 \end{bmatrix} \begin{bmatrix} f_{crop} \\ f_{ltsoil} \\ f_{dksoil} \end{bmatrix} . \tag{9-30}$$

We'll call this the *augmented* mixing equation. It can now be solved *exactly* for the fractions at each pixel,

$$\begin{bmatrix} f_{crop} \\ f_{ltsoil} \\ f_{dksoil} \end{bmatrix} = \begin{bmatrix} E_{crop3} & E_{ltsoil3} & E_{dksoil3} \\ E_{crop4} & E_{ltsoil4} & E_{dksoil4} \\ 1 & 1 & 1 \end{bmatrix}^{-1} \begin{bmatrix} DN_3 \\ DN_4 \\ 1 \end{bmatrix} . \tag{9-31}$$

For our example, the endmembers are defined in two ways, first as the extreme pixels of the scatterplot ("data-defined") and second as "virtual" endmembers, which do not actually exist in the data, but whose convex hull will enclose all data in the scatterplot. These virtual endmembers are assumed to represent 100% pure pixels in their respective classes. The DN values are tabulated in Table 9-7, with the associated E and E^{-1} matrices in Table 9-8. With the E^{-1} matrix determined, Eq. (9-31) is applied to each multispectral pixel vector (augmented with the third, unit-valued "band") to find the fractions of each endmember class. The results in Fig. 9-30 show little visual and numerical difference for the two endmember selection methods with these data. For the "data-defined" endmembers, only 29 pixels (out of 97,500 total) in the "crop" class and 85 pixels in the "light soil" class were less than zero, and then by no more than -0.08; 112 pixels had fractions that summed to more than 1, again by no more than 1.08.

TABLE 9-7. Endmember DN values for the 2-D unmixing example.

endmember type	band	crop	light soil	dark soil
data-defined	3	21	84	18
	4	84	72	14
virtual	3	14	93	15
	4	90	77	6

In traditional hard classification, an "unknown" label is often included to accommodate everything not represented by the training sets. Any pixel that falls outside the decision threshold for any trained class is assigned to the "unknown" class. In the linear mixing model, however, we cannot define an "unknown" endmember, because such a spectral signature is not defined as a single spectral vector or even a small cluster of spectral vectors. A "shade" endmember is sometimes used to accommodate the variations in signatures arising from topography and subpixel surface texture (Adams *et al.*, 1993).

TABLE 9-8. Augmented matrices for unmixing.

endmember type	E	E^{-1}
data-defined	$\begin{bmatrix} 21 & 84 & 18 \\ 84 & 72 & 14 \\ 1 & 1 & 1 \end{bmatrix}$	$\begin{bmatrix} -0.013045 & +0.014845 & +0.026991 \\ +0.015744 & -0.00067476 & -0.27395 \\ -0.0026991 & -0.01417 & +1.247 \end{bmatrix}$
virtual	$\begin{bmatrix} 14 & 93 & 15 \\ 90 & 77 & 6 \\ 1 & 1 & 1 \end{bmatrix}$	$\begin{bmatrix} -0.01072 & +0.011777 & +0.09014 \\ +0.012683 & +0.00015099 & -0.19115 \\ -0.0019629 & -0.011928 & +1.101 \end{bmatrix}$

The constrained, linear unmixing example worked previously is simple and exact. More interesting unmixing challenges are presented when the number of spectral bands is greater than the number of classes, as in hyperspectral imagery. The linear mixing model then becomes an overdetermined set of K equations (K spectral bands) with L variables (L endmember fractions), where K is much greater than L. The endmember matrix is then $K \times L$ and cannot be inverted. A solution for the fraction $\hat{f}$ can be obtained via the *pseudoinverse* (see Chapter 7),

$$\hat{f}_{ij} = (E^T E)^{-1} E^T \cdot DN_{ij},$$ (9-32)

which minimizes the mean-squared error,

$$\|\varepsilon_{ij}\| = min[\varepsilon_{ij}^T \varepsilon_{ij}] = (DN_{ij} - E\hat{f}_{ij})^T (DN_{ij} - E\hat{f}_{ij}),$$ (9-33)

in the fit of the estimated fraction mixture to the data. A map of $\|\varepsilon_{ij}\|$ at each pixel is called a *residual image*. This least-squares algorithm is quite intensive for high-dimensional hyperspectral data because the matrix-vector product in Eq. (9-32) must be calculated for every pixel. The matrix E is fixed for a given problem, so Eq. (9-32) is equivalent to a rotational transform of the DN, as in the PCT and TCT transforms (Chapter 5). If the additional constraints of Eq. (9-25) and Eq. (9-26) are included in the analysis, it is known as the *Constrained Least Squares (CLS)* method and requires additional mathematical techniques for solution (Shimabukuro and Smith, 1991).[14]

---

14. Note the similarity to the least-squares solution for the polynomial coefficients that best fit a set of GCPs (Chapter 7). We've assumed the noise is uncorrelated in feature space (for unmixing) and among GCPs (for warping), and that its covariance matrix is the diagonal identity matrix. If this assumption does not hold, the least-squares solution includes a dependence on C_ε (Settle and Drake, 1993).

crop   light soil   dark soil

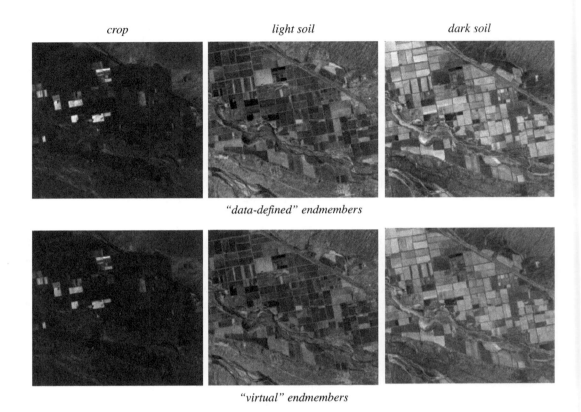

"data-defined" endmembers

"virtual" endmembers

FIGURE 9-30. *Class fraction maps produced by the data-defined endmembers (middle triangle, Fig. 9-29)
and "virtual" endmembers (outer triangle, Fig. 9-29). All pictures have the same display LUT, so they are
directly comparable.*

Relation of fractions to neural network output

As noted earlier, there is a connection between the fractions produced by linear unmixing and the
soft output of neural network classifiers. In Plate 9-1, we show the fraction maps for the three
classes of our example combined in a color composite, with crop as red, light soil as green and dark
soil as blue. Clearly, this representation correlates well with the original image and the soft ANN
output map. Now, if we look at the scatterplot between each class in the fraction map and in the
ANN map, we see their relationship in an empirical way. The ANN output value is proportional to
the fraction of the class at a pixel, and it appears that the ANN sigmoid soft decision function is a
factor in this relation to fractions. Further analysis for this example has not been done, however. It

is also known that ANN outputs can approximate the *a posteriori* class probabilities of Bayes classifiers under certain conditions. The reader is referred to Paola and Schowengerdt (1995b), Moody
et al. (1996), Foody (1996), Schowengerdt (1996), and Haykin (1999) for more discussion.

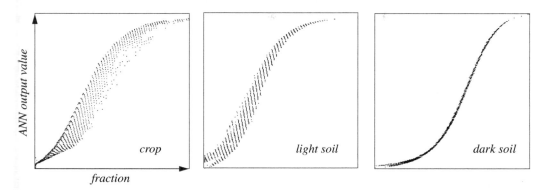

FIGURE 9-31. *Scatterplots between the ANN output node values and linear unmixing fractions of the image
in Plate 9-1. Both variables are scaled [0, 255]. Note the overall sigmoid-like shape for each class. The
variations within the curve envelopes indicate the two measures do not have an exactly one-to-one
relationship.*

Endmember specification

Specification of the endmembers can be a major problem in unmixing analysis. Clearly, accurate
unmixing relies on accurate endmembers. It may be impossible to find completely pure pixels in the
image data, i.e., pixels that contain only one endmember spectrum. Since we are considering subpixel content, how can we know if a pixel is "pure" or not? Even if such pixels exist, identifying
them in high-dimensional hyperspectral imagery is especially difficult. The following methods for
endmember specification have been proposed:

- Laboratory or field reflectance spectra (Boardman, 1990). In this case, the image data
  must be calibrated to reflectance (Chapter 7).

- Comparison of remote sensing data to libraries of material spectra. There are a number of
  spectral reflectance libraries available on the Internet for different materials and applications in remote sensing. Image pixel spectra can be modeled as mixtures of reflectance
  spectra from such libraries (Smith *et al.*, 1990; Roberts *et al.*, 1993). Again, the data must
  be calibrated.

- Automated techniques based on data transforms (Full *et al.*, 1982) and projection tech-
  niques such as *N-FINDR* (Winter, 1999; Winter, 2004; Plaza and Chang, 2005) , *Pixel
  Purity Index (PPI)* (Boardman *et al.*, 1995), and *Vertex Component Analysis (VCA)*
  (Nascimento and Dias, 2005).

- *K*-dimensional interactive visualization tools (Bateson and Curtiss, 1996).

It is possible to allow the types and numbers of pure spectral endmembers to vary from pixel-to-pixel for unmixing (Roberts *et al.*, 1998). Furthermore, while it is common to think of endmembers as invariant, i.e., that a single endmember represents a spectral class, it is also useful to allow for variability within endmembers, for example with tree canopy spectra (Bateson *et al.*, 2000). The variable endmembers, or "bundles," produce a range of fractions rather than a single fraction when unmixing for each category.

Finally, we note that the linear unmixing approach described here is not the only tool available for unmixing hyperspectral data. For example, *Independent Component Analysis (ICA)* has also been used for this task (Bayliss *et al.*, 1997; Tu, 2000; Shah *et al.*, 2004; Nascimento and Dias, 2005).

9.8.2 Fuzzy Set Classification

The fuzzy set concept, in which an entity may have partial membership in more than one category, is a natural model for the unmixing problem. Two fuzzy algorithm tools for remote-sensing classification are described here: *Fuzzy C-Means (FCM)* clustering and fuzzy supervised classification.

Fuzzy C-Means (FCM) clustering

This algorithm is much like the *K*-means unsupervised clustering algorithm described earlier. The essential difference is that the feature space is partitioned into fuzzy regions (Bezdek *et al.*, 1984; Cannon *et al.*, 1986). A *membership grade* matrix U is created for N pixels and C clusters,

$$U = \begin{bmatrix} u_{11} & \cdots & u_{1N} \\ \vdots & & \vdots \\ u_{C1} & \cdots & u_{CN} \end{bmatrix}. \tag{9-34}$$

Each column of U represents the membership values of the image pixels in each of the C clusters. The following constraints also apply,

$$\sum_{n=1}^{N} u_{ln} > 0 \ , \ \sum_{l=1}^{C} u_{ln} = 1 \ , \ 0 \le u_{ln} \le 1 \ , \tag{9-35}$$

which are similar to the constraints on fractions of endmembers in unmixing and on *a posteriori* probabilities in maximum-likelihood classification (Foody, 1992).

K-means hard clustering can be shown to minimize the squared-error function, for a fixed number of clusters C,

$$\varepsilon^2 = \sum_{n=1}^{N} \sum_{l=1}^{C} \left\| DN_n - \mu_l^* \right\|^2 \tag{9-36}$$

where

$$\left\| DN_n - \mu_l^* \right\|^2 = (DN_n - \mu_l^*)^T (DN_n - \mu_l^*) \tag{9-37}$$

is the square of the Euclidean distance L_2 from a pixel vector to the current *fuzzy mean vector* μ_l^* for cluster l (Jain and Dubes, 1988). To achieve a fuzzy partitioning of feature space, the function to be minimized incorporates the membership values,

$$J_m = \sum_{n=1}^{N} \sum_{l=1}^{C} u_{ln}^m \left\| DN_n - \mu_l \right\| \ , \ m \geq 1 \ . \tag{9-38}$$

The parameter m determines the "fuzziness" of the partitioning; m equal to one results in a hard clustering, and values around two are typically used. The iterative adjustment process used for the K-means algorithm is followed (Sect. 9.4.2), with the following formulas to update the cluster means and membership values,

$$\mu_l^* = \left[\sum_{n=1}^{N} u_{ln}^m DN_n \right] / \sum_{n=1}^{N} u_{ln}^m \tag{9-39}$$

$$u_{ln} = 1 / \sum_{j=1}^{C} \left[\left\| DN_n - \mu_l^* \right\| / \left\| DN_n - \mu_j^* \right\| \right]^{-2/(m-1)} . \tag{9-40}$$

These rather obscure equations are actually simple in concept. The fuzzy cluster means are simply calculated from the data samples weighted by their fuzzy membership values, and the membership values are updated by the normalized distance to the cluster means.

The Marana TM image is used to demonstrate the FCM algorithm (with an m value of two) in Fig. 9-32. Only three clusters are requested, so the crop areas are not distinct and there is considerable within-cluster data variation. The FCM algorithm shows similar overall cluster assignments as K-means, but also indicates the membership likelihood at each pixel (displayed on the same greyscale so they can be compared). The active crop fields do not appear in any one fuzzy cluster, but have a nonzero likelihood of belonging to all three clusters. Fields with low vegetation density in the upper left have nonzero membership values in clusters 1 and 2.

Fuzzy supervised classification

This method is described in Wang (1990a, 1990b). A membership matrix U is defined, as in FCM. The fuzzy mean is calculated from weighted training data as in Eq. (9-39), and the fuzzy covariance is calculated as

$$C_l^* = \left[\sum_{n=1}^{N} u_{ln} (DN_n - \mu_l^*)(DN - \mu_l^*)^T \right] / \sum_{n=1}^{N} u_{ln} . \tag{9-41}$$

cluster 1   cluster 2   cluster 3

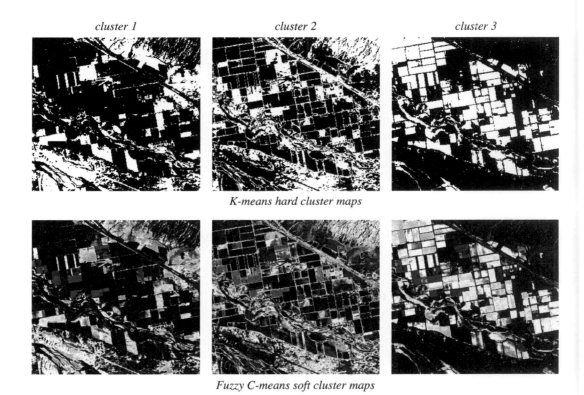

K-means hard cluster maps

Fuzzy C-means soft cluster maps

FIGURE 9-32. Comparison of hard and fuzzy clustering results. (Software code and example courtesy of Te-shen Liang and Ho-yuen Pang, University of Arizona.)

These parameters are then used to define the fuzzy membership function as a modified Gaussian distribution,

$$U_l = P_i^*(\boldsymbol{DN}) / \sum_{l=1}^{C} P_i^*(\boldsymbol{DN}) \qquad (9\text{-}42)$$

where,

$$P_l^*(\boldsymbol{DN}) = \frac{1}{\left|\boldsymbol{C}_l^*\right|^{1/2}(2\pi)^{K/2}} e^{-(\boldsymbol{DN}-\boldsymbol{\mu}_l^*)^T \boldsymbol{C}_l^{*-1}(\boldsymbol{DN}-\boldsymbol{\mu}_l^*)/2} , \qquad (9\text{-}43)$$

which is the normal distribution in *K* dimensions (Chapter 4).

Training data for a fuzzy supervised classification need not be homogeneous in a single class; if some pixels have known mixtures, the membership function can be used to calculate weighted fuzzy means and covariances. On the other hand, if the training data are pure in each class, the fuzzy means and covariances are the same as those in a conventional hard classification. The only difference then is that we interpret Eq. (9-42), given Eq. (9-43), as the fuzzy membership partitions of feature space. Fuzzy membership grades for each pixel can therefore be calculated from the *a posteriori* probabilities of a conventional hard maximum-likelihood classification.

9.9   Hyperspectral Image Analysis

While any of the multispectral classification techniques may be directly extended to hyperspectral imagery, that may not be desirable because

- computation costs in K dimensions are high
- more training data are required
- traditional classifiers do not necessarily exploit the greater information content of hyperspectral data

In this section we look at some of the unique tools that have been developed for hyperspectral image analysis.

9.9.1 Visualization of the Image Cube

Visualization is a persistent problem in the analysis of hyperspectral image data. The data are not only voluminous, but multidimensional (Chapter 1). One technique is to extract a line of pixels from a hyperspectral image and display the spectral values of each pixel as greylevels, resulting in an array where the *x*-coordinate is the pixel number within the line and the *y*-coordinate is the wavelength band (Fig. 9-33). This display is a *spatial spectrogram*. These at-sensor radiance data have been normalized to make the spectral information visible from 400 to 2400 nm (see Chapter 7).

Visualization of class spectral signatures is likewise difficult for hyperspectral data. An effective tool for second-order statistics is the *statistics image* (Benediktsson and Swain, 1992; Lee and Landgrebe, 1993). Examples from the Palo Alto AVIRIS data (Plate 1-3) are shown in Plate 9-3. The class correlation matrix is displayed as a pseudo-colored map; the colors indicate the degree of correlation between bands, from negative to positive values. The mean spectrum with the standard deviation in each band for the training data is also displayed. The differences among the class means and correlation matrices are clearly evident. The correlation matrix can help distinguish among classes with similar mean signatures and some studies indicate that the maximum-likelihood classifier may be superior to the nearest-mean classifier for hyperspectral data because of information contained in the covariances (Lee and Landgrebe, 1993).

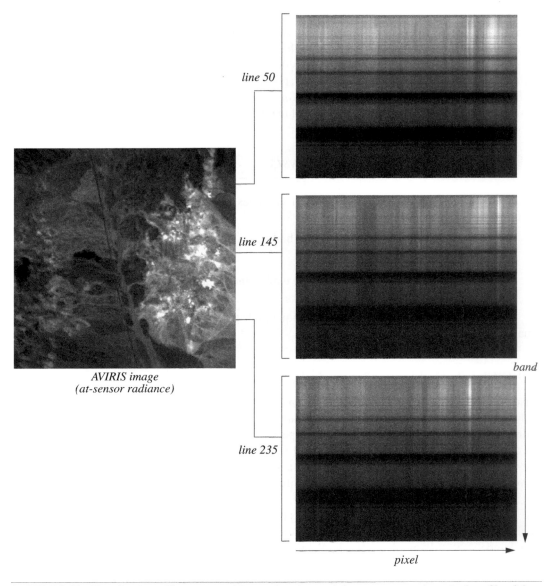

AVIRIS image
(at-sensor radiance)

FIGURE 9-33. *Visualization of hyperspectral image data by spatial spectrograms. A vertical profile of the display is the spectrum measured for the pixel in that column of the given line. These displays are equivalent to horizontal slices through the hyperspectral image cube (Fig. 1-7). Similar displays could be made along image columns or any arbitrary image profile. The dark horizontal lines are the atmospheric absorption bands. The display is much like the visualization of a single line of BIL-formatted data (Fig. 1-23). The same format is also produced directly by a 2-D array imaging spectrometer.*

The large volume of data available permits unique dynamic visualizations, such as *spectral movies*, where the image bands are displayed sequentially in rapid succession. An eight-second "movie" results from 240 bands displayed at 30 bands/second (video frame rate). The main benefits from such a display are in visual detection of unique signatures and in rapid previewing of the data for bad bands.

9.9.2 Training for Classification

For any classifier, a sufficient number of training pixels must be used to estimate the class signature properties accurately. If a maximum-likelihood classifier is used and Gaussian class distributions are assumed, the class sample mean vectors and covariance matrices must be calculated. If K spectral or other features are used, the training set for each class must contain at least $K + 1$ pixels in order to calculate the sample covariance matrix. To obtain reliable class statistics, however, 10 to 100 training pixels *per class, per feature* are typically needed (Swain and Davis, 1978). The number of training pixels required for a given signature accuracy increases if there is large within-class variability.

This requirement is not too severe for multispectral imagery, where we have only a small number of bands. With hyperspectral imagery, however, it may be impossible to meet for some classes of limited spatial extent. Also, as the dimensionality of feature space increases, an accuracy degradation known as the *Hughes phenomenon* can require more training samples (Landgrebe, 2003). Techniques for mitigating this situation have been proposed (Shahshahani and Landgrebe, 1994; Hoffbeck and Landgrebe, 1996).

9.9.3 Feature Extraction from Hyperspectral Data

The order of magnitude higher spectral resolution of hyperspectral imagery compared to multispectral imagery creates opportunities for new feature-extraction techniques, some examples of which are described here.

Image residuals

One of the earliest feature-extraction specifications for hyperspectral data was the calculation of image "residual" spectra for mineral detection and identification (Marsh and McKeon, 1983). The intent is to remove the external factor of topographic shading and to emphasize the absorption bands of different minerals relative to an average signature without absorption features (see the discussion in Chapter 7). The process is:

1. *Divide the spectrum of each pixel by a "reference" band*, chosen to be relatively free of surface absorption features. This normalization insures that every pixel's spectral value in the reference band will be one. It also achieves topographic shading suppression (Chapter 5), particularly in the SWIR where atmospheric scattering is not a factor.

2. Calculate the average normalized spectrum for the entire scene and subtract it from the normalized spectrum produced in step 1. Surface absorption features that are below the average spectrum will now appear as negative "residuals." This step will be effective if the occurrence of absorption minerals is a relatively small proportion of the full scene. Otherwise, the absorption features would influence the mean spectrum and not appear as residuals.

This feature-extraction technique is illustrated with AVIRIS data of Cuprite, Nevada (Plate 7-1). This area is highly mineralized as described in Chapter 7, containing several mineral types with distinguishing absorption bands at different wavelengths, particularly in the SWIR. For normalization, the band at 2.04μm is used as in Marsh and McKeon (1983). The residual spectra were calculated for each pixel as described above. A color composite of the residual images at 2.3, 2.2, and 2.1μm is shown in Plate 9-3. The different colors in the residual image correspond well to known occurrences of alunite, buddingtonite, and kaolinite (Kruse *et al.*, 1990; Abrams and Hook, 1995).

Absorption-band parameters

The high spectral resolution of imaging spectrometer systems creates the opportunity to identify some materials by their absorption-band characteristics. For single absorption bands, the parameters of depth, width,, and position can be defined (Fig. 9-34). The existence of an absorption band at a particular wavelength can be declared whenever the depth exceeds a certain threshold (Rubin, 1993); the width and position are then calculated, if desired. These features, as measured from hyperspectral imagery, can be compared to the same features derived from laboratory reflectance spectra for identification.

Spectral derivative ratios

The derivative of a function tends to emphasize changes and suppress the mean level. This fact has been used to develop a technique that reduces the effects of atmospheric scattering and absorption on spectral signatures (Philpot, 1991). With the aid of a simple radiometric model for remote-sensing data, it can be shown that the ratio of any-order derivative of the at-sensor radiance data at two wavelengths approximately equals the ratio of the same-order derivative of the spectral reflectance,

$$\frac{d^n L/d\lambda^n\big|_{\lambda 1}}{d^n L/d\lambda^n\big|_{\lambda 2}} \approx \frac{d^n \rho/d\lambda^n\big|_{\lambda 1}}{d^n \rho/d\lambda^n\big|_{\lambda 2}}. \tag{9-44}$$

For hyperspectral data collected in discrete wavelength bands, the first three continuous derivatives at wavelength λ can be approximated by the discrete derivatives,

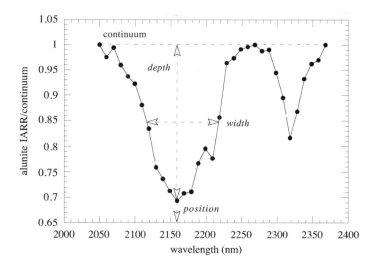

FIGURE 9-34. *The definition of three absorption-band parameters. Depth is measured from continuum-corrected Internal Average Relative Reflectance (IARR) data (see Chapter 7). The width is measured at half the band depth, and the position is the wavelength at the band minimum. Note the similarity to the water vapor band depth calculation in Fig. 7-40. (Diagram after Kruse, 1988.)*

$$dL/d\lambda \approx [L(\lambda) - L(\lambda^+)]/\Delta\lambda$$
$$d^2L/d\lambda^2 \approx [L(\lambda^-) - 2L(\lambda) + L(\lambda^+)]/(\Delta\lambda)^2 \qquad (9\text{-}45)$$
$$d^3L/d\lambda^3 \approx [L(\lambda^-) - 3L(\lambda) + 3L(\lambda^+) - L(\lambda^{++})]/(\Delta\lambda)^3$$

where λ^- is the nearest smaller wavelength, λ^+ is the nearest larger wavelength, λ^{++} is the next nearest larger wavelength, and $\Delta\lambda$ is the wavelength interval. Philpot determined the spectral ranges over which Eq. (9-44) was valid for sample water and leaf spectra. The value of the approach is that radiance data, uncalibrated for the atmosphere, may be used within the spectral regions of validity of this technique to find spectral derivative signatures that will match those from reflectance data.

Spectral fingerprints

An interesting feature-extraction technique for hyperspectral data was developed in a series of papers by Piech and Piech (1987, 1989). The basic concept is to locate the local points of inflection (maximum slope) in the spectral curve using scale-space filtering (Chapter 6). Absorption features lead to characteristic patterns in scale-space, termed "fingerprints." If the data have not been corrected for the atmosphere, then atmospheric absorption bands will be present in the fingerprints, along with bands due to material properties. If the data are corrected for atmospheric absorption, only material-related absorption features will be indicated.

A spectral fingerprint is calculated by convolving the spectrum with LoG filters of different widths and plotting the zero-crossings on a graph of the filter's sigma value (Eq. (6-44)) versus wavelength. As the scale (σ) increases, the number of zero-crossings cannot increase. This is a *scale-space* plot as originally defined in (Witkin, 1983); examples of the application of zero-crossings to boundary detection are shown in Chapter 6. The logarithm of the scale at which a fingerprint loop closes is linear with the logarithm of the area of the absorption feature in the original spectrum, a relationship reminiscent of fractals (Piech and Piech, 1987).

9.9.4 Classification Algorithms for Hyperspectral Data

In principle, any of the conventional algorithms described earlier for multispectral data can be directly applied to hyperspectral data. There are no theoretical limitations on the number of bands (or features) used by any of these algorithms. However, algorithms such as maximum-likelihood, even with efficiency improvements (Bolstad and Lillesand, 1991; Lee and Landgrebe, 1991; Jia and Richards, 1994), appear suddenly inefficient when applied to a 200-band hyperspectral image! A number of special classification tools have been developed specifically for hyperspectral imagery. Their structure is not only driven by a need for efficiency, but also by different types of pattern recognition made possible by the high-resolution spectral data.

Large, multiclass classifications, as commonly performed on TM or SPOT data, are computationally expensive for hyperspectral imagery. Therefore, three restricted classification modes have been described for imaging spectrometer data (Mazer *et al.*, 1988):

- Comparison of the spectrum of a single pixel, or the average of a group of pixels, to all pixels in the image. This is like a traditional supervised classification with a set of training pixels for each class. The data need not be calibrated to reflectance.

- Match the spectra of all image pixels to a single spectrum from a material reflectance library. This mode is used to locate all occurrences of a particular material in the data and requires calibration, or at least partial normalization, to reflectance.

- Match a single spectrum from the remote-sensing data to the spectra in the reflectance library. In this case, we want to find materials that match the given image spectrum. Remote-sensing data calibration is also required here.

It is desirable to exclude the atmospheric water vapor absorption bands from spectral classifications. This will reduce computation time and avoid classification degradation from low *SNR* bands. It is also routine practice to exclude any bands that are in the overlapping spectral regions of two spectrometers, as occurs in AVIRIS and Hyperion data (Fig. 3-24). Some software maintains a user-controlled "bad bands" list for this purpose.

reflectance fingerprints

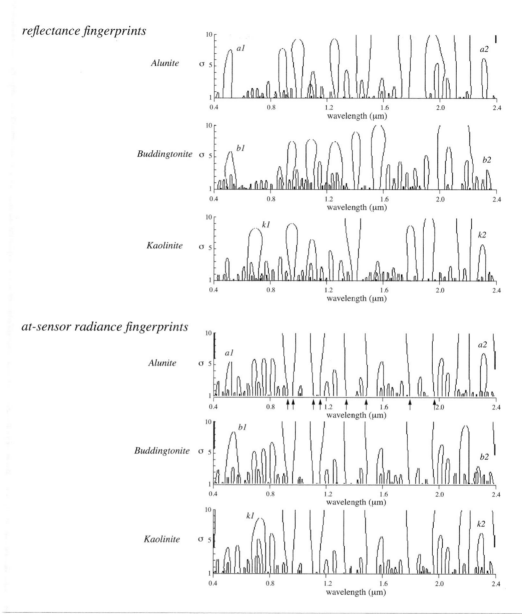

at-sensor radiance fingerprints

FIGURE 9-35.   *Spectral fingerprints for three mineral reflectance spectra (data from Fig. 1-8) and corresponding radiance data from the AVIRIS image of Cuprite in Plate 7-3. Some characteristic absorption signatures are noted for each mineral in reflectance and radiance. The closed loops indicate an absorption feature; loops that have not closed at a σ of 10 will close for a larger value of σ. The zero-crossings arising from atmospheric absorption bands in the radiance data are clearly evident as similar features in each fingerprint (major bands are indicated by arrows; compare their wavelengths with those in Fig. 2-4).*

Binary encoding

The large size of hyperspectral images (about 140 MB for an AVIRIS image) and memory limits on computers prompted early interest in techniques for simultaneous data reduction and pattern matching. Binary encoding of the spectral dimension was one approach (Mazer *et al.*, 1988; Jia and Richards, 1993).

To code the spectral radiance, a single DN threshold is specified and values above the threshold are coded as one and values below as zero. A single bit can then be used in each band to code the spectrum. To illustrate, the radiance spectra from an AVIRIS image (Plate 1-3) are used in Fig. 9-36. A threshold of 700 captures the major absorption features, but soil and the bright roof of a building are indistinguishable. To aid discrimination, the derivative of the spectrum can also be encoded to zero or one, depending on whether it is negative or positive. The coded features can be made less sensitive to external factors of solar irradiance and the atmosphere if the coding is done with respect to the local spectral mean (Mazer *et al.*, 1988). The signature representation is also improved by using multiple thresholds with, however, an increase in computation and the amount of data in the compressed features (Jia and Richards, 1993).

The coded spectra can be compared bit-by-bit using the *Hamming distance*, which is defined as the number of bits that are different in two binary numbers. The Hamming distances for our example are given in Table 9-9. We see that the building and soil signatures in fact differ by one bit, which is nevertheless probably below the noise level in the data. A minimum distance threshold is easily used to reduce noise effects in the classification.

Spectral-angle mapping

This classifier is like a nearest-mean classifier, using spectral-angle distance (the ANG distance measure defined in Table 9-3). Although originally developed for hyperspectral data (Kruse *et al.*, 1993), it does not use any of the special characteristics of that data and can be also applied to multispectral data. The spectral-angle distance is independent of the magnitude of the spectral vectors and, therefore, is insensitive to topographic variations (see the discussion in Sect. 4.7.1 and Fig. 4-37). The classifier can, therefore, be applied to remote-sensing data that have not been corrected for topography and facilitates comparison of these data to laboratory reflectance spectra. It has been compared to spectral unmixing using an AVIRIS spectral library (Dennison *et al.*, 2004).

A general 2-D classification example is shown in Fig. 9-37. Classes b and c might represent dark and light soil, or the same soil with variations in topographic shading. In any event, they cannot be distinguished because their mean signatures lie very close to, and the class distributions are aligned with, the decision boundary.

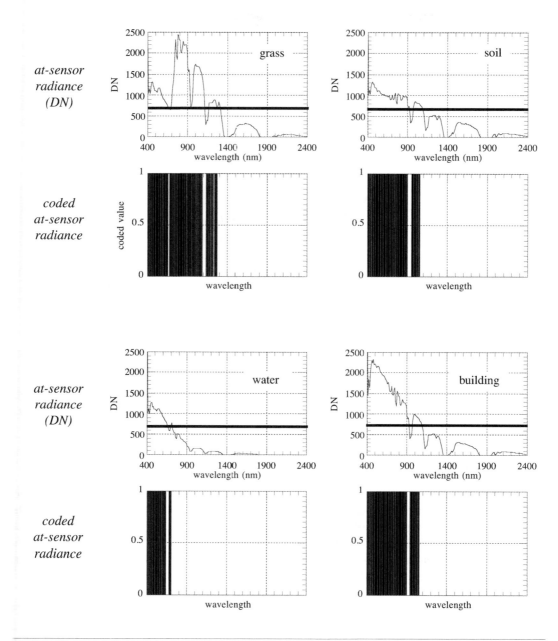

FIGURE 9-36. *Binary encoding of the spectral radiance for four classes in the AVIRIS Palo Alto scene of Plate 1-3. The DN threshold is 700.*

TABLE 9-9. *Hamming distance table for the binary-coded spectral classes of Fig. 9-36. The matrix is symmetric.*

	grass	soil	water	building
grass	0	20	58	19
soil		0	38	1
water			0	39
building				0

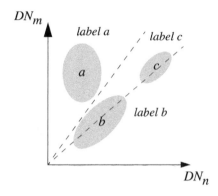

FIGURE 9-37. *The spectral-angle classifier decision boundaries. Classes b and c cannot be distinguished. A threshold on the angular distance can be used to "tighten" the classification and exclude outliers, just as in the level-slice and nearest-mean classifiers.*

Orthogonal Subspace Projection (OSP)

This technique, originally derived in the context of maximizing signal-to-noise in detection of spectral signatures (Harsanyi and Chang, 1994), is equivalent to traditional spectral unmixing (Settle, 1996). The L spectral signatures of interest constitute a $K \times L$ matrix E, which may be thought of as the endmember matrix described earlier (Eq. (9-24)). E is assumed to consist of two parts, the first $L - 1$ columns U containing $L - 1$ endmember vectors and the last column containing a particular spectral signature of interest d. The optimal classification operator is then given by,

$$q^T = d^T(I - UU^{\#})$$ (9-46)

where $U^{\#}$ is the pseudoinverse of U,

$$U^{\#} = (U^TU)^{-1}U^T,$$ (9-47)

as used earlier for unmixing (Eq. (9-32)). The matrix $I - UU^{\#}$ is a *projection matrix*, P. The classifier is applied as a matrix-vector operator on an unknown pixel vector DN,

$$\alpha_P = \beta q^T DN . \tag{9-48}$$

The classifier can be viewed as "projecting" the unknown data vector onto the particular vector of interest, d, while simultaneously "nullifying" the other class signatures (Harsanyi and Chang, 1994). Larger values of α_P indicate a better match between d and DN. The scalar β is a normalizing factor equal to

$$\beta = (d^T P d)^{-1} . \tag{9-49}$$

Although derived from a different viewpoint, this technique is mathematically equivalent to the unconstrained, least-squares fraction estimate, α_P, of the class whose endmember is d (Settle, 1996). Further discussion of the relationship between the two approaches is provided in Chang (1998).

9.10 Summary

A wide assortment of techniques for thematic classification of remote-sensing imagery have been presented. They range from heuristic, intuitive approaches to theoretically well-founded, statistical pattern-recognition techniques. The important aspects of this chapter are:

- Thematic classification methods are either heuristic (nonparametric) or based on statistical distribution assumptions (parametric).

- It is not possible to declare one classifier better than another for all applications; their performance is strongly data-dependent if there is any overlap of class signatures. If the class distributions are Gaussian, the maximum-likelihood classifier results in minimum total misclassification error.

- Soft classifications have notable advantages over hard classifications, among them the ability to discriminate physical class continua and to provide estimates of subpixel class component fractions.

- Hyperspectral imagery from imaging spectrometers offers opportunities for new classification algorithms that can exploit high spectral resolution information.

In closing the discussion on thematic classification, we note that there are a number of other, "higher" level approaches. These methods may use external (to the imagery) knowledge to improve labeling of scene components and often include spatial context information (Tilton *et al.*, 1982; Wharton, 1982; Clément *et al.*, 1993). Approaches range from rule-based, "expert" systems (Nagao and Matsuyama, 1980; Wang *et al.*, 1983; McKeown *et al.*, 1985; Wharton, 1987; Goldberg *et al.*, 1988; Mehldau and Schowengerdt, 1990; Srinivasan and Richards, 1990; Wang, 1993) to evidential reasoning, such as the Dempster-Shafer theory (Lee *et al.*, 1987); a review of some of the work in these areas is provided in Argialas and Harlow (1990). Some algorithms can be applied to a label map produced by conventional classification. Such classification schemes, although they tend to produce large and complex analysis systems, appear promising for many applications, particularly those involving higher resolution imagery. They have not been discussed in detail here because they use high-level descriptive models for image content, rather than the physical models emphasized throughout this book.

9.11   Exercises

Ex 9-1. Select a Level II category from the Anderson classification scheme (Table 9-2) and expand it to Level III, using your own ideas of appropriate classes.

Ex 9-2. Explain why the separability of two distributions in a 1-D feature space depends not only on their means, but also on their standard deviations.

Ex 9-3. Suppose you have a three-band image and classify it into two classes using the nearest-mean algorithm with the Euclidean distance measure. Show mathematically that the intersection of the decision plane with the plane of any two bands is a line, but not necessarily the same line as the decision boundary obtained if only those two bands were used in the classification. What is required to make the two lines the same?

Ex 9-4. Calculate the *DN*s for the two decision boundaries in Fig. 9-17.

Ex 9-5. Derive the explicit mathematical form for the maximum-likelihood decision boundary between two classes, under the normal distribution assumption, in a two-dimensional feature space. What family of mathematical functions does the solution belong to? Specify some reasonable parameter values and plot the decision boundaries for the cases of:

  • equal class means and unequal, diagonal class covariance matrices and

  • unequal class means and equal class covariance matrices.

Ex 9-6. Explain how Eq. (9-7) and Eq. (9-8) are determined.

Ex 9-7. Explain how the derivative of the activation function (Fig. 9-11) influences the iterative adjustment of weights in the back-propagation algorithm.

Ex 9-8. Compare Fig. 9-33 to Fig. 2-4 and identify each absorption band by its center wavelength.

Ex 9-9. Develop a simple *nonlinear* mixing model similar to the linear model in Fig. 9-27. Assume each *GIFOV* consists of a partially-transmitting vegetation canopy and a soil background.

APPENDIX A

Sensor Acronyms

TABLE A-1. *Some of the more common remote sensor system acronyms. The reference used is not necessarily the first for that sensor, nor the most complete, but is in either an archival journal or readily available conference proceedings .*

acronym	name	reference
ADEOS	ADvanced Earth Observing Satellite	Kramer, 2002
AIS	Airborne Imaging Spectrometer	Vane *et al.*, 1984
ALI	Advanced Land Imager	Ungar *et al.*, 2003
AOCI	Airborne Ocean Color Imager	Wrigley *et al.*, 1992
APT	Automatic Picture Transmission	Bonner, 1969
ASAS	Advanced Solid-State Array Spectroradiometer	Irons *et al.*, 1991
ASTER	Advanced Spaceborne Thermal Emission and Reflection Radiometer	Asrar and Greenstone, 1995
AVHRR	Advanced Very High Resolution Radiometer	Kramer, 2002
AVIRIS	Airborne Visible/InfraRed Imaging Spectrometer	Porter and Enmark, 1987
CZCS	Coastal Zone Color Scanner	Kramer, 2002
ETM+	Enhanced Thematic Mapper Plus	Kramer, 2002
GOES	Geostationary Operational Environmental Satellite	Kramer, 2002
HCMM	Heat Capacity Mapping Mission	Short and Stuart, 1982
HSI	HyperSpectral Imager	Kramer, 2002
HYDICE	HYperspectral Digital Imagery Collection Experiment	Basedow *et al.*, 1995
HyMap	Hyperspectral Mapper	Huang *et al.*, 2004
IRS	India Remote sensing Satellite	Kramer, 2002
JERS	Japanese Earth Resources Satellite (Fuyo-1 post-launch)	Nishidai, 1993
LISS	Linear Self Scanning Sensor	IRS
MAS	Modis Airborne Simulator	Myers and Arvesen, 1995
MERIS	MEdium Resolution Imaging Spectrometer	Curran and Steele, 2005
MISR	Muli-angle Imaging SpectroRadiometer	Diner *et al.*, 1989
MODIS	Moderate Resolution Imaging Spectroradiometer	Salomonson *et al.*, 1989
MSS	Multispectral Scanner System	Lansing and Cline, 1975
MTI	Multispectral Thermal Imager	Szymanski and Weber, 2005
RBV	Return Beam Vidicon	Kramer, 2002

TABLE A-1. *Some of the more common remote sensor system acronyms. The reference used is not necessarily the first for that sensor, nor the most complete, but is in either an archival journal or readily available conference proceedings (continued).*

acronym	name	reference
SeaWiFS	Sea-viewing Wide Field-of-view Sensor	Barnes and Holmes, 1993
SPOT	Systeme Probatoire d'Observation	Chevrel *et al.*, 1981
SSM/I	Special Sensor Microwave/Imager	Hollinger *et al.*, 1990
TIMS	Thermal Infrared Multispectral Scanner	Kahle and Goetz, 1983
TM	Thematic Mapper	Engel and Weinstein, 1983
TMS	Thematic Mapper Simulator	Myers and Arvesen, 1995
VHRR	Very-High Resolution Radiometer	Kramer, 2002
VIIRS	Visible/Infrared Imager/Radiometer Suite	NPOESS
WiFS	Wide Field Sensor	IRS-1C

APPENDIX B

1-D and 2-D Functions

Several functions were used in Chapter 3 to model components of the Point Spread Function (*PSF*) of a sensor. In particular, the *rect* function is defined as follows,

$$rect(x/W) = \begin{array}{ll} 0 & |x/W| > 1/2 \\ 1/2 & |x/W| = 1/2 \\ 1 & |x/W| < 1/2 \end{array} \tag{B-1}$$

which is a square pulse of width W and amplitude one (Fig. B-1). The defined value of the function at the edges is not critical (Bracewell, 2004); we have used the definition in Gaskill (1978).

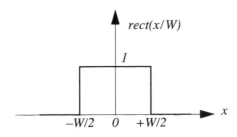

FIGURE B-1. The 1-D square pulse, or rectangle, function.

A separable function is defined as a two-dimensional function that is a product of two, one-dimensional functions in orthogonal directions,

$$f(x, y) = f_1(x)f_2(y) \tag{B-2}$$

The 2-D, separable *rect* function is therefore,

$$rect(x/A, y/B) = rect(x/A)rect(y/B) \ , \tag{B-3}$$

which is a rectangular "box" (Fig. B-2). The 2-D *rect* function is a reasonable model for the spatial response of a single sensor detector element, excluding any nonuniform sensitivity across the detector or detector-to-detector "cross-talk" between elements.

Another useful function is the Gaussian,

$$gaus(x/W) = e^{-x^2/2W^2} \tag{B-4}$$

and its 2-D, separable form,

$$gaus(x/A, y/B) = e^{-x^2/2A^2}e^{-y^2/2B^2} \ . \tag{B-5}$$

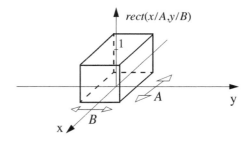

FIGURE B-2. The 2-D square pulse, or box, function.

This function is particularly interesting because it is not only separable, but also rotationally symmetric (isotropic) if A equals B. The *gaus* function is commonly used to model the optical spatial response of a sensor. In that case, it should be normalized to unit volume under the function, as in Chapter 3.

Other examples of 2-D functions and their characteristics in separable and isotropic form are discussed in relation to modeling of spatial correlation in Chapter 4. We note that the definition of a separable function is analogous to the definition of independent statistical variables, also discussed in Chapter 4. A multivariate distribution function is separable and equal to the product of the individual distribution functions if the constituent variables are independent.

Separable functions in the spatial domain are transformed to the Fourier frequency domain as separable functions (Chapter 6). This convenient property makes it possible to treat many common 2-D functions as the product of two 1-D functions in both domains. For example, the Fourier transform of the 2-D *rect* function in Eq. (B-3) is a 2-D separable *sinc* function,

$$|A||B|sinc(Au,Bv) \;=\; |A||B|sinc(Au)sinc(Bv) \;=\; |A||B|\left[\frac{\sin(\pi Au)}{\pi Au} \cdot \frac{\sin(\pi Bv)}{\pi Bv}\right] \qquad \text{(B-6)}$$

This 2-D *sinc* function has zeros at $u = \pm 1/A$, $\pm 2/A$, . . . and at $v = \pm 1/B$, $\pm 2/B$ . . . and extends to infinity in both directions. The 1-D *sinc* function is shown in Fig. 7-14 and the 2-D *sinc* function is shown in Fig. B-3. This example illustrates another fundamental property of Fourier transforms, namely that the width of a function in one domain is the reciprocal of the width in the other domain. A *rect* function A units wide in the spatial domain transforms to a *sinc* function 1/A units wide (to the first zeros) in the spatial frequency domain. This and other useful properties are listed in Table B-1.

In Chapter 6, we discussed the modeling of the ALI sensor spatial imaging performance in the Fourier spatial frequency domain. The various transfer functions used there are Fourier transforms of the respective spatial *PSF*. The primary components are the optics, modeled by a Gaussian *PSF*, and the detector and image motion, modeled by square pulse *PSF*s; the Fourier transform pair for the electronics component can be found in Gaskill (1978). We can use the properties of 2-D

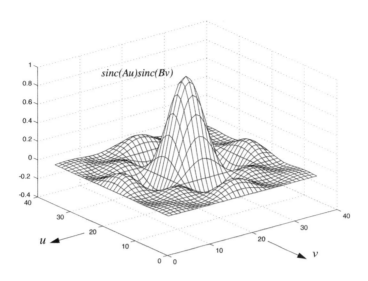

FIGURE B-3. The 2-D sinc function. The oscillatory behavior extends to infinity in both directions, but the amplitude decreases.

TABLE B-1. 2-D Fourier transform properties

name	f(x,y)	F(u,v)
separability	$f_1(x)f_2(y)$	$F_1(u)F_2(v)$
scaling	$f(x/a, y/b)$	$|a||b|F(au, bv)$
shifting	$f(x \pm a, y \pm b)$	$e^{\pm j2\pi(au + bv)}F(u, v)$
linearity (superposition)	$af_1(x, y) + bf_2(x, y)$	$aF_1(u, v) + bF_2(u, v)$
convolution	$f_1(x, y) * f_2(x, y)$	$F_1(u, v) \cdot F_2(u, v)$

functions and their Fourier transforms to generate Table B-2. The *MTF* is the magnitude of the complex *TF*, but for some models, such as the Gaussian used for the optics and the exponential used for the electronics, there is no difference between the *TF* and the *MTF* since the *TF* is a real, all-positive function. For the *sinc* function model used for the detector and image motion, however, the distinction between the *TF* and *MTF* is important.

For simplicity, a separable model for the electronics *PSF* and *TF* is assumed in Table B-2; an isotropic model is used by Hearn (2000),

$$TF(u, v) = e^{-\left|\rho/\rho_0\right|} = e^{-\left|\sqrt{u^2+v^2}/\rho_0\right|}. \tag{B-7}$$

The width of the ALI electronics *PSF* is much smaller than that of the other component *PSF*s, or equivalently, the width of the electronics *MTF* is much larger than that of the other component *MTF*s (see Chapter 6), so the difference between a separable and an isotropic function is negligible in this case. The general properties of separable, isotropic and anisotropic functions are discussed in Chapter 4 in the context of spatial statistics models.

TABLE B-2. *Spatial and frequency domain functions used to model the ALI sensor in Chapter 6.*

component	spatial domain PSF model	frequency domain TF model	MTF
optics	$\dfrac{1}{ab} \cdot e^{-x^2/a^2} e^{-y^2/b^2}$	$e^{-a^2 u^2} e^{-b^2 v^2}$	$\left\| e^{-a^2 u^2} e^{-b^2 v^2} \right\|$
detector	$\dfrac{rect(x/w)rect(y/w)}{w^2}$	$sinc(wu)sinc(wv)$	$\|sinc(wu)sinc(wv)\|$
image motion	$\dfrac{rect(y/s)}{s}$	$sinc(sv)$	$\|sinc(sv)\|$
electronics	$\dfrac{2v_0}{1+(2\pi v_0 x)^2} \cdot \dfrac{2v_0}{1+(2\pi v_0 y)^2}$	$e^{-\|u/v_0\|} e^{-\|v/v_0\|}$	$\left\| e^{-\|u/v_0\|} e^{-\|v/v_0\|} \right\|$

This brief Appendix only touches on the most important aspects of these topics for our purposes; they are covered in much greater detail in many electrical engineering, image processing, and optics textbooks. Several recommended resources are Gaskill (1978), Bracewell (2004), Gonzalez and Woods (2002), and Castleman (1996).

References

Color Plates

Barthel, K. U. (2005). *Color inspector 3d*. http://www.f4.fhtw-berlin.de/~barthel/ImageJ/ColorInspector//help.htm. Berlin.

Giglio, L., J. Descloitres, C. O. Justice and Y. J. Kaufman (2003). "An enhanced contextual fire detection algorithm for MODIS." *Remote Sensing of Environment* **87**: 273-282.

Goode, J. P. (1925). "The Homolosine projection: a new device for portraying the Earth's surface entire." *Annals of the Association of American Geographers* **15**: 119-125.

Justice, C. O., L. Giglio, S. Korontzi, J. Owens, J. T. Morisette, D. Roy, J. Descloitres, S. Alleaume, F. Petitcolin and Y. Kaufman (2002). "The MODIS fire products." *Remote Sensing of Environment* **83**: 244-262.

Paola, J. D. and R. A. Schowengerdt (1995b). "A detailed comparison of backpropagation neural network and maximum-likelihood classifiers for urban land use classification." *IEEE Transactions on Geoscience and Remote Sensing* **33**(4): 981-996.

Qu, Z., B. C. Kindel and A. F. H. Goetz (2003). "The high accuracy atmospheric correction for hyperspectral data (HATCH) model." *IEEE Transactions on Geoscience and Remote Sensing* **41**(6): 1223-1231.

Richter, R. and A. Muller (2005). "De-shadowing of satellite/airborne imagery." *International Journal of Remote Sensing* **26**(15): 3137-3148.

Shepherd, J. D. and J. R. Dymond (2003). "Correcting satellite imagery for the variance of reflectance and illumination with topography." *International Journal of Remote Sensing* **24**(17): 3503-3514.

Steinwand, D. R. (1994). "Mapping raster imagery to the Interrupted Goode Homolsine projection." *International Journal of Remote Sensing* **15**(17): 3463-3471.

CHAPTER 1 *The Nature of Remote Sensing*

Asrar, G. and R. Greenstone (1995). MTPE/EOS Reference Handbook. Greenbelt, MD, NASA/Goddard Space Flight Center, No. NP-215.

Avery, T. E. and G. L. Berlin (1992). *Fundamentals of Remote Sensing and Airphoto Interpretation*. Fifth Edition. New York, NY: Macmillan Publishing Company, 472 p.

Badhwar, G. D., J. G. Carnes and W. W. Austin (1982). "Use of Landsat-derived temporal profiles for corn-soybean feature extraction and classification." *Remote Sensing of Environment* **12**(1): 57-79.

Bowker, D. E., R. E. Davis, D. L. Myrick, K. Stacy, and W. T. Jones (1985). Spectral Reflectances of Natural Targets for Use in Remote Sensing Studies, NASA, No. Reference Publication 1139.

Campbell, J. B. (2002). *Introduction to Remote Sensing*. Third Edition. New York, NY: The Guilford Press, 620 p.

Clark, B. P. (1990). "Landsat Thematic Mapper data production: a history of bulk image processing." *Photogrammetric Engineering and Remote Sensing* **56**(4): 447-451.

Clark, R. N., G. A. Swayze, A. J. Gallagher, T. V. V. King, and W. M. Calvin (1993). The U. S. Geological Survey Digital Spectral Library: Version 1: 0.2 to 3.0 microns. Denver, CO, U. S. Geological Survey, No. 93-592.

Colwell, R. N., Editor. (1983). *Manual of Remote Sensing*. Falls Church, VA: American Society for Photogrammetry and Remote Sensing.

Curlander, J. C. and R. N. McDonough (1991). *Synthetic Aperture Radar - Systems and Signal Processing*. New York, NY: John Wiley & Sons, 647 p.

Elachi, C. (1988). *Spaceborne Radar Remote Sensing: Applications and Techniques*. New York, NY: IEEE Press, 255 p.

EOSAT (1993). Fast Format Document. Lanham, Maryland: EOSAT.

Filiberti, D., S. Marsh and R. Schowengerdt (1994). "Synthesis of high spatial and spectral resolution imagery from multiple image sources." *Optical Engineering* **33**(8): 2520-2528.

Fritz, L. W. (1996). "The era of commercial earth observation satellites." *Photogrammetric Engineering and Remote Sensing* **62**(1): 39-45.

Goetz, A., G. Vane, J. E. Solomon, and B. N. Rock (1985). "Imaging spectrometry for Earth remote sensing." *Science* **228**(4704): 1147-1153.

Haralick, R. M., C. A. Hlavka, R. Yokoyama, and S. M. Carlyle (1980). "Spectral-temporal classification using vegetation phenology." *IEEE Transactions on Geoscience and Remote Sensing* **GE-18**: 167-174.

Haydn, R., G. W. Dalke, J. Henkel, and J. E. Bare (1982). "Application of the IHS color transform to the processing of multisensor data and image enhancement." In *International Symposium on Remote Sensing of Arid and Semi-Arid Lands*, Cairo, Egypt, Environmental Research Institute of Michigan: 599-616.

Hollinger, J. P., R. Lo, G. Poe, R. Savage and J. Peirce (1987). Special Sensor Microwave/Imager User's Guide. Washington, D.C., Naval Research Laboratory.

Hollinger, J. P., J. L. Peirce, and G. A. Poe (1990). "SSM/I instrument evaluation." *IEEE Transactions on Geoscience and Remote Sensing* **28**(5): 781-790.

Hook, S. J. (1998). *JPL Spectral Library*. NASA/JPL.

Huete, A., K. Didan, T. Miura, E. P. Rodriguez, X. Gao and L. G. Ferreira (2002). "Overview of the radiometric and biophysical performance of the MODIS vegetation indices." *Remote Sensing of Environment* **83**: 195-213.

Hunt, G. R. (1979). "Near-infrared (1.3–2.4μm) spectra of alteration minerals - potential for use in remote sensing." *Geophysics* **44**(12): 1974-1986.

Jensen, J. R. (2004). *Introductory Digital Image Processing – A Remote Sensing Perspective*. Third Edition. Upper Saddle River, NJ: Prentice Hall, 544 p.

Justice, C. O., L. Giglio, S. Korontzi, J. Owens, J. T. Morisette, D. Roy, J. Descloitres, S. Alleaume, F. Petit-colin and Y. Kaufman (2002). "The MODIS fire products." *Remote Sensing of Environment* **83**: 244-262.

Knyazikhin, Y., J. V. Martonchik, R. B. Myneni, D. J. Diner and S. W. Running (1998). "Synergistic algorithm for estimating vegetation canopy leaf area index and fraction of absorbed photosynthetically active radiation from MODIS and MISR data." *Journal of Geophysical Research* **103**(D24): 32257-32275.

Kramer, H. J. (2002). *Observation of the earth and its environment: Survey of missions and sensors*. Fourth Edition. Berlin: Springer–Verlag, 1510 p.

Landgrebe, D. A. (1978). The Quantitative Approach: Concept and Rationale. *Remote Sensing: The Quantitative Approach*. P. H. Swain and S. M. Davis, (Eds.). New York, NY: McGraw-Hill, 1-20.

Landgrebe, D. (1997). "The evolution of Landsat data analysis." *Photogrammetric Engineering and Remote Sensing* **63**(7): 859-867.

Landgrebe, D. A. (2003). *Signal Theory Methods in Multispectral Remote Sensing*. Hoboken, NJ: John Wiley & Sons, Inc., 508 p.

Lillesand, T. M., R. W. Kiefer and J. W. Chipman (2004). *Remote Sensing and Image Interpretation*. Fifth Edition. New York: John Wiley & Sons, 763 p.

Markham, B. L. and J. L. Barker, (Eds.) (1985). *Special LIDQA Issue*. Photogrammetric Engineering and Remote Sensing, American Society for Photogrammetry and Remote Sensing.

Marsh, S. E. and R. J. P. Lyon (1980). "Quantitative relationship of near-surface spectra to Landsat radiometric data." *Remote Sensing of Environment* **10**(4): 241-261.

Mather, P. M. (1999). *Computer Processing of Remotely-Sensed Images: An Introduction*. Second Edition. Chichester, England: John Wiley & Sons, 292 p.

McDonald, R. A. (1995a). "Opening the Cold War sky to the public: declassifying satellite reconnaissance imagery." *Photogrammetric Engineering and Remote Sensing* **LXI**(4): 385-390.

McDonald, R. A. (1995b). "CORONA: Success for space reconnaissance, a look into the Cold War, and a revolution for intelligence." *Photogrammetric Engineering and Remote Sensing* **LXI**(6): 689-719.

Moik, J. G. (1980). *Digital Processing of Remotely Sensed Images*. Washington, D.C.: NASA, U.S. Government Printing Office, 330 p.

NASA (2000). *EOS data products handbook, volume 2*. Greenbelt, MD, NASA Goddard Space Flight Center.

NASA (2004). *EOS data products handbook, volume 1 (revised)*. Greenbelt, MD, NASA Goddard Space Flight Center.

Niblack, W. (1986). *An Introduction to Digital Image Processing*. Prentice Hall International (UK) Ltd, 215 p.

Nishida, K., R. R. Nemani, J. M. Glassy and S. W. Running (2003). "Development of an evapotranspiration index from Aqua/MODIS for monitoring surface moisture status." *IEEE Transactions on Geoscience and Remote Sensing* **41**(2): 493-501.

Platnick, S., M. D. King, S. A. Ackerman, W. P. Menzel, B. A. Baum, J. C. Riedi and R. A. Frey (2003). "The MODIS cloud products: Algorithms and examples from Terra." *IEEE Transactions on Geoscience and Remote Sensing* **41**(2): 459-473.

Rast, M., S. J. Hook, C. D. Elvidge, and R. E. Alley (1991). "An evaluation of techniques for the extraction of mineral absorption features from high spectral resolution remote sensing data." *Photogrammetric Engineering and Remote Sensing* **57**(10): 1303-1309.

Richards, J. A. and X. Jia (1999). *Remote Sensing Digital Image Analysis – An Introduction*. Third Edition. Berlin: Springer-Verlag, 356 p.

Rubin, T. D. (1993). "Spectral mapping with imaging spectrometers." *Photogrammetric Engineering and Remote Sensing* **59**(2): 215-220.

Sabins, F. F. J. (1997). *Remote Sensing – Principles and Interpretation*. Third Edition. New York: W. H. Freeman and Company, 432 p.

Salomonson, V. V., Editor. (1984). Special Issue on Landsat–4. *IEEE Transactions on Geoscience and Remote Sensing*, IEEE.

Salomonson, V. V., J. Barker, and E. Knight (1995). "Spectral characteristics of the Earth Observing System (EOS) Moderate Resolution Imaging Spectroradiometer (MODIS)." In *Imaging Spectrometry*, Orlando, FL, SPIE, vol. 2480: 142-152.

Schaaf, C. B., F. Gao, A. H. Strahler, W. Lucht, X. Li, T. Tsang, N. C. Strugnell, X. Zhang, Y. Jin, J.-P. Muller, P. Lewis, M. Barnsley, P. Hobson, M. Disney, G. Roberts, M. Dunderdale, C. Doll, R. P. d'Entremont, B. Hu, S. Liang, J. L. Privette and D. Roy (2002). "First operational BRDF, albedo nadir reflectance products from MODIS." *Remote Sensing of Environment* **83**: 135-148.

Schetselaar, E. M. (2001). "On preserving spectral balance in image fusion and its advantages for geological image interpretation." *Photogrammetric Engineering and Remote Sensing* **67**(8): 925-934.

Schott, J. R. (1996). *Remote Sensing: The Image Chain Approach*. New York, NY: Oxford University Press, 394 p.

Schowengerdt, R. A. (1983). *Techniques for Image Processing and Classification in Remote Sensing*. Orlando, FL: Academic Press, 249 p.

Slater, P. N. (1980). *Remote Sensing - Optics and Optical Systems*. Reading, MA: Addison-Wesley, 575 p.

SPOTImage (1991). Reston, VA: SPOT Image Corporation.

Storey, J. (2005). Personal communication.

Storey, J. C., M. J. Choate and D. J. Meyer (2004). "A geometric performance assessment of the eo-1 advanced land imager." *IEEE Transactions on Geoscience and Remote Sensing* **42**(3): 602-607.

Swain, P. H. and S. M. Davis, (Eds.) (1978). *Remote Sensing: The Quantitative Approach*. New York, NY: McGraw-Hill, 396 p.

Townshend, J., C. Justice, W. Li, C. Gurney, and J. McManus (1991). "Global land cover classification by remote sensing: present capabilities and future possibilities." *Remote Sensing of Environment* **35**: 243-255.

Twomey, S., C. Bohren and J. Mergenthaler (1986). "Reflectance and albedo differences between wet and dry surfaces." *Applied Optics* **25**(3): 431-437.

USGS (2000). *Landsat-7 Level-0 and Level-1 Data Sets Document*. U.S. Geological Survey.

Vane, G. and A. F. H. Goetz (1988). "Terrestrial imaging spectroscopy." *Remote Sensing of Environment* **24**: 1-29.

Vermote, E. F., N. El Saleous, C. O. Justice, Y. J. Kaufman, J. L. Privette, L. Remer, J. C. Roger and D. Tanré (1997). "Atmospheric correction of visible to middle-infrared EOS-MODIS data over land surfaces: Background, operational algorithm and validation." *Journal of Geophysical Research - Atmospheres* **102**(D14): 17131-17141.

Vetter, R., M. Ali, M. Daily, J. Gabrynowicz, S. Narumalani, K. Nygard, W. Perrizo, P. Ram, S. Reichenbach, *et al.* (1995). "Accessing earth system science data and applications through high-bandwidth networks." *IEEE Journal on Selected Areas in Communications* **13**(5): 793-805.

Way, J. and E. A. Smith (1991). "The evolution of synthetic aperture radar systems and their progression to the EOS SAR." *IEEE Transactions on Geoscience and Remote Sensing* **29**(6): 962-985.

Welch, R. and M. Ehlers (1988). "Cartographic feature extraction with integrated SIR-B and Landsat TM images." *International Journal of Remote Sensing* **9**(5): 873-889.

Wolfe, R. E., M. Nishihama, A. J. Fleig, J. A. Kuyper, D. P. Roy, J. C. Storey and F. S. Patt (2002). "Achieving sub-pixel geolocation accuracy in support of MODIS land science." *Remote Sensing of Environment* **83**: 31-49.

Wong, F. H. and R. Orth (1980). "Registration of SEASAT-LANDSAT composite images to UTM coordinates." In *Proc. Sixth Canadian Symposium on Remote Sensing*, Halifax, Nova Scotia: 161-164.

CHAPTER 2 *Optical Radiation Models*

Berk, A., L. S. Bernstein and D. C. Robertson (1989). MODTRAN: A Moderate Resolution Model for LOWTRAN 7, U. S. Air Force Geophysics Laboratory, No. GL-TR-89-0122.

Boyd, D. S. and F. Petitcolin (2004). "Remote sensing of the terrestrial environment using middle infrared radiation (3.0–5.0 μm)." *International Journal of Remote Sensing* **25**(17): 3343-3368.

Chavez, P. S., Jr. (1988). "An improved dark-object subtraction technique for atmospheric scattering correction of multispectral data." *Remote Sensing of Environment* **24**: 459-479.

Curcio, J. A. (1961). "Evaluation of atmospheric aerosol particle size distribution from scattering measurement in the visible and infrared." *Journal of the Optical Society of America* **51**: 548-551.

Diner, D. J., C. J. Bruegge, J. V. Martonchik, T. P. Ackerman, R. Davies, S. A. W. Gerstl, H. R. Gordon, P. J. Sellers, J. Clark, *et al.* (1989). "MISR: A Multi-angle Imaging SpectroRadiometer for geophysical and climatological research from EOS." *IEEE Transactions on Geoscience and Remote Sensing* **27**(2): 200-214.

Dubayah, R. O. and J. Dozier (1986). "Orthographic terrain views using data derived from digital elevation models." *Photogrammetric Engineering and Remote Sensing* **52**(4): 509-518.

Gao, B.-C., A. F. H. Goetz and W. J. Wiscombe (1993). "Cirrus cloud detection from airborne imaging spectrometer data using the 1.38μm water vapor band." *Geophysical Research Letter* **20**(4): 301-304.

Giles, P. T., M. A. Chapman and S. E. Franklin (1994). "Incorporation of a digital elevation model derived from stereoscopic satellite imagery in automated terrain analysis." *Computers & Geosciences* **20**(4): 441-460.

Goel, N. S. (1988). "Models of vegetation canopy reflectance and their use in estimation of biophysical parameters from reflectance data." *Remote Sensing Reviews* **4**: 1-212.

Horn, B. K. P. (1981). "Hill shading and the reflectance map." *Proceedings of the IEEE* **69**(1): 14-47.

Liang, S. (2004). *Quantitative Remote Sensing of Land Surfaces*. Hoboken, New Jersey: John Wiley & Sons, 534 p.

Mushkin, A., L. K. Balick and A. R. Gillespie (2005). "Extending surface temperature and emissivity retrieval to the mid-infrared (3–5 μm) using the Multispectral Thermal Imager (MTI)." *Remote Sensing of Environment* **98**: 141-151.

Proy, C., D. Tanre and P. Y. Deschamps (1989). "Evaluation of topographic effects in remotely sensed data." *Remote Sensing of Environment* **30**: 21-32.

Schott, J. R. (1996). *Remote Sensing: The Image Chain Approach*. New York, NY: Oxford University Press, 394 p.

Schowengerdt, R. A. (1982). "Enhanced thermal mapping with Landsat and HCMM digital data." In *Proc. Annual American Society of Photogrammetry Convention*, Denver, CO, American Society for Photogrammetry and Remote Sensing: 414-422.

Sjoberg, R. W. and B. K. P. Horn (1983). "Atmospheric effects in satellite imaging of mountainous terrain." *Applied Optics* **22**(11): 1702-1716.

Slater, P. N. (1980). *Remote Sensing–Optics and Optical Systems*. Reading, MA: Addison-Wesley, 575 p.

Slater, P. N. (1996). personal communication.

Teillet, P. M. and G. Fedosejevs (1995). "On the dark target approach to atmospheric correction of remotely sensed data." *Canadian Journal of Remote Sensing* **21**(4): 374-387.

Vermote, E. F., D. Tanre, J. L. Deuze, M. Herman and J.-J. Morcrette (1997). "Second simulation of the satellite signal in the solar spectrum, 6S: An overview." *IEEE Transactions on Geoscience and Remote Sensing* **35**(3): 675-686.

CHAPTER 3 Sensor Models

Anuta, P. E. (1973). Geometric correction of ERTS-1 digital multispectral scanner data, Laboratory for Applications of Remote Sensing, Purdue University, No. Information Note 103073.

Anuta, P. E., L. A. Bartolucci, M. E. Dean, D. F. Lozano, E. Malaret, C. D. McGillem, J. A. Valdes, and C. R. Valenzuela (1984). "LANDSAT-4 MSS and Thematic Mapper data quality and information content analysis." *IEEE Transactions on Geoscience and Remote Sensing* **GE-22**(3): 222-236.

Bachmann, M. and J. Bendix (1992). "An improved algorithm for NOAA-AVHRR image referencing." *International Journal of Remote Sensing* **13**(16): 3205-3215.

Bernstein, R., J. B. Lospiech, H. J. Myers, H. G. Kolsky, and R. D. Lees (1984). "Analysis and processing of Landsat-4 sensor data using advanced image processing techniques and technologies." *IEEE Transactions on Geoscience and Remote Sensing* **GE-22**(3): 192-221.

Blonski, S., M. A. Pagnutti, R. E. Ryan and V. Zanoni (2002). "In-flight edge response measurements for high-spatial-resolution remote sensing systems." In *Earth Observing Systems VII*, San Diego, CA, SPIE, vol. 4814: 317-326.

Borgeson, W. T., R. M. Batson and H. H. Kieffer (1985). "Geometric accuracy of Landsat-4 and Landsat-5 Thematic Mapper images." *Photogrammetric Engineering and Remote Sensing* **51**(12): 1893-1898.

Brush, R. J. H. (1985). "A method for real-time navigation of AVHRR imagery." *IEEE Transactions on Geoscience and Remote Sensing* **GE-23**(6): 876-887.

Brush, R. J. H. (1988). "The navigation of AVHRR imagery." *International Journal of Remote Sensing* **9**(9): 1491-1502.

Bryant, N. A., A. L. Zobrist, R. E. Walker, and B. Gokhman (1985). "An analysis of Landsat Thematic Mapper P-product internal geometry and conformity to Earth surface geometry." *Photogrammetric Engineering and Remote Sensing* **51**(9): 1435-1447.

Chavez, P. S., Jr. (1989). "Use of the variable gain settings on SPOT." *Photogrammetric Engineering and Remote Sensing* **55**(2): 195-201.

Chen, L. C. and L. H. Lee (1993). "Rigorous generation of digital orthophotos from SPOT images." *Photogrammetric Engineering and Remote Sensing* **59**(5): 655-661.

Delvit, J.-M., D. Leger, S. Roques and C. Valorge (2004). "Modulation transfer function estimation from non-specific images." *Optical Engineering* **43**: 1355-1365.

Dereniak, E. L. and G. D. Boreman (1996). *Infrared Detectors and Systems*. New York, NY, Wiley-Interscience, 560 p.

Desachy, J., G. Begni, B. Boissin, and J. Perbos (1985). "Investigation of Landsat-4 Thematic Mapper line-to-line and band-to-band registration and relative detector calibration." *Photogrammetric Engineering and Remote Sensing* **51**(9): 1291-1298.

Ehlers, M. and R. Welch (1987). "Stereocorrelation of Landsat TM images." *Photogrammetric Engineering and Remote Sensing* **53**: 1231–1237.

Emery, W. J., J. Brown and Z. P. Nowak (1989). "AVHRR image navigation: summary and review." *Photogrammetric Engineering and Remote Sensing* **55**(8): 1175-1183.

Emery, W. J. and M. Ikeda (1984). "A comparison of geometric correction nethods of AVHRR imagery." *Canadian Journal of Remote Sensing* **10**: 46-56.

Evans, W. E. (1974). "Marking ERTS images with a small mirror reflector." *Photogrammetric Engineering* **XL**(6): 665-671.

Forrest, R. B. (1981). "Simulation of orbital image–sensor geometry." *Photogrammetric Engineering and Remote Sensing* **47**(8): 1187-1193.

Friedmann, D. E., J. P. Friedel, K. L. Magnussen, R. Kwok, and S. Richardson (1983). "Multiple scene precision rectification of spaceborne imagery with very few ground control points." *Photogrammetric Engineering and Remote Sensing* **49**(12): 1657-1667.

Fritz, L. W. (1996). "The era of commercial earth observation satellites." *Photogrammetric Engineering and Remote Sensing* **62**(1): 39-45.

Fusco, L., U. Frei and A. Hsu (1985). "Thematic Mapper: Operational activities and sensor performance at ESA/Earthnet." *Photogrammetric Engineering and Remote Sensing* **51**(9): 1299-1314.

Ganas, A., E. Lagios and N. Tzannetos (2002). "An investigation into the spatial accuracy of the IKONOS 2 orthoimagery within an urban environment." *International Journal of Remote Sensing* **23**(17): 3513-3519.

Helder, D., T. Choi and M. Rangaswamy (2004). "In-flight characterization of spatial quality using point spread functions." *Post-launch calibration of satellite sensors*. S. A. Morain and A. M. Budge, Editors. International Society for Photogrammetry and Remote Sensing. **2**: 149-170.

Helder, D. L., M. Coan, K. Patrick and P. Gaska (2003). "IKONOS geometric characterization." *Remote Sensing of Environment* **88**: 69-79.

Helder, D. L. and E. Micijevic (2004). "Landsat-5 Thematic Mapper outgassing effects." *IEEE Transactions on Geoscience and Remote Sensing* **42**(12): 2717-2729.

Ho, D. and A. Asem (1986). "NOAA AVHRR image referencing." *International Journal of Remote Sensing* **7**: 895-904.

Iwasaki, A. and H. Fujisada (2005). "ASTER geometric performance." *IEEE Transactions on Geoscience and Remote Sensing* **43**(12): 2700-2706.

Jovanovic, V. M., M. A. Bull, M. M. Smyth and J. Zong (2002). "MISR in-flight camera geometric model calibration and georectification performance." *IEEE Transactions on Geoscience and Remote Sensing* **40**(7): 1512-1519.

Justice, C. O., B. L. Markham, J. R. G. Townshend, and R. L. Kennard (1989). "Spatial degradation of satellite data." *International Journal of Remote Sensing* **10**(9): 1539-1561.

Kohm, K. (2004). "Modulation transfer function measurement method and results for the OrbView-3 high resolution imaging satellite." In *XXth ISPRS Congress, Commission 1*, Istanbul, Turkey, International Society for Photogrammetry and Remote Sensing, pp. 7-12.

Krasnopolsky, V. M. and L. C. Breaker (1994). "The problem of AVHRR image navigation revisited." *International Journal of Remote Sensing* **15**(5): 979-1008.

Kratky, V. (1989). "On-line aspects of sterophotogrammetric processing of SPOT images." *Photogrammetric Engineering and Remote Sensing* **55**(3): 311-316.

Lee, D. S., J. C. Storey, M. J. Choate and R. W. Hayes (2004). "Four years of Landsat-7 on-orbit geometric calibration and performance." *IEEE Transactions on Geoscience and Remote Sensing* **42**(12): 2786-2795.

Legeckis, R. and J. Pritchard (1976). Algorithm for correcting the VHRR imagery for geometric distortions due to the earth curvature, earth rotation and spacecraft roll attitude errors. Washington, D.C., NOAA, No. NESS 77.

Leger, D., F. Viallefont, E. Hillairet and A. Meygret (2003). "In-flight refocusing and MTF assessment of Spot5 HRG and HRS cameras." In *Sensors, Systems, and Next-Generation Satellites VI*, Crete, Greece, SPIE, vol. 4881: 224-231.

Maling, D. H. (1992). *Coordinate Systems and Map Projections*. Second Edition. Oxford, England: Pergamon Press, 476 p.

Markham, B. L. (1985). "The Landsat sensors' spatial responses." *IEEE Transactions on Geoscience and Remote Sensing* **GE-23**(6): 864-875.

Markham, B. L. and J. L. Barker (1983). "Spectral characterization of the Landsat-4 MSS sensors." *Photogrammetric Engineering and Remote Sensing* **49**(6): 811-833.

Mendenhall, J. A., C. F. Bruce, C. J. Digenis, D. R. Hearn and D. E. Lencioni (2002). *EO-1 Advanced Land Imager Technology Validation Report*. Lincoln Laboratory, Massachusetts Institute of Technology.

Moik, J. G. (1980). *Digital Processing of Remotely Sensed Images*. Washington, D.C.: NASA, U.S. Government Printing Office, 330 p.

Moreno, J. F., S. Gandía, and J. Meliá (1992). "Geometric integration of NOAA AVHRR and SPOT data: low resolution effective parameters from high resolution data." *IEEE Transactions on Geoscience and Remote Sensing* **30**(5): 1006-1014.

Moreno, J. F. and J. Meliá (1993). "A method for accurate geometric correction of NOAA AVHRR HRPT data." *IEEE Transactions on Geoscience and Remote Sensing* **31**(1): 204-226.

NASA (2006). *Landsat-7 Science Data Users Handbook*. Greenbelt, MD, NASA Goddard Space Flight Center.

Nelson, N. R. and P. S. Barry (2001). "Measurement of Hyperion MTF from on-orbit scenes." In *Geoscience and Remote Sensing Symposium IGARSS '01*, IEEE, pp. 2967 - 2969.

Nishihama, M., R. Wolfe, D. Solomon, F. Patt, J. Blanchette, A. Fleig and E. Masuoka (1997). *MODIS Level 1A Earth Location: Algorithm Theoretical Basis Document Version 3.0*. NASA Goddard Space Flight Center.

Palmer, J. M. (1984). "Effective bandwidths for LANDSAT-4 and LANDSAT-D' multispectral scanner and Thematic Mapper subsystems." *IEEE Transactions on Geoscience and Remote Sensing* **GE-22**(3): 336-338.

Parada, M., A. Millan, A. Lobato and A. Hermosilla (2000). "Fast coastal algorithm for automatic geometric correction of AVHRR images." *International Journal of Remote Sensing* **21**(11): 2307-2312.

Park, S. K., R. Schowengerdt, and M.-A. Kaczynski (1984). "Modulation-transfer-function analysis for sampled imaging systems." *Applied Optics* **23**(15): 2572-2582.

Park, S. K. and R. A. Schowengerdt (1982). "Image sampling, reconstruction, and the effect of sample–scene phasing." *Applied Optics* **21**(17): 3142-3151.

Pearlman, J., C. Segal, L. Liao, S. Carman, M. Folkman, B. Browne, L. Ong and S. Ungar (2004). "Development and operations of the EO-1 Hyperion imaging spectrometer." In *Earth Observing Systems V*, SPIE, vol. 4135: 243-253.

Puccinelli, E. F. (1976). "Ground location of satellite scanner data." *Photogrammetric Engineering and Remote Sensing* **42**(4): 537-543.

Radhadevi, P. V., R. Ramachandran and A. S. R. K. V. M. Mohan (1998). "Restitution of IRS-1C pan data using an orbit attitude model and minimum control." *ISPRS Journal of Photogrammetry and Remote Sensing* **53**: 262-271.

Rauchmiller, R. F. and R. A. Schowengerdt (1988). "Measurement of the Landsat Thematic Mapper modulation transfer function using an array of point sources." *Optical Engineering* **27**(4): 334-343.

Reichenbach, S. E., D. E. Koehler and D. W. Strelow (1995). "Restoration and reconstruction of AVHRR images." *IEEE Transactions on Geoscience and Remote Sensing* **33**(4): 997-1007.

Richards, J. A. and X. Jia (1999). *Remote Sensing Digital Image Analysis – An Introduction*. Third Edition. Berlin: Springer-Verlag, 356 p.

Robinet, F., D. Leger, H. Cerbelaud and S. Lafont (1991). "Obtaining the MTF of a CCD imaging system using an array of point sources: Evaluation of performances." In *Geoscience and Remote Sensing Symposium IGARSS '91*, IEEE, pp. 1357 - 1361.

Rojas, F., R. A. Schowengerdt and S. F. Biggar (2002). "Early results on the characterization of the Terra MODIS spatial response." *Remote Sensing of Environment* **83**(1-2): 50-61.

Ryan, R. E., B. Baldridge, R. A. Schowengerdt, T. Choi, D. L. Helder and S. Blonski (2003). "IKONOS spatial resolution and image interpretability characterization." *Remote Sensing of Environment* **88**(1-2): 37-52.

Salamonowicz, P. H. (1986). "Satellite orientation and position for geometric correction of scanner imagery." *Photogrammetric Engineering and Remote Sensing* **52**(4): 491-499.

Sawada, N., M. Kidode, H. Shinoda, H. Asada, M. Iwanaga, S. Watanabe, K.-I. Mori, and M. Akiyama (1981). "An analytic correction method for satellite MSS geometric distortions." *Photogrammetric Engineering and Remote Sensing* **47**(8): 1195-1203.

SBRC (1984). *Thematic Mapper - Design Through Flight Evaluation*. Goleta, CA: Hughes Santa Barbara Research Center, No. NAS5-24200.

Schott, J. R. (1996). *Remote Sensing: The Image Chain Approach*. New York: Oxford University Press, 394 p.

Schowengerdt, R. A. and P. N. Slater (1972). "Determination of inflight MTF of orbital earth resources sensors." In *ICO IX Congress on Space Optics*, Santa Monica, CA, pp. 693-703.

Schowengerdt, R. A., S. K. Park, and R. T. Gray (1984). "Topics in the two–dimensional sampling and reconstruction of images." *International Journal of Remote Sensing* **5**(2): 333-347.

Schowengerdt, R. A., C. Archwamety, and R. C. Wrigley (1985). "Landsat Thematic Mapper image-derived MTF." *Photogrammetric Engineering and Remote Sensing* **51**(9): 1395-1406.

Schowengerdt, R. A., R. W. Basedow and J. E. Colwell (1996). "Measurement of the HYDICE system MTF from flight imagery." In *Hyperspectral Remote Sensing and Applications*, Denver, CO, SPIE, vol. 2821: 127-136.

Schowengerdt, R. A. (2002). "Spatial response of the EO-1 Advanced Land Imager (ALI)." In *IEEE Int. Geoscience and Remote Sensing Symposium (IGARRS '02)*, Toronto, pp. 3121-3123.

Seto, Y. (1991). "Geometric correction algorithms for satellite imagery using a bi–directional scanning sensor." *IEEE Transactions on Geoscience and Remote Sensing*: 292-299.

Slater, P. N. (1979). "A re-examination of the Landsat MSS." *Photogrammetric Engineering and Remote Sensing* **45**(1): 1479-1485.

Slater, P. N. (1980). *Remote Sensing - Optics and Optical Systems*. Reading, MA: Addison-Wesley, 575 p.

Steiner, D. and M. E. Kirby (1976). "Geometrical referencing of Landsat images by affine transformation and overlaying of map data." *Photogrammetria* **33**: 41-75.

Storey, J. C. (2001). "Landsat 7 on-orbit modulation transfer function estimation." In *Sensors, Systems, and Next-Generation Satellites V*, Toulouse, France, SPIE, vol. 4540: 50-61.

Tilton, J. C., B. L. Markham, and W. L. Alford (1985). "Landsat-4 and Landsat-5 MSS coherent noise: characterization and removal." *Photogrammetric Engineering and Remote Sensing* **51**(9): 1263-1279.

Toutin, T. (2004). "Review article: Geometric processing of remote sensing images: Models, algorithms and methods." *International Journal of Remote Sensing* **25**(10): 1893-1924.

Turker, M. and A. O. Gacemer (2004). "Geometric correction accuracy of IRS-1D pan imagery using topographic map versus GPS control points." *International Journal of Remote Sensing* **25**(6): 1095-1104.

USGS/NOAA (1984). Landsat 4 Data Users Handbook. Alexandria, VA: U. S. Geological Survey and National Oceanic and Atmospheric Administration.

Verdebout, J., S. Jacquemoud, and G. Schmuck (1994). Optical properties of leaves: modelling and experimental studies. *Imaging Spectrometry - A Tool for Environmental Observations*. J. Hill and J. Megier, (Eds.). Dordrecht, The Netherlands: Kluwer Academic Publishers. **4**: 335p.

Walker, R. E., A. L. Zobrist, N. A. Bryant, B. Gohkman, S. Z. Friedman, and T. L. Logan (1984). "An analysis of Landsat-4 Thematic Mapper geometric properties." *IEEE Transactions on Geoscience and Remote Sensing* **GE-22**(3): 288-293.

Welch, R., T. R. Jordan, and M. Ehlers (1985). "Comparative evaluations of the geodetic accuracy and cartographic potential of Landsat-4 and Landsat-5 Thematic Mapper image data." *Photogrammetric Engineering and Remote Sensing* **51**(11): 1799-1812.

Welch, R. and E. L. Usery (1984). "Cartographic accuracy of Landsat-4 MSS and TM image data." *IEEE Transactions on Geoscience and Remote Sensing* **GE-22**(3): 281-288.

Wessman, C. A. (1994). Estimating canopy biochemistry through imaging spectrometry. *Imaging Spectrometry - A Tool for Environmental Observations*. J. Hill and J. Megier, (Eds.). Dordrecht, The Netherlands: Kluwer Academic Publishers. **4**: .

Westin, T. (1990). "Precision rectification of SPOT imagery." *Photogrammetric Engineering and Remote Sensing* **56**(2): 247-253.

Westin, T. (1992). "Inflight calibration of SPOT CCD detector geometry." *Photogrammetric Engineering and Remote Sensing* **58**(9): 1313-1319.

Wolf, P. R. and B. A. Dewitt (2000). *Elements of Photogrammetry with Applications in GIS*. Third Edition. McGraw-Hill, 624 p.

Wolfe, R. E., M. Nishihama, A. J. Fleig, J. A. Kuyper, D. P. Roy, J. C. Storey and F. S. Patt (2002). "Achieving sub-pixel geolocation accuracy in support of MODIS land science." *Remote Sensing of Environment* **83**: 31-49.

Wong, K. W. (1975). "Geometric and cartographic accuracy of ERTS-1 imagery." *Photogrammetric Engineering and Remote Sensing* **41**: 621-635.

Wrigley, R. C., C. A. Hlavka, D. H. Card, and J. S. Buis (1985). "Evaluation of Thematic Mapper interband registration and noise characteristics." *Photogrammetric Engineering and Remote Sensing* **51**(9): 1417-1425.

Xu, Q. and R. A. Schowengerdt (2003). "Urban targets for image quality analysis of high resolution satellite imaging systems." In *Visual Information Processing XII*, Orlando, SPIE, vol. 5108: 31-38.

CHAPTER 4 Data Models

Adams, J. B., M. O. Smith, and A. R. Gillespie (1993). "Imaging Spectroscopy: Interpretation Based on Spectral Mixture Analysis." *Remote Geochemical Analysis: Elemental and Mineralogical Composition*. C. M. Pieters and P. A. Englert, (Eds.) Cambridge: Cambridge University Press: 145-166.

Atkinson, P. M. (1993). "The effect of spatial resolution on the experimental variogram of airborne MSS imagery." *International Journal of Remote Sensing* **14**(5): 1005-1011.

Carr, J. R. (1995). *Numerical Analysis for the Geological Sciences*. Englewood Cliffs, NJ: Prentice Hall, 592 p.

Carr, J. R. and F. P. d. Miranda (1998). "The semivariogram in comparison to the co-occurrence matrix for classification of image texture." *IEEE Transactions on Geoscience and Remote Sensing* **36**(6): 1945-1952.

Castleman, K. R. (1996). *Digital Image Processing*. Englewood Cliffs, NJ: Prentice Hall, 667 p.

Civco, D. L. (1989). "Topographic normalization of Landsat Thematic Mapper digital imagery." *Photogrammetric Engineering and Remote Sensing* **55**(9): 1303-1309.

Conese, C., M. A. Gilabert, F. Maselli, and L. Bottai (1993). "Topographic normalization of TM scenes through the use of an atmospheric correction method and digital terrain models." *Photogrammetric Engineering and Remote Sensing* **59**(12): 1745-1753.

Cornsweet, T. N. (1970). *Visual Perception*. New York, NY: Academic Press, 475 p.

Curran, P. J. (1988). "The semivariogram in remote sensing: an introduction." *Remote Sensing of Environment* **24**: 493-507.

Curran, P. J. and J. L. Dungan (1989). "Estimation of signal-to-noise: a new procedure applied to AVIRIS data." *IEEE Transactions on Geoscience and Remote Sensing* **27**(5): 620-628.

Djamdji, J.-P. and A. Bijaoui (1995). "Disparity analysis: A wavelet transform approach." *IEEE Transactions on Geoscience and Remote Sensing* **33**(1): 67-76.

Dymond, J. R. and J. D. Shepherd (2004). "The spatial distribution of indigenous forest and its composition in the wellington region, new zealand, from ETM+ satellite imagery." *Remote Sensing of Environment* **90**(1): 116-125.

Eliason, P. T., L. A. Soderblom and J. P. S. Chavez (1981). "Extraction of topographic and spectral albedo information from multispectral images." *Photogrammetric Engineering and Remote Sensing* **48**(11): 1571-1579.

Feder, J. (1988). *Fractals*. New York, NY: Plenum Press, 283 p.

Feng, J., B. Rivard and A. Sanchez-Azofeifa (2003). "The topographic normalization of hyperspectral data: Implications for the selection of spectral end members and lithologic mapping." *Remote Sensing of Environment* **85**(2): 221-231.

Fukunaga, K. (1990). *Introduction to Statistical Pattern Recognition*. Second Edition. San Diego: Academic Press, 591 p.

Gaddis, L. R., L. A. Soderblom, H. H. Kieffer, K. J. Becker, J. Torson, and K. Mullins (1996). "Decomposition of AVIRIS spectra: Extraction of surface-reflectance, atmospheric, and instrumental components." *IEEE Transactions on Geoscience and Remote Sensing* **34**(1): 163-178.

Gohin, F. and G. Langlois (1993). "Using geostatistics to merge in situ measurements and remotely-sensed observations of sea surface temperature." *International Journal of Remote Sensing* **14**(1): 9-19.

Gonzalez, R. C. and R. E. Woods (1992). *Digital Image Processing*. Reading, MA: Addison-Wesley, 703 p.

Gu, D. and A. Gillespie (1998). "Topographic normalization of Landsat TM images of forest based on sub-pixel sun–canopy–sensor geometry." *Remote Sensing of Environment* **64**: 166-175.

Gu, D., A. R. Gillespie, J. B. Adams and R. Weeks (1999). "A statistical approach for topographic correction of satellite images by using spatial context information." *IEEE Transactions on Geoscience and Remote Sensing* **37**(1): 236-246.

Haralick, R. M., K. Shanmugan, and I. Dinstein (1973). "Textural features for image classification." *IEEE Transactions on Systems, Man, and Cybernetics* **SMC-3**: 610-621.

Holben, B. N. and C. O. Justice (1980). "The topographic effect on spectral response from nadir pointing sources." *Photogrammetric Engineering and Remote Sensing* **46**(9): 1191-1200.

Holben, B. N. and C. O. Justice (1981). "An examination of spectral band ratioing to reduce the topographic effect on remotely sensed data." *International Journal of Remote Sensing* **2**(2): 115-133.

IRARS (1995). *Multispectral Imagery Interpretability Rating Scale Reference Guide.* Image Resolution Assessment and Reporting Standards (IRARS) Committee.

IRARS (1996). *Civil NIIRS Reference Guide.* Image Resolution Assessment and Reporting Standards (IRARS) Committee.

Isaaks, E. H. and R. M. Srivastava (1989). *An Introduction to Applied Geostatistics.* New York, Oxford University Press, 561 p.

Itten, K. I. and P. Meyer (1993). "Geometric and radiometric correction of TM data of mountainous forested areas." *IEEE Transactions on Geoscience and Remote Sensing* **31**(4): 764-770.

Iwasaki, A. and H. Tonooka (2005). "Validation of a crosstalk correction algorithm for ASTER/SWIR." *IEEE Transactions on Geoscience and Remote Sensing* **43**(12): 2747-2751.

Jain, A. K. (1989). *Fundamentals of Digital Image Processing.* Englewood Cliffs, NJ: Prentice Hall, 569 p.

Jasinski, M. F. and P. S. Eagleson (1989). "The structure of red-infrared scattergrams of semivegetated land-scapes." *IEEE Transactions on Geoscience and Remote Sensing* **27**(4): 441-451.

Jasinski, M. F. and P. S. Eagleson (1990). "Estimation of subpixel vegetation cover using red-infrared scatter-grams." *IEEE Transactions on Geoscience and Remote Sensing* **28**(2): 253-267.

Journel, A. G. and C. J. Huijbregts (1978). *Mining Geostatistics.* London: Academic Press.

Jupp, D. L. B., A. H. Strahler, and C. E. Woodcock (1989a). "Autocorrelation and regularization in digital images I. basic theory." *IEEE Transactions on Geoscience and Remote Sensing* **26**: 463-473.

Jupp, D. L. B., A. H. Strahler, and C. E. Woodcock (1989b). "Autocorrelation and regularization in digital images II. simple image models." *IEEE Transactions on Geoscience and Remote Sensing* **27**: 247-258.

Justice, C. O., S. W. Wharton, and B. N. Holben (1981). "Application of digital terrain data to quantify and reduce the topographic effect on Landsat data." *International Journal of Remote Sensing* **2**(3): 213-230.

Kawata, Y., S. Ueno, and T. Kusaka (1988). "Radiometric correction for atmospheric and topographic effects on Landsat MSS images." *International Journal of Remote Sensing* **9**(4): 729-748.

Lacaze, B., S. Rambal, and T. Winkel (1994). "Identifying spatial patterns of Mediterranean landscapes from geostatistical anlaysis of remotely-sensed data." *International Journal of Remote Sensing* **15**(12): 2437-2450.

Leachtenauer, J. C., W. Malila, J. Irvine, L. Colburn and N. Salvaggio (1997). "General image-quality equation: GIQE." *Applied Optics* **36**(32): 8322-8328.

Mandelbrot, B. B. (1967). "How long is the coast of Britain? Statistical self-similarity and fractal dimension." *Science* **155**: 636-638.

Mandelbrot, B. B. (1983). *The Fractal Geometry of Nature*. New York, NY: W. H. Freeman and Company, 468 p.

Miranda, F. P., L. E. N. Fonseca and J. R. Carr (1998). "Semivariogram textural classifcation of JERS-1 (Fuyo-1) SAR data obtained over a flooded area of the Amazon rainforest." **19**(3): 549-556.

Peli, T. (1990). "Multiscale fractal theory and object characterization." *Journal of the Optical Society of America* **7**(6): 1101-1112.

Pelig, S., J. Naor, R. Hartley, and D. Avnir (1984). "Multiple resolution texture analysis and classification." *IEEE Transactions on Pattern Analysis and Machine Intelligence* **PAMI-6**(4): 518-523.

Pentland, A. P. (1984). "Fractal-based description of natural scenes." *IEEE Transactions on Pattern Analysis and Machine Intelligence* **PAMI-6**(6): 661-674.

Pratt, W. K. (1991). *Digital Image Processing*. Second Edition. New York, NY: John Wiley & Sons, 698 p.

Press, W. H., B. P. Flannery, S. A. Teukolsky, and W. T. Vetterling (1992). *Numerical Recipes in C–The Art of Scientific Computing*. Second Edition. Cambridge: Cambridge University Press, 994 p.

Proy, C., D. Tanre, and P. Y. Deschamps (1989). "Evaluation of topographic effects in remotely sensed data." *Remote Sensing of Environment* **30**: 21-32.

Richards, J. A. and X. Jia (1999). *Remote Sensing Digital Image Analysis – An Introduction*. Third Edition. Berlin: Springer-Verlag, 356 p.

Richter, R. (1997). "Correction of atmospheric and topographic effects for high spatial resolution satellite imagery." *International Journal of Remote Sensing* **18**(5): 1099-1111.

Richter, R. and D. Schlapfer (2002). "Geo-atmospheric processing of airborne imaging spectrometry data. Part 2: Atmospheric/topographic correction." *International Journal of Remote Sensing* **23**(13): 2631 –2649.

Rossi, R. E., J. L. Dungan, and L. R. Beck (1994). "Kriging in shadows: geostatistical interpolation for remote sensing." *Remote Sensing of Environment* **48**: 1-25.

Ryan, R. E., B. Baldridge, R. A. Schowengerdt, T. Choi, D. L. Helder and S. Blonski (2003). "IKONOS spatial resolution and image interpretability characterization." *Remote Sensing of Environment* **88**(1-2): 37-52.

Schott, J. R. (1996). *Remote Sensing: The Image Chain Approach*. New York, NY: Oxford University Press, 394 p.

Shepherd, J. D. and J. R. Dymond (2003). "Correcting satellite imagery for the variance of reflectance and illumination with topography." *International Journal of Remote Sensing* **24**(17): 3503-3514.

St-Onge, B. A. and F. Cavayas (1997). "Automated forest structure mapping from high resolution imagery based on directional semivariogram estimates." *Remote Sensing of Environment* **61**: 82-95.

Teillet, P. M., B. Guindon, and D. G. Goodenough (1982). "On the slope-aspect correction of multispectral scanner data." *Canadian Journal of Remote Sensing* **8**(2): 84-106.

Townshend, J. R. G., C. O. Justice, C. Gurney, and J. McManus (1992). "The impact of misregistration on change detection." *IEEE Transactions on Geoscience and Remote Sensing* **30**(5): 1054-1060.

Wald, L. (1989). "Some examples of the use of structure functions in the analysis of satellite images of the ocean." *Photogrammetric Engineering and Remote Sensing* **55**(10): 1487-1490.

Warner, T. A. and X. Chen (2001). "Normalization of Landsat thermal imagery for the effects of solar heating and topography." *International Journal of Remote Sensing* **22**(5): 773-788.

Woodcock, C. E., A. H. Strahler, and D. L. B. Jupp (1988a). "The use of variograms in remote sensing: I scene models and simulated images." *Remote Sensing of Environment* **25**: 324-348.

Woodcock, C. E., A. H. Strahler, and D. L. B. Jupp (1988b). "The use of variograms in remote sensing: II real digital images." *Remote Sensing of Environment* **25**: 349-379.

CHAPTER 5 Spectral Transforms

Avery, T. E. and G. L. Berlin (1992). *Fundamentals of Remote Sensing and Airphoto Interpretation*. Fifth Edition. New York, NY: Macmillan Publishing Company, 472 p.

Byrne, G. F., P. F. Crapper, and K. K. Mayo (1980). "Monitoring land-cover change by principal component analysis of multitemporal Landsat data." *Remote Sensing of Environment* **10**: 175-184.

Chavez, P. S., Jr., G. L. Berlin, and L. B. Sowers (1982). "Statistical method for selecting Landsat MSS ratios." *Journal of Applied Photographic Engineering* **8**(1): 23-30.

Coppin, P., I. Jonckheere, K. Nackaerts, B. Muys and E. Lambinint (2004). "Digital change detection methods in ecosystem monitoring: A review." *International Journal of Remote Sensing* **25**(9): 1565-1596.

Crist, E. P. and R. C. Cicone (1984). "A physically-based transformation of Thematic Mapper data — the TM Tasseled Cap." *IEEE Transactions on Geoscience and Remote Sensing* **GE-22**: 256-263.

Crist, E. P., R. Laurin, and R. C. Cicone (1986). "Vegetation and soils information contained in transformed Thematic Mapper data." in: *IGARSS' 86*, Zurich, ESA Publications Division, vol. ESA SP-254: 1465-1472.

Crist, E. P. (1996). personal communication.

Dallas, W. J. and W. Mauser (1980). "Preparing pictures for visual comparison." *Applied Optics* **19**(21): 3586-3587.

Du, Y., P. M. Teillet and J. Cihlar (2002). "Radiometric normalization of multitemporal high-resolution satellite images with quality control for land cover change detection." *Remote Sensing of Environment* **82**: 123-134.

Durand, J. M. and Y. H. Kerr (1989). "An improved decorrelation method for the efficient display of multispectral data." *IEEE Transactions on Geoscience and Remote Sensing* **27**(5): 611-619.

Eastman, J. R. and M. Fulk (1993). "Long sequence time series evaluation using standardized principal components." *Photogrammetric Engineering and Remote Sensing* **59**(6): 991-996.

Fahnestock, J. D. and R. A. Schowengerdt (1983). "Spatially-variant contrast enhancement using local range modification." *Optical Engineering* **22**(3): 378-381.

Fukunaga, K. (1990). *Introduction to Statistical Pattern Recognition*. Second Edition. San Diego: Academic Press, 591 p.

Fung, T. and E. LeDrew (1987). "Application of principal components analysis to change detection." *Photogrammetric Engineering and Remote Sensing* **53**(12): 1649-1658.

Gao, X., A. R. Huete, W. Ni and T. Miura (2000). "Optical–biophysical relationships of vegetation spectra without background contamination." *Remote Sensing of Environment* **74**: 609-620.

Gillespie, A. R., A. B. Kahle, and R. E. Walker (1986). "Color enhancement of highly correlated images. I. Decorrelation and HSI contrast stretches." *Remote Sensing of Environment* **20**: 209-235.

Gonzalez, R. C. and R. E. Woods (2002). *Digital Image Processing*. Second Edition. Upper Saddle River, NJ: Prentice-Hall, 793 p.

Green, A. A., M. Berman, P. Switzer, and M. D. Craig (1988). "A transformation for ordering multispectral data in terms of image quality with implications for noise removal." *IEEE Transactions on Geoscience and Remote Sensing* **26**(1): 65-74.

Haeberli, P. and D. Voorhies (1994). "Image processing by interpolation and extrapolation." *Silicon Graphics IRIS Universe Magazine*.

Haydn, R., G. W. Dalke, J. Henkel, and J. E. Bare (1982). "Application of the IHS color transform to the processing of multisensor data and image enhancement." in: *International Symposium on Remote Sensing of Arid and Semi-Arid Lands*, Cairo, Egypt: Environmental Research Institute of Michigan: 599-616.

Huang, C., B. Wylie, L. Yang, C. Homer and G. Zylstra (2002). "Derivation of a tasselled cap transformation based on Landsat 7 at-satellite reflectance." *International Journal of Remote Sensing* **23**(8): 1741-1748.

Huete, A. R. (1988). "A Soil Adjusted Vegetation Index (SAVI)." *Remote Sensing of Environment* **25**: 295-309.

Huete, A. R. and R. D. Jackson (1987). "Suitability of spectral indices for evaluating vegetation characteristics on arid rangelands." *Remote Sensing of Environment* **23**: 213-232.

Huete, A., K. Didan, T. Miura, E. P. Rodriguez, X. Gao and L. G. Ferreira (2002). "Overview of the radiometric and biophysical performance of the MODIS vegetation indices." *Remote Sensing of Environment* **83**: 195-213.

Ingebritsen, S. E. and R. J. P. Lyon (1985). "Principal components analysis of multitemporal image pairs." *International Journal of Remote Sensing* **6**: 687-696.

Jackson, R. D. (1983). "Spectral indices in N-space." *Remote Sensing of Environment* **13**(5): 409-421.

Justice, C. O., J. R. G. Townshend, and V. L. Kalb (1991). "Representation of vegetation by continental data sets derived from NOAA-AVHRR data." *International Journal of Remote Sensing* **12**(5): 999-1021.

Kauth, R. J. and G. S. Thomas (1976). "The Tasselled Cap – A graphic description of the spectral–temporal development of agricultural crops as seen by Landsat." in: *Symposium on Machine Processing of Remotely Sensed Data*, IEEE, vol. 76CH 1103–1MPRSD: 41-51.

Knyazikhin, Y., J. V. Martonchik, R. B. Myneni, D. J. Diner and S. W. Running (1998). "Synergistic algorithm for estimating vegetation canopy leaf area index and fraction of absorbed photosynthetically active radiation from MODIS and MISR data." *Journal of Geophysical Research* **103**(D24): 32257-32275.

Kruse, F. A., K. S. Kierein-Young, and J. W. Boardman (1990). "Mineral mapping at Cuprite, Nevada with a 63-channel imaging spectrometer." *Photogrammetric Engineering and Remote Sensing* **56**(1): 83-92.

Lavreau, J. (1991). "De-hazing Landsat Thematic Mapper images." *Photogrammetric Engineering and Remote Sensing* **57**(10): 1297-1302.

Lee, J. B., A. S. Woodyatt, and M. Berman (1990). "Enhancement of high spectral resolution remote-sensing data by a noise-adjusted principal components transform." *IEEE Transactions on Geoscience and Remote Sensing* **28**(3): 295-304.

Liang, S. (2004). *Quantitative Remote Sensing of Land Surfaces*. Hoboken, New Jersey: John Wiley & Sons, 534 p.

Lu, D., P. Mausel, E. Brondizio and E. Moran (2004). "Change detection techniques." *International Journal of Remote Sensing* **25**(12): 2365-2407.

Myneni, R. B., R. R. Nemani and S. W. Running (1997). "Estimation of global leaf area index and absorbed par using radiative transfer models." *IEEE Transactions on Geoscience and Remote Sensing* **35**(6): 1380-1393.

Pizer, S. M., E. P. Amburn, J. D. Austin, R. Cromartie, A. Geselowitz, T. Greer, B. H. Romeny, J. B. Zimmerman, and K. Zuiderveld (1987). "Adaptive histogram equalization and its variations." *Computer Vision, Graphics and Image Processing* **39**: 355-368.

Pun, T. (1981). "Entropic thresholding, a new approach." *Computer Graphics and Image Processing* **16**: 210-239.

Ready, P. J. and P. A. Wintz (1973). "Information extraction, SNR improvement and data compression in multispectral imagery." *IEEE Transactions on Communications* **COM-21**(10): 1123-1131.

Richards, J. A. (1984). "Thematic mapping from multitemporal image data using the principal components transformation." *Remote Sensing of Environment* **16**: 35-46.

Richards, J. A. and X. Jia (1999). *Remote Sensing Digital Image Analysis – An Introduction*. Third Edition. Berlin: Springer-Verlag, 356 p.

Richardson, A. J. and C. L. Wiegand (1977). "Distinguishing vegetation from soil background information." *Photogrammetric Engineering and Remote Sensing* **43**: 1541-1552.

Rothery, D. A. and G. A. Hunt (1990). "A simple way to perform decorrelation stretching and related techniques on menu-drive image processing systems." *International Journal of Remote Sensing* **11**(1): 133-137.

RSI (2005). *ENVI*. Boulder, CO.

Schetselaar, E. M. (2001). "On preserving spectral balance in image fusion and its advantages for geological image interpretation." *Photogrammetric Engineering and Remote Sensing* **67**(8): 925-934.

Schowengerdt, R. A. (1983). *Techniques for Image Processing and Classification in Remote Sensing*. Orlando, FL: Academic Press, 249 p.

Singh, A. and A. Harrison (1985). "Standardized principal components." *International Journal of Remote Sensing* **6**(6): 883-896.

Smith, A. R. (1978). "Color gamut transform pairs." in: *ACM-SIGGRAPH*, ACM, vol. 12: 12-19.

Thompson, D. R. and O. A. Whemanen (1980). "Using Landsat digital data to detect moisture stress in corn-soybean growing regions." *Photogrammetric Engineering and Remote Sensing* **46**(8): 1087-1093.

Townshend, J. R. G. and C. O. Justice (1986). "Analysis of the dynamics of African vegetation using the normalized difference vegetation index." *International Journal of Remote Sensing* **7**(11): 1435-1445.

Tucker, C. J. and P. C. Sellers (1986). "Satellite remote sensing of primary production." *International Journal of Remote Sensing* **7**(11): 1395-1416.

CHAPTER 6 Spatial Transforms

Barnes, W. L., T. S. Pagano and V. V. Salomonson (1998). "Prelaunch characteristics of the moderate resolution imaging spectroradiometer (MODIS) on EOS-AM1." *IEEE Transactions on Geoscience and Remote Sensing* **36**: 1088-1100.

Brigham, E. O. (1988). *The Fast Fourier Transform and Its Applications*. Englewood Cliffs, NJ: Prentice Hall, 448 p.

Burrus, C. S., R. A. Gopinath and H. Guo (1998). *Introduction to Wavelets and Wavelet Transforms - A Primer*. Upper Saddle River, NJ: Prentice Hall, 268 p.

Burt, P. J. (1981). "Fast filter transforms for image processing." *Computer Graphics and Image Processing* **16**: 20-51.

Burt, P. J. and E. H. Adelson (1983). "The laplacian pyramid as a compact image code." *IEEE Transactions on Communications* **31**(4): 532-540.

Castleman, K. R. (1996). *Digital Image Processing*. Englewood Cliffs, NJ: Prentice-Hall, 667 p.

Cohen, L. (1989). "Time-frequency distributions—A review." *Proceedings of the IEEE* **77**(7): 941-981.

Daubechies, I. (1988). "Orthonormal bases of compactly supported wavelets." *Communications on Pure and Applied Mathematics* **41**: 909-996.

Davis, L. S. (1975). "A survey of edge detection techniques." *Computer Graphics and Image Processing* **4**: 248-270.

Dougherty, E. R. and R. A. Lotufo (2003). *Hands-On Morphological Image Processing*. Bellingham, WA: SPIE, 290 p.

Friedmann, D. E. (1980). "Forum: A re-examination of the Landsat MSS." *Photogrammetric Engineering and Remote Sensing* **46**(12): 1541-1542.

Gaskill, J. D. (1978). *Linear Systems, Fourier Transforms, and Optics*. New York, NY: John Wiley & Sons, 554 p.

Gardina, C. R. and E. R. Dougherty (1988). *Morphological Methods in Image and Signal Processing*. Englewood Cliffs, NJ: Prentice Hall, 321 p.

Grossman, A. and J. Morlet (1984). "Decomposition of Hardy functions into square integrable wavelets of constant shape." *SIAM Journal of Applied Mathematics* **15**: 723-736.

Hearn, D. R. (2000). *EO-1 Advanced Land Imager modulation transfer functions*. Lincoln Laboratory, Massachusetts Institute of Technology. Technical Report 1061.

Hunt, B. R. and T. M. Cannon (1976). "Nonstationary assumptions for Gaussian models of images." *IEEE Transactions on Systems, Man & Cybernetics* **SMC-6**: 876-882.

Jain, A. K. (1989). *Fundamentals of Digital Image Processing*. Englewood Cliffs, NJ: Prentice-Hall, 569 p.

Jensen, J. R. (2004). *Introductory Digital Image Processing – A Remote Sensing Perspective*. Third Edition. Upper Saddle River, NJ: Prentice Hall, 544 p.

Lee, J.-S. (1980). "Digital image enhancement and noise filtering by use of local statistics." *IEEE Transactions on Pattern Analysis and Machine Intelligence* **PAMI-2**(2): 165-168.

Levine, M. D. (1985). *Vision in Man and Machine*. New York, NY: McGraw-Hill.

Mallat, S. (1991). "Zero-crossings of a wavelet transform." *IEEE Transactions on Information Theory* **37**(4): 1019-1033.

Mallat, S. G. (1989). "A theory for multiresolution signal decomposition: the wavelet representation." *IEEE Transactions on Pattern Analysis and Machine Intelligence* **11**(7): 674-693.

Markham, B. L. (1985). "The Landsat sensors' spatial responses." *IEEE Transactions on Geoscience and Remote Sensing* GE-**23**(6): 864-875.

Marr, D. (1982). *Vision*. New York, NY: W. H. Freeman and Company, 397 p.

Marr, D. and E. Hildreth (1980). "Theory of edge detection." *Proceedings of the Royal Society of London* **207**: 187-217.

McDonnell, M. J. (1981). "Box-filtering techniques." *Computer Graphics and Image Processing* **17**(1): 65-70.

Oppenheim, A. V. and J. S. Lim (1981). "The importance of phase in signals." *Proceedings of the IEEE* **69**(5): 529-541.

Park, S. K., R. Schowengerdt and M.-A. Kaczynski (1984). "Modulation-transfer-function analysis for sampled imaging systems." *Applied Optics* **23**(15): 2572-2582.

Pratt, W. K. (1991). *Digital Image Processing*. Second Edition. New York, NY: John Wiley & Sons, 698 p.

Press, W. H., B. P. Flannery, S. A. Teukolsky, and W. T. Vetterling (1992). *Numerical Recipes in C–The Art of Scientific Computing*. Second Edition. Cambridge: Cambridge University Press, 994 p.

Reichenbach, S. E., D. E. Koehler and D. W. Strelow (1995). "Restoration and reconstruction of AVHRR images." *IEEE Transactions on Geoscience and Remote Sensing* **33**(4): 997-1007.

Robinson, G. S. (1977). "Detection and coding of edges using directional masks." *Optical Engineering* **16**(5).

Rojas, F., R. A. Schowengerdt and S. F. Biggar (2001). "Modulation transfer analysis of the moderate resolution imaging spectroradiometer (MODIS)." In *Earth Observing Systems VI*, San Diego, CA, SPIE, vol. 4483: 222-230.

Schalkoff, R. J. (1989). *Digital Image Processing and Computer Vision*. New York, NY: John Wiley & Sons, 489 p.

Schott, J. R. (1996). *Remote Sensing: The Image Chain Approach*. New York, NY: Oxford University Press, 394 p.

Serra, J. (1982). *Image Analysis and Mathematical Morphology*. New York, NY: Academic Press.

Shensa, M. J. (1992). "The discrete wavelet transform: Wedding the a trous and Mallat algorithms." *IEEE Transactions on Signal Processing* **40**: 2464-2482.

Slater, P. N. (1979). "A re-examination of the Landsat MSS." *Photogrammetric Engineering and Remote Sensing* **45**(1): 1479-1485.

Soille, P. (2002). *Morphological Image Analysis*. Second Edition. Springer, 391 p.

Storey, J. C. (2001). "Landsat 7 on-orbit modulation transfer function estimation." In *Sensors, Systems, and Next-Generation Satellites V*, Toulouse, France, SPIE, vol. 4540: 50-61.

Wallis, R. (1976). "An approach to the space variant restoration and enhancement of images." In *Proc. Symposium on Current Mathematical Problems in Image Science*, Monterey, CA: Naval Post-graduate School.

CHAPTER 7 Correction and Calibration

Abramson, S. B. and R. A. Schowengerdt (1993). "Evaluation of edge–preserving smoothing filters for digital image mapping." *ISPRS Journal of Photogrammetry and Remote Sensing* **48**(2): 2-17.

Adler-Golden, S. M., M. W. Matthew, L. S. Bernstein, R. Y. Levine, A. Berk, S. C. Richtsmeier, P. K. Acharya, G. P. Anderson, G. Feldeb, J. Gardner, M. Hokeb, L. S. Jeong, B. Pukall, J. Mello, A. Ratkowski and H.-H. Burke (1999). "Atmospheric correction for short-wave spectral imagery based on MODTRAN4." In *Imaging Spectrometry V*, Denver, CO, SPIE, vol. 3753: 61-69.

Ahern, F. J., D. G. Goodenough, S. C. Jain, V. R. Rao, and G. Rochon (1977). "Use of clear lakes as standard reflectors for atmospheric measurements." in: *Eleventh International Symposium on Remote Sensing of Environment*, Ann Arbor, MI: Environmental Research Institute of Michigan, : 583-594.

Algazi, V. R. and G. E. Ford (1981). "Radiometric equalization of nonperiodic striping in satellite data." *Computer Graphics and Image Processing* **16**: 287-295.

Andrews, H. C. and B. R. Hunt (1977). *Digital Image Restoration*. Englewood Cliffs, NJ: Prentice-Hall, 238 p.

Anuta, P. E. (1973). Geometric correction of ERTS-1 digital multispectral scanner data, Laboratory for Applications of Remote Sensing, Purdue University, No. Information Note 103073.

Basedow, R. W., W. S. Aldrich, J. E. Coiwell and W. D. Kinder (1996). "HYDICE system performance – an update." In *Hyperspectral Remote Sensing and Applications*, Denver, CO, SPIE, vol. 2821: 76-84.

Bates, R. H. T. and M. J. McDonnell (1986). *Image Restoration and Reconstruction*. Oxford: Clarendon Press, 288 p.

Bernstein, R., J. B. Lospiech, H. J. Myers, H. G. Kolsky, and R. D. Lees (1984). "Analysis and processing of Landsat-4 sensor data using advanced image processing techniques and technologies." *IEEE Transactions on Geoscience and Remote Sensing* **GE-22**(3): 192-221.

Bindschadler, R. and H. Choi (2003). "Characterizing and correcting Hyperion detectors using ice-sheet images." *IEEE Transactions on Geoscience and Remote Sensing* **41**(6): 1189-1193.

Bugayevskiy, L. M. and J. Snyder (1995). *Map Projections - A Reference Manual*. London, UK: Taylor & Francis Ltd, 328 p.

Büttner, G. and A. Kapovits (1990). "Characterization and removal of horizontal striping from SPOT panchromatic imagery." *International Journal of Remote Sensing* **11**(2): 359-366.

Canty, M. J., A. A. Nielsen and M. Schmidt (2004). "Automatic radiometric normalization of multitemporal satellite imagery." *Remote Sensing of Environment* **91**: 441-451.

Carrere, V. and J. E. Conel (1993). "Recovery of atmospheric water vapor total column abundance from imaging spectrometer data around 940 nm−sensitivity analysis and application to Airborne Visible/Infrared Imaging Spectrometer (AVIRIS) data." *Remote Sensing of Environment* **44**: 179-204.

Castleman, K. R. (1996). *Digital Image Processing*. Englewood Cliffs, NJ: Prentice Hall, 667 p.

Centeno, J. A. S. and V. Haertel (1995). "Adaptive low-pass fuzzy filter for noise removal." *Photogrammetric Engineering and Remote Sensing* **61**(10): 1267-1272.

Chavez, P. S., Jr. (1975). "Simple high-speed digital image processing to remove quasi-coherent noise patterns." in: *41st Annual Meeting*, Washington, D.C., American Society of Photogrammetry: 595-600.

Chavez, P. S., Jr. (1988). "An improved dark-object subtraction technique for atmospheric scattering correction of multispectral data." *Remote Sensing of Environment* **24**: 459-479.

Chavez, P. S., Jr. (1989). "Radiometric calibration of Landsat Thematic Mapper multispectral images." *Photogrammetric Engineering and Remote Sensing* **55**(9): 1285-1294.

Chavez, P. S., Jr. (1996). "Image-based atmospheric corrections−revisited and improved." *Photogrammetric Engineering and Remote Sensing* **62**(9): 1025-1036.

Chen, L.-C. and L.-H. Lee (1992). "Progressive generation of control frameworks for image registration." *Photogrammetric Engineering and Remote Sensing* **58**(9): 1321-1328.

Chin, R. T. and C.-L. Yeh (1983). "Quantitative evaluation of some edge-preserving noise-smoothing techniques." *Computer Vision, Graphics and Image Processing*(23): 67-91.

Clark, R. N. and T. L. Roush (1984). "Reflectance Spectroscopy: Quantitative Analysis Techniques for Remote Sensing Applications." *Journal of Geophysical Research* **89**: 6329-6340.

Colby, J. D. (1991). "Topographic normalization in rugged terrain." *Photogrammetric Engineering and Remote Sensing* **57**(5): 531-537.

Craig, M. D. and A. A. Green (1987). "Registration of distorted images from airborne scanners." *The Australian Computer Journal* **19**(3): 148-153.

Crippen, R. E. (1989). "A simple spatial filtering routine for the cosmetic removal of scan-line noise from Landsat TM P-tape imagery." *Photogrammetric Engineering and Remote Sensing* **55**(3): 327-331.

Curran, P. J. and J. L. Dungan (1989). "Estimation of signal-to-noise: a new procedure applied to AVIRIS data." *IEEE Transactions on Geoscience and Remote Sensing* **27**(5): 620-628.

Devereux, B. J., R. M. Fuller, L. Carter, and R. J. Parsell (1990). "Geometric correction of airborne scanner imagery by matching Delaunay triangles." *International Journal of Remote Sensing* **11**(12): 2237-2251.

Dikshit, O. and D. P. Roy (1996). "An empirical investigation of image resampling effects upon the spectral and textural supervised classification of a high spatial resolution mulitspectral image." *Photogrammetric Engineering and Remote Sensing* **62**(9): 1085-1092.

Eliason, E. and A. S. McEwen (1990). "Adaptive box filter for removal of random noise from digital images." *Photogrammetric Engineering and Remote Sensing* **56**(4): 453-458.

EOSAT (1993). Fast Format Document. Lanham, Maryland, EOSAT.

Filho, C. R. d. S., S. A. Drury, A. M. Denniss, R. W. T. Carlton, and D. A. Rothery (1996). "Restoration of corrupted optical Fuyo-1 (JERS-1) data using frequency domain techniqeus." *Photogrammetric Engineering and Remote Sensing* **62**(9): 1037-1047.

Fischel, D. (1984). "Validation of the Thematic Mapper radiometric and geometric correction algorithms." *IEEE Transactions on Geoscience and Remote Sensing* **GE-22**(3): 237-242.

Fusco, L., U. Frei, D. Trevese, P. N. Blonda, G. Pasquariello, and G. Milillo (1986). "Landsat TM image forward/reverse scan banding: characterization and correction." *International Journal of Remote Sensing* **7**(4): 557-575.

Gao, B.-C. and A. F. H. Goetz (1990). "Column atmospheric water vapor and vegetation liquid water retrievals from airborne imaging spectrometer data." *Journal of Geophysical Research* **95**: 3549-3564.

Gao, B.-C., K. B. Heidebrecht, and A. F. H. Goetz (1993). "Derivation of scaled surface reflectances from AVIRIS data." *Remote Sensing of Environment* **44**: 165-178.

Gilbert, E. N. (1974). "Distortion in maps." *SIAM Review* **16**(1): 47-62.

Goetz, A. F. H., B. C. Kindel, M. Ferri and Z. Qu (2003). "HATCH: Results from simulated radiances, AVIRIS and Hyperion." *IEEE Transactions on Geoscience and Remote Sensing* **41**(6): 1215-1222.

Gonzalez, R. C. and R. E. Woods (2002). *Digital Image Processing*. Second Edition. Upper Saddle River, NJ: Prentice-Hall, 793 p.

Gu, D. and A. Gillespie (1998). "Topographic normalization of Landsat TM images of forest based on sub-pixel sun–canopy–sensor geometry." *Remote Sensing of Environment* **64**: 166-175.

Hadjimitsis, D. G., C. R. I. Clayton and V. S. Hope (2004). "An assessment of the effectiveness of atmospheric correction algorithms through the remote sensing of some reservoirs." *International Journal of Remote Sensing* **25**(18): 3651–3674.

Helder, D. L., B. K. Quirk, and J. J. Hood (1992). "A technique for the reduction of banding in Landsat Thematic Mapper images." *Photogrammetric Engineering and Remote Sensing* **58**(10): 1425-1431.

Helder, D. L. and T. A. Ruggles (2004). "Landsat Thematic Mapper reflective-band radiometric artifacts." *IEEE Transactions on Geoscience and Remote Sensing* **42**(12): 2704-2716.

Holben, B., E. Vermote, Y. J. Kaufman, D. Tanre, and V. Kalb (1992). "Aerosol retrieval over land from AVHRR data - application for atmospheric correction." *IEEE Transactions on Geoscience and Remote Sensing* **30**(2): 212-222.

Horn, B. K. P. and R. J. Woodham (1979). "Destriping Landsat MSS images by histogram modification." *Computer Graphics and Image Processing* **10**(1): 69-83.

Huang, C., J. R. G. Townshend, S. Liang, S. N. V. Kalluri and R. S. DeFries (2002). "Impact of sensor's point spread function on land cover characterization: Assessment and deconvolution." *Remote Sensing of Environment* **80**: 203-212.

Hummer-Miller, S. (1990). "Techniques for noise removal and registration of TIMS data." *Photogrammetric Engineering and Remote Sensing* **56**(1): 49-53.

Itten, K. I. and P. Meyer (1993). "Geometric and radiometric correction of TM data of mountainous forested areas." *IEEE Transactions on Geoscience and Remote Sensing* **31**(4): 764-770.

Jain, A. K. and R. C. Dubes (1988). *Algorithms for Clustering Data*. Englewood Cliffs, NJ: Prentice Hall, 320 p.

Jain, A. K. (1989). *Fundamentals of Digital Image Processing*. Englewood Cliffs, NJ: Prentice-Hall, 569 p.

Jensen, J. R. (2004). *Introductory Digital Image Processing – A Remote Sensing Perspective*. Third Edition. Upper Saddle River, NJ: Prentice Hall, 544 p.

Keys, R. G. (1981). "Cubic convolution interpolation for digital image processing." *IEEE Transactions on Acoustics, Speech, and Signal Processing* **ASSP-29**: 1153-1160.

Khan, B., L. W. B. Hayes, and A. P. Cracknell (1995). "The effects of higher-order resampling on AVHRR data." *International Journal of Remote Sensing* **16**(1): 147-163.

Kieffer, H. H. (1996). "Detection and correction of bad pixels in hyperspectral sensors." In *Hyperspectral Remote Sensing and Applications*, SPIE, vol. 2821: 93-108.

King, M. D., W. P. Menzel, Y. J. Kaufman, D. Tanré, B.-C. Gao, S. Platnick, S. A. Ackerman, L. A. Remer, R. Pincus and P. A. Hubanks (2003). "Cloud and aerosol properties, precipitable water, and profiles of temperature and water vapor from MODIS." *IEEE Transactions on Geoscience and Remote Sensing* **41**(2): 442-458.

Kruse, F. A. (1988). "Use of Airborne Imaging Spectrometer data to map minerals associated with hydrothermally altered rocks in the northern Grapevine Mountains, Nevada and California." *Remote Sensing of Environment* **24**(1): 31-51.

Kruse, F. A., K. S. Kierein-Young, and J. W. Boardman (1990). "Mineral mapping at Cuprite, Nevada with a 63-channel imaging spectrometer." *Photogrammetric Engineering and Remote Sensing* **56**(1): 83-92.

Lavreau, J. (1991). "De-hazing Landsat Thematic Mapper images." *Photogrammetric Engineering and Remote Sensing* **57**(10): 1297-1302.

Lee, J. (1983). "Digital image smoothing and the sigma filter." *Computer Vision, Graphics and Image Processing* **24**: 255-269.

Lee, J.-S. (1980). "Digital image enhancement and noise filtering by use of local statistics." *IEEE Transactions on Pattern Analysis and Machine Intelligence* **PAMI-2**(2): 165-168.

Lee, J.-S. (1981). "Speckle analysis and smoothing of synthetic aperture radar images." *Computer Graphics and Image Processing* **17**: 24-32.

Leprieur, C., V. Carrere, and X. F. Gu (1995). "Atmospheric corrections and ground reflectance recovery for Airborne Visible/Infrared Imaging Spectrometer (AVIRIS) data: MAC Europe'91." *Photogrammetric Engineering and Remote Sensing* **61**(10): 1233-1238.

Markham, B. L., K. J. Thome, J. A. Barsi, E. Kaita, D. L. Helder, J. L. Barker and P. L. Scaramuzza (2004). "Landsat-7 ETM+ on-orbit reflective-band radiometric stability and absolute calibration." *IEEE Transactions on Geoscience and Remote Sensing* **42**(12): 2810-2820.

Marsh, S. E. and J. B. McKeon (1983). "Integrated analysis of high-resolution field and airborne spectroradiometer data for alteration mapping." *Economic Geology* **78**(4): 618-632.

Mastin, G. A. (1985). "Adaptive filters for digital image noise smoothing: an evaluation." *Computer Vision, Graphics and Image Processing*(31): 103-121.

Matthew, M. W., S. M. Adler-Golden, A. Berk, S. C. Richtsmeier, R. Y. Levine, L. S. Bernstein, P. K. Acharya, G. P. Anderson, G. W. Felde, M. P. Hoke, A. Ratkowski, H.-H. Burke, R. D. Kaiser and D. P. Miller (2000). "Status of atmospheric correction using a MODTRAN4-based algorithm." In *Algorithms for Multispectral, Hyperspectral, and Ultraspectral Imagery VI*, SPIE, vol. 4049: 199-207.

Miller, C. J. (2002). "Performance assessment of ACORN atmospheric correction algorithm." In *Algorithms and Technologies for Multispectral, Hyperspectral, and Ultraspectral Imagery VIII*, SPIE, vol. 4725: 438-449.

Moik, J. G. (1980). *Digital Processing of Remotely Sensed Images*. Washington, D.C.: NASA, U.S. Government Printing Office, 330 p.

Montes, M. J., B.-C. Gao and C. O. Davis (2003). "Tafkaa atmospheric correction of hyperspectral data." In *Imaging Spectrometry IX*, SPIE, vol. 5159: 188-197.

Moran, M. S., R. D. Jackson, P. N. Slater, and P. M. Teillet (1992). "Evaluation of simplified procedures for retrieval of land surface reflectance factors from satellite sensor output." *Remote Sensing of Environment* **41**: 169-184.

Moran, M. S., R. Bryant, K. Thome, W. Ni, Y. Nouvellon, M. P. Gonzalez-Dugo, J. Qi and T. R. Clarke (2001). "A refined empirical line approach for reflectance factor retrieval from Landsat-5 TM and Landsat-7 ETM+." *Remote Sensing of Environment* **78**: 71-82.

Nagao, M. and T. Matsuyama (1979). "Edge preserving smoothing." *Computer Graphics and Image Processing* **9**: 394-407.

NASA (2006). *Landsat-7 Science Data Users Handbook*. Greenbelt, MD, NASA Goddard Space Flight Center.

Neville, R. A., L. Sunb and K. Staenz (2003). "Detection of spectral line curvature in imaging spectrometer data." In *Algorithms and Technologies for Multispectral, Hyperspectral, and Ultraspectral Imagery IX*, SPIE, vol. 5093: 144-154.

Ouaidrari, H. and E. F. Vermote (1999). "Operational atmospheric correction of Landsat TM data." *Remote Sensing of Environment* **70**: 4-15.

Pan, J.-J. (1989). "Spectral analysis and filtering techniques in digital spatial data processing." *Photogrammetric Engineering and Remote Sensing* **55**(8): 1203-1207.

Pan, J.-J. and C.-I. Chang (1992). "Destriping of Landsat MSS images by filtering techniques." *Photogrammetric Engineering and Remote Sensing* **58**(10): 1417-1423.

Parada, M., A. Millan, A. Lobato and A. Hermosilla (2000). "Fast coastal algorithm for automatic geometric correction of AVHRR images." *International Journal of Remote Sensing* **21**(11): 2307-2312.

Park, S. K. and R. A. Schowengerdt (1983). "Image reconstruction by parametric cubic convolution." *Computer Vision, Graphics and Image Processing* **20**(3): 258-272.

Pearlman, J. S., P. S. Barry, C. C. Segal, J. Shepanski, D. Beiso and S. L. Carman (2003). "Hyperion, a space-based imaging spectrometer." *IEEE Transactions on Geoscience and Remote Sensing* **41**(6): 1160-1173.

Perkins, T., S. Adler-Golden, M. Matthew, A. Berk, G. Anderson, J. Gardner and G. Felde (2005). "Retrieval of atmospheric properties from hyper- and multi-spectral imagery with the FLAASH atmospheric correction algorithm." In *Remote Sensing of Clouds and the Atmosphere X*, SPIE, vol. 5979: 5979OE-1-5979OE-11.

Peros, D. J. and C. J. Peterson (1985). "Methods for destriping Landsat Thematic Mapper images - A feasibility study for an online destriping process in the Thematic Mapper Image Processing System (TIPS)." *Photogrammetric Engineering and Remote Sensing* **51**(9): 1371-1378.

Potter, J. F. (1984). "The channel correlation method for estimating aerosol levels from multispectral scanner data." *Photogrammetric Engineering and Remote Sensing* **50**: 43-52.

Potter, J. F. and M. Mendolowitz (1975). "On the determination of the haze levels from Landsat data." in: *10th International Symposium on Remote Sensing of Environment*, Ann Arbor, MI, Environmental Research Institute of Michigan, : 695-703.

Pratt, W. K. (1991). *Digital Image Processing*. Second Edition. New York, John Wiley & Sons, 698 p.

Qu, Z., B. C. Kindel and A. F. H. Goetz (2003). "The high accuracy atmospheric correction for hyperspectral data (HATCH) model." *IEEE Transactions on Geoscience and Remote Sensing* **41**(6): 1223-1231.

Quarmby, N. C. (1987). "Noise removal for SPOT imagery." *International Journal of Remote Sensing* **8**: 1229-1234.

Rao, C. R. N. and J. Chen (1994). Post-Launch Calibration of the Visible and Near Infrared Channels of the Advanced Very High Resolution Radiometer on NOAA-7, -9 and -11 Spacecraft. Washington, D.C.: National Oceanic and Atmospheric Administration, No. NESDIS 78.

Rast, M., S. J. Hook, C. D. Elvidge, and R. E. Alley (1991). "An evaluation of techniques for the extraction of mineral absorption features from high spectral resolution remote sensing data." *Photogrammetric Engineering and Remote Sensing* **57**(10): 1303-1309.

Reichenbach, S. E., D. E. Koehler and D. W. Strelow (1995). "Restoration and reconstruction of AVHRR images." *IEEE Transactions on Geoscience and Remote Sensing* **33**(4): 997-1007.

Reichenbach, S. E. and J. Shi (2004). "Two-dimensional cubic convolution for one-pass image restoration and reconstruction." In *Geoscience and Remote Sensing Symposium, IGARSS '04*, IEEE, pp. 2074-2076a.

Richards, J. A. and X. Jia (1999). *Remote Sensing Digital Image Analysis – An Introduction*. Third Edition. Berlin: Springer-Verlag, 356 p.

Richards, M. E. (1985). "A comparison of two nonlinear destriping procedures." in: *Architectures and Algorithms for Digital Image Processing II*, SPIE, vol. 534: 128-134.

Richter, R. (1996a). "A spatially adaptive fast atmospheric correction algorithm." *International Journal of Remote Sensing* **17**(6): 1201-1214.

Richter, R. (1996b). "Atmospheric correction of satellite data with haze removal including a haze/clear transition region." *Computers & Geosciences* **22**(6): 675-681.

Richter, R. and D. Schlapfer (2002). "Geo-atmospheric processing of airborne imaging spectrometry data. Part 2: Atmospheric/topographic correction." *International Journal of Remote Sensing* **23**(13): 2631–2649.

Richter, R. and A. Muller (2005). "De-shadowing of satellite/airborne imagery." *International Journal of Remote Sensing* **26**(15): 3137-3148.

Rindfleisch, T. C., J. A. Dunne, H. J. Frieden, W. D. Stromberg, and R. M. Ruiz (1971). "Digital processing of the Mariner 6 and 7 pictures." *Journal of Geophysical Research* **76**(2): 394-417.

Ripley, B. D. (1981). *Spatial Statistics*. New York, NY: John Wiley & Sons, 252 p.

Rojas, F., R. A. Schowengerdt and S. F. Biggar (2002). "Error and correction for MODIS-AM's spatial response on the NDVI and EVI science products." In *Earth Observing Systems VII*, SPIE, vol. 4814: 447-456.

Rose, J. F. (1989). "Spatial interference in the AVIRIS imaging spectrometer." *Photogrammetric Engineering and Remote Sensing* **55**(9): 1339-1346.

Roy, D. P. and O. Dikshit (1994). "Investigation of image resampling effects upon the textural information content of a high spatial resolution remotely sensed image." *International Journal of Remote Sensing* **15**(5): 1123-1130.

Ruiz, C. P. and F. J. A. Lopez (2002). "Restoring SPOT images using PSF-derived deconvolution." *International Journal of Remote Sensing* **23**(12): 2379-2391.

Salama, M. S., J. Monbaliu and P. Coppin (2004). "Atmospheric correction of advanced very high resolution radiometer imagery." *International Journal of Remote Sensing* **25**(7-8): 1349–1355.

Sanders, L. C., J. R. Schott and R. Raqueno (2001). "A VNIR/SWIR atmospheric correction algorithm for hyperspectral imagery with adjacency effect." *Remote Sensing of Environment* **78**: 252-263.

Schott, J. R., C. Salvaggio and W. J. Volchok (1988). "Radiometric scene normalization using pseudoinvariant features." *Photogrammetric Engineering and Remote Sensing* **26**: 1-16

Schowengerdt, R. A. (1983). *Techniques for Image Processing and Classification in Remote Sensing*. Academic Press, 249 p.

Schürmann, J. (1996). *Pattern Classification–A Unified View of Statistical and Neural Approaches*. New York, NY: John Wiley & Sons, 373 p.

Sethmann, R., B. A. Burns and G. C. Heygster (1994). "Spatial resolution improvement of SSM/I data with image restoration techniques." *IEEE Transactions on Geoscience and Remote Sensing* **32**(6): 1144-1151.

Shepherd, J. D. and J. R. Dymond (2003). "Correcting satellite imagery for the variance of reflectance and illumination with topography." *International Journal of Remote Sensing* **24**(17): 3503-3514.

Shetler, B. and H. Kieffer (1996). "Characterization and reduction of stochastic and periodic anomalies in an hyperspectral imaging sensor system." In *Hyperspectral Remote Sensing and Applications*, SPIE, vol. 2821: 109-126.

Simpson, J. J. and S. R. Yhann (1994). "Reduction of noise in AVHRR channel 3 data with minimum distortion." *IEEE Transactions on Geoscience and Remote Sensing* **32**(2): 315-328.

Simpson, J. J., J. R. Stitt and D. M. Leath (1998). "Improved finite impulse response filters for enhanced destriping of geostationary satellite data." *Remote Sensing of Environment* **66**: 235–249.

Singh, S. M. and A. P. Cracknell (1986). "The estimation of atmospheric effects for SPOT using AVHRR channel-1 data." *International Journal of Remote Sensing* **7**(3): 361-377.

Smith, J. A., T. L. Lin, and K. Ranson (1980). "The lambertian assumption and Landsat data." *Photogrammetric Engineering and Remote Sensing* **46**(9): 1183-1189.

Srinivasan, R., M. Cannon, and J. White (1988). "Landsat data destriping using power spectral filtering." *Optical Engineering* **27**(11): 939-943.

Steiner, D. and M. E. Kirby (1976). "Geometrical referencing of Landsat images by affine transformation and overlaying of map data." *Photogrammetria* **33**: 41-75.

Steinwand, D. R. (1994). "Mapping raster imagery to the Interrupted Goode Homolsine projection." *International Journal of Remote Sensing* **15**(17): 3463-3471.

Steinwand, D. R., J. A. Hutchinson, and J. P. Snyder (1995). "Map projections for global and continental data sets and an analysis of pixel distortion caused by reprojection." *Photogrammetric Engineering and Remote Sensing* **61**(12): 1487-1497.

Storey, J. (2006). Personal communication.

Swann, R., D. Hawkins, A. Westwell-Roper, and W. Johnstone (1988). "The potential for automated mapping from geocoded digital image data." *Photogrammetric Engineering and Remote Sensing* **54**(2): 187-193.

Switzer, P., W. S. Kowalik, and R. J. Lyon (1981). "Estimation of atmospheric path radiance by the covariance matrix method." *Photogrammetric Engineering and Remote Sensing* **47**: 1469-1476.

Tanre, D., C. Deroo, P. Duhaut, M. Herman, J. J. Morcrette, J. Perbos, and P. Y. Deschamps (1990). "Description of a computer code to simulate the satellite signal in the solar spectrum: 5S code." *International Journal of Remote Sensing* **11**: 659-668.

Teillet, P. M. and G. Fedosejevs (1995). "On the dark target approach to atmospheric correction of remotely sensed data." *Canadian Journal of Remote Sensing* **21**(4): 374-387.

Thome, K. J., S. F. Biggar, D. I. Gellman, and P. N. Slater (1994). "Absolute-radiometric calibration of Landsat-5 Thematic Mapper and the proposed calibration of the Advanced Spaceborne Thermal Emission and Reflection Radiometer." in: *IGARSS-94*, Pasadena, CA: IEEE, vol. 4: 2295-2297.

Tilton, J. C., B. L. Markham, and W. L. Alford (1985). "Landsat-4 and Landsat-5 MSS coherent noise: characterization and removal." *Photogrammetric Engineering and Remote Sensing* **51**(9): 1263-1279.

Vermote, E. F., D. Tanre, J. L. Deuze, M. Herman and J.-J. Morcrette (1997). "Second simulation of the satellite signal in the solar spectrum, 6S: An overview." *IEEE Transactions on Geoscience and Remote Sensing* **35**(3): 675-686.

Wegener, M. (1990). "Destriping multiple sensor imagery by improved histogram matching." *International Journal of Remote Sensing* **11**(5): 859-875.

Weinreb, M. P., R. Xie, J. H. Lienesch, and D. S. Crosby (1989). "Destriping GOES images by matching empirical distribution functions." *Remote Sensing of Environment* **29**: 185-195.

Westin, T. (1990). "Filters for removing coherent noise of period 2 in SPOT imagery." *International Journal of Remote Sensing* **11**(2): 351-357.

Westin, T. (1990). "Precision rectification of SPOT imagery." *Photogrammetric Engineering and Remote Sensing* **56**(2): 247-253.

Wolberg, G. (1990). *Digital Image Warping*. Los Alamitos, CA: IEEE Computer Society Press, 318 p.

Wolf, P. R. (1983). *Elements of Photogrammetry*. Second International Student Edition. Singapore: McGraw-Hill, 628 p.

Wood, L., R. A. Schowengerdt and D. Meyer (1986). "Restoration for sampled imaging systems." In *Applications of Digital Processing IX*, San Diego, CA, SPIE, vol. 697: 333-340.

Wrigley, R. C., D. H. Card, C. A. Hlavka, J. R. Hall, F. C. Mertz, C. Archwamety, and R. A. Schowengerdt (1984). "Thematic Mapper image quality: Registration, noise, and resolution." *IEEE Transactions on Geoscience and Remote Sensing* **GE-22**(3): 263-271.

Wrigley, R. C., M. A. Spanner, R. E. Slye, R. F. Pueschel, and H. R. Aggarwal (1992). "Atmospheric correction of remotely sensed image data by a simplified model." *Journal of Geophysical Research* **97**(D-7): 18797-18814.

Wu, H.-H. P. and R. A. Schowengerdt (1993). "Improved fraction image estimation using image restoration." *IEEE Transactions on Geoscience and Remote Sensing* **31**(4): 771-778.

Zagolski, F. and J. P. Gastellu-Etchegorry (1995). "Atmospheric corrections of AVIRIS images with a procedure based on the inversion of the 5S model." *International Journal of Remote Sensing* **16**(16): 3115-3146.

CHAPTER 8 *Image Registration and Fusion*

Anuta, P. (1970). "Spatial registration of multispectral and multitemporal digital imagery using Fast Fourier Transform techniques." *IEEE Transactions on Geoscience Electronics* **GE-8**(4): 353-368.

Barnea, D. I. and H. E. Silverman (1972). "A class of algorithms for fast digital image registration." *IEEE Transactions on Computers* **C-21**(2): 179-186.

Bernstein, R. (1976). "Digital image processing of earth observation sensor data." *IBM Journal of Research and Development* **20**(1): 40-57.

Biemond, J., R. Lagendijk and R. Mersereau (1990). "Iterative methods for image deblurring." *Proceedings of the IEEE* **78**(5): 856-883.

Brockelbank, D. C. and A. P. Tam (1991). "Stereo elevation determination techniques for SPOT imagery." *Photogrammetric Engineering and Remote Sensing* **57**(8): 1065-1073.

Brown, L. (1992). "A survey of image registration techniques." *ACM Computing Surveys* **24**(4).

Carper, W. J., T. M. Lillesand, and R. W. Kiefer (1990). "The use of intensity-hue-saturation transformations for merging SPOT panchromatic and multispectral image data." *Photogrammetric Engineering and Remote Sensing* **56**(4): 459-467.

Chavez, P. S., Jr. (1986). "Digital merging of Landsat TM and digitized NHAP data for 1:24,000-scale image mapping." *Photogrammetric Engineering and Remote Sensing* **52**(10): 1637-1646.

Chavez, P. S., Jr., S. C. Sides, and J. A. Anderson (1991). "Comparison of three different methods to merge multiresolution and multispectral data: Landsat TM and SPOT panchromatic." *Photogrammetric Engineering and Remote Sensing* **57**(3): 295-303.

Chen, H.-M., M. K. Arora and P. K. Varshney (2003). "Mutual information-based image registration for remote sensing data." *International Journal of Remote Sensing* **24**(18): 3701-3706.

Cliche, G., F. Bonn, and P. Teillet (1985). "Integration of the SPOT panchromatic channel into its multispectral mode for image sharpness enhancement." *Photogrammetric Engineering and Remote Sensing* **51**(3): 311-316.

Craig, M. D. and A. A. Green (1987). "Registration of distorted images from airborne scanners." *The Australian Computer Journal* **19**(3): 148-153.

Dai, X. and S. Khorram (1999). "A feature-based image registration algorithm using improved chain-code representation combined with invariant moments." *IEEE Transactions on Geoscience and Remote Sensing* **37**(5): 2351-2362.

Daily, M. I., T. Farr, C. Elachi, and G. Schaber (1979). "Geologic interpretation from composited radar and Landsat imagery." *Photogrammetric Engineering and Remote Sensing* **45**(8): 1109-1116.

Djamdji, J. P., A. Bijaoui, and R. Maniere (1993b). "Geometrical registration of images: the multiresolution approach." *Photogrammetric Engineering and Remote Sensing* **59**(5): 645-653.

Eckert, S., T. Kellenberger and K. Itten (2005). "Accuracy assessment of automatically derived digital elevation models from aster data in mountainous terrain." *International Journal of Remote Sensing* **26**(9): 1943–1957.

Ehlers, M. (1991). "Multisensor image fusion techniques in remote sensing." *ISPRS Journal of Photogrammetry and Remote Sensing* **46**: 19-30.

Ehlers, M. and R. Welch (1987). "Stereocorrelation of Landsat TM images." *Photogrammetric Engineering and Remote Sensing* **53**: 1231–1237.

Filiberti, D., S. Marsh, and R. Schowengerdt (1994). "Synthesis of high spatial and spectral resolution imagery from multiple image sources." *Optical Engineering* **33**(8): 2520-2528.

Filiberti, D. P. and R. A. Schowengerdt (2004). "Improving multisource image fusion using thematic content." In *Visual Information Processing XIII*, Orlando, FL, SPIE, vol. 5438: 111-119.

Flusser, J. and T. Suk (1994). "A moment-based approach to registration of images with affine geometric distortion." *IEEE Transactions on Geoscience and Remote Sensing* **32**(2): 382-387.

Fonseca, L. M. G. and B. S. Manjunath (1996). "Registration techniques for multisensor remotely sensed imagery." *Photogrammetric Engineering and Remote Sensing* **62**(9): 1049-1056.

Garguet-Duport, B., J. Girel, J.-M. Chassery, and G. Pautou (1996). "The use of multiresolution analysis and wavelets transform for merging SPOT panchromatic and multispectral image data." *Photogrammetric Engineering and Remote Sensing* **62**(9): 1057-1066.

Gonzalez, R. C. and R. E. Woods (1992). *Digital Image Processing*. Reading, MA, Addison-Wesley, 703 p.

Gonzalez-Audicana, M., X. Otazu, O. Fors and A. Seco (2005). "Comparison between Mallat's and the 'a` trous' discrete wavelet transform based algorithms for the fusion of multispectral and panchromatic images." *International Journal of Remote Sensing* **26**(3): 595-614.

Goshtasby, A. (1988). "Registration of images with geometric distortions." *IEEE Transactions on Geoscience and Remote Sensing* **26**(1): 60-64.

Goshtasby, A. (1993). "Correction to "Registration of images with geometric distortions"." *IEEE Transactions on Geoscience and Remote Sensing* **31**(1): 307.

Goshtasby, A., G. Stockman, and C. Page (1986). "A region-based approach to digital image registration with subpixel accuracy." *IEEE Transactions on Geoscience and Remote Sensing* **GE-24**(3).

Greenfield, J. S. (1991). "An Operator–Based Matching System." *Photogrammetric Engineering and Remote Sensing* **57**(8): 1049-1055.

Hall, E. L. (1979). *Computer Image Processing and Recognition*. New York, NY: Academic Press, 584 p.

Harris, J. R., R. Murray, and T. Hirose (1990). "IHS transform for the integration of radar imagery with other remotely sensed data." *Photogrammetric Engineering and Remote Sensing* **56**(12): 1631-1641.

Haydn, R., G. W. Dalke, J. Henkel, and J. E. Bare (1982). "Application of the IHS color transform to the processing of multisensor data and image enhancement." in: *International Symposium on Remote Sensing of Arid and Semi-Arid Lands*, Cairo, Egypt: Environmental Research Institute of Michigan: 599-616.

Henderson, T. C., E. E. Triendl, and R. Winter (1985). "Edge- and shape-based geometric registration." *IEEE Transactions on Geoscience and Remote Sensing* **GE-23**(3): 334-341.

Hong, T. D. and R. A. Schowengerdt (2005). "A robust technique for precise registration of radar and optical satellite images." *Photogrammetric Engineering and Remote Sensing* **71**(5): 585-593.

Hood, J., L. Ladner, and R. Champion (1989). "Image processing techniques for digital orthophotoquad production." *Photogrammetric Engineering and Remote Sensing* **55**(9): 1323-1329.

Kennedy, R. E. and W. B. Cohen (2003). "Automated designation of tie-points for image-to-image coregistration." *International Journal of Remote Sensing* **24**(17): 3467–3490.

Konecny, G., P. Lohmann, H. Engel, and E. Kruck (1987). "Evaluation of SPOT imagery on analytical photogrammetric instruments." *Photogrammetric Engineering and Remote Sensing* **53**(9): 1223-1230.

Le Moigne, J., W. J. Campbell and R. F. Cromp (2002). "An automated parallel image registration technique based on the correlation of wavelet features." *IEEE Transactions on Geoscience and Remote Sensing* **40**(8): 1849-1864.

Lemeshewsky, G. P. (2005). "Sharpening Advanced Land Imager multispectral data using a sensor model." In *Visual Information Processing XIV*, Orlando, FL, SPIE, vol. 5817: 336-346.

Li, H., B. S. Manjunath, and S. K. Mitra (1995). "A contour-based approach to multisensor image registration." *IEEE Transactions on Image Processing* **4**(3): 320-334.

Liang, T. and C. Heipke (1996). "Automatic relative orientation of aerial images." *Photogrammetric Engineering and Remote Sensing* **62**(1): 47-55.

Lillo-Saavedra, M., C. Gonzalo, A. Arquero and E. Martinez (2005). "Fusion of multispectral and panchromatic satellite sensor imagery based on tailored filtering in the fourier domain." *International Journal of Remote Sensing* **26**(6): 1263-1268.

Liu, J. G. (2000). "Smoothing filter-based intensity modulation: A spectral preserve image fusion technique for improving spatial details." *International Journal of Remote Sensing* **21**(18): 3461–3472.

Mikhail, E. M., J. S. Bethel and J. C. McGlone (2001). *Introduction to Modern Photogrammetry*. John Wiley & Sons, 496 p.

Moran, S. M. (1990). "A window-based technique for combining Landsat thematic mapper thermal data with higher-resolution multispectral data over agricultural lands." *Photogrammetric Engineering and Remote Sensing* **56**(3): 337-342.

Muller, J.-P., A. Mandanayake, C. Moroney, R. Davies, D. J. Diner and S. Paradise (2002). "MISR stereoscopic image matchers: Techniques and results." *IEEE Transactions on Geoscience and Remote Sensing* **40**(7): 1547-1559.

Munechika, C. K., J. S. Warnick, C. Salvaggio, and J. R. Schott (1993). "Resolution enhancement of multispectral image data to improve classification accuracy." *Photogrammetric Engineering and Remote Sensing* **59**(1): 67-72.

Panton, D. J. (1978). "A flexible approach to digital stereo matching." *Photogrammetric Engineering and Remote Sensing* **44**(12): 1499-1512.

Park, J. H. and M. G. Kang (2004). "Spatially adaptive multi-resolution multispectral image fusion." *International Journal of Remote Sensing* **25**(23): 5491–5508.

Pellemans, A. H. J. M., R. W. L. Jordans, and R. Allewijn (1993). "Merging multispectral and panchromatic SPOT images with respect to the radiometric properties of the sensor." *Photogrammetric Engineering and Remote Sensing* **59**(1): 81-87.

Pradines, D. (1986). "Improving SPOT images size and multispectral resolution." in: *Earth Remote Sensing using the Landsat Thematic Mapper and SPOT systems*, SPIE, vol. 660: 98-102.

Pratt, W. K. (1974). "Correlation techniques of image registration." *IEEE Transactions on Aerospace and Electronic Systems* **AES-10**(3): 353-358.

Pratt, W. K. (1991). *Digital Image Processing*. Second Edition. New York, NY: John Wiley & Sons, 698 p.

Price, J. C. (1987). "Combining panchromatic and multispectral imagery from dual resolution satellite instruments." *Remote Sensing of Environment* **21**: 119-128.

Quam, L. H. (1987). Hierarchical Warp Stereo. *Readings in Computer Vision: Issues, Problems, Principles, and Paradigms*. M. A. Fischler and O. Firschein, (Eds.). Los Altos, CA: Morgan Kaufmann Publishers, Inc.: 800.

Ramapriyan, H. K., J. P. Strong, Y. Hung, and J. Charles W. Murray (1986). "Automated matching of pairs of SIR-B images for elevation mapping." *IEEE Transactions on Geoscience and Remote Sensing* **GE-24**(4): 462-472.

Ranchin, T., B. Aiazzi, L. Alparone, S. Baronti and L. Wald (2003). "Image fusion—the arsis concept and some successful implementation schemes." *ISPRS Journal of Photogrammetry and Remote Sensing* **58**: 4-18.

Rao, T. C. M., K. V. Rao, A. R. Kumar, D. P. Rao, and B. L. Deekshatula (1996). "Digital Terrain Model (DTM) from Indian Remote Sensing (IRS) satellite data from the overlap area of two adjacent paths using digital photogrammetric techniques." *Photogrammetric Engineering and Remote Sensing* **62**(6): 727-731.

Reddy, B. S. and B. N. Chatterji (1996). "An FFT-based technique for translation, rotation, and scale-invariant image registration." *IEEE Transactions on Image Processing* **5**(8): 1266-1271.

Redriguez, V., P. Gigord, A. C. d. Gaujac, P. Munier, and G. Begni (1988). "Evaluation of the stereoscopic accuracy of the SPOT satellite." *Photogrammetric Engineering and Remote Sensing* **54**(2).

Rosenfeld, A. and A. C. Kak (1982). *Digital Picture Processing*. Second Edition. Orlando, FL: Academic Press, 349 p.

Scambos, T. A., M. J. Dutkiewicz, J. C. Wilson, and R. A. Bindschadler (1992). "Application of image cross–correlation to the measurement of glacier velocity using satellite image data." *Remote Sensing of Environment* **42**: 177-186.

Schenk, T., J. C. Li, and C. Toth (1991). "Towards an Autonomous System for Orienting Digital Stereopairs." *Photogrammetric Engineering and Remote Sensing* **57**(8): 1057-1064.

Schowengerdt, R. A. (1980). "Reconstruction of multispatial, multispectral image data using spatial frequency content." *Photogrammetric Engineering and Remote Sensing* **46**(10): 1325-1334.

Schowengerdt, R. A. and D. Filiberti (1994). "Spatial frequency models for multispectral image sharpening." in: *Algorithms for Multispectral and Hyperspectral Imagery*, Orlando, Florida, SPIE, vol. 2231: 84-90.

Shettigara, V. K. (1992). "A generalized component substitution technique for spatial enhancement of multispectral images using a higher resolution data set." *Photogrammetric Engineering and Remote Sensing* **58**(5): 561-567.

SPOTImage, SPOTView Digital Ortho-Image Data Sampler CD-ROM, 1995.

Tateishi, R. and A. Akutsu (1992). "Relative DEM production from SPOT data without GCP." *International Journal of Remote Sensing* **13**(14): 2517-2530.

Tom, V. T., M. J. Carlotto, and D. K. Scholten (1985). "Spatial sharpening of Thematic Mapper data using a multiband approach." *Optical Engineering* **24**(6): 1026-1029.

Ton, J. and A. K. Jain (1989). "Registering Landsat images by point matching." *IEEE Transactions on Geoscience and Remote Sensing* **27**(5).

Toutin, T. (2002). "DEM from stereo Landsat 7 ETM+ data over high relief areas." *International Journal of Remote Sensing* **23**(10): 2133 –2139.

Toutin, T. (2004). "DSM generation and evaluation from QuickBird stereo imagery with 3D physical modelling." *International Journal of Remote Sensing* **25**(22): 5181–5193.

USGS (1995). Capital Cities of the United States. *Digital Raster Graphic Data CD-ROM*, U.S. Geological Survey.

Ventura, A. D., A. Rampini, and R. Schettini (1990). "Image registration by the recognition of corresponding structures." *IEEE Transactions on Geoscience and Remote Sensing* **28**(3): 305-387.

Welch, R. and M. Ehlers (1987). "Merging multiresolution SPOT HRV and Landsat TM data." *Photogrammetric Engineering and Remote Sensing* **53**(3): 301-303.

Welch, R. and M. Ehlers (1988). "Cartographic feature extraction with integrated SIR-B and Landsat TM images." *International Journal of Remote Sensing* **9**(5): 873-889.

Westin, T. (1990). "Precision rectification of SPOT imagery." *Photogrammetric Engineering and Remote Sensing* **56**(2): 247-253.

Wisniewski, W. T. and R. A. Schowengerdt (2005). "Information in the joint aggregate pixel distribution of two images." In *Visual Information Processing XIV*, Orlando, FL, pp. 167-178.

Wolf, P. R. and B. A. Dewitt (2000). *Elements of Photogrammetry with Applications in GIS*. Third Edition. McGraw-Hill, 624 p.

Wong, F. H. and R. Orth (1980). "Registration of SEASAT-LANDSAT composite images to UTM coordinates." in: *Sixth Canadian Symposium on Remote Sensing*, Halifax, Nova Scotia: 161-164.

Yocky, D. A. (1996). "Multiresolution wavelet decomposition image merger of Landsat Thematic Mapper and SPOT panchromatic data." *Photogrammetric Engineering and Remote Sensing* **62**(9): 1067-1074.

Zheng, Q. and R. Chellappa (1993). "A computational vision approach to image registration." *IEEE Transactions on Image Processing* **2**: 311-326.

Zhou, J., D. L. Civco and J. A. Silander (1998). "A wavelet transform method to merge Landsat TM and SPOT panchromatic data." *International Journal of Remote Sensing* **19**(4): 743-757.

Zhukov, B., D. Oertel, F. Lanzl and G. Reinhackel (1999). "Unmixing-based multisensor multiresolution image fusion." *IEEE Transactions on Geoscience and Remote Sensing* **37**(3): 1212-1226.

Zitova, B. and J. Flusser (2003). "Image registration methods: A survey." *Image and Vision Computing* **21**(11): 977-1000.

CHAPTER 9 Thematic Classification

Abrams, M. and S. J. Hook (1995). "Simulated Aster data for geologic studies." *IEEE Transactions on Geoscience and Remote Sensing* **33**(3): 692-699.

Abuelgasim, A. A., S. Gopal, J. R. Irons, and A. H. Strahler (1996). "Classification of ASAS multiangle and multispectral measurements using artificial neural networks." *Remote Sensing of Environment* **57**: 79-87.

Adams, J. B., M. O. Smith, and A. R. Gillespie (1993). Imaging Spectroscopy: Interpretation Based on Spectral Mixture Analysis. *Remote Geochemical Analysis: Elemental and Mineralogical Composition*. C. M. Pieters and P. A. Englert, (Eds.), Cambridge: Cambridge University Press: 145-166.

Anderberg, M. R. (1973). *Cluster Analysis for Applications*. New York, NY: Academic Press, 353 p.

Anderson, J. R., E. E. Hardy, J. T. Roach, and R. E. Witmer (1976). A Land Use and Land Cover Classification System for Use with Remote Sensor Data. Washington, D.C., U. S. Geological Survey, No. Professional Paper 964.

Arai, K. (1992). "A supervised Thematic Mapper classification with a purification of training samples." *International Journal of Remote Sensing* **13**(11): 2039-2049.

Argialas, D. P. and C. A. Harlow (1990). "Computational image interpretation models: an overview and a perspective." *Photogrammetric Engineering and Remote Sensing* **56**(6): 871-886.

Arriaza, J. A. T., F. G. Rojas, M. P. López and M. Cantón (2003). "An automatic cloud-masking system using backpro neural nets for AVHRR scenes." *IEEE Transactions on Geoscience and Remote Sensing* **41**(4): 826-831.

Asner, G. P., C. A. Wessman and J. L. Privette (1997). "Unmixing the directional reflectances of AVHRR sub-pixel landcovers." *IEEE Transactions on Geoscience and Remote Sensing* **35**(4): 868-878.

Augusteijn, M. F., L. E. Clemens, and K. A. Shaw (1995). "Performance evaluation of texture measures for ground cover identification in satellite image by means of a neural network classifier." *IEEE Transactions on Geoscience and Remote Sensing* **33**(3): 616-626.

Avery, T. E. and G. L. Berlin (1992). *Fundamentals of Remote Sensing and Airphoto Interpretation*. Fifth Edition. New York, NY: Macmillan Publishing Company, 472 p.

Ball, G. and D. Hall (1967). "A clustering technique for summarizing multivariate data." *Behavioral Science* **12**: 153-155.

Ballard, D. H. and C. M. Brown (1982). *Computer Vision*. Englewood Cliffs, NJ: Prentice Hall, 523 p.

Baraldi, A. and F. Parmiggiani (1995). "A neural network for unsupervised categorization of multivalued input patterns: an application to satellite image clustering." *IEEE Transactions on Geoscience and Remote Sensing* **33**(2): 305-316.

Bateson, A. and B. Curtiss (1996). "A method for manual endmember selection and spectral unmixing." *Remote Sensing of Environment* **55**: 229-243.

Bateson, C. A., G. P. Asner and C. A. Wessman (2000). "Endmember bundles: A new approach to incorporating endmember variability into spectral mixture analysis." *IEEE Transactions on Geoscience and Remote Sensing* **38**(2): 1083-1094.

Bayliss, J., J. A. Gualtieri and R. F. Cromp (1997). "Analyzing hyperspectral data with independent component analysis." In *26th AIPR Workshop: Exploiting New Image Sources and Sensors*, Washington DC, SPIE, vol. 3240: 133-143.

Benediktsson, J. A., J. R. Sveinsson, and K. Arnason (1995). "Classification and feature extraction of AVIRIS data." *IEEE Transactions on Geoscience and Remote Sensing* **33**(5): 1194-1205.

Benediktsson, J. A. and P. H. Swain (1992). "Consensus theoretic classification methods." *IEEE Transactions on Systems, Man & Cybernetics* **22**(4): 688-704.

Benediktsson, J. A., P. H. Swain, and O. K. Ersoy (1990a). "Neural network approaches versus statistical methods in classification of multisource remote sensing data." *IEEE Transactions on Geoscience and Remote Sensing* **28**(4): 540-551.

Benie, G. B. and K. P. B. Thomson (1992). "Hierarchical image segmentation using local and adaptive similarity rules." *International Journal of Remote Sensing* **13**(8): 1559-1570.

Benjamin, S. and L. Gaydos (1990). "Spatial resolution requirements for automated cartographic road extraction." *Photogrammetric Engineering and Remote Sensing* **56**(1): 93-100.

Bezdek, J. C., R. Ehrlich, and W. Full (1984). "FCM: The fuzzy c-means clustering algorithm." *Computers and Geosciences* **10**(2-3): 191-203.

Billingsley, F. C. (1982). "Modeling misregistration and related effects on multispectral classification." *Photogrammetric Engineering and Remote Sensing* **48**(3): 421-430.

Bischof, H., W. Schneider, and A. J. Pinz (1992). "Multispectral classification of Landsat–images using neural networks." *IEEE Transactions on Geoscience and Remote Sensing* **30**(3): 482-490.

Boardman, J. (1990). "Inversion of high spectral resolution data." SPIE, vol. 1298: 222-233.

Boardman, J. W., F. A. Kruse, and R. O. Green (1995). "Mapping target signatures via partial unmixing of AVIRIS data." in: *Fifth JPL Airborne Earth Science Workshop*, Pasadena, CA: JPL, vol. 95-1: 23-26.

Bolstad, P. V. and T. M. Lillesand (1991). "Rapid maximum likelihood classification." *Photogrammetric Engineering and Remote Sensing* **57**(1): 67-74.

Borel, C. C. and S. A. W. Gerstl (1994). "Nonlinear spectral mixing models for vegetative and soil surfaces." *Remote Sensing of Environment* **47**: 403-416.

Bouman, C. A. and M. Shapiro (1994). "A multiscale random field model for bayesian image segmentation." *IEEE Transactions on Image Processing* **3**(2): 162-177.

Bruce, L. M., C. Morgan and S. Larsen (2001). "Automated detection of subpixel hyperspectral targets with continuous and discrete wavelet transforms." *IEEE Transactions on Geoscience and Remote Sensing* **39**(10): 2217-2226.

Bruzzone, L., D. F. Prieto and S. B. Serpico (1999). "A neural-statistical approach to multitemporal and multisource remote-sensing image classification." *IEEE Transactions on Geoscience and Remote Sensing* **37**(3): 1350-1359.

Bryant, J. (1979). "On the clustering of multidimensional pictorial data." *Pattern Recognition* **11**(2): 115-125.

Bryant, J. (1989). "A fast classifier for image data." *Pattern Recognition* **22**: 45-48.

Bryant, J. (1990). "AMOEBA clustering revisited." *Photogrammetric Engineering and Remote Sensing* **56**(1): 41-47.

Buchheim, M. P. and T. M. Lillesand (1989). "Semi-automated training field extraction and analysis for efficient digital image classification." *Photogrammetric Engineering and Remote Sensing* **55**(9): 1347-1355.

Camps-Valls, G., L. Gómez-Chova, J. Calpe-Maravilla, J. D. Martín-Guerrero, E. Soria-Olivas, L. Alonso-Chordá and J. Moreno (2004). "Robust support vector method for hyperspectral data classification and knowledge discovery." *IEEE Transactions on Geoscience and Remote Sensing* **42**(7): 1530-1542.

Cannon, R. L., J. V. Dave, J. C. Bezdek, and M. M. Trivedi (1986). "Segmentation of a Thematic Mapper image using the fuzzy c-means clustering algorithm." *IEEE Transactions on Geoscience and Remote Sensing* **GE-24**(3): 400-408.

Carleer, A. P., O. Debeir and E. Wolff (2005). "Assessment of very high spatial resolution satellite image segmentations." *Photogrammetric Engineering and Remote Sensing* **71**(11): 1285-1294.

Carpenter, G. A., M. N. Gjaja, S. Gopal and C. E. Woodcock (1997). " ART neural networks for remote sensing: Vegetation classification from Landsat TM and terrain data." *IEEE Transactions on Geoscience and Remote Sensing* **35**(2): 308-325.

Carpenter, G. A., S. Gopal, S. Macomber, S. Martens and C. E. Woodcock (1999). "A neural network method for mixture estimation for vegetation mapping." *Remote Sensing of Environment* **70**: 138-152.

Chang, C.-I. (1998). "Further results on relationship between spectral unmixing and subspace projection." *IEEE Transactions on Geoscience and Remote Sensing* **36**(3): 1030-1032.

Chen, K. S., Y. C. Tzeng, C. F. Chen, and W. L. Kao (1995). "Land-cover classification of multispectral imagery using a dynamic learning neural network." *Photogrammetric Engineering and Remote Sensing* **61**(4): 403-408.

Civco, D. L. (1993). "Artificial neural networks for land–cover classification and mapping." *International Journal of Geographical Information Systems* **7**(2): 173-186.

Clément, V., G. Giraudon, S. Houzelle, and F. Sandakly (1993). "Interpretation of remotely sensed images in a context of multisensor fusion using a multispecialist architecture." *IEEE Transactions on Geoscience and Remote Sensing* **31**(4): 779-791.

Coleman, G. B. and H. C. Andrews (1979). "Image segmentation by clustering." *Proceedings of the IEEE* **67**(5): 773-785.

Cross, A. M., J. J. Settle, N. A. Drake, and R. T. M. Paivinen (1991). "Subpixel measurment of tropical forest cover using AVHRR data." *International Journal of Remote Sensing* **12**(5): 1119-1129.

Dennison, P. E., K. Q. Halligan and D. A. Roberts (2004). "A comparison of error metrics and constraints for multiple endmember spectral mixture analysis and spectral angle mapper." *Remote Sensing of Environment* **93**(3): 359-367.

Dozier, J. (1981). "A method for satellite identification of surface temperature fields of subpixel resolution." *Remote Sensing of Environment* **11**: 221-229.

Dreyer, P. (1993). "Classification of land cover using optimized neural nets on SPOT data." *Photogrammetric Engineering and Remote Sensing* **59**(5): 617-621.

Duda, R. D., P. E. Hart and D. G. Stork (2001). *Pattern Classification and Scene Analysis*. Second Edition. New York: John Wiley & Sons, 482 p.

Dymond, J. R. (1993). "An improved Skidmore/Turner classifier." *Photogrammetric Engineering and Remote Sensing* **59**(5): 623-626.

Evans, C., R. Jones, I. Svalbe and M. Berman (2002). "Segmenting multispectral Landsat TM images into field units." *IEEE Transactions on Geoscience and Remote Sensing* **40**(5): 1054-1064.

Fang, H. and S. Liang (2003). "Retrieving leaf area index with a neural network method: Simulation and validation." *IEEE Transactions on Geoscience and Remote Sensing* **41**(9): 2052-2062.

Felzer, B., P. Hauff, and A. F. H. Goetz (1994). "Quantitative reflectance spectroscopy of buddingtonite from the Cuprite mining district, Nevada." *Journal of Geophysical Research* **99**(B2): 2887-2895.

Foody, G. M. (1992). "A fuzzy sets approach to the representation of vegetation continua from remotely sensed data: an example from lowland heath." *Photogrammetric Engineering and Remote Sensing* **58**(2): 221-225.

Foody, G. M. (1995). "Using prior knowledge in artificial neural network classification with a minimal training set." *International Journal of Remote Sensing* **16**(2): 301-312.

Foody, G. M. (1996). "Relating the land-cover composition of mixed pixels to artificial neural network classification output." *Photogrammetric Engineering and Remote Sensing* **62**(5): 491-499.

Foody, G. M. and D. P. Cox (1994). "Sub-pixel land cover composition estimation using a linear mixture model and fuzzy membership functions." *International Journal of Remote Sensing* **15**(3): 619-631.

Foody, G. M., M. B. McCulloch, and W. B. Yates (1995). "Classification of remotely sensed data by an artificial neural network: issues related to training data characteristics." *Photogrammetric Engineering and Remote Sensing* **61**(4): 391-401.

Foody, G. M. (2004). "Supervised image classification by MLP and RBF neural networks with and without an exhaustively defined set of classes." *International Journal of Remote Sensing* **25**(15): 3091–3104.

Foody, G. M. and A. Mathur (2004). "A relative evaluation of multiclass image classification by support vector machines." *IEEE Transactions on Geoscience and Remote Sensing* **42**(6): 1335-1343.

Frans, E. P. and R. A. Schowengerdt (1997). "Spatial-spectral unmixing using the sensor PSF." In *Imaging Spectrometry III*, San Diego, CA, pp. 241-249.

Frans, E. and R. Schowengerdt (1999). "Improving spatial-spectral unmixing with the sensor spatial response function." *Canadian Journal of Remote Sensing* **25**(2): 131-151.

Fukunaga, K. (1990). *Introduction to Statistical Pattern Recognition.* Second Edition. San Diego: Academic Press, 591 p.

Full, W. E., R. Ehrlich, and J. C. Bezdek (1982). "Fuzzy Qmodel - a new model approach for linear unmixing." *Mathematical Geology* **14**(3): 259-270.

Ghosh, J. K. (2004). "Automated interpretation of sub-pixel vegetation from IRS LISS-II images." *International Journal of Remote Sensing* **25**(6): 1207-1222.

Go, J., G. Han, H. Kim and C. Lee (2001). "Multigradient: A new neural network learning algorithm for pattern classification." *IEEE Transactions on Geoscience and Remote Sensing* **39**(5): 986-993.

Goldberg, M., D. G. Goodenough, and G. Plunkett (1988). "A knowledge-based approach for evaluating forestry-map congruency with remotely sensed imagery." *Philosophical Transactions of the Royal Society of London* **324**: 447-456.

Gross, H. N. and J. R. Schott (1998). "Application of spectral mixture analysis and image fusion techniques for image sharpening." *Remote Sensing of Environment* **63**(2): 85-94.

Gyer, M. S. (1992). "Adjuncts and alternatives to neural networks for supervised classification." *IEEE Transactions on Geoscience and Remote Sensing* **22**(1): 35-46.

Hara, Y., R. G. Atkins, R. T. Shin, J. A. Kong, S. H. Yueh, and R. Kwok (1995). "Application of neural networks for sea ice classification in polarimetric SAR images." *IEEE Transactions on Geoscience and Remote Sensing* **33**(3): 740-748.

Hara, Y., R. G. Atkins, S. H. Yueh, R. T. Shin, and J. A. Kong (1994). "Application of neural networks to radar image classification." *IEEE Transactions on Geoscience and Remote Sensing* **32**(1): 100-109.

Haralick, R. M. and I. h. Dinstein (1975). "A spatial clustering procedure for multi-image data." *IEEE Transactions on Circuits and Systems* **CAS-22**(5): 440-450.

Hardin, P. J. (1994). "Parametric and nearest-neighbor methods for hybrid classification: a comparison of pixel assignment accuracy." *Photogrammetric Engineering and Remote Sensing* **60**(12): 1439-1448.

Hardin, P. J. and C. N. Thomson (1992). "Fast nearest neighbor classification methods for multispectral imagery." *The Professional Geographer* **44**(2): 191-201.

Harsanyi, J. C. and C.-I. Chang (1994). "Hyperspectral image classification and dimensionality reduction: an orthogonal subspace projection approach." *IEEE Transactions on Geoscience and Remote Sensing* **32**(4): 779-785.

Hartigan, J. A. (1975). *Clustering Algorithms.* New York, NY: John Wiley & Sons, 351 p.

Heermann, P. D. and N. Khazenie (1992). "Classification of multispectral remote sensing data using a back–propagation neural network." *IEEE Transactions on Geoscience and Remote Sensing* **30**(1): 81-88.

Hlavka, C. A. and M. A. Spanner (1995). "Unmixing AVHRR imagery to assess clearcuts and forest regrowth in Oregon." *IEEE Transactions on Geoscience and Remote Sensing* **33**(3): 788-795.

Hoffbeck, J. P. and D. A. Landgrebe (1996). "Covariance matrix estimation and classification with limited training data." *IEEE Transactions on Pattern Analysis and Machine Intelligence* **18**(7): 763-767.

Holben, B. N. and Y. E. Shimabukuro (1993). "Linear mixing models applied to coarse spatial resolution data from multispectral satellite sensors." *International Journal of Remote Sensing* **14**(11): 2231-2240.

Horwitz, H. M., R. F. Nalepka, P. D. Hyde, and J. P. Morgenstern (1971). "Estimating the proportions of objects within a single resolution element of a multispectral scanner." in: *Seventh International Symposium on Remote Sensing of Environment*, Ann Arbor, Michigan, Environmental Research Institute of Michigan, vol. 2: 1307-1320.

Hu, X., C. V. Tao and B. Prenzel (2005). "Automatic segmentation of high-resolution satellite imagery by integrating texture, intensity, and color features." *Photogrammetric Engineering and Remote Sensing* **71**(12): 1399-1406.

Huete, A. R. and R. Escadafal (1991). "Assessment of biophysical soil properties through spectral decomposition techniques." *Remote Sensing of Environment* **35**: 149-159.

Irons, J. R., B. L. Markham, R. F. Nelson, D. L. Toll, D. L. Williams, R. S. Latty, and M. L. Stauffer (1985). "The effects of spatial resolution on the classification of Thematic Mapper data." *International Journal of Remote Sensing* **6**(8): 1385-1403.

Jain, A. K. and R. C. Dubes (1988). *Algorithms for Clustering Data*. Englewood Cliffs, NJ: Prentice Hall, 320 p.

James, M. (1985). *Classification Algorithms*. London: Collins, 209 p.

Jasinski, M. F. (1996). "Estimation of subpixel vegetation density of natural regions using satellite multispectral imagery." *IEEE Transactions on Geoscience and Remote Sensing* **34**(3): 804-813.

Jensen, J. R. (2004). *Introductory Digital Image Processing – A Remote Sensing Perspective*. Third Edition. Upper Saddle River, NJ: Prentice Hall, 544 p.

Jenson, S. K., T. R. Loveland, and J. Bryant (1982). "Evaluation of AMOEBA: a spectral-spatial classification method." *Journal of Applied Photographic Engineering* **8**(3): 159-162.

Jia, X. and J. A. Richards (1993). "Binary coding of imaging spectrometry data for fast spectral matching and classification." *Remote Sensing of Environment* **43**: 47-53.

Jia, X. and J. A. Richards (1994). "Efficient maximum likelihood classification for imaging spectrometer data sets." *IEEE Transactions on Geoscience and Remote Sensing* **32**: 274-281.

Kanellopoulos, I., A. Varfis, G. G. Wilkinson, and J. Mégier (1991). "Neural network classification of multidate satellite imagery." in: *11th Annual International Geoscience and Remote Sensing Symposium*, Espoo, Finland: IEEE : 2215-2218.

Kanellopoulos, I., A. Varfis, G. G. Wilkinson, and J. Mégier (1992). "Land cover discrimination in SPOT imagery by artificial neural network–a twenty class experiment." *International Journal of Remote Sensing* **13**(5): 917-924.

Kauth, R. J., A. P. Pentland, and G. S. Thomas (1977). "BLOB, an unsupervised clustering approach to spatial preprocessing of MSS imagery." in: *Eleventh International Symposium on Remote Sensing of Environment*, Ann Arbor, MI: Environmental Research Institute of Michigan: 1309-1317.

Kepuska, V. Z. and S. O. Mason (1995). "A hierarchical neural network system for signalized point recognition in aerial photographs." *Photogrammetric Engineering and Remote Sensing* **61**(7): 917-925.

Kettig, R. L. and D. A. Landgrebe (1976). "Classification of multispectral image data by extraction and classification of homogeneous objects." *IEEE Transactions on Geoscience Electronics* **GE-4**(1): 19-26.

Key, J., J. A. Maslanik, and S. A. J. (1989). "Classification of merged AVHRR and SMMR arctic data with neural networks." *Photogrammetric Engineering and Remote Sensing* **55**(9): 1331-1338.

Kim, K. and M. M. Crawford (1991). "Adaptive parametric estimation and classification of remotely sensed imagery using a pyramid structure." *IEEE Transactions on Geoscience and Remote Sensing* **29**(4): 481-493.

Kruse, F. A. (1988). "Use of Airborne Imaging Spectrometer data to map minerals associated with hydrothermally altered rocks in the northern Grapevine Mountains, Nevada and California." *Remote Sensing of Environment* **24**(1): 31-51.

Kruse, F. A., K. S. Kierein-Young, and J. W. Boardman (1990). "Mineral mapping at Cuprite, Nevada with a 63-channel imaging spectrometer." *Photogrammetric Engineering and Remote Sensing* **56**(1): 83-92.

Kruse, F. A., A. B. Lefkoff, J. W. Boardman, K. B. Heidebrecht, A. T. Shapiro, P. J. Barloon, and A. F. H. Goetz (1993). "The Spectral Image Processing System (SIPS) - interactive viualization and analysis of imaging spectrometer data." *Remote Sensing of Environment* **44**: 145-163.

Landgrebe, D. A. (1980). "The development of a spectral-spatial classifier for earth observational data." *Pattern Recognition* **12**: 165-175.

Landgrebe, D. A. (2003). *Signal Theory Methods in Multispectral Remote Sensing*. Hoboken, NJ: John Wiley & Sons, Inc., 508 p.

Le Moigne, J. and J. C. Tilton (1995). "Refining image segmentation by integration of edge and region data." *IEEE Transactions on Geoscience and Remote Sensing* **33**(3): 605-615.

Lee, C. and D. A. Landgrebe (1991). "Fast likelihood classification." *IEEE Transactions on Geoscience and Remote Sensing* **29**(4): 509-517.

Lee, C. and D. A. Landgrebe (1993). "Analyzing high-dimensional multispectral data." *IEEE Transactions on Geoscience and Remote Sensing* **31**(4): 792-800.

Lee, T., J. A. Richards, and P. H. Swain (1987). "Probabilistic and evidential approaches for multisource data analysis." *IEEE Transactions on Geoscience and Remote Sensing* **GE-25**: 283-293.

Li, Z., A. Khananian, R. H. Fraser and J. Cihlar (2001). "Automatic detection of fire smoke using artificial neural networks and threshold approaches applied to AVHRR imagery." *IEEE Transactions on Geoscience and Remote Sensing* **39**(9): 1859-1870.

Lillesand, T. M., R. W. Kiefer and J. W. Chipman (2004). *Remote Sensing and Image Interpretation*. Fifth Edition. New York: John Wiley & Sons, 763 p.

Lippmann, R. P. (1987). "An introduction to computing with neural nets." *IEEE ASSP Magazine*(April): 4-22.

Liu, Z. K. and J. Y. Xiao (1991). "Classification of remotely-sensed image data using artificial neural networks." *International Journal of Remote Sensing* **12**(11): 2433-2438.

Lobell, D. B. and G. P. Asner (2004). "Cropland distributions from temporal unmixing of MODIS data." *Remote Sensing of Environment* **93**: 412-422.

Lobo, A. (1997). "Image segmentation and discriminant analysis for the identification of land cover units in ecology." *IEEE Transactions on Geoscience and Remote Sensing* **35**(5): 1136-1145.

Markham, B. L. and J. R. G. Townshend (1981). "Land cover classification accuracy as a function of sensor spatial resolution." in: *Fifteenth International Symposium on Remote Sensing of Environment*, Ann Arbor, MI, Environmental Research Institute of Michigan, vol. III: 1075-1090.

Marsh, S. E. and J. B. McKeon (1983). "Integrated analysis of high-resolution field and airborne spectroradiometer data for alteration mapping." *Economic Geology* **78**(4): 618-632.

Marsh, S. E., P. Switzer, and R. J. P. Lyon (1980). "Resolving the percentage component terrains within single resolution elements." *Photogrammetric Engineering and Remote Sensing* **46**(8): 1079-1086.

Maxwell, E. L. (1976). "Multivariate systems analysis of multispectral imagery." *Photogrammetric Engineering and Remote Sensing* **42**(9): 1173-1186.

Mazer, A. S., M. Martin, M. Lee, and J. E. Solomon (1988). "Image processing software for imaging spectrometry data analysis." *Remote Sensing of Environment* **24**: 201-211.

McIntire, T. J. and J. J. Simpson (2002). "Arctic sea ice, cloud, water, and lead classification using neural networks and 1.6-μm data." *IEEE Transactions on Geoscience and Remote Sensing* **40**(9): 1956-1972.

McKeown, D. M., Jr., J. Wilson A. Harvey, and J. McDermott (1985). "Rule-based interpretation of aerial imagery." *IEEE Transactions on Pattern Analysis and Machine Intelligence* **PAMI-7**(5): 570-585.

Mehldau, G. and R. A. Schowengerdt (1990). "A C-extension for rule-based image classification systems." *Photogrammetric Engineering and Remote Sensing* **56**(6): 887-892.

Moody, A., S. Gopal and A. H. Strahler (1996). "Artificial neural network response to mixed pixels in coarse-resolution satellite data." *Remote Sensing of Environment* **58**: 329-343.

Moreno, J. F. and R. O. Green (1996). "Surface and atmospheric parameter retrieval from AVIRIS data: The importance of non-linear effects." In *AVIRIS Airborne Geoscience Workshop*, Pasadena, CA, NASA Jet Propulsion Laboratory, pp. 175-184.

Muchoney, D. and J. Williamson (2001). "A gaussian adaptive resonance theory neural network classification algorithm applied to supervised land cover mapping using multitemporal vegetation index data." *IEEE Transactions on Geoscience and Remote Sensing* **39**(9): 1969-1977.

Nagao, M. and T. Matsuyama (1980). *A Structural Analysis of Complex Aerial Photographs*. New York, NY: Plenum Press, 199 p.

Nalepka, R. P. and P. D. Hyde (1972). "Classifying unresolved objects from simulated space data." in: *Eighth International Symposium on Remote Sensing of Environment*, Ann Arbor, MI: Environmental Research Institute of Michigan, vol. 2: 935-949.

Nascimento, J. M. P. and J. M. B. Dias (2005). "Does independent component analysis play a role in unmixing hyperspectral data?" *IEEE Transactions on Geoscience and Remote Sensing* **43**(1): 175-187.

Nascimento, J. M. P. and J. M. B. Dias (2005). "Vertex component analysis: A fast algorithm to unmix hyper-spectral data." *IEEE Transactions on Geoscience and Remote Sensing* **43**(4): 898-910.

Novo, E. M. and Y. E. Shimabukuro (1994). "Spectral mixture analysis of inland tropical waters." *International Journal of Remote Sensing* **15**(6): 1351-1356.

Painter, T. H., J. Dozier, D. A. Roberts, R. E. Davis and R. O. Green (2003). "Retrieval of subpixel snow-covered area and grain size from imaging spectrometer data." *Remote Sensing of Environment* **85**(64-77).

Pal, N. R. and S. K. Pal (1993). "A review on image segmentation techniques." *Pattern Recognition* **26**(9): 1277-1294.

Pao, Y. H. (1989). *Adaptive Pattern Recognition and Neural Networks*. Reading, MA: Addison-Wesley, 299 p.

Paola, J. D. and R. A. Schowengerdt (1995a). "A review and analysis of backpropagation neural networks for classification of remotely-sensed multi-spectral imagery." *International Journal of Remote Sensing* **16**(16): 3033-3058.

Paola, J. D. and R. A. Schowengerdt (1995b). "A detailed comparison of backpropagation neural network and maximum-likelihood classifiers for urban land use classification." *IEEE Transactions on Geoscience and Remote Sensing* **33**(4): 981-996.

Paola, J. D. and R. A. Schowengerdt (1997). "The effect of neural network structure on a multispectral land-use/land-cover classification." *Photogrammetric Engineering and Remote Sensing*.

Pesaresi, M. and J. A. Benediktsson (2001). "A new approach for the morphological segmentation of high-resolution satellite imagery." *IEEE Transactions on Geoscience and Remote Sensing* **39**(2): 309-320.

Philpot, W. D. (1991). "The derivative ratio algorithm: avoiding atmosheric effects in remote sensing." *IEEE Transactions on Geoscience and Remote Sensing* **29**(3): 350-357.

Piech, M. A. and K. R. Piech (1987). "Symbolic representation of hyperspectral data." *Applied Optics* **26**(18): 4018-4026.

Piech, M. A. and K. R. Piech (1989). "Hyperspectral interactions: invariance and scaling." *Applied Optics* **28**(3): 481-489.

Plaza, A. and C.-I. Chang (2005). "An improved N-FINDR algorithm in implementation." In *Algorithms and Technologies for Multispectral, Hyperspectral, and Ultraspectral Imagery XI*, Orlando, FL, SPIE, vol. 5806: 298-306.

Pratt, W. K. (1991). *Digital Image Processing*. Second Edition. New York, NY: John Wiley & Sons, 698 p.

Qiu, F. and J. R. Jensen (2004). "Opening the black box of neural networks for remote sensing image classification." *International Journal of Remote Sensing* **25**(9): 1749–1768.

Ray, T. W. and B. C. Murray (1996). "Nonlinear spectral mixing in desert vegetation." *Remote Sensing of Environment* **55**: 59-64.

Richards, J. A. and X. Jia (1999). *Remote Sensing Digital Image Analysis – An Introduction*. Third Edition. Berlin: Springer-Verlag, 356 p.

Richter, R. (1996a). "A spatially adaptive fast atmospheric correction algorithm." *International Journal of Remote Sensing* **17**(6): 1201-1214.

Ritter, N. D. and G. F. Hepner (1990). "Application of an artificial neural network to land–cover classification of Thematic Mapper imagery." *Computers & Geosciences* **16**(6): 873-880.

Roberts, D. A., M. O. Smith, and J. B. Adams (1993). "Green vegetation, nonphotosynthetic vegetation. and soils in AVIRIS data." *Remote Sensing of Environment* **44**: 255-269.

Roberts, D. A., M. Gardner, R. Church, S. Ustin, G. Scheer and R. O. Green (1998). "Mapping chaparral in the Santa Monica Mountains using multiple endmember spectral mixture models." *Remote Sensing of Environment* **65**(3): 267-279.

Rubin, T. D. (1993). "Spectral mapping with imaging spectrometers." *Photogrammetric Engineering and Remote Sensing* **59**(2): 215-220.

Rumelhart, D. E., G. E. Hinton, and R. J. Williams (1986). Learning internal representations by error propagation. *Parallel DIstributed Processing: Explorations in the Microstruction of Cognition*. D. E. Rumelhart and J. L. McClelland, (Eds.). Cambridge, MA, The MIT Press. **I**: 318-362.

Ryan, T. (1985). "Image segmentation algorithms." in: *Architectures and Algorithms for Digital Image Processing II*, SPIE, vol. 534: 172-178.

Salu, Y. and J. Tilton (1993). "Classification of multispectral image data by the binary neural network and by nonparametric, pixel-by-pixel methods." *IEEE Transactions on Geoscience and Remote Sensing* **31**(3): 606-617.

Salvato, P. J. (1973). "Iterative techniques to estimate signature vectors for mixture processing of multispectral data." in: *Symposium on Machine Processing of Remotely Sensed Data*, IEEE, vol. 73CH0834-2GE: 3B:48-62.

Sarkar, A., M. K. Biswas, B. Kartikeyan, V. Kumar, K. L. Majumder and D. K. Pal (2002). "A MRF model-based segmentation approach to classification for multispectral imagery." *IEEE Transactions on Geoscience and Remote Sensing* **40**(5): 1102-1113.

Sayood, K. (2005). *Introduction to Data Compression*. Third Edition. San Francisco, CA: Morgan Kaufmann, 704 p.

Schowengerdt, R. A. (1983). *Techniques for Image Processing and Classification in Remote Sensing*. Orlando, FL: Academic Press, 249 p.

Schowengerdt, R. A. (1996). "On the estimation of spatial-spectral mixing with classifier likelihood functions." *Pattern Recognition Letters*, **17**(13): 1379-1387.

Schürmann, J. (1996). *Pattern Classification–A Unified View of Statistical and Neural Approaches*. New York, NY: John Wiley & Sons, 373 p.

Serpico, S. B. and F. Roli (1995). "Classification of multisensor remote-sensing images by structured neural networks." *IEEE Transactions on Geoscience and Remote Sensing* 33(3): 562-577.

Settle, J. J. (1996). "On the relationship between spectral unmixing and subspace projection." *IEEE Transactions on Geoscience and Remote Sensing* 34(4): 1045-1046.

Settle, J. J. and N. A. Drake (1993). "Linear mixing and the estimation of ground cover proportions." *International Journal of Remote Sensing* 14(6): 1159-1177.

Shah, C. A., M. K. Arora and P. K. Varshney (2004). "Unsupervised classification of hyperspectral data: An ICA mixture model based approach." *International Journal of Remote Sensing* 25(2): 481-487.

Shahshahani, B. M. and D. A. Landgrebe (1994). "The effect of unlabeled samples in reducing the small sample size problem and mitigating the Hughes phenomenon." *IEEE Transactions on Geoscience and Remote Sensing* 32(5): 1087-1095.

Shimabukuro, Y. E. and J. A. Smith (1991). "The least-squares mixing models to generate fraction images derived from remote sensing multispectral data." *IEEE Transactions on Geoscience and Remote Sensing* 29(1): 16-20.

Skidmore, A. K. and B. J. Turner (1988). "Forest mapping accuracies are improved using a supervised nonparametric classifier with SPOT data." *Photogrammetric Engineering and Remote Sensing* 54(10): 1415–1421.

Smith, M. O., S. L. Ustin, J. B. Adams, and A. R. Gillespie (1990). "Vegetation in deserts: I. A regional measure of abundance from multispectral images." *Remote Sensing of Environment* 31: 1-26.

Sohn, Y. and R. M. McCoy (1997). "Mapping desert shrub rangeland using spectral unmixing and modeling spectral mixtures with TM data." *Photogrammetric Engineering and Remote Sensing* 63(6): 707-716.

Song, M. and D. Civco (2004). "Road extraction using SVM and image segmentation." *Photogrammetric Engineering and Remote Sensing* 70(12): 1365-1371.

Srinivasan, A. and J. A. Richards (1990). "Knowledge-based techniques for multi-source classification." *International Journal of Remote Sensing* 11: 505-525.

Strahler, A. H. (1980). "The use of prior probabilities in maximum likelihood classification of remotely sensed data." *Remote Sensing of Environment* 10: 135-163.

Sunar Erbek, F., C. Ozkan and M. Taberner (2004). "Comparison of maximum likelihood classification method with supervised artificial neural network algorithms for land use activities." *International Journal of Remote Sensing* 25(9): 1733–1748.

Swain, P. H. and S. M. Davis, (Eds.) (1978). *Remote Sensing: The Quantitative Approach*. New York, NY: McGraw-Hill, 396 p.

Tilton, J. C., S. B. Vardeman, and P. H. Swain (1982). "Estimation of context for statistical classification of multispectral image data." *IEEE Transactions on Geoscience and Remote Sensing* GE-20(4): 445-452.

Townshend, J. R. G., C. O. Justice, C. Gurney, and J. McManus (1992). "The impact of misregistration on change detection." *IEEE Transactions on Geoscience and Remote Sensing* 30(5): 1054-1060.

Tu, T.-M. (2000). "Unsupervised signature extraction and separation in hyperspectral images: A noise-adjusted fast independent component analysis approach." *Optical Engineering* 39(4): 897-906.

Valdes, M. and M. Inamura (2000). "Spatial resolution improvement of remotely sensed images by a fully interconnected neural network approach." *IEEE Transactions on Geoscience and Remote Sensing* 38(5): 2426-2430.

Visa, A. and J. Iivarinen (1997). "Evolution and evaluation of a trainable cloud classifier." *IEEE Transactions on Geoscience and Remote Sensing* **35**(5): 1307-1315.

Wang, F. (1990a). "Improving remote sensing image analysis through fuzzy information representation." *Photogrammetric Engineering and Remote Sensing* **56**(8): 1163-1169.

Wang, F. (1990b). "Fuzzy supervised classification of remote sensing images." *IEEE Transactions on Geoscience and Remote Sensing* **28**(2): 194-201.

Wang, F. (1993). "A knowlege-based vision system for detecting land changes at urban fringes." *IEEE Transactions on Geoscience and Remote Sensing* **31**(1): 136-145.

Wang, S., D. B. Elliott, J. B. Campbell, R. W. Erich, and R. M. Haralick (1983). "Spatial reasoning in remotely sensed data." *IEEE Transactions on Geoscience and Remote Sensing* **GE-21**(1): 94-101.

Weiss, J. M., S. A. Christopher and R. M. Welch (1998). "Automatic contrail detection and segmentation." *IEEE Transactions on Geoscience and Remote Sensing* **36**(5): 1609-1619.

Wharton, S. (1987). "A spectral-knowledge-based approach for urban and land-cover discrimination." *IEEE Transactions on Geoscience and Remote Sensing* **GE-25**(3): 272-282.

Wharton, S. W. (1982). "A contextural classification method for recognizing land use patterns in high resolution remotely sensed data." *Pattern Recognition* **15**(4): 317-324.

Winter, M. E. (1999). "N-FINDR: An algorithm for fast autonomous spectral end-member determination in hyperspectral data." In *Imaging Spectrometry V*, Denver, CO, SPIE, vol. 3753: 266-275.

Winter, M. E. (2004). "A proof of the N-FINDR algorithm for the automated detection of end-members in a hyperspectral image." In *Algorithms and Technologies for Multispectral, Hyperspectral, and Ultraspectral Imagery X*, Orlando, FL, SPIE, vol. 5425: 31-41.

Witkin, A. P. (1983). "Scale-space filtering." in: *Ninth International Joint Conference on Artificial Intelligence*, Karlsruhe, West Germany: Morgan Kaufmann Publishers, 1019-1022.

Woodcock, C. and V. J. Harward (1992). "Nested-hierarchical scene models and image segmentation." *International Journal of Remote Sensing* **13**(16): 3167-3187.

Wu, H.-H. P. and R. A. Schowengerdt (1993). "Improved fraction image estimation using image restoration." *IEEE Transactions on Geoscience and Remote Sensing* **31**(4): 771-778.

Yhann, S. R. and J. J. Simpson (1995). "Application of neural networks to AVHRR cloud segmentation." *IEEE Transactions on Geoscience and Remote Sensing* **33**(3): 590-604.

Yoshida, T. and S. Omatu (1994). "Neural network approach to land cover mapping." *IEEE Transactions on Geoscience and Remote Sensing* **32**(5): 1103-1109.

Appendix A Sensor Acronyms

Asrar, G. and R. Greenstone (1995). MTPE/EOS Reference Handbook. Greenbelt, MD: NASA/Goddard Space Flight Center, No. NP-215.

Barnes, R. A. and A. W. Holmes (1993). "Overview of the SeaWiFS ocean sensor." In *Sensor Systems for the Early Earth Observing System Platforms*, Orlando, FL, SPIE, vol. 1939: 224-232.

Basedow, R. W., D. C. Carmer, and M. E. Anderson (1995). "HYDICE system, implementation and performance." in: *Imaging Spectrometry*, Orlando, FL: SPIE, vol. 2480: 258-267.

Bonner, W. D. (1969). "Gridding scheme for APT satellite pictures." *Journal of Geophysical Research* **74**(18): 4581-4587.

Chevrel, M., M. Courtois, and G. Weill (1981). "The SPOT satellite remote sensing mission." *Photogrammetric Engineering and Remote Sensing* **47**(8): 1163-1171.

Curran, P. J. and C. M. Steele (2005). "MERIS: The re-branding of an ocean sensor." *International Journal of Remote Sensing* **26**(9): 1781-1798.

Diner, D. J., C. J. Bruegge, J. V. Martonchik, T. P. Ackerman, R. Davies, S. A. W. Gerstl, H. R. Gordon, P. J. Sellers, J. Clark, et al. (1989). "MISR: A Multi-angle Imaging SpectroRadiometer for geophysical and climatological research from EOS." *IEEE Transactions on Geoscience and Remote Sensing* **27**(2): 200-214.

Engel, J. L. and O. Weinstein (1983). "The Thematic Mapper - An overview." *IEEE Transactions on Geoscience and Remote Sensing* **GE-21**(3): 258-265.

Hollinger, J. P., J. L. Peirce, and G. A. Poe (1990). "SSM/I instrument evaluation." *IEEE Transactions on Geoscience and Remote Sensing* **28**(5): 781-790.

Huang, Z., B. J. Turner, S. J. Dury, I. R. Wallis and W. J. Foley (2004). "Estimating foliage nitrogen concentration from HyMap data using continuum removal analysis." *Remote Sensing of Environment* **93**(1): 18-29.

Irons, J. R., K. J. Ranson, D. L. Williams, R. R. Irish, and F. G. Huegel (1991). "An off-nadir-pointing imaging spectroradiometer for terrestrial ecosystem studies." *IEEE Transactions on Geoscience and Remote Sensing* **29**(1): 66-74.

Kahle, A. B. and A. F. H. Goetz (1983). "Mineralogic Information from a new airborne thermal infrared multi-spectral scanner." *Science* **222**(4619): 24-27.

Kramer, H. J. (2002). *Observation of the Earth and Its Environment: Survey of Missions and Sensors*. Fourth Edition. Berlin: Springer–Verlag, 1510 p.

Lansing, J. C., Jr. and R. W. Cline (1975). "The four- and five-band multispectral scanners for Landsat." *Optical Engineering* **14**: 312.

Myers, J. and J. Arvesen (1995). Sensor Systems of the NASA Airborne Science Program. Moffett Field, CA: NASA/Ames Research Center.

Nishidai, T. (1993). "Early results from 'Fuyo-1' Japan's Earth Resources Satellite (JERS-1)." *International Journal of Remote Sensing* **14**(9): 1825-1833.

Porter, W. M. and H. T. Enmark (1987). "A system overview of the Airborne Visible/Infrared Imaging Spectrometer (AVIRIS)." in: *31st Annual International Technical Symposium*, SPIE, vol. 834: 22-31.

Salomonson, V. V., W. L. Barnes, P. W. Maymon, H. E. Montgomery, and H. Ostrow (1989). "MODIS: Advanced facility instrument for studies of the earth as a system." *IEEE Transactions on Geoscience and Remote Sensing* **27**(2): 145-153.

Short, N. M. and J. Locke M. Stuart (1982). *The Heat Capacity Mapping Mission (HCMM) Anthology*, Washington, D.C.: NASA, No. SP-465, 264 p.

Szymanski, J. J. and P. G. Weber (2005). "Multispectral Thermal Imager: Mission and applications overview." *IEEE Transactions on Geoscience and Remote Sensing* **43**(9): 1943-1949.

Ungar, S. G., J. S. Pearlman, J. A. Mendenhall and D. Reuter (2003). "Overview of the Earth Observing One (EO-1) mission." *IEEE Transactions on Geoscience and Remote Sensing* **41**(6): 1149-1159.

Vane, G., A. F. H. Goetz and J. B. Wellman (1984). "Airborne Imaging Spectrometer: a new tool for remote sensing." *IEEE Transactions on Geoscience and Remote Sensing* **GE-22**(6): 546-549.

Wrigley, R. C., R. E. Slye, S. A. Klooster, R. S. Freedman, M. Carle, and L. F. McGregor (1992). "The Airborne Ocean Color Imager: system description and image processing." *Journal of Imaging Science and Technology* **36**(5): 423-430.

Appendix B   1-D and 2-D Functions

Bracewell, R. N. (2004). *Fourier Analysis and Imaging*. Springer, 704 p.

Castleman, K. R. (1996). *Digital Image Processing*. Englewood Cliffs, NJ: Prentice-Hall, 667 p.

Gaskill, J. D. (1978). *Linear Systems, Fourier Transforms, and Optics*. New York, NY: John Wiley & Sons, 554 p.

Gonzalez, R. C. and R. E. Woods (2002). *Digital Image Processing*. Second Edition. Upper Saddle River, NJ: Prentice-Hall, 793 p.

Hearn, D. R. (2000). *EO-1 Advanced Land Imager modulation transfer functions*. Lincoln Laboratory, Massachusetts Institute of Technology. Technical Report 1061.

Index

A

affine transform 184, 289
Artificial Neural Network (ANN) 407, 426, 440
    decision boundaries 407, 413
        comparison to maximum-likelihood 426
    degrees-of-freedom 412
    hidden layer 407
        number of nodes 412
    processing element 407
    relation of output to fractions 440
    sigmoid function 409
atmosphere
    absorption bands 8, 48, 450
    correction 56, 61, 189, 192, 337, 343
    path radiance 48, 49, 54, 191, 332, 337, 350
    transmittance 51, 53
        angular dependence 54
        correction for hyperspectral data 341
        MWIR and LWIR 66
        VSWIR 50
        water vapor bands 344
at-sensor radiance
    solar and thermal in MWIR 47
at-sensor radiance L 53

B

Back-propagation algorithm 409
Bayes
    classification 418
    classifier 441
    decision rule 418
    rule 417

Bi-directional Reflectance Distribution Function (BRDF) 40, 53

C

calibration
    coefficients 334
    detector 30, 320
    hyperspectral 341
    normalization techniques 344
    radiometric 332
    scene 109
    sensor 109, 334
    target 61
camera equation 104
CDF 375
classification
    as a compression tool 388
    decision boundaries
        ANN 415
        ANN compared to maximum-likelihood 414, 426
        level-slice 413
        maximum-likelihood 424, 426
        nearest-mean 424
    fuzzy c-means clustering 442
    fuzzy supervised classification 443
    hard versus soft 394
    K-means clustering 400
    nonparametric 405
        ANN 407
        histogram estimation 406
        level-slice, box, parallelepiped 405
        nearest-neighbor 407

parametric 417
    maximum-likelihood 417
    nearest-mean 421
  thematic hierarchy 392
  training 388
    supervised 396
    unsupervised 399
cluster
  example 137
  in unsupervised training 399
clustering
  examples 400
  ISODATA algorithm 400
  K-means algorithm 400
color
  contrast enhancement 219
    decorrelation stretch 222
  natural 39
  scatterplots 140
Color IR (CIR) 39
Color-Space Transform (CST) 223
  application to color enhancement 223
  application to image fusion 374
  hexcone algorithm 224
Constrained Least Squares (CLS)
  application in unmixing 439
contrast
  decrease for display 209
  definitions 146
  enhancement 130, 206
  in segmentation 431
  in Sigma filter 315
  modification by blending algorithm 226
  ratio 81
  relation to NEDr 152
  relation to SNR 147
  spatial texture feature 165
  spectral reflectance ratio 186
  stretch 208
contrast enhancement
  CDF reference stretch 210
  color
    decorrelation stretch 222
  color images 219
  gain 208
  gaussian stretch 213

  global transforms 208
  histogram equalization stretch 209
  Local Range Modification (LRM) 217
  min-max stretch 209
  normalization stretch 210
  piecewise-linear stretch 209
  saturation stretch 209
  statistical pixel matching 213
  thresholding 216
convex hull 435
convolution
  by Fourier transforms 255
  cascaded filters 241
  circular 237
  circular from Fourier transforms 252
  discrete 232
  filter 230
  for debanding 330
  for image pyramid 265
  in MTF compensation 309
  in resampling 300, 378
  in system modeling 259
  in wavelet transform 278
  sum of weights equals MTF(0,0) 257
Convolution Theorem 255, 464
co-occurrence matix
  texture features 165
co-occurrence matrix 164
  effect of resampling on 306
coordinates
  conventions for spatial 128
  geocentric 114
  geodetic 114
correlation
  spatial 357
    correlation coefficient 360
    spatial texture measure 165
  spectral 138
    correlation coefficient 138
    correlation matrix 138
correlation length
  relation to power spectrum 162
  relation to semivariogram 155
correlation matrix
  in PCT 199
  statistics image display 445

covariance matrix 137
    eigenvalues 193
    eigenvectors 195
    in maximum-likelihood classifier 419
    in nearest-mean classifier 421
    in PCT 193
    trace in PCT 195
Cumulative Distribution Function (CDF) 131
    in destriping 325
    in fusion 375
    in histgram equalization 209
    in reference stretch 210

D

Dark Object Subtraction (DOS) 337
data format
    Band SeQuential (BSQ) 36
    Band-Interleaved-by-Line (BIL) 36
    Band-Interleaved-by-Pixel (BIP) 36
    Band-Interleaved-by-Sample (BIS) 36
decision boundaries 393, 394, 422
    Artificial Neural Network (ANN) classifier 415
    level-slice classifier 406
    maximum-likelihood classifier 421
deconvolution 309
    simultaneous with resampling 311
Delaunay triangulation 298
diffuse spectral reflectance r 52
Digital Elevation Model (DEM) 58
    calculation from stereo imagery 366
    in orthorectification 363
    shaded relief image 59
Digital Number (DN) 23
    quantization 107
    relation to Grey Level (GL) 36
digital orthophotoquads 363
Dirichlet tessellation 298
discriminant function 418
    comparison of ANN and maximum-likelihood 426
    gaussian maximum-likelihood 419
    nearest-mean 422
distortion
    geometric 110
    in airborne imagery 112
    in line and whiskbroom scanners 119
    in pushbroom scanners 119
    in scanners 113
    in wide FOV sensors 119, 121
    topographic 121

E

emissivity 64
endmember 434
    bundle 442
    in linear mixing model 437
    specification 441

F

Field Of View (FOV) 19
filter
    spatial
        box filter algorithm 240
        Difference-of-Gaussians (DoG) 274
        directional 236
        High-Boost Filter (HBF) 234, 235
        High-Pass Filter (HPF) 330
        Identity Filter (IF) 233
        Laplacian-of-Gaussian (LoG) 269
        Low-Pass Filter (LPF) 378
        Nagao-Matsuyama 317
        notch 327
    spectral 104
formation flying 35
Fourier
    analysis and synthesis 246
    Discrete Fourier Transform (DFT) 249
    transform 246, 253
fractals 166

G

geocoding 286
geometric correction 286
    coordinate transformation 296
    Ground Control Points (GCPs) 291
    Ground Points (GPs) 293
    piecewise polynomial model 296
    polynomial models 287
    resampling 300
geostatistics 152

Grey Level (GL) 36
greybody 64
Ground Sample Distance (GSD) 20
Ground-projected Field Of View (GFOV) 19
Ground-projected Instantaneous Field Of View
  (GIFOV) 20
        effect on spatial statistics 174
        effect on spectral feature space 178
Ground-projected Sample Interval (GSI) 3, 20
        definition 21
        pixel growth 21
        relation to GIFOV 22

H

histogram 129
        K-dimensional 134
hyperspectral
        calibration and atmospheric correction 341
        classification algorithms 450
        data analysis 8
        feature extraction 447
        image analysis 445
        normalization 344
        sensor 3
        spectral resolution 82
        subpixel classification 430
        training for classification 447
        visualization 445

I

image
        fusion 371
                feature space 374
                high frequency modulation 376
                scale-space 380
                sharpening with a sensor model 378
                spatial domain 375
        multitemporal 356
        registration 286
                area correlation 357
                Hierarchical Warp Stereo (HWS) 366
                scale-space 362
        restoration 309
        restoration and fusion 378
Instantaneous Field Of View (IFOV) 22

Instantaneous Field of View (IFOV) 22
irradiance 46
ISODATA clustering algorithm 400

K

K-means clustering algorithm 400
kurtosis 132

L

Lambertian surface 52
        comparison to Minnaert model 339
Laplacian pyramid 266
Laplacian-of-Gaussian (LoG) filter 269, 362
learning rate 411
least-squares solution
        equivalence to orthogonal subspace projection
          455
        geometric correction 293
        linear spectral unmixing 439
Long-Wave InfraRed (LWIR)
        at-sensor radiance 68
        definition 10
        Thermal IR (TIR) 10
Look-Up Table (LUT) 36
        atmospheric characterization 338
        histogram estimation classifier 406
        image coding 388
        level-slice classifier 405
        reference stretch 210

M

mean 131
median 132
microwave
        atmospheric absorption 8
        brightness temperature 10
        wavelength-frequency nomograph 13
        wavelengths for remote sensing 12
Mid-Wave InfraRed (MWIR)
        at-sensor radiance 64
        definition 10
        optical properties 61
mixing
        effect on spectral scattergram 180
        intimate 435

linear model 434
    nonlinear 434
mode 132
modulation
    definition 146
    relation to contrast 147
Modulation Transfer Function (MTF)
    compensation 309
    image filter 257
    system model 259
momentum 411

N

National Imagery Interpretability Scale (NIIRS) 149
    General Image-Quality Equation (GIQE) 150
Near InfraRed (NIR)
    definition 10
noise
    amplification by MTF compensation 309
    banding 328
    coherent 320
    comparison between pushbroom and whisk-
        broom sensors 32
    comparison of random and striping 151
    detection by spectral correlation 318
    effect on spatial statistics 174
    estimation by semivariogram 155
    models 140
    periodic 320
    photographic granularity 142
    reduction by spatial filter masking 327
    reduction techniques 315
    spectral crosstalk 145
    striping 143, 323
Noise Equivalent Radiance (NER) 152
normal (Gaussian) distribution
    1-D 130
    K-D 138

O

Orthogonal Subspace Projection (OSP) 454
orthorectification 363
    collinearity equations 363
    Hierarchical Warp Stereo (HWS) 366
outliers 132

in classification 405
in contrast stretch 209
in distortion correction 296
removal by median filter 242

P

phased-array 98
Planck's blackbody equation 46
Point Spread Function (PSF)
    detector 88
    electronics 90
    image motion 88
    in fusion 378
    in spatial statistics 160
    in spectral mixing 180
    measurement 95
    net 90
    net spatial response 85
    optical 86
    sensor comparison 90
    sensor modeling 259
    useful models 462
power spectrum 263
Principal Components Transform (PCT)
    application to color contrast enhancement 222
    application to image fusion 374
    application to noise detection 318
probability
    a posteriori 417
    a priori 417
    class-conditional 417
pseudo-color 39

R

radiance 23
    at-sensor 47, 53, 334
        in camera equation 104
        retrieval by sensor calibration 109
        spectral derivative 448
        total emitted 68
        total solar 55
    emitted 64, 65
    path 48, 54, 332
    path-emitted 63
    surface 337, 339

rectification 286
region growing 427
    application to classifier training 396
registration 286, 356
    area correlation 359
    automated 357
    spatial features 362
resampling 286, 300
    bilinear 301
    combined with MTF compensation 311
    cubic 302
    effect on mixing 430
    effect on spectral feature space 308
    in fusion 375
    nearest-neighbor 300
    Parametric Cubic Convolution (PCC) 306
resolution 2, 76
    hyperspectral 448
    radiometric 23
    relation to classification 390
    spatial 2, 3, 77
    spectral 82
    temporal 32
restoration 309
    amplitude gain 310
    deconvolution 309
    inverse filter 309
    simultaneous with image resampling 311
    Wiener Filter 310

S

sample-scene phase 81
    phased target array 98
scale of aerial photographs 390
scale-space transfom
    in fusion 380
scale-space transform 263
    in registration 362
    spectral fingerprints 449
    wavelet transform 278
scanner
    2-D pushbroom array 23
    line 19
    paddlebroom 19
    pushbroom 19

    whiskbroom 19
scattergram 134
    spatial co-occurrence matrix 153, 164
    TM examples 141
scattering
    atmospheric 49
    Mie 51
    Rayleigh 51
scatterplot 133
    reduction from 2-D to 1-D 135
    reduction from 3-D to 2-D 135
semivariogram 153
    correlation length 155
    nugget 155
    range 155
    sill 155
sensor gain 11, 29, 70, 106
    detector equalization 325
sensor offset 106
    detector equalization 325
separability
    analysis for classification 396
    interclass measures 397
    scanner imaging model 86
    spatial statistics model 160
shaded relief 58, 171, 172
shadow
    de-shadowing 344
    in TIR image 69, 71
    projected shadows 58
    self-shadowed 58
Short Wave InfraRed (SWIR)
    definition 10
Short-Wave InfraRed (SWIR)
    ASTER spectral crosstalk 145
    comparison of MODIS, ASTER, and ETM+
      bands 106
    liquid water absorption bands 15
    mineral absorption bands 346
    mineral discrimination 447
Side-Looking Airborne Radar (SLAR) 317
Signal-to-Noise Ratio (SNR) 147
simulation
    imaging system 91
    random image noise 149

sensor characteristics and spectral scattergrams
178
sensor characteristics on spatial statistics 174
TM imaging 91
topography and spectral scattergrams 169
skewness 132
skylight 48
soil line 187
solar
irradiance 46
path radiance 47
path transmittance 48
spectrum 46
spatial statistics
covariance 153
Markov covariance model 155
power spectrum 162
semivariogram 153
spatial-spectral segmentation 427
effect on spectral feature space 432
spectral 46
flux density 47
radiant exitance 46
reflectance 15
signature 13
standard deviation 132
Standardized Principal Components (SPC) 199
stereo
base-to-height ratio 124
parallax 123
subpixel
object detection 78, 79
surface radiance 52
swath width 19
Synthetic Aperture Radar (SAR) 10

T

Tasseled-Cap Transform (TCT) 204
comparison to PCT 205
Landsat coefficients 204
temperature
approximately linear relation to radiant exitance 65
blackbody 46
effective earth 64

effective solar 47
relation to wavelength of maximum radiant exitance (Wien's displacement law) 46
separation from emissivity effect 64
Thermal InfraRed (TIR) 10
Long-Wave InfraRed (LWIR) 10
transfer function 255, 257

U

unmixing 435
and image restoration 311, 435
endmember specification 441
equivalence to Orthogonal Subspace Projection (OSP) 454
examples 437
Independent Component Analysis (ICA) 442
relation to ANN classification 440

V

Vegetation Index (VI) 188
Enhanced Vegetation Index (EVI) 189
Normalized Difference Vegetation Index (NDVI) 188
Perpendicular Vegetation Index (PVI) 189
Ratio Vegetation Index (RVI) 188
relation to physical models 190
Soil-Adjusted Vegetation Index (SAVI) 189
Transformed Vegetation Index (TVI) 189
Visible (V) 10
Voronoi diagram 298

W

wavelet transform 278
application to GCP location 362
application to image fusion 380
Wien's displacement law 46

Z

zero-crossings 267
image algorithm 273
in HP-filtered data 234
spectral fingerprints 450